HVAC Maintenance and Operations Handbook

Other McGraw-Hill HVAC Books of Interest

CHEN and DEMSTER • *Variable Air Volume Systems for Environmental Quality*

GLADSTONE and HUMPHREYS • *Mechanical Estimating Guidebook for HVAC*

GLADSTONE et al. • *HVAC Testing, Adjusting, and Balancing Field Manual*

GRIMM and ROSALER • *HVAC Systems and Components Handbook*

HAINES and WILSON • *HVAC Systems Design Handbook*

HARTMAN • *Direct Digital Controls for HVAC Systems*

LEVENHAGEN and SPETHMANN • *HVAC Controls and Systems*

MULL • *HVAC Principles and Applications Manual*

PARMLEY • *HVAC Design Data Sourcebook*

RISHEL • *HVAC Pump Handbook*

SUN • *Air Handling Systems Design*

WANG et al. • *Handbook of Air Conditioning and Refrigeration*

To order or receive additional information on these or any other McGraw-Hill titles, in the United States please call 1-800-722-4726. In other countries, contact your local McGraw-Hill representative.

HVAC Maintenance and Operations Handbook

Robert C. Rosaler, P.E.

McGraw-Hill

New York   San Francisco   Washington, D.C.   Auckland   Bogotá
Caracas   Lisbon   London   Madrid   Mexico City   Milan
Montreal   New Delhi   San Juan   Singapore
Sydney   Tokyo   Toronto

Library of Congress Cataloging-in-Publication Data

HVAC maintenance and operations handbook / [edited by] Robert C.
Rosaler.
     p.   cm.
   Includes index.
   ISBN 0-07-052169-7 (alk. paper)
    1. Heating—Handbooks, manuals, etc. 2. Ventilation—Handbooks,
manuals, etc. 3. Air conditioning—Handbooks, manuals, etc.
I. Rosaler, Robert C.
TH7011.H82   1997
697—dc21   97-31235
                                              CIP

McGraw-Hill

*A Division of The **McGraw·Hill** Companies*

1 2 3 4 5 6 7 8 9 0 DOC/DOC 9 0 2 1 0 9 8 7

ISBN 0-07-052169-7

*The sponsoring editor for this book was Harold B. Crawford, the editing
supervisor was Suzanne Ingrao, and the production supervisor was Tina
Cameron. It was set in Century Schoolbook by Pro-Image Corporation.*

Printed and bound by R. R. Donnelley & Sons Company.

McGraw-Hill books are available at special quantity discounts to use as pre-
miums and sales promotions, or for use in corporate training programs. For
more information, please write to the Director of Special Sales, McGraw-Hill,
11 West 19 Street, New York, NY 10011. Or contact your local bookstore.

This book is printed on recycled, acid-free paper containing a minimum of 50% recycled de-inked fiber.

These latest years have seen an explosive growth in the global usage of HVAC systems. HVAC has made the outdoors tolerable and the indoors comfortable and beneficial.

This book is dedicated to the unheralded HVAC engineers and technicians who have made it all possible.

Contents

Preface

Publication of this book reflects the world-wide growth of the HVAC industry, particularly the recognition of the increasing importance of keeping HVAC installations operating reliably and efficiently.

It has been long established that the costs of accomplishing this far exceed that of design and engineering. It is our objective to provide practical, useful information that will identify those maintenance and operations procedures that will keep their financial burdens to a minimum.

Acknowledgments

The editor wishes to acknowledge the valuable assistance and guidance of McGraw-Hill supervising editor Harold B. Crawford.

Robert C. Rosaler

Contributors

Don Batz   *Comm Air Mechanical Services Company, Oakland, California* (CHAP. 19.1)

Mike Batz   *Comm Air Mechanical Services Company, Oakland, California* (CHAP. 19.1)

Bayley Fan, Lau Industries   *Lebanon, Indiana* (CHAP. 16.1)

Michael Bilecky   *von Otto & Bilecky, Washington, DC* (CHAPS. 4–8)

John Brewster   *Pro Tec Technical Services, Willowdale, Ontario, Canada* (CHAP. 11)

Roland W. Brown   *Power-Flame, Inc., Parsons, Kansas* (CHAP. 18.2)

William A. Butler   *von Otto & Bilecky, Washington, DC* (CHAPS. 4–8)

Cleaver-Brookes, Division of Aqua-Chem, Inc.   *Milwaukee, Wisconsin* (CHAP. 18.1)

Giffels Associates   *Southfield, Minnesota* (CHAP. 3)

Joe Gosmano   *Marley Cooling Tower, Brea, California* (CHAP. 19.5)

Ben Harstine   *Joy / Green Fan Div., New Philadelphia Fan Co., New Philadelphia, Ohio* (CHAP. 16.2)

M. B. Herbert   *Consulting Engineer, Willow Grove, Pennsylvania* (CHAP. 2)

Terry Hoffman   *Johnson Controls, Inc., Milwaukee, Wisconsin* (CHAP. 21)

Billy Langley   *Consulting Engineer, Azle, Texas* (CHAPS. 19.2, 20.2)

Howard J. McKew   *Wm. A. Berry & Son, Danvers, Massachusetts* (CHAP. 12)

Mechanical Contracting Foundation, The   *Rockville, Maryland* (CHAP. 15.1)

Ron Moore   *R.M. Group, The, Knoxville, Tennessee* (CHAPS. 9, 10, 14)

Donald A. Morton   *System Planning Corp., Arlington, Virginia* (CHAP. 13)

Michael S. Palazzolo   *Safety King Corp., Utica, Michigan* (CHAP. 15.2)

George M. Player   *Wm. A. Berry & Son, Danvers, Massachusetts* (CHAP. 12)

George H. Raaf   *von Otto & Bilecky, Washington, DC* (CHAPS. 4–8)

Robur Corporation   *Evansville, Indiana* (CHAP. 1)

Willis Schroader   *Robur Corp., Evansville, Indiana* (CHAP. 19.4)

David O. Seaward   *Alfa Laval Thermal, Inc., Richmond, Virginia* (CHAP. 18.3)

George E. Taber   *Taco, Inc., Cranston, Rhode Island* (CHAP. 17)

C. Curtis Trent   *Trent Technologies, Inc., Tyler, Texas* (CHAP. 22)

Warren C. Trent   *Trent Technologies, Inc., Tyler, Texas* (CHAP. 22)

Travis West   *Building Air Quality, The Woodlands, Texas* (CHAP. 20.1)

George White   *Aggreko Co., Venetia, California* (CHAP. 19.3)

Introduction

Glossary of HVAC Terms

Robur Corporation
Evansville, Indiana

Absorption: The process by which one material extracts other substances from a mixture of gases or liquids.

Access Door: A door or panel provided in any structure, as in a duct, wall, etc., or in a cooling or heating unit, to permit inspection and adjustment of the inside components.

Air, Ambient: The air surrounding an object.

Air Change: The amount of air required to completely replace the air in a room or building; not to be confused with recirculated air.

Air Circulation: Natural or forced movement of air.

Air Cleaner: A mechanical, electrical or chemical device (usually a filter) for removing dust, gas, vapor, fumes, smoke and other impurities from air.

Air Conditioner: A machine that controls the temperature, moisture, cleanliness and distribution of air.

Air Conditioner, Unitary: An air conditioner consisting of several factory-made components within an insulated casing, normally including an evaporator (cooling) coil, a condenser and compressor combination, and controls and fans for distributing conditioned air; some also contain filters, heaters and dampers.

Air Conditioning: The process of treating air to control its temperature, moisture, cleanliness and distribution.

Air Conditioning System, Year-Round: A system that controls the total environment during all seasons under all comfort requirements.

Air Diffuser: An outlet which discharges supply air in a spreading pattern.

Air Handler: A fan or blower for moving air within a distribution system.

Air Handling Unit: An air handler, heating element and cooling coil and other components in a cabinet or casing.

Air, Recirculated: Air drawn from a space and passed through the conditioner and discharged again into the conditioned space.

Air, Ventilation: The quantity of supply air drawn from outdoors needed to maintain the desired amount of oxygen and the quality of air within a designated space.

Anemometer: An instrument for measuring the velocity of air.

Atomize: To break up a liquid into a fine spray, as in oil burner nozzles and atomizing water humidifiers.

Baffle: A surface used to deflect or direct air flow, usually in the form of a plate or wall.

Barometer: A device for measuring atmospheric pressure.

Blow (Throw):   In air distribution, the distance an air stream travels from an outlet to a point where air motion is reduced to a specific flow measured in feet (meters) per minute (terminal velocity).

Blower (Fan):   An air-handling device for moving air in a distribution system.

British Thermal Unit (Btu):   A unit for measuring heat quantity equal approximately to the amount of heat produced by the burning of an ordinary wooden kitchen match. More specifically, it is the amount of heat required to raise the temperature of one pound of water one degree Fahrenheit.

Capacity Reducer: A device, usually placed at the energy source, to reduce the Btuh capacity of the unit; usually used to decrease a furnace's output without decreasing its air-handling capacity; frequently required in well-insulated, tightly constructed buildings.

Capillary Tube: A small-diameter tube (metering device), whose bore and length are designed to permit the passage of a specific amount of liquid refrigerant or other liquid at a specific pressure drop.

Centigrade: A metric scale for measuring temperature. Zero degrees Centigrade (C) is $+32°$ Fahrenheit (F), and $100°C$ is $+212°F$.

Change of Air: Introduction of new air to a conditioned space, measured by the number of complete enclosed-space changes per unit time, usually per hour.

Charge, Refrigerant: The amount of refrigerant in a system; to put in the refrigerant charge.

Chill: To reduce temperature moderately without freezing.

Chimney Effect: The tendency of air or gas to rise in a duct or other vertical passage when heated.

Coil: A heating or cooling element made of pipe or tubing, with or without extensions or fins.

Coil, Cooling: A commonly used term meaning the evaporator.

Comfort Zone: The range of temperatures, humidities and air velocities at which the greatest percentage of people feel comfortable.

Compression: In a compression refrigeration system, a process by which the pressure of the refrigerant is increased.

Compression, ratio of: The ratio of absolute pressures after and before compression.

Compressor: A device for increasing the pressure of heat-laden refrigerant vapors thus increasing the heat level (super-heating) within those vapors so that the heat contained can be released into the outside atmosphere.

Condensate: The liquid formed by the condensation of a vapor, e.g. moisture on cold window glass, or dew.

Condensation: The process whereby a vapor is changed to a liquid by removal of heat after its dew point (condensation temperature) is reached.

Condenser: A vessel, or arrangement of pipes or tubing, in which vapor is liquified by removal of heat.

Condensing Unit: A refrigerating machine consisting of one or more power-driven compressors, condensers, liquid receivers and other components.

Conduction: The process of transferring heat along the elements of a substance, as from a tube to a fin.

Conduit: A tube or pipe used for conveying liquid or gas; also, a tube or pipe in which wire or other pipes may be inserted for protection.

Connecting Rod: A device connecting a piston to a crankshaft used in reciprocating compressors.

Control: A device for regulating a system, or part of a system, in normal operation. Such a device can be either manually operated or automatic.

Convection: The transfer of heat by movement of fluid or air.

Cycle, Refrigeration (Absorption):   The complete cycle of a refrigerating agent through a system whereby the agent absorbs heat, transports it to a point where a second, unrelated, substance extracts the heat from the initial agent which is returned through the system to absorb more heat and repeat the process. Energy source for such a cycle can be one of several different types.

Cycle, Refrigeration (Mechanical):   Complete cycle of a refrigerant from a condensing unit (compressor-condenser) through a system of lines or tubing to the evaporator coil and back into the condensing unit; a cycle in which heat is absorbed in the evaporator and given up in the condenser.

Damper: A device for adjusting the amount of air flowing through an outlet, inlet or duct.

Damper, Fire: A damper placed in a duct or at an outlet to prevent or retard the emission of smoke and/or heat into a room or space during a fire. Such devices are also designed to prevent or retard the spread of fire throughout the structure, thus confining it to a specific area.

Defrosting: The removal of accumulated ice or frost from a cooling element or coil.

Dehumidification: The reduction of water vapor in air by cooling the air below the dew point; removal of water vapor from air by chemical means, refrigeration, etc.

Dehydrate: To remove water in all forms from other matter.

Dew Point: The temperature at which water vapor turns to liquid.

Draft: A current caused by the movement of air from an area of high pressure to an area of low pressure; usually considered objectionable.

Drier: A device placed in a refrigeration system to attract and collect unwanted moisture which may be in the system.

Dry: To separate or remove liquid or vapor from another substance, such as moisture from a refrigerant.

Duct: A pipe or closed conduit made of sheet metal or other suitable material used for conducting air to and from an air-handling unit.

Entrainment: The induction of room air into an airstream from an outlet. (Also, see *Induction*.)

Evaporator: That part (heat exchanger) of a cooling system in which refrigerant is vaporized.

Exfiltration: Air flow outward from an enclosed space through a wall, leak, membrane, etc.

Exhaust Opening: An opening through which air is exhausted from a room which is being cooled, heated, or ventilated.

Fan Coil Unit: An air-handling unit especially designed to condition small spaces.

Filter: A device for removing dust particles from air or unwanted elements from liquids.

Filter Drier: A combination of a liquid refrigerant line strainer and dehydrator.

Fluid: Any liquid or gas.

Freezing Point: The temperature at which liquids will solidify or freeze upon removal of heat. (e.g., the freezing point of water is +32°F or 0°C)

Furnace: That part of an environmental system which converts gas, oil, electricity, or other fuel into heat for distribution within a structure.

Gage (Gauge):   An instrument for measuring temperatures, pressures, or liquid levels. Also, an arbitrary scale of measurement for sheet metal thicknesses, wire and drill diameters, etc., liquid levels within boiler pipes, tanks and other enclosures.

Gas, Fuel: An expanded hydrocarbon fluid used to provide heat in environmental systems.

Gas, Inert: A gas that neither undergoes nor causes chemical reaction nor change of form.

Gravity, Specific: Density measured on the basis of the known density of a given substance, usually air or water.

Grille: A covering for an opening through which air passes.

Heat: Energy which can be transferred as a result of temperature differences.

Heat, Latent: Heat energy which changes the form of a substance (e.g., ice to water) without changing its temperature.

Heat Sensible: Heat which changes the temperature of a substance without changing its form.

Heat Exchanger: Usually considered to be that part of a furnace that transfers heat from burning fuel or electricity into air or other medium. Also, it is any device in which heat is absorbed by one fluid from another fluid, such as a condenser, evaporator or boiler.

Heat Pump: A reverse-cycle refrigeration system designed to perform both heating and cooling operations.

Heat Transmission: Any time-rate flow of heat; usually refers to conduction, convection and radiation.

High Side: That part of a refrigeration system where refrigerant is under the greatest amount of pressure where heat is rejected (condenser section). This is area of highest temperatures and highest pressures.

Humidifier: A device that adds moisture to warm air being circulated or directed into a space.

Humidistat: A device designed to regulate humidity input by reacting to changes in the moisture content of the air.

Humidity, Relative: The percentage of moisture in the air measured against the amount of moisture the air could hold at a given temperature. (e.g., Since cold air is capable of holding less moisture than warm air, if the temperature drops and the moisture volume remains constant, the relative humidity will increase. Conversely, if the temperature rises and the moisture volume remains constant, the relative humidity will decrease.)

Induction: The entrainment of room air into an air stream from an air outlet.

Infiltration: Air flow inward into a space through walls, leaks around doors and windows or through the building material used in the structure.

Insulation, Thermal: A material having a relatively high resistance to heat flow, used primarily to reduce heat loss and heat gain.

Liquid Line: The tube or piping carrying liquid refrigerant from the condenser or receiver of a refrigeration system to a pressure-reducing device.

Louver: A series of vanes that permit directional adjustment of air flow. They are usually installed on outside grilles or intake openings to impede water entry into ducts, elbows, and to minimize turbulence; they may also include indoor grilles to eliminate light penetration.

Low Side: Those parts of a refrigeration system where heat is absorbed at evaporator pressure; the cooling coil or evaporator with associated components; also commonly referred to as the suction side; this is the area of lowest temperature and lowest pressure.

Main (Trunk Line):   Pipe or duct for distributing fluids such as air, water, or steam to various branch ducts and collecting fluids from various branches.

Make-Up Air Unit: Unitary equipment which introduces outside air into a space and conditions it to offset air exhausted from that space.

Manometer: An instrument for measuring pressures by the difference in liquid volume.

Meter: An instrument for measuring rates of flow of energy or fuel over a given period. Also, in the metric system of measurement, it is a distance of 39.37 inches.

Motor, Air (Pneumatic Motor):   An air-operated device normally used for opening and closing dampers and valves.

Multizone System: An air conditioning system designed to serve one or more areas having different heating/cooling/humidification requirements. Such a system can consist of more than one heating/cooling/humidification unit, or can consist of a single unit controlled by dampers, by-passes and thermostats.

Muffler, Noise Reducing: A device for the prevention or reduction of sound or vibration being transmitted from one space to another via air conditioning distribution systems or equipment.

Oil Separator: A device for separating oil or oil vapor from refrigerant.

Plenum Chamber: An air compartment connecting one or more ducts with the heating or cooling unit. Any enlarged section within an air duct with several duct, grille, or diffuser connections.

Pneumatic: Any device actuated by air pressure is said to be pneumatic.

Power Roof Ventilator: A motor-driven exhaust fan mounted above the roof on a roof curb.

Pressure, Absolute: That pressure which will register on a pressure gauge, plus atmospheric pressure present, e.g. a gauge pressure of 68.5 pounds per square inch (psi) (5 bars) at sea level would represent an absolute pressure of 83.2 (68.5 plus 14.7 equals 83.2) [6 bars]. In the United States, absolute pressure is expressed as pounds per square inch absolute, or as psia. Elsewhere, in bars or pascals.

Pressure, Atmospheric: Pressure resulting from the weight of the atmosphere which is 14.696 pounds per square inch of surface (1 bar, 100 kPa); or, that pressure in the outdoors or within a space that is present as a result of the forces of nature with no artificially induced pressure change.

Pressure Drop: That pressure lost between any two points of a piping or duct system due to friction, leakage or other reasons.

Pressure, Static: Force (per unit area) expended against the walls of a container such as an air duct. Commonly, force is measured in heating and air-conditioning in pounds per square inch, or inches of water pressure.

Pressure, Total: A combination of both static and velocity pressures.

Pressure, Velocity: Force of air as it moves in an air duct.

Psychrometer: A thermometer-like instrument for measuring wet bulb and dry bulb temperatures simultaneously.

Radiation: Transmission of heat or energy by electromagnetic waves, such as heat transmitted by an electric heater and absorbed by a disjoined substance.

Radiator: A mechanical device that transmits heat by the process of radiation.

Refrigerant: The fluid in a refrigeration cycle that absorbs heat at low temperatures and rejects heat at higher temperatures.

Register: A combination grille and damper assembly covering an air-supply outlet, designed to distribute air into a room.

Return Air: Air recirculated through a return air system to an air-handling unit or furnace.

Saturation: A condition which occurs when air contains all the moisture that it can possibly hold at a given temperature.

Sight Glass: A glass tube used to indicate the liquid condition and level in pipes, tanks, bearings, and similar equipment.

Steam: Water that has turned to vapor as a result of the application of heat at a given pressure, or the reduction of pressure at a given temperature.

Strainer: A device for withholding foreign matter from a flowing liquid or gas.

Sun Effect: Heat transmitted into space through glass and building materials exposed to the sun.

System, Duct: A series of tubular or rectangular sections, elbows and connectors fabricated as a channel to carry air from one point to another.

Temperature: The heat content of a substance in degrees Fahrenheit or Centigrade.

Temperature Drop: A measurement of the difference in heat between two points of a system, such as at the furnace plenum and at the outlet grille.

Temperature, Dry-Bulb: The temperature of the air at any given location indicated by an accurate thermometer and not influenced by outside interferences such as radiation or water.

Ton of Refrigeration: Removal of 12,000 Btus of heat per hour, or 200 Btus per minute (360 watts), from a given area. The heat required to melt one ton of ice (288,000 Btus) in a period of 24 hours.

Unit: An assembly for heating, cooling, dehumidifying, and/or ventilating.

The HVAC Design Factor

M. B. Herbert, P.E.
Consulting Engineer
Willow Grove, Pennsylvania

2.1 Introduction

Heating, ventilating, and air-conditioning (HVAC) systems are designed to provide control of space temperature, humidity, air contaminants, differential pressurization, and air motion. Usually an upper limit is placed on the noise level that is acceptable within the occupied spaces. To be successful, the systems must satisfactorily perform the tasks intended.

Most heating, ventilating, and air-conditioning systems are designed for human comfort; many industrial applications have objectives other than human comfort. If human comfort can be achieved while the demands of industry are satisfied, the design will be that much better.

Automatic control of the HVAC system is required to maintain desired environmental conditions. The method of control is dictated by the requirements of the space. The selection and the arrangement of the system components are determined by the method of control. Controls are necessary because of varying weather conditions and internal loads. These variations must be understood before the system is designed. Control equipment is discussed in Chap. 21.

The proliferation of affordable computers has made it possible for most offices to automate their design efforts. Each office should eval-

---

Condensed from Chapter 1, "Conceptual and Preliminary Design," the *HVAC Systems & Components Handbook*.

uate its needs, choose from the available computer programs on the market, and then purchase a compatible computer and its peripherals.

2.1.1 Equipment selection

From the calculations and the method of control, the capacity and operating conditions may be determined for each component of the system. Manufacturers' catalogs give extensive tables and sometimes performance curves for their equipment. All equipment that moves or is moved vibrates and generates noise. In most HVAC systems, noise is of utmost importance.

Many equipment test codes have been written by ASHRAE, American Refrigeration Institute (ARI), Air Moving and Conditioning Association (AMCA), and other societies and manufacturer groups. A comprehensive list of these codes is contained in ASHRAE handbooks. Manufacturer's catalogs usually contain references to codes by which their equipment has been rated.

2.1.2 Equipment location

Mechanical and electrical equipment must be serviced periodically and eventually replaced when its useful life has expired. To achieve this end, *every piece of equipment must be accessible and have a planned means of replacement.*

Ceiling spaces should not be used for locating equipment. Servicing equipment in the ceiling entails erecting a ladder at the proper point and removing a ceiling tile or opening an access door, to gain access to the equipment. Crawling over the ceiling is dangerous and probably violates OSHA regulations. No matter how careful maintenance personnel are, eventually the ceiling will become dirty, tiles will be broken, and if water is involved, the ceiling will be stained.

Also, the equipment will suffer from lack of proper maintenance, because no one on a ladder can work efficiently. This work in the occupied space is disruptive to the normal activities of that space.

Equipment should be located in spaces specifically designed to house them. Sufficient space should be provided so that workers can walk around pieces of equipment, swing a wrench, rig a hoist, or replace an electric motor, fan shaft, or fan belts. Do not forget to provide space for the necessary electrical conduits, piping, and air ducts associated with this equipment. Boilers and other heat exchangers require space for replacing tubes. Valves in piping should be located so that they may be operated without resorting to a ladder or crawling through a tight space. If equipment is easily reached, it will be maintained. Adequate space also provides for good housekeeping, which is a safety feature.

Provision of adequate space in the planning stage can be made only after the types and sizes of systems have been estimated. Select equipment based on the estimated loads. Lay out each piece to a suitable scale. Arrange the equipment room with cutout copies of the equipment. Allow for air ducts, piping, electrical equipment, access aisles, and maintenance workspace. Cutouts permit several arrangements to be prepared for study.

In locating the equipment rooms, be sure each piece of equipment can be brought into and removed from the premises at any time during the construction. A strike may delay the delivery of a piece of equipment beyond its scheduled delivery date. This delay should not force construction to be halted, as it would if the chiller or boiler had to be set in place before the roof or walls were constructed.

2.1.3 Distribution systems

HVAC distribution systems are of two kinds: air ducts and piping. Air ducts are used to convey air to and from desired locations. Air ducts include supply air, return-relief air, exhaust air, and air-conveying systems (see Chap. 15.1). Piping is used to convey steam and condensate, heating hot water, chilled water, brine, cooling tower water, refrigerants, and other heat-transfer fluids (see Chap. 15.2). Energy is required to force the fluids through these systems. This energy should be considered when systems are evaluated or compared.

Building Design and Equipment Location

Giffels Associates, Inc.
Southfield, Michigan

The adage "out of sight out of mind" applies to maintenance. Equipment that a designer knows should be periodically checked and maintained may get neither when access is difficult. Maintenance instructions are available from equipment manufacturers; the system designer should be acquainted with these instructions, and the design should include reasonable access, including walk space and headroom, for ease of maintenance.

Penthouse and rooftop equipment should be serviceable via stairs or elevators and via roof walkways (to protect the roofing). Ship's ladders are inadequate when tools, parts, chemicals, etc., are to be carried. Rooftop air handlers, especially those used in cold climates, should have enclosed service corridors. If heavy rooftop replacement parts, filters, or equipment are expected to be skidded or rolled across a roof, the architect must be advised of the loading to permit proper roof system design.

Truss-mounted air handlers, unit heaters, valves, exhaust fans, etc., should be over aisles (for servicing from mechanized lifts and rolling platforms) when catwalks are impractical. Locate isolated valves and traps within reach of building columns and trusses to provide a degree of stability for service personnel on ladders.

Condensed from Chapter 2.2, "Applications of HVAC Systems" of the *HVAC Systems and Components Handbook*.

It is important that access to ceiling spaces be coordinated with the architect. Lay-in ceilings provide unlimited access to the space above, except possibly at lights, speakers, sprinklers, etc. When possible, locate valves, dampers, air boxes, coils, etc., above corridors and janitor closets so as to disturb the client's operations the least.

Piping-system diagrams and valve charts are important and should be provided by the construction documents. Piping should be labeled with service and flow arrows, and valves should be numbered, especially when not within easy view of the source (such as steam piping not being within easy view of the boiler).

The air-handling, refrigeration, and heating equpment could be located either within an enclosed mechanical-equipment room or on the building roof in the form of unitary self-contained equipment. For larger systems, of 200 tons (703 kW) of refrigeration or more, the mechanical-equipment room offers distinct advantages from the standpoint of maintenance; however, the impact on building cost must be evaluated carefully. An alternate approach to the enclosed equipment room is a custom-designed factory-fabricated equipment room. These are shipped to the jobsite in preassembled, bolted-together, ready-to-run modules. For small offices and retail stores, the most appropriate approach would be roof-mounted, packaged, self-contained, unitary equipment. It will probably be found that this is the lowest in first cost, but it will not fare well in a life-cycle analysis because of increased maintenance costs after 5 to 10 years of service.

Commissioning

Introduction to Commissioning

Michael S. Bilecky, PE, CEM
George H. Raaf, Consultant
William A. Butler, Consultant
Von Otto & Bilecky, Prof. Corp.
Washington, D.C.

4.1 What Are the Objectives of Commissioning?

Commissioning is quite simply assuring that mechanical systems are installed and function in accordance with design intent. Achieving this assurance is anything but simple. It involves rigorous efforts to exercise mechanical systems through their full range of operations in attempts to verify that the systems will perform as intended by the designer and as expected by the owner. Commissioning is the final stage in any quality assurance program; and though it can be an add-on to an already completed project, commissioning provides the best results when it has been integrated into the project from the earliest stages of design.

Commissioning as described in this book is presented as being performed by an independent commissioning agent; that is a person not responsible for the design, installation, or operating and maintenance of the mechanical systems. Employing the services of a commissioner acting as an independent agent for the owner provides the commissioner autonomy, allowing for objective and unbiased evaluations and assessments in the commissioning process. The services of an independent agent, though providing the optimum, are not absolutely required. Commissioning can be performed by the design architect/engineer (A/E), by the installing contractor, by the owner, or by the owner's operating and maintenance personnel. Commissioning by each of these entities brings the potential for individual bias and self-interest to skew the commissioning process. The A/E may attempt to mask design deficiencies and avoid embarrassing change orders by accepting system performance at levels less than had been intended and expected by the owner, but not suitably provided by the contract documents. The installing contractor may assert that systems are functioning up to that level provided in contract documents; or more accurately, his assessment that the installation meets specified performance. The installing contractor has the greatest potential for exercising bias in commissioning his installations, this being the proverbial "fox guarding the chickens". The bias that can be exercised by the owner or the owner's operating and maintenance personnel can be one of high expectations not substantiated by contract documents. In this case the contractual work can get bogged down in claims and counterclaims among the design team, installing contractor, and owner. Though not a spelled out contract service of the independent commissioning agent, his or her independence should provide all parties with an unbiased source of dispute resolution; telling the A/E, the contractor, or the owner that, in his or her opinion, the work in dispute is or is not provided in the contract documents. Remember, the commissioning is an assurance of what has been purchased by the owner has in fact been provided. Getting the owner what he or she wants starts with the design process, and the design documents are the basis from which all subsequent assessments are made.

The initial role of the A/E is to provide the owner with mechanical systems which will serve the owner's needs. These systems are defined by the contract documents which normally consist of plans and specifications. It is wise to have the commissioning agent be a team member as early as the design phase of the project. *Commissioning is not intended to provide a design review; it is assumed that the design of mechanical systems will be competently provided by the design engineer.* The role of the commissioning agent through the design phase is primarily one of observer, to gain insight into design intent and expectations and to be cognizant of design decisions and trade-offs which ultimately will affect systems performance and operations. The commissioning agent also serves as an advisor, offering guidance in the preparation of documents to assure that all necessary system components and test procedures are provided to facilitate the commissioning.

4.2 Why Perform Commissioning?

Full commissioning is not accomplished in the conventional services offered by the A/E design team or by the installing contractor. The

A/E performs their design, does shop drawing review during construction, periodically visits the site during construction to perform an often cursory review of project progress or to resolve questions and conflicts between design documents and field conditions, and performs punch list and final inspections which usually entail very little actual exercising of the mechanical systems through their full range of operational sequences.

The installing mechanical contractor is looking to get off of the job as soon as possible as this generally will represent his largest profit. This does not necessarily mean that the contractor has short-changed the job as far as the quality of materials and workmanship put into assembling the pieces, which will comprise the total mechanical installation. But what *is* often short-changed is the testing of systems to assure proper long term operation; not just operation that appears to be proper for the few minutes during which a system is demonstrated to the owner or the A/E. Further, today's mechanical systems are becoming more and more complex, particularly with the pervasive influx of computer-based control systems. It is not uncommon for mechanical contractors to rely on their subcontractors for testing of systems for which they were responsible and accepting at face value statements from the subcontractors that their work is complete and functioning properly. This may well be true when a subsystem is viewed as its own distinct entity. However, in the reality of the building mechanical systems as a whole, what may have appeared to be proper operation of a subsystem may in fact be incomplete or even improper when the subsystem is integrated with other interactive mechanical operations. The mechanical contractor's self-testing of the mechanical systems as a whole is seldom performed.

The owner or the owner's operating and maintenance personnel are usually in the weakest position to commission a building's mechanical system. They are normally handed a completed product, given a few hours or days of instructions on equipment operations and maintenance, and left to fend for themselves over the warranty period and eventually through the life of the system. Operating and maintenance personnel perform a hands-on type of commissioning by responding to system failures and complaints from building occupants. During the warranty period, the failures and complaints are usually passed back to the contractor who returns and "fixes" the problem; sometimes these fixes are made by totally subverting the original design intent, taking the shortest path to silencing complaints. What often results is the masking of design flaws or installation short-comings which will be costly over the life of the building.

Further, at the time the building is turned over from the contractor to the owner, the limited exposure given to operating and maintenance personnel is seldom sufficient for them to understand the underlying design concepts and intricacies of the systems installed. They are provided operating and maintenance manuals which provide little insight as to how interactive subsystems are to function to create the whole of the mechanical system. They are relegated to attempting to decipher manufacturer's printed literature for equipment and extrapolate on how systems are supposed to operate based upon information on pieces of equipment. What often results is that operating and maintenance personnel, primarily from lack of adequate training and instruction from the contractor, resort to jumping the controls out or making modifications to equipment and systems to get them to function in a way that they understand; which is not necessarily the way the designer intended.

The real-world process of designing, installing, operating, and maintaining mechanical systems has created the need for commissioning. Someone has to be responsible for a holistic assessment of the building's mechanical systems. The assessment of the whole, completed project is most effectively achieved by the independent commissioning agent, who by project's end probably possesses the greatest knowledge of the component systems and the interaction of these components which constitute the whole of mechanical systems.

4.3   Effects of Commissioning

To architects, engineers, and contractors who have not been through a mechanical commissioning process, the independent commissioning agent is often initially viewed as a threat. The perception is that the commissioner is the agent of the owner, policing the performance of the design team and the contractors serving to put light on errors and omissions. Though the perception is largely correct, viewing the commissioning agent as a threat normally subsides over the life of the construction project; and usually results in resounding support of the commissioning process. After all, all parties are, in theory, working toward the best interests of the owner; and the theory should be supported by the reality that the rewards gleaned from a good project are far greater than any short term gain (and often losses) from a poorly designed or constructed project. Once it is realized that the commissioning agent shares the common goal of a good project which satisfies the owner and from which all parties walk away with their fair profits, support for the commissioning process grows.

By being involved with a project from the earliest design phases, the commissioning agent may identify design errors or omissions which would subsequently become requests for information from the contractors and subsequent change orders, proving both costly in man-hours and possibly damaging, or at least embarrassing, to the design team. By assuring that the work is installed by the contractors as intended by the contract documents, the commissioning agent helps assure that a properly working building is turned over to the owner. This minimizes the amount of contractor call-backs to correct deficiencies which eat into contractors' profits. Having assured a properly functioning mechanical system, one of the final acts of the commissioning agent is to participate in the demonstrations and training sessions provided to the owner's operating and maintenance personnel; and this involvement should result in a better and more comprehensive understanding of systems operations by the operating and maintenance personnel. Thus, the commissioning process should result in a win-win-win proposition for the design team, the contractors, and the owner.

The Commissioning Process

Michael S. Bilecky, PE, CEM
George H. Raaf, Consultant
William A. Butler, Consultant
Von Otto & Bilecky, Prof. Corp.
Washington, D.C.

5.1 When to Start

When should mechanical systems commissioning start? Actually, the process of commissioning starts with the outset of design when building and mechanical system types are being discussed in the early planning phase. Decisions made by owners, architects, and engineers in the early stages of planning will determine the mechanical systems to be designed, installed, and ultimately commissioned. Ideally, the commissioning agent would be a participant in the design process. By being involved with project design, the commissioning agent gleans an understanding of the decisions that were made to establish the type of mechanical system to be installed. This understanding will assist in making on-site observations during construction and, more importantly, during observations of actual systems performance. The commissioning agent can also be an active contributor during the early design process by bringing mechanical engineering expertise to the table in addition to and complementary of the project A&E. During design, the commissioning agent offers a second set of eyes to review contract documents for completeness and accuracy, helping to ensure that all of the necessary pieces and procedures are in place by design and specification to properly commission the mechanical systems.

Perhaps the most important contribution of the commissioning agent during the design process is assisting in the incorporation of a written commissioning plan in the contract specifications. This is an essential component to achieve successful commissioning of the building. The commissioning process requires time and man-power of the A/E, the mechanical contractor and subcontractors, and of the owner's operating and maintenance personnel which extends beyond that normally provided when commissioning is not performed. Particularly, contractors must be placed on notice, via the commissioning specification, that they will be required to demonstrate various tests and systems operations to the commissioning agent. Without this notice being given in the contract specifications, the contractor can rightfully become less than cooperative when asked to demonstrate (and redemonstrate) mechanical tests and operations because the contractor did not allocate the necessary man-hours in his bid. Demonstrations and instructions to owner's operating and maintenance personnel in a commissioned building also extend beyond those normally provided by the contractor; and as such, the owner's personnel must also be on notice that more time will be required of them to accept the building.

The commissioning plan should be a distinct section of the mechanical specifications (normally Section 15995). The specification should delineate the responsibilities of the various participants during construction through the completion of the project and acceptance of the building by the owner; a sample specification is included in Chapter 6. As with any good specification, the commissioning plan should be written specifically applicable to the mechanical systems being installed on the project; it does little good to enumerate the procedures for commissioning an air cooled chiller when a water cooled machine is designed for the project.

The involvement of the commissioning agent during design, though important to the commissioning process, does not require an extensive amount of time; and should represent about 15 percent of the total man-hours and fee for commissioning.

5.2   Construction Phase

5.2.1   Construction schedules and the commissioning plan

After a set of construction documents has been prepared, bid, and a contract for construction awarded, the commissioning agent starts to play a more active role. Usually, a pre-construction meeting is held at which time the general contractor presents a construction schedule. This schedule will contain milestones for the general construction of the project and should identify major mechanical milestones such as scheduled delivery and installation of major pieces of equipment (air handling units, boilers, chillers, etc.). The mechanical contractor should also present his construction schedule which more specifically identifies the timing of the mechanical systems (distribution systems installation, receipt and setting of equipment, insulation application, equipment start-up, testing, adjusting, balancing, demonstrations, etc.). The mechanical contractor will normally possess a construction schedule for his own use, but may not be in the habit of producing his schedule as a record document. The project commissioning specifications should delineate milestones necessary for the commissioning process, and put the mechanical contractor on notice that he needs to present a formal construction schedule responsive to the specified milestones. The construction schedules are essential to allow the commissioning agent to keep abreast of the project's progress and the installation of mechanical systems. The commissioning agent needs to periodically observe the installation of piping and duct distribution systems, witness pressure tests, observe the setting of major pieces of equipment and witness start-ups, etc. Utilizing the construction schedules, the commissioning agent should prepare and distribute to all appropriate personnel a written commissioning plan either reinforcing or amplifying the plan contained in the project specifications, integrated with the construction schedule. The commissioning plan should give brief and concise descriptions of the work to be performed by the commissioning agent which in turn provides the contractors with knowledge of whom to have present for the various acts of commissioning. This restating or reinforcing of the commissioning plan is then used to record achievement of the mechanical milestones throughout the project's construction.

For example, projects involving water piping will normally contain specifications for flushing and cleaning the piping. This is an event that should be witnessed by the commissioning agent and identified as such in the project commissioning specification section. The mechanical contractor is placed on notice that the project schedule should identify flushing and cleaning as a milestone. Keeping current with updated progress schedules, the commissioning agent and contractors are aware of an impending act which requires their coordination. The reality is that, even with progress schedules updated as often as bi-weekly, the actual day of occurrence often cannot be clearly defined and communication from the contractor to the commissioning agent is necessary. Per the specifications and commissioning plan, the contractor should be given the responsibility of notifying the commissioning agent at least 48 hours in advance of items which can only be observed when they occur, such as the flushing of piping.

5.2.2 Construction documents and correspondence

Via the contract specifications, and reinforced at the pre-construction meeting, the commissioning agent is to be placed on the distribution

list for receipt of equipment submittals, shop drawings, and correspondence related to the mechanical systems being commissioned; i.e.,
if the commissioning is only for the heating, ventilating, and air conditioning systems, the commissioning agent does not need to be furnished documentation of plumbing systems, fire protection, etc. The
equipment submittals and shop drawings sent to the commissioning
agent should be those that bear the A/E's review stamp as the review
and approval of shop drawings and submittals for compliance with
contract documents is the responsibility of the design A/E. The review
of submittals by the commissioning agent is complementary to that of
the A/E and is necessary for the commissioning agent to be fully
aware of the actual equipment that is to be furnished on the project
(substitutions, design alternates, etc.). Should the commissioning
agent note some discrepancies between submittals and contract requirements not cited by the A/E, his duty is to notify the A/E of the
discrepancy. Ultimate resolution of acceptance or rejection of submittals resides with the A/E.

The commissioning agent should also be copied on all correspondence of a mechanical nature throughout project construction, most
notably contractor generated requests for information or clarification,
the consulting engineer's responses, and mechanical change orders.
The commissioning agent plays no role in evaluating change orders,
but must be aware of any changes from the contract documents in the
mechanical work.

The commissioning agent is required to generate correspondence
throughout the commissioning process documenting observations and
milestones passed. This correspondence should be copied to all concerned parties; i.e., the owner, architect, mechanical engineer, general
contractor and mechanical contractor. It is the mechanical contractor's
responsibility to further copy any concerned subcontractors.

5.2.3 Construction—through approximately 90 percent completion

As construction progresses, the commissioning agent makes periodic
site visits to observe the quality of work being installed and to stay
abreast of work progress and any contractual changes made. The frequency and duration of the site visits intensify as the project nears
completion. The commissioning agent should record his visits and observations in a log and issue brief summary reports for record. Any
discrepancies noted during construction should be brought to the attention of the A/E as the A/E bears the responsibility of enforcing the
tenets of the contract documents.

During the early phases of construction, the mechanical work will largely consist of the installation of piping and duct distribution systems; and the observations made by the commissioning agent would primarily be to ascertain that materials, valves, fittings, supports, and joining materials and methods are in accordance with project specifications. Prior to the application of piping and duct insulation installation, various pressure and leak tests are to be performed (per specifications); and the commissioning agent should witness most, if not all, tests.

As the project progresses, equipment will start to arrive on site and be installed by the contractor. The commissioning agent cannot observe the actual installation of all equipment, but should observe the installation of a representative sampling. Observing an installation does not mean that the commissioning agent needs to lord over a mechanic from uncrating through mounting and fitting; but rather, through periodic site visits, get an assessment of the workmanship. In the case of terminal equipment such as unit ventilators, fan coil units, variable volume air terminals, etc., the observations should be made as the first few of each type of terminal device are installed so that any problems or deficiencies noted can be brought to the attention of the engineer and contractor before the problems or deficiencies are repeated throughout the project. In the case of main air handling units and central plant equipment, such as chillers, boilers, pumps, and cooling towers, the commissioning agent should observe the site prior to setting of the equipment to ascertain if specified supports and vibration isolation are in place, and again after equipment has been set and leveled. Further site visits and observations should be made as piping and distribution systems are connected. The commissioning agent needs to be present during all equipment start-ups performed by factory representatives (such as boilers, chillers, variable frequency drives, etc.).

From the start of the project through the final connections of equipment and distribution systems, the commissioning agent will have made numerous unscheduled independent visits to the project observing the work as well as several scheduled visits coordinated with the contractor for specific milestone observations. After systems have been flushed and cleaned, balancing of air and water commence along with the calibration and tuning of controls systems. This marks a major milestone in the commissioning process. Up to this point, the commissioning agent has primarily been an observer during construction. Once balancing and control systems operations commence, the commissioning agent becomes more proactive with the greatest demand of time and effort lying ahead.

The involvement of the commissioning agent during construction, up to the point where balancing and controls calibration commence, should represent about 35 percent of the total man-hours and fee for commissioning. Thus, the commissioning process is now 50 percent complete (including the 15 percent during design).

5.2.4 Construction completion

Herein resides the irony which has created the need for mechanical systems commissioning. By this time in the project, the commissioning agent has 50 percent of his work ahead of him, but the mechanical contractor normally will have assessed his work to be 90 to 95 percent complete and invoiced accordingly. Yet systems which have not been properly balanced and/or control systems which do not function properly are probably the cause of 90 percent of comfort complaints, inefficient operation, and even systems failures. In many instances, the failures of balancing and control systems are not due to the poor performance or lack of diligence of the design engineer or of the contractors, but more so *due to the lack of the presence of a single overseer able to devote the time to fully exercise systems through their full range of control sequences.* The mechanical contractor relies mostly on the subcontractors, accepting their assertions that they have properly completed their work and systems are functioning in accordance with design. The mechanical contractor seldom performs his own comprehensive testing of operations to validate the assertions of his subcontractors. The individual subcontractors may feel that their work is completed and checked out, but it is only done so to the extent of the work for which they were responsible.

To illustrate, the following scenario is all too common: The controls subcontractor determines from the contract plans that an air handling unit is to provide minimum outdoor air at 20 percent of the total volume of the unit. He sets his outdoor air damper control signal to be 20 percent of the full range from closed to open and contends that, by moving the damper 20 percent, he has provided 20 percent outdoor air. The balancing subcontractor comes along and disconnects control damper linkages to set damper positions for minimum outdoor air based upon measured air volumes, marks the damper position, and then reconnects the linkage; he has completed his job. However, unless the controls subcontractor revisits the air handling unit to verify that his 20 percent control signal in fact moves the damper to the position which the balancing subcontractor marked, the air handling unit will introduce outdoor air in some indeterminate quantity based upon where the control signal has placed the damper. It is not unusual to see a 20 percent control signal not open the damper at all or to open the damper to a position which admits far more than the 20 percent outdoor air minimum. In the former case, minimum ventilation rates are not met. In the latter case, admittance of excessive outdoor air may exceed the heating or cooling capacity of the air handling equipment resulting in spatial discomfort and, even if discomfort is not experienced, energy is wasted.

In the illustration, both controls and balancing subcontractors would have reported to the mechanical contractor that they have completed their work, and most certainly, the mechanical contractor did not himself go back and exercise the controls to view damper operation.

The consulting engineer may perform this test, but most likely will not. The consulting engineer cannot be relied upon to be the omniscient overseer. The engineer performs a "punch out," but this does not approach the depth that independent commissioning does. The crass reality behind this is that the fee paid for engineering services does not cover the costs of the manhours which are required to exercise all systems' controls through their full range of sequences and observe the results. However, even if the engineer were given an adequate fee to encompass commissioning, the engineer may be willing to accept systems deficiencies because they are the result of his own design errors or omissions; and hope that the deficiencies do not come to the owner's attention.

Once balancing and controls calibrations commence, the commissioning agent becomes a more visible presence on site. Initially, he is there to physically observe the performance of work of the balancing and controls subcontractors. This literally involves following the workmen around for a few hours at a time. The purpose is as much to observe their work as it is to enforce in their minds that a serious look at the results of their work is impending.

Balancing, controls calibration, and making controls sequences operational can take contractors a few days or several weeks depending on the size of the project. In the case of balancing, on-site work is normally completed weeks before a testing and balancing report is submitted. Often, the building is actually turned over to the owner and occupied before the balancing report is submitted. This occurs because balancing is one of the last processes performed; and, more often than not, work on the building continues up to the completion date unless the building is delivered ahead of schedule. This can get into contractual grey areas unless it has been clearly written in the contract documents that the owner may take beneficial occupancy of the building without accepting mechanical systems as complete until the commissioning process has recommended acceptance of the mechanical systems. Commissioning can proceed ahead of submission of the written balance report, relying upon an assertion from the balancing subcontractor that balancing has been satisfactorily completed; credence being added to the subcontractor's assertion by the commissioning agent having viewed the balancing process.

By the time the controls subcontractor has completed calibration and performed his own test of sequences of operation, the mechanical contractor asserts that the installation is complete and ready for check-out. This is when the commissioning agent assumes a dominant role and exerts his most intensive efforts. Up to this point, all previous work of the commissioning agent required the efforts of a single person (though not necessarily the same person). Now, a team of people comprise the commissioning agent. At a minimum, at least two people are required equipped with radios. This is due to the fact that it is necessary to simultaneously view operations of interactive mechanical components that are remote from each other; a simple example being a room mounted thermostat and a rooftop air handling unit. It takes one person to exercise the thermostat while another observes the correlating operation of the rooftop unit. This action/reaction testing is the foundation of the commissioning process and is required to be performed on all major mechanical systems (central plant equipment and air handling units). It is normally performed on only a representative sampling of terminal equipment such as variable air volume boxes, fan coil units, cabinet heaters, etc. Control and operations of terminal units are repetitive, and any endemic problems will be discovered provided that a suitable number of each type of typical terminal unit is observed; usually a quarter to a third of each of all types of units installed is sufficient.

In addition to the team representing the commissioning agent, the presence of contractors are required during this stage in the commissioning process, and coordination is required. The general contractor may wish to be present, but may delegate his authority to the mechanical contractor. A representative of the mechanical contractor, usually the project manager or superintendent, is required to be with the commissioning agent throughout this stage. The controls subcontractor is needed through most of this stage as exercising and verifying operation of controls is the predominant activity taking place. Other subcontractors or equipment suppliers are necessary predicated on the nature of the work; for example, a factory representative is needed when equipment with factory furnished controls are exercised. The presence of representatives of the A/E and the owner are at their discretion.

The time involved with verifying the operation of controls and systems is tremendously indeterminate; based entirely upon the expertise of and diligence exercised by the contractors during installation. A system that is initially found to perform properly in accordance with design intent obviously takes less time to evaluate than one which has malfunctions. No amount of overseeing by the commissioning agent during construction can assure a flawless operational test. The reality is that mistakes will have been made, through oversight, misinterpretation of contract documents, errors, omissions, etc. It is the mission of the commissioning agent to unearth the flaws and have them corrected. Further, it is often necessary to revisit a subsystem when some other interrelated component is evaluated and found unsatisfactory. A simple illustration would be the case of a variable air volume terminal unit not being able to satisfy its space loads. The VAV terminal may be acting properly, but the air handling unit may not be delivering air at a sufficient temperature or quantity. An initial review of the air handling unit may have established that the unit appeared to work properly in response to the temperature reset or static pressure signals which it received from components elsewhere in the system. The apparent deficiency of the VAV terminal would necessitate a reevaluation of the air handling unit. Another more dramatic illustration of the difficulty in estimating time associated with tasks would be if the commissioning agent went to site to perform observations of systems operations in the cooling mode, expecting chiller operations to be producing 45°F (7°C) water, and finding that the chiller is generating 52°F (11°C) water due to chiller malfunctions. The commissioning could be terminated until proper chiller operations were established or commissioning could proceed working around the insufficient chilled water temperature; but in any case, another visit to the site would be required once chiller problems have been rectified.

Buildings furnished with computer based direct digital controls and energy management systems are considerably easier to commission than those not so equipped. The historical recording capability of automation systems permits substantial diagnosis of operations prior to actually visiting the site for inspection. Once on-site, use of the computer allows simulation of wide ranges of operational sequences. A common failure in commissioning of the automated control systems is reliance upon what the computer says that the system is doing; successful commissioning dictates visual observation of equipment to assure that the signal sent from or received by the computer correlates to the actual operation of the field device. The use of an energy management automation system, though a valuable tool in commissioning the mechanical systems, creates another whole layer in the commissioning process that requires a very detailed specification and poses unique demands upon the commissioning agent. The automation system essentially becomes the brains of the mechanical systems and warrants such extensive checking, that Chap. 8 is provided solely to address the energy management systems.

It is incumbent upon the commissioning agent to have all mechanical systems demonstrated to function in accordance with design intent as provided by the contract documents. During the demonstrations, problems found resultant from faulty installation must be rectified by the contractor, and the system reinspected. Problems that exist after all installation work checks out properly need to be brought to the attention of the A/E for rectification.

To arrive at verification that systems perform in accordance with contract documents and design intent represents approximately 38 percent of the total commissioning man-hours and fee; thus by this time, commissioning is 88 percent complete. The commissioning agent will have issued documentation of the results of the site visits made during this check-out process, citing systems that were found to perform acceptably as well as deficiencies.

5.3 Close Out

The commissioning agent reviews all close-out documents required of the contractor, including as-built drawings, operation and maintenance manuals, warranty statements, and all other certifications required for submission by project specifications. As-built drawings, including control drawings, should be furnished and used by the commissioning agent during the field check-out process. This assures their accuracy.

The review of close-out documents by the commissioning agent is complementary to those performed by the A/E.

The commissioning agent is present at all demonstration and training periods given by the contractor to owner's operating and maintenance personnel.

Upon satisfactory completion of all of the foregoing, the commissioning agent provides a letter stating that the mechanical systems are ready for acceptance by the owner. It is at this point that warranty periods should begin. This phase represents approximately 2 percent of the commissioning man-hours and fee; commissioning is now 90 percent complete.

5.4 Follow-up

As the acceptance of the mechanical systems usually occurs either in summer or winter, i.e., in the cooling or heating mode; a subsequent season inspection is required. Satisfactory operation of systems in the cooling mode is no guarantee that systems will perform properly in the heating mode and vice-versa. Therefore, trends generated by the automation system are again reviewed (if the building is so equipped) and the commissioning team needs to revisit the site to exercise systems in the remaining mode of operation: This process constitutes approximately 10 percent of the commissioning man-hours and fee; and when satisfactorily completed, the commissioning process is 100 percent complete.

5.5 Practical Commissioning

Though the benefits of mechanical systems commissioning are most fully realized when the commissioning agent is brought on-board as a team member at the earliest stages of design; commissioning can commence at any phase in the project, with commensurate reductions in what commissioning achieves.

For example, perhaps the commissioning agent is retained at about the time that the contractor has concluded his work; all systems being operational, balancing and controls calibration and set-up having been completed. The commissioning agent essentially performs a test and verification of systems operations. Before the commissioning agent visits the site, he must perform catch-up work to understand the mechanical systems design and intent. He must be provided with copies of contract drawings and specifications, construction correspondence (requests for information, clarifications, change orders, etc.), mechanical equipment and controls shop drawings, test and balance reports, and as-built drawings for review. If the building has a computer based automation system, trends need to be provided for review. Once the commissioning agent is satisfied with his understanding of mechanical systems design intent, and if trends indicate apparent proper systems operation when an automation system is involved, the commissioning team is ready to visit the site to perform tests and observations. Critical components of the commissioning process are omitted by this late stage commissioning; particularly, the commissioning agent will not have witnessed systems flushing and cleaning, start-up of major pieces of equipment, testing and balancing procedures, and controls set-up. Many deficiencies manifested by errors or inadequacies in systems installation and set-up may still be discovered by the commissioning agent, though certainly not to the extent had the commissioning agent been involved with the project through construction; and more importantly, having been involved during construction, deficiencies which are discovered and require remedial action by the contractor may have completely been avoided by early detection.

To illustrate, suppose that in checking out an air handling system, the commissioning agent has found that the air handling unit appears to be operating properly. The filters and coils are clean, the test and balance report indicates that air and water flows are in accordance with design, controls appear to function properly, chilled water is being delivered at proper temperature, yet the space which is served by the air handler cannot maintain comfort cooling conditions. Initial perceptions may be that the system is inadequate for the load imposed upon it; maybe the coil or fan are too small, i.e., a design error. Before asserting a design error (usually the first response from contractors), the commissioning agent must perform further diagnostics. In doing so, he may unearth installation or set-up failures by the contractors. It may be that when the system was balanced, full water flow was documented at the cooling coil; but during subsequent operation, some slag lodged in the control valve, preventing it from opening fully. All outward appearances would indicate that the valve has stroked fully, but checking water and air temperatures across the coil would indicate that full flow was not being realized. Diagnostics may have also revealed, that the controls contractor's signal to open the air handling unit's outdoor air damper to its minimum position, in fact opens the damper well *beyond* the minimum setpoint, introducing hot, humid outdoor air in excess of the design cooling coil's capacity. Due to the contractor's inadequate flushing of the water system or lack of coordination between balancing and controls subcontractors, latent problems arose which then require remedial work (possibly draining systems, pipe fitters, controls work, balancing work, etc.). The system then must be retested by the commissioning agent.

Some of the problems of the foregoing illustration may be avoided by bringing the commissioning agent into the project at *earlier construction stages;* but there is no optimal time other than at commencement of design. Fees for commissioning are much more difficult to establish when the commissioning agent is not a part of the team from the beginning. The commissioning agent may offer a fixed fee based upon his assessment of the complexity of the mechanical systems and the time which he will have to devote, depending on when he poorly installed project becomes involved with the project. However, his actual time to complete the commissioning is greatly governed by the quality of work performed by the contractor. As indicated by the illustration of the air handling unit that did not maintain space temperature, the commissioning agent was required to spend additional time performing diagnostics and re-inspection beyond that time which he may have initially allocated to check a properly operating air handling unit. Therefore, if the commissioning agent is to be locked into a fixed fee, he may inflate it to protect from losing money on a poorly installed project. If the fixed fee proposed by the commissioning agent is too low to accommodate the poorly installed project, he has the choice of taking a loss, to provide less than the full service he had intended, or to solicit extra compensation from the owner. For these reasons, a fee proposal which offers compensation for hourly services with perhaps an upset limit is more practical, fair and equitable for both the owner and the commissioning agent. Also, a means by which the owner may extract recompense from the contractor for extra time spent by the commissioning agent for a poorly installed project, may instill greater diligence by the contractor in performing the installation.

6

The Commissioning Specification

Michael S. Bilecky, PE, CEM
George H. Raaf, Consultant
William A. Butler, Consultant
Von Otto & Bilecky, Prof. Corp.
Washington, D.C.

6.1 General

The Commissioning Specification is critical to the success of the commissioning process. It serves multiple purposes, both objectively and subjectively. It objectively defines the scope of work to be performed by all parties involved with the building: the design team, the builders, the owner's operating and maintenance personnel, and the commissioning agent. Via the overt statements of responsibilities and expectations, it places everyone involved on notice that an independent oversight of design and installation will be performed, instilling a reality that business will not be conducted as usual. Going through the commissioning process entails demonstration and verification of mechanical systems operations and performance not normally performed by conventional design and construction processes. This realization should result in a greater dedication to detail and workmanship by the designers and contractors.

6.2 Related Documents

The Commissioning Specification does not stand alone. It must be written as an integral part of the complete project specifications; most particularly to be in sync with Division 1, General Requirements, and

Division 15, Mechanical. Division 1 will contain the general dictates and tenets and Division 15 will contain the specific dictates and tenets for submittals, shop drawings, as-builts, record documents, warranties, etc. Division 15 will further provide the specifics of the mechanical equipment and systems. It should be assured that items such as warranties, certificates, materials testing, start-up procedures, as-built drawings, inspections, demonstrations, and close-out submittals are clearly spelled out and covered. These issues are not provided in the commissioning section of the specification, but are enforced and verified by the commissioning section. The following offer several examples:

The requirements for flushing, cleaning, and testing of piping systems is to be specified in Division 15, but not in the commissioning section. The commissioning section specifies that the commissioning agent will observe the flushing, cleaning, and testing.

Testing, adjusting, and balancing (TAB) is a complete section of Division 15 and spells out procedures for the complete testing, adjusting, and balancing of the mechanical systems. The commissioning specification cites that the commissioning agent will observe TAB procedures and allows for limited amounts of retesting to be performed to demonstrate that TAB has been performed correctly. The TAB and commissioning sections of the specification have to be coordinated so that what is expected by the commissioning agent is provided in the TAB section.

Demonstrations and instructions for operating and maintenance personnel should be included in the General Requirements section of Division 15; spelling out a procedure and giving a minimum number of hours that will be required of the contractor to perform demonstrations and instructions. The commissioning section of the specification will assert that the commissioning agent will be present during the demonstrations and instructions.

The foregoing are but a few examples illustrating that much of what the commissioning agent performs is observance of compliance with requirements contained elsewhere in the project specifications. The presence of a commissioning authority does not relieve the A/E from their obligations to prepare comprehensive specifications for the project.

6.3   The Specification and Check Lists

The following sample specification and check lists are offered as a guide. They should be prepared by the independent commissioning agent to be incorporated with the project specifications prior to the project being issued for bids. The specification and check lists could also be prepared by the project mechanical consulting engineer if a commissioning agent has not been involved with the project through design. As with any good contract document, the specification and check lists should be written tailored to the specific requirements of the project to which they will be applied. It is impossible to provide guide specifications and check lists within the confines of this book that would encompass all mechanical system types and the work required to properly commission them. The sample specification and check lists presented should offer a basis from which project specific documents may be written. To facilitate this process, explanations and amplifications are given to selective paragraphs in italic type in the specifications.

For convenience in reproduction for the reader's own use, the checklists are shown separately in Chapter 7 following.

Sample Specification

Part 1—General

1.01   INTENT: The intended result of the HVAC Commissioning process is to assure the Owner that the HVAC systems are installed and operate in accordance with contract drawings and specifications prior to the Owner's acceptance of the building.

1.02   SCOPE:

   A. Work Included: The HVAC commissioning shall provide substantial verification that systems and equipment are installed and performing in accordance with the contract documents and design intent. This independent commissioning shall be complementary to the construction period services performed by the Architect and Mechanical Consulting Engineer.

   B. Work Not Included: It shall not be incumbent upon the Commissioning Agent to verify adequacy of HVAC systems to accommodate the heating, cooling, and/or ventilating loads imposed upon them, i.e. to evaluate design. Systems installed and performing in accordance with plans and specifications which do not achieve and/or maintain spatial conditions in accordance with design intent will be so noted when observed. Commissioning of plumbing, fire protection sprinkler, and electrical systems are excluded from the HVAC commissioning process except as may be incidental to the operations of the HVAC system; i.e. note would be made if condensate drainage was observed to be inadequate, if a fan turned backwards due to improper power connections, etc.

1.03   RELATED DOCUMENTS: The contract drawings and requirements of DIVISION 1, GENERAL REQUIREMENTS, DIVISION 15, MECHANICAL, AND DIVISION 16, ELECTRICAL apply to this Section of the Specifications.

(This is important as the commissioning section of the specification essentially only provides the procedural methods which will be employed to verify that the requirements of drawings and specifications have been met.)

1.04   COMMISSIONING AGENT AUTHORITY: Throughout the commissioning process, the commissioning agent's role is primarily one of an observer/witness; monitoring the installation, start-up, and operation of the mechanical heating, ventilating, and air conditioning (HVAC) systems. The commissioning agent shall have no authority to alter design or installation procedures. If acceptable performance cannot be achieved, it will be the commissioning agent's responsibility to appraise the Owner, design engineer, and/or contractor of the deficiency. Corrective actions shall be the responsibility of the Owner, design engineer, and/or contractor; and not that of the commissioning agent. The commissioning agent shall have the authority to require tests and demonstrations to verify proper performance.

1.05   ARCHITECT/ENGINEER RESPONSIBILITY:

A.   In addition to their normal performance of Construction Period Services, the Architect/Engineer will furnish to the Commissioning Agent one copy of all approved HVAC systems Shop Drawings and Submittals and place the Commissioning Agent on the mailing list for all communications regarding the HVAC systems.

B.   The Architect/Engineer shall respond in writing to deficiencies cited in correspondence issued by the Commissioning Agent.

(This has been found to be a necessary means of documenting that issues raised by the commissioning agent have been addressed to minimize the amount of time required of the commissioning agent to re-verify systems.)

1.06   CONTRACTOR'S RESPONSIBILITY:

A.   The General Contractor shall be responsible for assuring that the commissioning agent is provided with all relevant correspondence, submittals, notifications, and assistance as may be required to satisfactorily complete the commissioning process using whatever personnel, time and resources that are required. This Section provides minimum commissioning requirements, however, the Contractor shall exceed those requirements whenever necessary to achieve the intent of HVAC Commissioning.

B.   The General Contractor shall include in his Bid the cost of furnishing the material requested and manpower necessary for the verification of proper HVAC system installation and operation as specified in this Section.

C.   The Contractor shall respond in writing to deficiencies cited in correspondence issued by the Commissioning Agent.

(*This has been found to be a necessary means of documenting that remedial work has been performed, hopefully properly, to minimize the amount of time required of the commissioning agent to re-verify systems. Enforcing Contractor documentation assures that issues raised by the commissioning agent are not ignored or swept under the carpet.*)

Part 2—Products

2.01   NOT APPLICABLE TO THIS SPECIFICATION

Part 3—Execution

3.01   COMMISSIONING TEAM: At a minimum, the following are members of the commissioning team:

A.   The Owner's Authorized Representatives

(*This should include operating and maintenance personnel that will ultimately be responsible for the building*)

B.   The Architect

C.   The Consulting Mechanical Engineer

D.   The General Contractor

(*Though it is primarily the mechanical HVAC systems which are being commissioned, the General Contractor bears ultimate responsibility for the successful completion of the project and needs to stay abreast of the work of the subcontractors. In addition, it is not unusual to unearth construction related problems when commissioning the HVAC systems. For example, it may be found that an HVAC system is installed and performing in accordance with design intent, but it cannot maintain proper space comfort conditions. While diagnosing the problem, it was found that an opening in an exterior wall was not properly sealed, allowing leakage into the building. This becomes an issue for the General Contractor to rectify.*)

E.   The Electrical Subcontractor

(The involvement of the Electrical Subcontractor can become necessary if power connections to HVAC equipment or control wiring interlocks with starters have been found to be missing or improperly connected. Also, in many installations involving computer based automation systems, much of the wiring installation may have be subbed to the project electrical subcontractor by the controls subcontractor. A common example of a failure in coordination between mechanical / electrical work is when the mechanical has specified a two speed fan and the electrical has provided a single speed starter.)

F.   The Mechanical Subcontractor (and all of his subcontractors performing HVAC work with particularly critical involvement of the Testing, Adjusting, and Balancing Subcontractor and Temperature Controls Subcontractor)

(At one point or another in the project, the commissioning agent will have been in contact with nearly all of the Mechanical Contractor's subcontractors. The Mechanical Contractor may perform only the piping components of the project with in-house personnel and have subbed out the sheet metal, insulation, controls, testing and balancing, etc. Additionally, all major pieces of equipment usually are specified to be initially started by factory authorized technicians. The commissioning agent will observe the work of all of these subcontractors and factory technicians; and the Mechanical Contractor is responsible for placing all of his subcontractors and equipment suppliers on notice to respond to the authority of the commissioning agent.)

G.   The Commissioning Agent

(The following paragraph is necessary to place all concerned parties on notice that time and effort will be required of them to accommodate HVAC commissioning.)

3.02   RESPONSIBILITIES: Each member of the commissioning team has responsibilities to the successful completion of the commissioning process as follows:

A.   The Owner's Representatives shall perform their normal construction contract administration functions.

(Depending upon who the Owner is, this work may be delegated to the Architect / Engineer if it is a small project with an Owner that does not have a construction management team. On government or institutional projects, the Owner may have their own Contract Administrator, field construction manager, or teams of people representing different divisions within the Owning agency such as a boiler maintenance shop, an HVAC shop, an energy management division, etc. Whoever is involved representing the

Owner, their duties are essentially unaltered by that of the commissioning agent.)

B.   The Architect shall provide adequate support to the Consulting Mechanical Engineer as related to his duties in the commissioning process. It shall also be the Architect's responsibility, either directly or through his assignee, to assure that the commissioning agent is:

    1.   Provided copies of approved shop drawings as they are returned to the Contractor.

    2.   Notified of time, date, and place of all regularly scheduled progress meetings, and of any special meetings that may be called regarding HVAC systems.

    3.   Copied on all correspondence pertinent to the HVAC systems including but not limited to minutes of progress meetings, responses to contractor requests for information, change order documentation.

(As the work being performed is HVAC commissioning, the Architect's duties are only moderately affected by the commissioning process. The Architect needs to stay abreast of the commissioning process as it is normally the Architect who bears ultimate responsibility for the design team and construction management. All correspondence for the project generally flows through the Architect and he needs to assure that the commissioning agent is in the loop.)

C.   The Consulting Mechanical Engineer shall perform his normal construction contract administration functions.

(The normal contract administration functions include attendance at progress meetings, shop drawing review, answering questions, issuing clarifications or change orders, site visits to monitor project progress or address conflicts, punch-out, review of As-Built drawings, review of operating and maintenance manuals and other close-out submittals. The presence of a commissioning agent does not relieve the mechanical consulting engineer of any of these duties. Any responses from the consulting engineer necessitated by observations of the commissioning agent fall within the normal contractual obligations of the consulting engineer. The Architect may delegate some of the correspondence and document transfer between the A / E team and the commissioning agent to the consulting engineer; i.e. directly providing shop drawings and permitting direct correspondence with record copies being furnished to the Architect.)

D.   The General Contractor shall, in addition to his normal responsibilities for construction of the project, assure that his subcontractors recognize the authority of the commissioning agent and perform responsive to the requirements of the commissioning process. He shall assure that proper notification, at least 48 hours in advance, is provided to the commissioning agent of the milestones of the mechanical systems installation, at a minimum as follows:

1. Pressure testing of piping systems
2. Flushing and cleaning of piping systems
3. Factory start-up of central plant equipment
4. Factory start-up of rooftop equipment
5. Calibration of Automatic Temperature Controls
6. Start date of Air and Water Balancing
7. Date of punch-out inspections
8. Date of instructions to Owner's operating personnel regarding operations of the HVAC system

(The foregoing establishes that the General Contractor, who has ultimate responsibility for the construction sequencing of the project, must also take responsibility for proper notifications of the commissioning agent. He may delegate this responsibility to the Mechanical Subcontractor at his discretion. The milestones listed are minimum requirements, and there may be others specific to the needs of the project which should be identified here and again restated in the commissioning plan presented at an early progress meeting.)

E.   The Electrical Subcontractor shall perform his normal contract obligations and be responsive to the authority of the commissioning agent.

(This places the Electrical Subcontractor on notice that he may be required to perform demonstrations and in other ways be responsive to the commissioning process.)

F.   The Mechanical Subcontractor shall, in addition to his normal responsibilities for construction of the project, assure that proper notification of the milestones of the mechanical systems installation as cited in Paragraph D. are provided to the General Contractor. The Mechanical Subcontractor shall assure that his subcontractors recognize the authority of the commissioning agent and perform responsive to the reqirements of the commissioning process, particularly the Testing, Adjusting and Balancing (TAB) and Temperature Controls and/or Automation System subcontractors.

(This places the Mechanical Subcontractor on notice that he, and all of his subcontractors, will be required to perform demonstrations and in other ways be responsive to the commissioning process.)

G.   The Commissioning Agent will follow the procedures as set forth in Paragraph 3.03 to execute his responsibility for:
1.   Verifying that the mechanical systems are installed and operating in accordance with contract documents and specifications.
2.   Assuring that Owner's Operating and Maintenance personnel are fully trained on systems operation and maintenance.
3.   Assuring that Close Out documentation is properly provided to the Owner.

3.03   COMMISSIONING AGENT PROCEDURES: The Commissioning Agent will perform systems commissioning following the procedures listed herein, and all members of the Commissioning Team shall cooperate fully with the execution of these procedures. Initial HVAC commissioning shall be performed while the central plant is in operation, either in the heating or mechanical cooling mode. Initial commissioning will not take place during "swing" seasons when neither boilers or chillers are operating. If commissioning occurs during the mechanical cooling operation, subsequent commissioning shall be performed during the next "swing" season and at the changeover to the heating season; and similarly, if commissioning occurs in the heating season, subsequent commissioning shall be performed during next "swing" season and during the next cooling season. The subsequent seasonal systems commissioning shall consist of observing the central plant equipment start-up, reviewing complete sets of automation system operational histories (*on projects with automation systems*), and follow-up site visits to spot check automatic temperature controls systems.

(The foregoing paragraph may be modified if such complete service is not retained by the Owner or if the mechanical systems are not designed for such distinct modes of operation. The full and complete commissioning of HVAC systems necessitates observations of systems in all modes of operation; meaning mechanical cooling, heating, and natural cooling modes. Some HVAC systems designs, particularly two-pipe heating and cooling systems have a "swing" season of operation where neither mechanical cooling or heating is provided; and the systems designed operate primarily in an economizer mode. An example of this could be a school which shuts down the chiller plant in late September and does not start the boiler plant until late October. There could be a 3–4 week period where the mechanical systems only operate in the natural ventilation and cooling mode. On the other hand, the school could be designed as a four-pipe system to operate without economizer cycles. In this case, only the heating and mechanical cooling modes need ver-

ification. In all cases, it is essential that the HVAC systems be commissioned in both heating and mechanical cooling modes as operaton in the heating mode may not reveal problems that exist in the cooling mode, and vice-versa.)

To achieve HVAC systems commissioning, the Commissioning Agent shall:

A. Attend periodic construction progress meetings and perform unscheduled walks through the building to observe and keep abreast of mechanical systems installation progress, means and methods. Commissioning agent's presence at meetings and in the building will be for his benefit in preparing to commission the building and shall in no way be construed as superseding the authority of the project architect/engineers.

(This reinforces that the role of the commissioning agent is primarily one of observer and that he has no authority to alter design.)

B. Perform a complementary review of HVAC Shop Drawings after approval of the Project Consulting Mechanical Engineer. The purpose of this review is primarily one of familiarization with equipment to be furnished on the project for on site verification by the commissioning agent and in no way relieves the Consulting Mechanical Engineer of his duties for shop drawing review.

(This reinforces that the Consulting Mechanical Engineer is responsible for approving what gets installed on the project, and that the role of the commissioning agent is primarily one of verification.)

C. Issue Commissioning Check Lists relevant to the project for Contractor monitoring and verification of installation progress (See Sample Check Lists). Establish milestones in the HVAC system installation at which time interim commissioning status reports will be prepared and issued by the Commissioning Agent.

(The commissioning check lists are essential to not only document completion of milestones, but also provide an amplification of the scope of work which will be required of various parties in the commissioning process. It is recommended that a complete set of commissioning check lists be included with this specification section, specific to the project, to assist contractors in allotting time for commissioning.)

D. Observe piping system pressure tests, flushing, and cleaning.

(The wording of this paragraph leaves open to the discretion of the commissioning agent as to how many pressure tests, flushing, and cleaning he will observe. On a small project, he may observe all such operations. On large projects, particularly those done in phases, he may opt to observe only selected systems once he is satisfied that work is being performed properly.)

E.   Observe representative sampling of duct systems, plenums, coils, and filters for cleanliness, damage, or leakage.

(The extent of a representative sampling is left up to the discretion of the commissioning agent, but places contractors on notice that such systems will be inspected.)

F.   Be present to observe the start-up of central plant equipment; i.e. chillers, cooling towers, and boilers. This shall be the start-up which is supervised and certified by the equipment manufacturer's authorized agent. Commissioning agent will observe the start-up of a representative number of self-contained refrigeration units (rooftops, split-systems) and verify that the operations have been certified by manufacturers' representatives.

(This paragraph should be further amplified specific to the project to include any other types of equipment which are specified to be started or certified by factory authorized representatives, such as variable frequency drives, ice storage systems, energy recovery equipment, etc.)

G.   Visit the project periodically during testing, adjusting, and balancing of the air and water systems to observe the actual work being performed, and review the certified reports submitted by the TAB agency.

(Again, the review of the TAB report by the commissioning agent is complementary to that of the mechanical consulting engineer. It is the consulting engineer's duty to address discrepancies or deficiencies; and it is the commissioning agent's responsibility to assure that this work is performed. It is the commissioning agent's duty to appraise the engineer of discrepancies or deficiencies that may have been missed by the consulting engineer in his review.)

H.   Periodically observe the installation of the automatic temperature controls system: observe the operation of the main air compressor (running time, pressure, moisture), verify calibration of a representative sampling of sensing devices, observe the operation of all air handling units and exercise their respective control sequences including safeties, exercise the controls and observe operation of a representative sampling of incremental units, observe the operation and controlling sequence of all central system piping isolation, changeover, and modulating control valves.

I. Verify the commissioning of the Engery Management Automation System by reviewing point histories and by being present during hardware and software punch outs at the facility. The operational and monitoring capabilities of the EMS will be employed in the commissioning of the mechanical systems.

(Paragraphs H. and I. need to be tailored to be harmonious with one another and specific to the project. Though most modern day projects are being designed with computer based control systems, not all are done using full direct digital controls and electronic / electric actuating devices. There still exist some projects with only pneumatic / electric controls and often times some hybrid form of controls exist where pneumatics are used for actuators while the rest of the control system is computer based. Commissioning of computer based automation controls systems requires substantial elaboration and unique processes which are amplified further in Chapter 8.)

J. Be present during Mechanical Contractor instructions to the Owner's operating personnel regarding operations of the HVAC system.

K. Review Contractor prepared Operating and Maintenance Manuals, As-Built drawings, and all certifications and warranties required for submission by project specifications.

(This review is complementary to that of the consulting mechanical engineer. Certificatons can include such items as equipment compliance with specified standards such as ASME and ARI, water treatment, etc. Warranties can be those for manufactured equipment such as compressors, variable frequency drives, etc. as well as the contractors' warranties.)

L. Furnish a written report and recommend acceptance of the HVAC system upon satisfactorily completing the commissioning process. Recommendations for approval, when appropriate, will be forwarded to the project Architect/Engineer for inclusion in their final submission of project close out documentation to the Owner.

("...recommend acceptance of the HVAC system upon satisfactorily completing the commissioning process." is a key phrase that needs to be defined in other procedural sections of the project specifications. Normally, acceptance of the project has occurred at what has been viewed as project completion; that is, systems have been inspected and found to be operating satisfactorily, and the contractors start their warranty periods. Problems that sub-

sequently arise are viewed as warranty issues. The fallacy in this is that the mechanical systems are normally accepted after having been viewed in only one mode of operation, either heating or cooling, depending on what time of year the General Contractor has completed the work; and, as stated previously, the mechanical systems need to be inspected in all modes of operation to satisfactorily pass the commissioning process. A mechanism may be inserted in the contract documents to afford the contractors their due in allowing a building to be accepted and warranty periods to commence after this single season inspection with a proviso that acceptance of the mechanical systems is given conditionally in anticipation that the systems will perform acceptably in the next seasonal change-over. Should the systems not perform acceptably, the initial acceptance is rescinded until such time as the subsequent seasonal operations are accepted. The warranty of the mechanical systems then commences at the new date of acceptance. For example, suppose that the building was accepted while the mechanical systems were operating in the cooling mode; and the date of acceptance and commencement of the warranty period was September 1. The subsequent inspection of systems in the heating mode revealed failures of the controls system to properly operate or an inability to establish proper water flow in a heating loop. The contractor should not be let off the hook for providing a full year's warranty (or two year warranty if so specified) for the mechanical systems; i.e. the Owner is purchasing a timed warranty of the mechanical systems. If successful commissioning of the mechanical systems in the heating mode did not occur until December 1, it should be the contractor who needs to carry the warranty an extra three months in lieu of the Owner having to accept a nine month warranty.

This methodology can effectively work with the warranty of the mechanical subcontractor, provided that it has been clearly stated as such in the specifications; however, issues of fairness should also be incorporated. For example, if the building is accepted by the Owner while systems were operating in the cooling mode, manufacturer's warranties for equipment such as chillers and direct expansion rooftop equipment should be permitted to remain in effect from the date of acceptance regardless of the results of the subsequent seasonal commissioning. After all, failure of the heating system should not adversely affect the manufacturer's warranty of the chiller since the chiller had been proven to operate properly and was accepted by the Owner. The review of warranty statements is part of the commissioning agent's responsibilities per paragraph K. above.)

3.04   CONTRACTOR COMMISSIONING PROCEDURES: The General Contractor and all relevant Sub-Contractors shall, in addition to being responsive to the procedures cited for execution by the Commissioning Agent in paragraph 3.03, perform as follows to achieve satisfactory HVAC systems commissioning. The Contractor shall:

A.   Demonstrate the performance of each piece of equipment to the Commissioning Agent and Owner's Representative after completion of construction. Schedule the TAB, HVAC controls, energy management and other sub-Trade Representatives as may apply to demonstrate the performance of the equipment and systems.

(It is critical to the success of the commissioning process that all parties responsible for the installation of a specific system be present during the demonstration. Invariably, a failed performance test will be blamed on the individual not present at the demonstration. For example, in testing an air handling unit, at a minimum, the mechanical contractor and the controls contractor need to be present for the initial test as the system is operated through its various control modes. If satisfactory performance cannot be achieved, and the cause of the failure not readily recognized and remedied, responsibility for the failure may be laid on the balancing contractor, the consulting engineer, the equipment supplier, etc. A subsequent visit will be required which should additionally involve all parties that had a hand in putting the system together. As much time and effort can be wasted in repeated calls back to revisit a poorly operating system, it behooves all parties concerned to get it right the first time; and failing that, assuredly get it right the second time. This becomes a cumbersome but essential process of scheduling the right people at the right time. Similarly, one does not want to waste the time of a lot of people by calling forces en-masse to the site to needlessly wait while systems for which they had no responsibility are checked; i.e. the boiler representative need not be on site while chiller operations are being inspected.

The commissioning agent can save all parties a lot of time by listing those parties that should be present during any particular test in the commissioning plan presented at the outset of construction.)

B.   At a minimum, the performance and operation demonstrations of the following equipment and/or systems will be required:

(Here is where the scope of work is defined, specific to the project. All systems to be commissioned need to be mentioned. The nature

of the tests and demonstrations to be performed would be further amplified by the line items shown on the Commissioning Check Lists. The systems, equipment, and percentages listed herein after are offered as examples.)

1. Incremental equipment (Fan Coil Units, Cabinet Heaters, Unit Ventilators, etc.); approximately 20 percent of units installed.

2. Air balancing: major trunk duct flow and pressure checks, air terminals, variable volume boxes, outdoor air damper settings. The TAB Trade Representative shall identify all places where temperature, pressure and/or velocity readings were taken in major duct systems; and performance shall be demonstrated on up to 20 percent of the locations. Up to 5 percent of air terminals shall have performance demonstrated. Outdoor air settings shall be verified on all major air handling units and on up to 10 percent of incremental equipment and self contained air conditioning units. Flows shall be verified on up to 20 percent of variable air volume terminals.

3. Water balancing: flow settings at pumps and in distribution piping, up to 20 percent of the locations of flow fittings.

4. Fire and smoke damper installation and operation, up to 30 percent.

5. AHU Coil performance; all coils during both cooling and heating.

6. Fan and motor performance; all major air handling units and up to 30 percent of exhaust fans.

7. Pump performance.

8. Chiller system performance.

9. Boiler performance.

10. HVAC Controls System; complete control sequence of central plant equipment and major air handling units, and up to 20 percent of incremental and terminal equipment (unit ventilators, fan coils, cabinet heaters, variable air volume boxes).

C. In addition to the foregoing, the Contractor shall repeat any other measurement contained in the TAB report where required by the Commissioning Agent for verification or diagnostic purposes. Should any verification test reveal operation or performance not in accordance contract documents, the Contractor shall rectify the deficiency, and re-inspection shall be performed. Should operation or performance still not be as specified on the re-inspection, the time and expenses of the Commissioning Agent to make further re-inspections shall be considered as additional cost to the Owner. The total sum of such costs shall be deducted from the final payment to the Contractor.

(This can get very dicey. There needs to be some mechanism to protect the commissioning agent from losing his shirt due to shoddy workmanship by the contractor. The commissioning agent will have entered into a contractual agreement with the Owner to provide a service. The fee for that service is not open ended, nor is the time that the commissioning agent gives in performing that service. If the commissioning agent is required to return to the project time and again to test systems that the contractor has not made to function properly, the commissioning agent cannot be expected to absorb the costs of these revisits under his base fee. To obtain recompense for exceptional expenditures of time, the failures of the systems have to be well documented to indicate that the commissioning agent was not frivolous in his rejection of systems performance in order to gain a higher fee. Also, the Owner should not bear the additional expenses incurred by the commissioning agent due to failures of the contractor. The hourly rates of the commissioning agent should be a matter of record with the Owner and the contractor so that the costs of additional services come as no surprise if invoked.)

3.05 DEFICIENCY RESOLUTION:

A. Deficiencies identified during the commissioning process shall be corrected in a timely fashion. The Commissioning Agent has no authority to dictate ways and means of deficiency resolution other than enforcing the dictates of Contract Drawings and Specifications. Resolution of deficiencies that require interpretations or modifications to the contract documents shall be the responsibility of the Architect and Engineers. Project completion date shall not be delayed due to lack of timely resolution of deficiencies unless authorized contract extensions have been executed.

B. Written responses shall be made to deficiencies correspondence issued by the Commissioning Agent. The Commissioning Agent shall issue such correspondence as deemed appropriate during the commissioning process with original provided to Owner and copies to the General Contractor, Architect, and Consulting Mechanical Engineer. The General Contractor, Architect, and/or Consulting Mechanical Engineer shall provide the Owner with a written response to each deficiency item cited by the Commissioning Agent as to corrective actions implemented. The written response shall be provided to the Owner within two (2) weeks of the date of the Commissioning Agent's deficiency correspondence; copies shall be provided to the Commissioning Agent, General Contractor, Architect, and Consulting Mechanical Engineer. Deficiencies which have not been fully resolved within the two week period shall be noted as such with explanation of intended resolution; and subsequent status reports of the continued deficiency resolution shall be made in writing at two week intervals until such time as the deficiency has been fully rectified. The Owner reserves the right to withhold partial payment for construction contract or professional services until satisfactory resolution of mechanical deficiencies have been documented and verified.

(The foregoing will go a long way in minimizing repeated observations of failures and deficiencies. By dictating that written acknowledgment be made, tracking of resolutions becomes a matter of record; and a lot of wasted time can be avoided. On large projects, this is essential as there will be many observations made, and it will be impossible to keep track of resolutions informally. For example, suppose at a point in the construction of the project, the commissioning agent observes that a balancing fitting has been omitted in a piping loop. He makes this observation known to all concerned parties and awaits a response. Should there not be a requirement for written acknowledgement and response, the contractor could overlook or ignore the notification. Perhaps the balancing subcontractor makes the mechanical contractor again aware of the missing fitting; but by this time the mechanical contractor has had systems tested and filled and is driving toward project completion. With completion date approaching, the contractor asserts that the has completed his job and is ready for inspection; hoping that the missing balancing fitting was not really necessary anyway. The commissioning agent may have repeatedly written about the missing fitting as a continuing deficiency, but he has no enforcement capability until such time as he performs operation tests. The commissioning agent goes to the job site and finds that an air handling unit is not performing in accordance with design. The commissioning agent may or may not be able to assess the deficiency to a lack of adequate water flow. In any event, subsequent visits will be required, and may eventually lead back to the absence of the balancing fitting.

By having written responses, it becomes more difficult for the contractor to ignore or overlook a deficiency; and without suitable written documentation from the contractor that the missing balancing fitting has been installed or acknowledgment from the design engineer that systems performance will not be adversely affected by the omission; the commissioning agent may refuse to test any systems affected by the missing fitting until appropriate notification from the contractor or the engineer are in hand.)

3.06   SATISFACTORY COMPLETION: The Contractor's personnel shall be made available to execute all aspects of the commissioning process until the Owner accepts final results. Commissioning tasks and meetings may be repeated until the Owner is satisfied and will not be fixed as one-time, one-chance events for the Contractor.

(This precludes the contractor from taking the position that he responded to the deficiency cited, remedied it, and sees no need in revisiting the issue. Actually, in some cases, this may be an entirely satisfactory response. For example, suppose an exhaust fan was found to be inoperative by the commissioning agent. The contractor found a broken fan belt, replaced it, and stated that the fan now runs fine. The commissioning agent need not convene a meeting to verify the fan's operation, but can do so himself incidental to one of his site visits.

On the other hand, there are many instances where fixing one problem may subsequently reveal another which could not previously be discerned as it was masked by the first. For example, suppose the fan with a broken belt was a return fan on a variable air volume system controlled by a variable frequency drive. Replacing the fan belt and making the fan operational in no way satisfies the need to establish that the fan runs properly under automatic controls and the variable frequency drive. Suppose that upon inspecting the operating return fan, it was found that the VFD is not properly tracking the input signal from the controls system. The controls or VFD problems need to be rectified and the system re-visited.

The key to this paragraph is that the contractor is responsible for as many visits as may be necessary to get it right; and the final arbiter of when enough is enough is the Owner, not the contractor.)

3.07   PHASING OF CONSTRUCTION AND COMMISSIONING: Where project completion is performed in stages, the commissioning plan will take into account the staged start-up of each phase as shown on the Drawings and/or Specified.

3.08   CLOSE OUT SUBMITTALS: Close out documents, consisting of but not limited to As-Built Drawings, Certificates of Inspections, Warranties, Final Testing and Balancing Reports, Operating and Maintenance Manuals as submitted to the Architect/

Engineer shall be copied to the commissioning agent for concurrent review.

3.09 DEMONSTRATIONS AND TRAINING: Commissioning agent shall be notified in advance of the dates and times of demonstrations for and training of Owner's operating and maintenance personnel, and the commissioning agent shall be present at all sessions.

3.10 COMMISSIONING REPORTS

A. The Commissioning Agent shall document commissioning milestones with reports. The documents shall acknowledge acceptance at the milestone or separately list deficiencies observed or discovered. The document shall be distributed to Commissioning Team members.

(It is equally as important to acknowledge successful completion of commissioning milestone as it is to cite deficiencies.)

B. The Commissioning Agent shall prepare a final formal report to the Owner which will include a narrative in the form of an Executive Summary of the results of commissioning process, impressions of the demonstration and training sessions, and a certification that the verification of each item is complete and all systems are operating as intended.

Commissioning Check Lists

Introduction

Michael S. Bilecky, PE, CEM
George H. Raaf, Consultant
William A. Butler, Consultant
Von Otto & Bilecky, Prof. Corp.
Washington, D.C.

COMMISSIONING CHECK LISTS: (See Tables 7.1–7.5)
(Two options are given for this paragraph; only one should be used in the project specifications. The first paragraph applies where all of the detailed check lists are included with the specifications. This is the most desirable approach as it most clearly defines scope and expectations. However, recognizing that there may be occurrences where, for one reason or another, completely detailed check lists cannot be provided at the time that the project specifications are printed and the project bid, the latter paragraph can be used to achieve a functional equivalent. Though the check lists should be as all encompassing as possible while being practical, there are always cases where something may have been missed and other cases where items are specified that are not included in the project. Therefore, an escape clause is provided to allow the commissioning agent and Owner flexibility in the use and application of the check lists. Check lists should be developed specific to each project.)
    (Option 1)
       The following Commissioning Check Lists are to be completed by the Contractor and certified by the Commissioning Agent. Com-

pletion of the requirements of the check lists shall be viewed as a minimum requirement in satisfying the commissioning process. The check lists shall not be used to exclude any additional work required to commission systems and equipment as herein before specified or as deemed necessary by the commissioning agent and/or Owner.

(*Option 2*)

The following Commissioning Check Lists applies to a pump. This is a sample sheet, and the Contractor will be provided Check Lists by the Commissioning Agent specific to the project at the first Project Progress Meeting. Sheets are to be completed by the Contractor and certified by the Commissioning Agent. Completion of the requirements of the check lists shall be viewed as a minimum requirement in satisfying the commissioning process. The check lists shall not be used to exclude any additional work required to commission systems and equipment as herein before specified or as deemed necessary by the commissioning agent and/or Owner.

Sample Commissioning Check Lists

TABLE 7.1

COMMISSIONING CHECK LIST

PUMP

Number: _____   Serving: _____

	GPM	Head (Ft.)	Hp	v/ph/fr	RPM	Type	Manufacturer	Model No.	Comments
Specified									
Installed									

Comments:
1.
2.
3.

Installation Check	Designed or Specified		Provided		Installation Check	Designed or Specified		Provided	
	Yes	No	Yes	No		Yes	No	Yes	No
Pressure Gauges					Suction Diffuser				
Suction					Triple Duty Valve				
Dicsharge					Strainer				
Isolation Valves					Drain				
Balancing Valve					Thermometers				
Check Valve					Inertial Base				
Flexible Connectors					Vibration Pads				
Unions									

Comments:
1.
2.
3.

Operational Check	Contractor		Commissioner		Comments
	Recorded	Date	Observed	Date	
Pump Discharge Pressure (PSIG)					
Pump Suction Pressure (PSIG)					
Pump Rotation					
Alignment					
Motor Amps					

Comments:
1.
2.
3.

TABLE 7.2

COMMISSIONING CHECK LIST

BOILER

Number: _____   Serving: _____

		Fuel				Output							
	Duty: Steam HW	Gas, Oil Elec, Dual	Gas Input MBH	Oil Input GPH	Elec Input KW	Net IBR	MBH	Steam Pres. PSIG	Water GPM	EWT °F	LWT °F	Mfr.	Model
Specified													
Installed													

Comments:
1.
2.
3.

Installation Check	Designed or Specified		Provided		Installation Check	Designed or Specified		Provided	
	Yes	No	Yes	No		Yes	No	Yes	No
Gas Train (UL, FM, etc.)					Isolation Valves				
Gas Train Vented					Flues/Vents				
Oil Transfer Pumps					Flues/Vents Dampers				
Oil Preheater					Drain Valves				
Releif Valve					ASME Stamp				
Releif Valve Discharge to Drain					Step Electric Control				
Releif Valve Discharge to Atmos					Step Burner Control				
Combustion Air Intake					Modulating Burner Control				

Comments:
1.
2.
3.

Operational Check	Contractor		Commissioner		Comments
	Recorded	Date	Observed	Date	
Start-Up					
Burner Set-Up/Test/Adjustment					
Flue Gas Analysis					
Combustion Air Intake Operation					
Boiler Sequencing					
Burner Sequencing					
Safeties/Alarms					
Emergency Shut Off					

Comments:
1.
2.
3.

TABLE 7.3

HYDRONIC SYSTEM ACCESSORIES

Number: _____   Serving: _____

	Air Separator	Expansion Tank	Strainer				
Specified							
Installed							

Comments:
1.
2.
3.

Installation Check	Designed or Specified		Provided		Installation Check	Designed or Specified		Provided	
	Yes	No	Yes	No		Yes	No	Yes	No
Air Separator					Make-Up Water				
With Strainer					Pressure Regulating Valve				
Without Strainer					Back-flow Preventer				
Pressure Gauges					Isolation Valves				
Isolation Valves					Strainer				
Drain					By-Pass				
Expansion Tank									
Tank Fitting					Pressure Relief Valve				
Site Glass									
Drain					Unions				

Comments:
1.
2.
3.

Operational Check	Contractor		Commissioner		Comments
	Recorded	Date	Observed	Date	
Air Separator Pressure Drop (PSIG)					
System Fill Pressure (PSIG)					
Relief Valve Setting (PSIG)					

Comments:
1.
2.
3.

TABLE 7.4

COMMISSIONING CHECK LIST

AIR HANDLING UNIT
CONSTANT VOLUME, HEATING/COOLING

Number: _____   Serving: _____

	Supply Fan									Unit Mtg.
	Total CFM	Min. OA Cfm	E.S.P. In.WG	RPM	Drive Belt/ Direct	HP	Power V/Ph/Freq	Mfr.	Model	Style-Horiz,Vert, Draw/Blow-thru AF/FC etc.	Floor, Hung, Roof
Specified											
Installed											

	Cooling Coil (Water or Glycol)										
	E.A.T. °F db/wb	L.A.T. °F db/wb	Air P.D. In.WG	GPM	Water P.D. Ft.	Rows/Fins	FV FPM	Total Clg MBH	Sens Clg MBH	E.W.T. °F	L.W.T. °F
Specified											
Installed											

	Heating Coil (Water, Steam, or Electric)										
	E.A.T. °F	L.A.T. °F	Air P.D. In.WG	GPM, Lbs/Hr or KW	Water P.D. Ft.	Rows/Fins	FV FPM	Total MBH	E.W.T. °F	L.W.T. °F	Ent. Steam PSIG	Electric Steps & V/ph/Freq
Specified												
Installed												

	Cooling Coil (Direct Expansion)										
	E.A.T. °F db/wb	L.A.T. °F db/wb	Air P.D. In.WG	FV FPM	No. Ckts., Type	Rows/Fins	Total Clg MBH	Sens Clg MBH	Refrig Type	Sat Suction Temp°F	Refrig Temp °F
Specified											
Installed											

	Prefilter				Final Filter				Coil Circulating Pump				
	Type	% Eff	Mfr	Model	Type	% Eff	Mfr	Model	GPM	Head	HP	V/Ph/Fr	Model
Specified													
Installed													

	Return/Relief Fan									Comments
	CFM	E.S.P. In.WG	RPM	Drive B/D	HP	Power V/Ph/Freq	Mfr.	Model	Style AF/FC etc.	
Specified										
Installed										

Comments:
1.
2.
3.
4.

TABLE 7.4 *(Continued)*

COMMISSIONING CHECK LIST (continued)

AIR HANDLING UNIT
CONSTANT VOLUME, HEATING/COOLING

Number: _____   Serving: _____

Installation Check	Designed or Specified		Provided		Installation Check	Designed or Specified		Provided	
	Yes	No	Yes	No		Yes	No	Yes	No
Supply Fan					Prefilter Section				
Isolation (External/Internal)					Type (Flat, Angle. Bag, etc.)				
Access Door					Access				
Discharge Flex Connection					Differential Pressure Gauge				
Return/Relief Fan					Frame				
Isolation (External/Internal)					Final Filter Section				
Access Door					Type (Flat, Angle, Bag, etc.)				
Inlet Flex Connection					Access				
Water/Glycol Cooling Coil					Differential Pressure Gauge				
Control Valve					Frame				
Face & By-Pass Dampers					Outdoor Air Intake				
Condensate Drain					Wall Louver, Size				
Coil Pull Access					Roof Intake, Size				
Reheat Coil (Water, Steam, Elec)					Ducted, Size				
Control Valve					Damper				
Face & By-Pass Dampers					Mixing Box				
Coil Pull Access					Factory Furnished				
Steam Trap					Field Constructed				
Electric Control (Step/SCR)					Dampers				
Preheat Coil (Water, Steam, Elec)					Return Air Size				
Control Valve					Relief/Exhaust Air Size				
Face & By-Pass Dampers					Outdoor Air Intake Size				
Coil Pull Access					Exhaust/Relief Air				
Steam Trap					Wall Louver, Size				
Electric Control (Step/SCR)					Roof Outlet, Size				
Coil Circulating Pump					Ducted, Size				
Access Section					Motor Operated Damper				
Freezestats					Barometric/Gravity Damper				
Smoke Detectors					Starters				
Supply Air					Vibration Isolation				
Return Air					Base (Spring, Pad, Etc.)				
Casing					Hanger (Spring, Rubber, Etc.)				
Double Wall					Curb Isolators				
Single Wall					Roof Curb				
Insulated					Spare Belts				
					Spare Filters				

TABLE 7.4 (*Continued*)

COMMISSIONING CHECK LIST (continued)

AIR HANDLING UNIT
CONSTANT VOLUME, HEATING/COOLING

Number: _____   Serving: _____

Operational Check	Contractor		Commissioner		Comments
	Recorded	Date	Observed	Date	
Dampers Operation					
Normally Open/Normally Closed					
Minimum Damper Positions					
Dampers Tracking					
Safeties (Freeze, Fire, Low Limit)					
Leakage					
Economizer Cycle					
Filters					
Clean					
Differential Pressure Clean					
Differential Pressure Loaded					
Control Valves					
Normally Open/Normally Closed					
Safeties					
Face & By-Pass Dampers					
Normally Open/Normally Closed					
Safeties					
Condensate Drain, Trapped/Operating					
Clean Coils					
Smoke Detectors Operation					
Freezestat Operation					
Low-Limit Operation					
Temperature Controls Operations					
Space Temperature					
Discharge Air					
Mixed Air					
Return Air					
Start/Stop Control					
Heating Sequence - Occupied					
Heating Sequence - Unoccupied					
Cooling Sequence - Occupied					
Cooling Sequence - Unoccupied					
Economizer Cycle					
Purge Cycle					

Comments:
1.
2.
3.

TABLE 7.5

COMMISSIONING CHECK LIST

MASTER

Number: _____ Serving: _____

Specified								
Installed								

Comments:
1.
2.
3.

Installation Check	Designed or Specified		Provided		Installation Check	Designed or Specified		Provided	
	Yes	No	Yes	No		Yes	No	Yes	No

Comments:
1.
2.
3.

Operational Check	Contractor		Commissioner		Comments
	Recorded	Date	Observed	Date	

Comments:
1.
2.
3.

Commissioning Computer-Based Control Systems

Michael S. Bilecky, PE, CEM
George H. Raaf, Consultant
William A. Butler, Consultant
Von Otto & Bilecky, Prof. Corp.
Washington, D.C.

8.1 General

The control system is at the heart of the successful operation of all mechanical systems, yet it is perhaps the least understood. Control design, installation, and operation are of such paramount importance that assuring properly operating controls consumes more time than any other single activity in the commissioning of mechanical systems. With the advent of computer based direct digital controls and energy management automation systems (EMS), controls systems have become more complex while simultaneously simplifying their commissioning. To commission a control system that is not equipped with central intelligence capable of storing data and printing trends of operations, much more leg work is required to exercise controls through their various sequences and modes of operations. The use of an energy management automation system greatly facilitates viewing mechanical systems operations and performance; however, to make full use of the automation system requires substantial work from the automation subcontractor and a deft understanding of the information that is presented in trended histories. The commissioning agent needs to assure that the project specifications are suitably written to provide for data acquisition appropriate to his needs in commissioning the mechanical systems. Further, the commissioning agent needs to be adept at verifying the automation system's energy and facility management operations in addition to the system's temperature control functions. The following sections of this chapter offer guidance in establishing that a project's automation system can be used to the fullest value in commissioning of the mechanical systems and also provides guidance for commissioning of the automation system itself.

8.2   The Contract Documents

The successful operation of mechanical systems starts with good design drawings and specifications. The design drawings show how the mechanical systems component parts are to fit together and be installed. The specifications further amplify quality of materials, installation methods, and performance. The controls system specifications state how everything is supposed to work.

Review and understanding of the controls systems specification by the commissioning agent is critical to the success of the project. Often, not all controls work is specified in the controls section of the specifications, but appears in a section for equipment where controls are to be furnished by the equipment manufacturer; i.e. the boiler controls specified with the boiler; chiller controller with the chiller, etc. These manufacturer furnished controls become an integral part of the overall building controls system, and it essential that work specified in the controls section of the specifications is harmonious with manufacturer furnished controls. For example, the controls section of the specification may call for the automation system to provide a chilled water reset signal to the chiller, but unless the chiller was specified to be capable of receiving and using the reset signal from the EMS, the manufacturer may ship a chiller incapable of operating with an externally generated reset signal.

Review and understanding of the contract drawings with regard for the specified control sequences is also critical; i.e., do the drawings provide the mechanical hardware and components to execute the operational sequences specified. A simple example of a conflict is where an air handling unit (AHU) was specified to have economizer cycle control but the drawings only provide an outdoor air intake duct to the AHU of sufficient size to introduce minimum outdoor air.

To get the fullest diagnostic benefit from an energy management automation system on projects so equipped, provisions must be made in the contract specifications to provide for the submission of operating trends for review (trends are also called histories). The importance of trends cannot be overstated. A tremendous amount of information can be provided that enables the commissioning agent to verify systems operation and even perform diagnostics without having to visit the site, saving great amounts of time in the commissioning process. However, unless the contract specifications have provided very detailed requirements for the accumulating and formatting of trends, the trends submitted by the controls contractor may be of little benefit. A common statement found in specifications may read "Contractor shall submit trends for all control points identified in the Points List." This is totally inadequate as it does not tell the contractor how many trend submittals are required, nor does it provide for format of the trends. What most likely will result from this specification is a long list of trends showing temperatures and occurrences of changes of values. Trends of this nature are difficult to diagnose as the simultaneous interaction of various mechanical systems components are not clear, and cause and effect relationships are not readily evident.

The place in the contract specifications for detailing the requirements of trend submittals is in the controls section where the EMS is specified. Should the controls section not provide for trending to the satisfaction of the commissioning agent, the trending requirements could be included in the commissioning section of the specifications. The requirements included in the controls section of the specification should be similar to the following sample, modified as necessary to the specific requirements of the project. If wording similar to the following is not provided, the commissioning specification should augment the controls specification to obtain the trends needed. Parenthetical italic statements are explanations and not part of the specification.

8.3 Sample Specification

Energy management automation system test and guarantee:

A. The Energy Management Automation System (EMS) shall be guaranteed free from all mechanical, electrical, and software defects for a period of two (2) years. During this two year period, the EMS Subcontractor shall be responsible for the proper adjustments of all systems, equipment, and apparatus installed by him and do all work necessary to insure efficient and proper functioning of the systems hardware and software (*This is a fairly common warranty statement*).

The Contractor (*General and / or Mechanical*) shall arrange to meet with the EMS Subcontractor, the Engineer, the Architect, the Owner, and the Commissioning Agent (*if so contracted*) within thirty (30) days prior to the specified end of the guarantee period for the purpose of compiling a list of items which require correction under specified guarantees. Should the Contractor fail to schedule the final meeting, then the Guarantee shall be automatically extended until such time as the meeting takes place; and the Contractor shall be fully responsible for correcting such deficiencies as if they occurred under the original guarantee period. (*This has tremendous value to the Owner. Over the two year warranty period, the operation of the building should be fairly well known by the Owner's operating and maintenance personnel. Habitual problems or idiosyncrasies of systems operations may have been discovered. This is the Owner's last chance to get systems operating satisfactorily prior to relieving the Contractors of their responsibilities. Some latent defects may require rectification under the terms of the contract, or some improvements to systems operations may be requested by change order.*)

B. Placing in service: Upon completion of the EMS installation, calibrate equipment and verify transmission media operation before system is placed on line. All testing, adjusting, and calibrating shall be completed and systems shall be properly operating prior to acceptance by Owner. Cross check each control point within the EMS by comparing the control command and the field controlled device and/or equipment.

C. Prior to final acceptance and authorization for final payment by the Owner, Energy Management Automation System "punch list" inspections shall be made by the Engineer, the Commissioning Agent, and the Owner's representatives. (*The Owner's representatives should include operating and maintenance personnel that will operate and service the mechanical equipment as well as those that are responsible for the operations of the automation system.*) This is not to preclude that punch lists shall be made by the Contractor to check the completion of his work prior to final "punch lists" inspection. (*This statement puts the Contractors on notice that they need to perform their own inspection of systems installation and operation prior to amassing forces on site for a formal punch-out.*) The punch lists inspections shall be in three parts.

1. Installation Punch List: An inspection shall be performed and punch list prepared regarding the physical installation of the EMS equipment, wiring, etc. (*This involves observation of installation techniques; i.e., is wiring properly strung and supported, are panels properly located and labeled, is control tubing the proper type and properly supported, has conduit been used as specified, are the control devices submitted in shop drawings those that are installed on the project, etc.*)

2. Site Programming Punch List: A separate punch list shall be prepared regarding the software programming at the site. At a

minimum, prior to requesting a Site Programming Punch List inspection, all specified software and control strategies shall have been loaded into the computers at the facility to be inspected.

3. Head-End Programming Punch List: A third punch list shall be prepared regarding the software programming at the head-end. Prior to requesting a Head-End Programming Punch List inspection, all required "head end" software shall be installed.

4. It is recognized that not all software programming may be fully "debugged" at the time of Punch List inspections; however, the EMS Contractor shall be able to demonstrate that all required software strategies are installed and that equipment is being controlled by those strategies. To facilitate punch outs of installations, the Contractor shall have their software completely loaded and functional at least one week prior to a scheduled software punch out and a complete set of histories shall have been submitted to the Engineer and the Commissioning Agent for review. The intent of this is to determine whether installed software is functioning as intended. As there may be more points required for trending than a single history program's capability, more than one run of histories may be necessary to provide all required data. Prior to submitting a trend, the Contractor shall perform a self review to identify and correct problems. Histories shall be presented at hourly intervals for a 24 hour period, unless directed otherwise.

(There are several important issues in the foregoing paragraph:

The contractor is required to have at least a week of system operations under the control of the programmed EMS. Often, the controls work is the last component of mechanical systems installation that is completed due to the failure of other trades. For example, the control contractor may have strung wiring and programmed the computers, but cannot demonstrate suitable operations because the electrical contractor has not completed motor starter installation. The Owner should not be expected to make allowances due to the Contractors inability to complete the project on time by assuming that the control sequence will function properly when the starters are finally installed.

On large projects, the amount of trending required may exceed the computer's memory capability. This cannot be an excuse for the EMS vendor to submit less than the required trends. Either a computer with sufficient memory needs to be provided or the contractor needs to make multiple runs of trends.

The review of trends can be quite time consuming. The contractor needs to perform a self-review of trends for proper operation to correct obvious malfunctions prior to submitting trends to the Engineer or

Commissioning Agent. This usually entails an initial trend run and in-house diagnosis by the Contractor. Failures revealed may require hardware or software corrections or both. Once the Contractor has made corrections, he needs to run another set of trends for his own review. The Contractor needs to go through this process as many times as necessary prior to submitted trends to Engineer or Commissioning Agent to assure that his system is performing properly. The purpose of trend review by the Engineer or Commissioning Agent should be that of verification, not de-bugging.

The requirement that trends be presented at hourly intervals for a 24 hour period tells the Contractor the format of the trends. This is far more valuable in diagnosing systems operations than a Change of Value format. Unless specified otherwise, the Contractor will most likely submit trends in the Change of Value format as it requires less work than the 24 hour format. The 24 hour format is a minimum time interval, some manufacturer's systems can present trends in this format for up to 36 hours.)

5. During punch out and/or if malfunctions are discerned in guarantee period histories, systems being monitored shall be operated with an occupancy schedule; i.e., indications that a system was scheduled off for the 24 hours of the history and remained off are of no value. *(For example, if a schedule places a facility in the unoccupied mode over a weekend, such a school, and trends are submitted for Saturday / Sunday operation; all equipment may be off and little knowledge of systems operations can be discerned. Yet the Contractor could contend that he fulfilled his obligation of submitting trends for review.)*

6. All system setpoints (both calculated and manual inputs) shall be provided with the trends. *(It is essential to know what a setpoint is to determine whether the controls are functioning properly. Trends may indicate that a space is being maintained at 72°F (22°C), and this may appear proper unless it is known that setpoint is 78°F (27°C). Also, software generated setpoints need to be presented in trends, such as heating water reset values. Diagnosis can then be made as to whether the reset setpoint is being properly calculated and whether control devices are properly tracking the setpoint.)*

7. Punch out of software shall not occur without the central plant operating in either the mechanical heating or cooling mode; if scheduled delivery of the project falls during a period in which the central plant is not operating, EMS punch out will be delayed until such time as the central plant has been operating under EMS control for the afore specified one week. *(This places*

the onus on the contractor to deliver the project on time. Suppose that a project was scheduled for delivery in August. It would be expected that a suitable punch-out could be performed with the central cooling plant in operation. But due to late completion of the work by the contractors, the facility is not ready for punch-out until late September. It is possible that the weather has cooled to the point that the cooling plant is no longer in operation, but it is not cold enough to run the heating plant. The Owner should not be expected to attempt to accept the control system without being able to witness the operations of the central plant.)

8. Initial acceptance of the EMS to start the warranty in either the heating or cooling mode shall be conditional, final acceptance shall be predicated on acceptance of the EMS in the subsequent seasonal operation; i.e., to maintain the warranty, the EMS must function properly in both heating and cooling modes. *(This places the Contractor on notice that the control system must operate properly in both the heating and cooling modes. Acceptance of operations in one mode does not automatically extend to acceptance of the other mode. The controls system is given conditional acceptance in anticipation that the systems will perform acceptably in the next seasonal change-over. Should the systems perform acceptably, the initial acceptance is honored as are the terms of the warranty period. Should the systems not perform acceptably, the initial acceptance is rescinded and a new date of acceptance established. The Warranty commences on the new acceptance date.)*

9. The following constitute those items which, at a minimum, shall be included in a trend submittal:

(The following is a sample. Listings should be specific to the project requirements. This type of list can also be used to cross check the controls specifications and points list to assure that all trended points required have been specified and / or shown on the points list.)

   a. Outdoor air:
      1) Drybulb temperature
      2) Wetbulb temperature and/or relative humidity
      3) Calculated enthalpy
   b. Phase, Voltage, Frequency: Histories should indicate time and duration of any occurrence of power interruption and normal power resumption.
   c. Control Air Compressor:
      1) Control air pressure (house side)
      2) Compressor run time

   d. Demand: Kw
   e. Boilers:
      1) Scheduled run status
      2) Actual run status
      3) Burner status
      4) Boiler water temperature (as sensed in the boiler shell)
   f. Heating Water System:
      1) Supply water temperature
      2) Return water temperature
      3) 3-way mixing valves position
      4) Calculated reset temperature
   g. Heating Water Pumps:
      1) Command
      2) Actual status
      3) Log alarmed outages
      4) Provide history for main and stand-by pumps
   h. Chilled Water System:
      1) Chiller command
      2) Chiller status
      3) Amperage
      4) Log alarmed outages
      5) Calculated chilled water reset temperature
      6) Reset signal
      7) Chilled water supply temperature
      8) Chilled water return temperature
   i. Condenser Water Pumps:
      1) Command
      2) Actual run status
      3) Log alarmed outages
      4) Provide history for main and stand-by pumps
      5) Flow status
   j. Condenser Water System:
      1) Cooling tower fans command
      2) Cooling tower fans status
      3) Amperage
      4) Log alarmed outages
      5) Condenser supply return temperature
      6) Condenser water return temperature
      7) Condenser water calculated reset temperature
      8) Mixing valve position
   k. System Water Pumps:
      1) Command
      2) Actual run status

       3) Log alarmed outages

       4) Provide history for main and stand-by pumps

  l. Water Cooling and/or Heating Coils:

       1) Space or discharge (controlling) temperature

       2) Valve position

  m. Air Handling Units:

       1) Supply fan command

       2) Supply fan status

       3) Supply air temperature

       4) Outdoor, return, and relief air damper positions

       5) Space air temperature

       6) Return air temperature and enthalpy when enthalpy optimization is performed

       7) Mixed air temperature

       8) Duct static pressure (on VAV units)

       9) Van position or variable frequency drive speed (on VAV units)

  n. Fan Powered Mixing Boxes:

       1) Supply fan command

       2) Supply fan status

       3) Space air temperature

       4) Air damper position

       5) Heating coil valve position

       6) Air flow

  o. Variable Air Volume Terminals

       1) Space air temperature

       2) Air valve position

       3) Heating coil valve position

  p. Incremental Units (Unit Ventilator, Fan Coil Units, etc.):

       1) Space temperature

       2) Fan command

       3) Fan status

       4) Valve position

       5) Damper position

  q. Direct Expansion Systems:

       1) Command (each step)

       2) Status (each step)

  r. Exhaust Fans:

       1) Scheduled run/enabled status

       2) Actual run status or enabled status

       3) Trends shall show fans operation as interlocked with other equipment/systems when controlled as such.

  s. Overrides: Run status

(The following paragraph could apply to a two-pipe heating and cooling system where there are distinct heating, cooling, and economizer modes of operation. The intent is to have the contractor furnish trends of the various seasonal modes of operations, once again reinforcing that controls must function properly in all modes.)

D. Over the two year guarantee period, a complete EMS trends shall be submitted by the Contractor four times a year to the Owner for review. The submission of trends may be increased or decreased pursuant to their evaluation; i.e., decreased if all is well, increased if problems exist. Prior to submitting a trend to the Owner, the Contractor shall perform a self review to identify and correct problems. The intent of trend reviews during the guarantee period by the Owner is fine tuning of the EMS, not debugging. During seasonal changeover of the mechanical systems from heating to cooling or vice versa, two 24 hour histories shall be provided; one containing information while the central plan was operational and one containing information with the central plant not in operation (ventilation mode). The histories shall contain all of the points cited hereinbefore for Punch-Out histories.

8.4 Close Out

As part of the commissioning process, the commissioning agent must review close-out submittals such as software documentation, operating and maintenance manuals, and as-built drawings. The commissioning agent is also required to be in attendance of demonstrations and training of the Owner's operating and maintenance personnel.

8.5 Bibliography

The following were used as sources in the preparation of the chapters on Mechanical Systems Commissioning:

Guideline for Commissioning of HVAC Systems, ASHRAE Guideline 1-1989, American Society of Heating, Refrigerating, and Air-Conditioning Engineers, Inc., 1791 Tullie Circle, NE, Atlanta, GA 30329; 1989.

Contractor Quality Control and Commissioning Program GUIDE-LINES & SPECIFICATION, Montgomery County (Maryland) Government, Montgomery Engineering Institute, Capital Projects Management Division, Department of Facilities & Services, 110 North Washington St., Third Floor, Rockville, MD 20850; December, 1993.

Management of Maintenance and Repair

Strategic Planning

Ron Moore
The RM Group
Knoxville, Tennessee

9.1 Introduction

Competitive pressures — regional, domestic, and, increasingly, global — are driving companies to change their historical ways of doing business, to become more competitive, to reduce costs, and to continuously seek ways to increase revenues and profits. As you might reasonably expect, HVAC systems and their operating efficiency, costs, etc. are coming under increasing scrutiny as to performance. After all, if the HVAC system fails, not much else happens in today's business environment. Because of this, many companies have taken a very strong risk averse position in their businesses, making certain that the HVAC system works better than 99.9 percent of the time (less than 8 hours per year of lost performance). This often requires things like redundant capacity, an extraordinary level of spare parts, mechanics on call, rapid, responsive, and often reactive maintenance. Given the risk of loss to the business and the probability of failure, this may be an appropriate strategy under historical paradigms. However, it may *not* be appropriate given the technology and methods that are now proven and available. This chapter outlines the basis for developing a maintenance strategy within your organization which has a strong reliability focus for assuring optimal performance. As you will see, the best companies have come to recognize, among other things, that maintenance is a *reliability* function, not a repair function. More on that below.

9.2   Benchmarking—Defining and
Measuring the Objectives of Your Strategy

One of the first steps necessary in developing your strategy is to define key objectives and develop the basis for measuring achievement of those objectives. One method used to do so is to seek out other organizations believed to be successful, and emulate them. This is often referred to as *benchmarking*. It is an increasingly common practice used to help companies improve their performance, and it is described below.

According to Dr. Jack Grayson of the National Center for Manufacturing Sciences,[1] benchmarking involves "seeking out another organization that does a process better than yours and learning from them, adapting and improving your own process. . ." Benchmarks, on the other hand, have come to be recognized as those specific performance measures which reflect a 'best in class' standard. *Best practices*, as the name implies, are those practices which are determined to be best for a given process, environment, etc., and allow a company to achieve a benchmark level of performance in a given category. Benchmarking, then, is a process for identifying benchmarks, or measures of best in class. Best practices are those practices which lead to benchmark performance. A good maintenance organization will be driven by a reliability strategy, and will define the key performance measures by which they should be judged, and will work diligently to assure that best practices are put in place which will lead to superior, or benchmark, performance. Caution: simply finding benchmark data and making arbitrary decisions without a clear understanding of the underlying practices, will often yield disastrous results—*don't do this!* Below we will explore the benchmarking and best practices relationship, and how to effectively use these principles to improve productivity, reduce costs, and support improved profitability.

9.3   Finding Benchmarks or Best
Performers

The first step in benchmarking is to define those processes for which benchmark metrics are desired, and which are believed to reflect the performance objectives of the company. While this may seem simple, it can often be quite complicated. You must first answer the question "What processes are of concern?" and "What measures best reflect my company's performance for those processes?" Then you must answer the question "What measures best reflect my department's performance which in turn supports organizational objectives?" and so on. These decisions will vary from industry to industry, and even from department to department. Let's suppose we have an overall objective to reduce the cost for maintaining HVAC systems, while concurrently assuring high reliability and availability. Performance measures might include cost per unit of 'facility', e.g., *square feet, horsepower,* (m^2, w^2) etc, and dollars lost from HVAC equipment downtime. Additional suggestions for the measures of success are provided in Table 9.1. *The key is to select those measures which reflect the performance objectives of the organization for a given process*, and subsequently to determine how the organization compares to others, generally within your industry. It is generally best when no more than five key metrics are used in each category of interest, and to make sure they can relate one to the other in a supportive, integrated relationship.

Comparing your company to other companies is the next step. Having selected the key metrics, you must now make sure that you are properly measuring these metrics within your company, that you understand the basis for the numbers being generated, and that you can equitably apply the same rationale to other information received. Developing comparable data from other companies may be somewhat difficult. You have a number of choices:

- Seek publicly available information through research.

- Set up an internal benchmarking group which will survey multiple facilities within a corporation, assuring fair and equitable treatment of the data. Benchmarks will then be defined in the context of the corporation.

- Seek the assistance of an outside company to survey multiple facilities, usually including your corporation, but also expanded to other companies within the company's industry.

- Seek the assistance of an outside company to survey multiple facilities, many of which may be outside your industry, in related, or even unrelated fields.

- Some combination of the above, or other alternative.

The Building Operation and Management Association (BOMA), provides an extensive database for certain performance measures, e.g., cost per square foot for various types of facilities, and may make a useful beginning. If the decision is made to actually do a benchmarking study, it is recommended that the Benchmarking Code of Conduct outlined in *The Benchmarking Management Guide*, published by Productivity Press, or similar standard, be followed. This assures that issues related to confidentiality, fairness, etc. are followed. For benchmarking outside the company, particularly within your industry, it is strongly recommended that you use an outside firm. This will help avoid any problems with statutes related to unfair trade practices, price fixing, etc. Benchmarking is an excellent tool to use to improve

TABLE 9.1   Sample Performance Measures

General Measures
Unit cost of supply (*$ / sq ft, $ / hp, $ / cu ft, $ / m², $ / w, $ / m³, etc*)
Productivity (*$ per person per asset, per sq ft, etc.*)
Equipment availability by *area, unit, component*
Equipment reliability—*average life or mean time between repair (MTBR)*
Energy consumed (*$ or units, e.g., KWH, BTU, etc.*)
Utilities consumed (*e.g., water, wastewater, nitrogen gas, distilled water, etc.*)
Spare Parts / MRO (maintenance repair order) inventory turns

Losses and "Bad Practice" Measures
Production or business losses from breakdowns (*$, quantity, no. of events*)
Overtime rate (*%$ or %hr*)
Rework rate (*$, units, number*)
Unplanned downtime
Reactive work order rate—*emergency, run-to-fail, breakdown, etc. (%, %hr, %$*)

Personnel Measures
OSHA injury rates (*Recordables, Lost time per 200K hr*)
Training (*time, $, certifications, etc. per craft*)
Personnel attrition rate (*staff turnover in % / yr*)

PM (preventive maintenance) Measures
Planned and scheduled work / total work (%)
PM / work order schedule compliance (% on schedule)
Hours covered by work orders (%)
PM work by operators (%)
PM's per month
Cost of PM's per month
"Wrench" time (%)
"Backlog" (weeks of work available)
Mean time to repair, including commissioning

Condition Monitoring Measures
Average vibration levels (*overall, balance, align, etc.*)
Average lube contamination levels
Schedule compliance for condition monitoring
PDM (predictive maintenance) effectiveness (*accuracy of predictive mainte-
   nance by technology*)
Process measures trended for *pressure, temperature, flow, current, etc.*

Census Measures
Mechanics, electricians, etc. per:
   Support person
   First line supervisor
   Planner
   Maintenance engineer
   Total site staff
Total number of crafts

Normalized Cost Measures
Maintenance cost / facility replacement value (%)
Facility replacement value per mechanic (*$ / mechanic*)
Stores value / facility replacement value (%)
Stores service level—stock out (%)
   Critical spares
   Normal spares
Contractor expenditure / total maintenance expenditure (%)

performance, but should not be used as a means to garner information otherwise not available on specific competitors, or to place your company at risk of violating the law.

Benchmarking has become associated with "marks", as opposed to finding someone who does something better than you do and doing what they do, which may more accurately be called best practices. Notwithstanding the semantics, benchmarking and application of best practices are powerful tools to help improve operational and financial performance. The ultimate benchmark, of course, is consistently providing the lowest cost per unit of supply, over the long term. Applying benchmarking and best practices can help assure that position.

Finally, you don't need to do benchmarking to make improvements, so don't let the lack of specific benchmarks stop the improvement process. Select those measures which you believe to be critical to your success, measure them, and then put in place practices which you believe will improve them, now.

9.4 Reliability and Best Practices—How to Achieve a Benchmark Level of Performance

We've discussed the process for benchmarking of key performance indicators. Now we're going to explore those practices which will support superior performance. Maintenance practices, when combined with good operations and design practices, will determine the reliability of HVAC equipment for a given facility. Therefore, it is incumbent upon us to assure good design, operation, and maintenance practices, and to use the experience of the maintenance function to improve operations and design, and vice versa. We must use an integrated approach built on teamwork to assure maximum reliability of our facility.

From a 1992 study,[2] the typical U.S. manufacturer was shown to have the following approximate levels of maintenance (reliability) practices in their manufacturing facilities:

- Reactive maintenance   50%
- Preventive maintenance   25%
- Predictive maintenance   15%
- Proactive maintenance   10%

Some definitions would be useful:

1. *Reactive maintenance* was defined by practices such as run-to-fail, breakdown, and emergency maintenance. Its common characteristics were that it was unanticipated and typically had some level of urgency.

2. *Preventive maintenance* were those practices that were time based, that is, on a periodic basis prescribed maintenance would be performed. Examples include: annual overhauls, quarterly calibrations, monthly lubrication, etc.

3. *Predictive maintenance* practices were those that were equipment condition based. Examples include changing a bearing long before it fails based on vibration analysis, changing lubricant based on oil analysis showing excess wear particles, replacing steam traps based upon ultrasonic analysis, etc. In the best facilities, it also included review of process parameters as part of the condition based approach.

4. *Proactive maintenance* practices were those that were root cause based, or that sought to extend machinery life. Examples were varied, and include modification of operation, design, or maintenance practices, but all sought to eliminate the root cause of particular problems. Specific examples might include root cause failure analysis, precision alignment and balancing of machinery, precision equipment installation commissioning, and improved design and vendor specifications. The overall goal was to ensure that the equipment was "fixed forever."

These typical facility maintenance practices were then compared to so-called benchmark facilities of the study. These benchmark facilities were characterized as those who had achieved extraordinary levels of improvement and/or performance in their operation.

The striking characteristics of these benchmark facilities, as compared to the typical facility, were twofold:

1. Reactive maintenance levels differed dramatically. The typical facility incurred some 50 percent reactive maintenance, while the benchmark facilities typically incurred less than 10 percent, principally because they were reliability focused in their maintenance practices.

2. The benchmarks facilities were driven by an integrated combination of preventive, predictive, and proactive methods, with a strong emphasis on predictive, or condition-based, maintenance practices. Further, preventive and predictive methods were used as tools to be proactive and eliminate the cause of failures.

These conclusions are reinforced by R. Ricketts in *Organizational Strategies for Managing Reliability*,[3] wherein he states that the best refineries were characterized by, among other things, the "religious pursuit of equipment condition assessment"; and that the worst were characterized by, among other things, "staffing. . . designed to accommodate rapid repair", and failures being "expected because they are the norm." In the same study, he also provides substantial data to support the fact that as reliability increases, maintenance costs decrease, and vice versa. This has also been the experience of the author.

9.5 "Worst Practices"

Reactive maintenance, at levels typically beyond 10–20 percent, should be considered a practice to avoid, a "worst practice". Reactive maintenance tends to cost more and lead to longer periods of downtime. In general, this is due to the ancillary damage which often results when machinery runs to failure; the frequent need for overtime; the application of extraordinary resources to "get the equipment back on line, NOW"; the frequent need to search for spares; the need for expedited (air freight) delivery of spares, etc. Further, in a reactive mode, the downtime period is often extended for these very same reasons. Moreover, in the rush to return the equipment to operation, many times no substantive effort will be made to verify the quality of the repair and equipment condition at start up. Hudachek and Dodd[4] report that reactive maintenance practices for general rotating machinery costs some 30 percent more than preventive maintenance practices, and about twice as much as predictive maintenance practices.

9.6 Best Practices

In the author's observation, organizations who employ best practices, that is those which facilitate operating at benchmark levels of performance — world class, best in class, etc. are those which have a *reliability culture*; whereas those who are mediocre and worse, have a *repair culture*.[5]

In a repair culture, the maintenance department is viewed as someone to call when things break. They can even become very good at crisis management and emergency repairs; they often have the better craft labor — after all, the crafts are called upon to perform miracles; but they will never rise to the level achieved in a reliability culture. They often even view themselves as second class employees, since they are viewed and treated as "grease monkeys", repair mechanics, and so on. They often complain of not being able to maintain the equipment properly, only to be admonished when it does break, and placed under extraordinary pressure to "get it back on line". They are rarely allowed to investigate the root cause of a particular problem and eliminate it, or to seek new technologies and methods for improving reliability—it's not in the budget is a familiar refrain. They are eager, sometimes desperate, to contribute more than a repair job, but are placed in an environment where it's just not possible. They are doomed to repeat the bad practices and history of the past, until the company can no longer afford them, or to stay in business.

In a reliability culture, reliability is the watchword. No failures is the mantra. Machinery failures are viewed as failures not of the machine, but of the system which allowed the failure to occur. In a reliability culture, preventive, predictive, and proactive maintenance practices are blended into an integral strategy. To the maximum extent possible, maintenance is performed based on condition of equipment and of process. Condition diagnostics are used to analyze the root cause of any failures, and methods are sought to avoid the failure in the future. The methods for achieving and supporting this culture are discussed in Chapter 10 following.

9.7   References

1. National Center for Manufacturing Sciences (NCMS) Newsletter, November 1991.
2. R. Moore, F. Pardue, A. Pride, J. Wilson, *The Reliability-Based Maintenance Strategy: A Vision for Improving Industrial Productivity*, September 1993, Computational Systems, Inc. Industry Report, Knoxville, TN, September, 1993.
3. R. Ricketts, Organization Strategies for Managing Reliability. National Petroleum Refiners Association, Washington, DC: New Orleans, LA, Annual Conference, May, 1994.
4. Progress and Payout of a Machinery Surveillance and Diagnostic Program, R; Hudachek and V. Dodd, American Society of Mechanical Engineers, New York, NY, 1985.
5. R. Moore, Reliability, Benchmarks, and Best Practices, *Reliability Magazine*, Knoxville, TN, December 1994.

9.8   Bibliography

J. Ashton, Key Maintenance Performance Measures, Dofasco Steel-Industry Survey, Hamilton, Ontario, Canada, 1993.
E. Jones, The Japanese Approach to Facilities Maintenance, Maintenance Technology, Barrington, IL, August 1991.
K. Kelly and P. Burrows, Motorola: Training for the Millennium, *Business Week,* March 1994.
Information from DuPont and Other Large Manufacturing Companies.

10

Preventive, Predictive, Proactive Maintenance

Ron Moore
The RM Group
Knoxville, Tennessee

10.1 Preventive Maintenance

Preventive maintenance takes on a new meaning when using a reliability based strategy. It does allow for time based maintenance, but only when the time periods are justifiable. Examples include instrument calibrations, strong wear related correlations and good statistical bases for time intervals, regulatory driven requirements, etc. Preventive maintenance is used principally as an analysis and planning function, and almost always includes comprehensive use of a computerized maintenance management system (CMMS). The maintenance function establishes and analyzes machinery histories, performs cost analyses of historical costs and impacts, uses these cost and machine histories to perform Pareto analyses to determine where resources should be allocated for maximum reliability. Preventive maintenance performs maintenance planning, assuring coordination with other functions, assuring that spare parts are available, that tools are available, that stores inventories and use histories are routinely reviewed to minimize stores, yet maximize the probability of assuring equipment uptime. Preventive maintenance staff have recognized that doing all maintenance on a strict interval basis is not optimal. They recognized that PM's are often based on arbitrary data, from the manufacturer or otherwise, and that most machines will exhibit wide variation in their mean time to failure. *Few machines are truly average,* a basic assumption of many PM schedules. Any group of equipment may follow a wide variation of failure modes, frequencies, and effects. Good preventive maintenance practices, then, include:

- Strong statistical base for those PM's done on a time interval.
- Exceptional planning and scheduling capability.
- Solid machinery history analysis capability.
- Strong and flexible cost analysis capability.
- Comprehensive link to stores and parts use histories.
- Routine analysis of stores and its support of cost effective reliability.
- Comprehensive training in the methods and technologies required for success.
- Comprehensive link to the predictive or condition based program.

10.2 Predictive Maintenance

Predictive or condition based maintenance is typically at the heart of a good reliability program. Knowing machinery condition, through the application of vibration, oil, infrared, ultrasonic, motor current, and particularly to process parameters pressure, temperature, flow, etc., drives world class reliability practices, and assures maximum reliability. It is generally best if all technologies are under single leadership, allowing for synergism and routine, informal communication teamwork, that focuses on maximizing reliability. For example, using modern vibration analysis tools we could identify a developing problem with a bearing long before it actually fails. This would allow the maintenance on the bearing (perhaps used in a fan, pump, etc.) to be planned and orderly. That is, we could check to assure we have the proper parts, tools, and resources, and we could work with operations to perform the maintenance so as to minimize the impact of the repair on the facility's requirements. Further, the vibration analysis could also preclude what might have historically resulted in a catastrophic failure. Finally, in a proactive use of vibration analysis, it could also facilitate a diagnosis of the root cause of the bearing failure. Other applications include using condition monitoring to plan overhauls more effectively, since it is known what repairs will be needed. Overhauls can be done for what is necessary, but only what is necessary. Therefore, they take less time—much of the work is done before shutdown; and only what is necessary is done. Condition monitoring allows for validation of the precision and quality of the repair at start up to assure the equipment is in like new condition. Condition monitoring allows better planning for spare parts needs, thus minimizing the need for excess inventory in stores. Good predictive maintenance practices include:

1. Routine use of equipment condition assessment.
2. Application of all appropriate and cost effective technologies.
3. Avoiding catastrophic failures and unplanned downtime by knowing of problems in machinery long before it becomes an emergency.
4. Diagnosing the root cause of problems and seeking to eliminate the cause.
5. Defining what maintenance jobs need to be done, when they need to be done — no more, no less.
6. Planning overhaul work more effectively, and doing as much of the work as possible before shutdown.
7. Setting commissioning standards and practices to verify equipment is in like new condition during start up.
8. Minimizing inventory, through knowledge of machine condition and subsequent planning of spare parts needs.
9. Comprehensive communications link to maintenance planning, equipment histories, stores, etc. for more effective teamwork.
10. Comprehensive training in the methods and technologies required for success.
11. An attitude that failures are failures of the design, operation, and maintenance processes, not the individual.
12. Continuously seeking ways to improve reliability and improve equipment performance.

10.3   Proactive Maintenance

Proactive maintenance at the best facilities, is the next step in reliability. At facilities which have a strong proactive program, they have gone beyond routine preventive maintenance, and beyond predicting when failures will occur. They aggressively seek the root cause of problems, actively communicate with other departments to understand and eliminate failures, employ methods for extending machinery life. Predictive maintenance, or more appropriately, condition monitoring, is an integral part of their function, because these methods and technologies provide the diagnostic capability to understand machinery behavior and condition. They have come to understand, for example, that alignment and balancing of rotating machinery can dramatically extend machinery life and reduce failure rates. They have also learned that doing the job perfectly in the facility is not sufficient, and that improved reliability also comes from their suppliers. Therefore, they have a set of supplier standards which require reliability tests and validation. In the net, they constantly seek to *design* equipment for reliability, to specify and *purchase* reliable equipment, to *store* it reliably, to *install* it reliably and verify the quality of the installation effort, to *operate* the equipment to maximize reliability, and finally to *maintain* the equipment for reliability. These issues are discussed further below.

Good *design* practice requires consideration of life cycle costs, including maintenance and operating costs, not just initial cost. It also requires maintainability, e.g., ease of access, isolation valves, lifting lugs, jacking bolts, inclusion of tools, etc.

Good *purchasing* practice requires good specifications from, and good communications with, maintenance and engineering in order to assure comprehensive and high quality supply. For example, a good motor specification might require that all motors: 1) be balanced to within 0.10 inches (25.5 mm) per second vibration at one times turning speed; 2) have no more than 5 percent difference in cross phase impedance at load; 3) have co-planar feet not to exceed 0.003 (×760 μm) inches.

Good *stores* practices requires a number of support functions which assure retaining the reliability of the equipment, and providing it when and where it is needed. Additional detail regarding the purchasing and stores function is provided in the chapter on stores and parts management.

Good *installation* requires precision in installation practices, and a process for validating the quality of the effort using specific measurement techniques to verify process quality (proper temperatures, flows, pressures, etc.), and well as equipment quality (proper vibration levels, oil quality, amperage levels, etc.). This should also include a process for resolving non-conformances.

Good *operation* practices assures that operators work with precision, and within the limits of the capability of the equipment. For example, not running pumps dead headed or in a cavitation mode, not trying to re-start a motor 10 times in a row without success (and burning out the motor), and so on. In the experience of the author, some 60–70 percent of equipment failures are a direct result of poor practices in areas other than maintenance which resulted in a maintenance or repair requirement. Good maintenance practices go well beyond simple maintenance issues.

Proactive *maintenance* practices which have yielded extraordinary gain include:

- Root cause failure analysis.
- Precision alignment and balancing.

- Supplier standards for reliability.
- Training of purchasing staff in reliability standards.
- Installation commissioning standards to verify proper installation.
- Precision operation — process control within all design limits.
- Comprehensive training in the methods and technologies required for success.
- Continuous communication with production, engineering, and purchasing to maximize reliability.

The key is to assure that the knowledge base within maintenance and operations which represent problems (otherwise called opportunities for improvement) are in fact used effectively for the continuous improvement process. The best facilities have developed and implemented these practices to assure success of their operation.

10.4 Key Performance Indicators

In order to apply the best practices described above, and to become a benchmark facility, you must have measures of your progress and success, i.e., you must have benchmarks by which you are judged. As Dr. Joseph Juran, a leading quality expert, said "If you don't measure it, you don't manage it." Put in a more positive sense, if you measure it, you will manage it, and it will improve. These measures, or benchmarks, by which you will judge your performance then, are critical for your success — you can hardly afford to be managing the wrong measures. The measures then should reflect operational and financial performance, and a listing of suggestions for consideration is provided in Table 9.1.

Table 10.1 provides a sampling of prospective benchmarks data. These are offered not as benchmarks which resulted from a comprehensive benchmarking study, but rather as based on a review of the literature, and the experience of the author. These measures, or benchmarks, will facilitate a measure of the effectiveness of the best practices espoused above.

Finally, overall success of any good HVAC reliability strategy requires:

Knowledge of equipment condition

Training to assure understanding

Focus on the right goals

Teamwork for synergy

Communication to assure understanding and cooperation

TABLE 10.1   Nominal Benchmarks for Various Measures[1]

Indicator	Definition	World class level
Planned maintenance	Planned maintenance/total maintenance	>90%
Reactive maintenance	Run to fail, emergency, etc. (1, 3)	<10%
Maintenance overtime	Maint OT/total maintenance time (1, 3)	<5%
Maintenance rework	Work orders reworked/total WO's (3)	~0%
Inventory turns	Turns ratio of spare parts (3, 4)	>2–3
Training	Workers receiving >40 hr/yr (3)	>90%
	Spending on worker training (% of payroll) (5)	~4%
Safety performance	OSHA injuries per 200,000 labor hr (6)	<0.5
Maintenance cost/yr	Percent of replacement value (4)	<3%
Hourly maintenance workers	Percent (%) of total (4)	15%

Benchmarks to provide metrics of performance and feedback on effectiveness

Leadership to create a continuous improvement environment and assure knowledge, training, focus, teamwork, communication, and benchmarks.

Reliability, benchmarking, and applying best practices are keys to your success. By learning what the best facilities are achieving and how you compare (benchmarking), by emulating how they are achieving superior results (best practices), and by judiciously applying the practices described above, adding a measure of your own common sense and management skill, your facility can exceed the performance levels of the current benchmarks. This will raise its overall effectiveness, lower operating and maintenance costs, improve reliability, and achieve world class performance.

10.5   References

1. R. Moore, Reliability, Benchmarks, and Best Practices, *Reliability Magazine,* Knoxville, TN, December 1994.

Maintenance: In-House versus Outsourcing*

John D. Brewster, P.E.
V.P., Maintenance & Engineering
Pro Tec Technical Services
Willowdale, Ontario, Canada

11.1 Contract Maintenance—What Is It?

Prior to installation of any plant system there ought to be a decision oriented toward how it is to be maintained. Will our own people maintain it, or will we outsource to a contractor? Immediately, when the word *contractor* is used, one tends to conjure up any number of evil thoughts about ability, costs, loss of control, dedication etc. These thoughts, in some cases, can be well deserved based upon past history with certain contractors. This chapter is included to assist in developing an understanding of the approach you may want to take with respect to outsourcing. Remember, however, that the success of outsourcing begins with your acceptance that, under a given set of conditions, your selected contractor can do the job while remaining transparent to your own in-house workforce activities. Perception has been and will remain the key stumbling block to acceptance of the use of outsourcing.

The decision to outsource should not be justified merely on a reduced labor rate or the ability to avoid a union environment. The perception of "loss of control" should never be considered. If you feel you will lack

---

*Note: see section 11.9 for explanation of all abbreviations.

control over maintenance and operation of the system through out-sourcing, the control, perhaps, never existed in the first place.

The cornerstone of successful contract maintenance is the *establish-ment of a long-term relationship between the owner and the contractor that will satisfy mutual goals where in both parties will prosper.* A partnering agreement, established between the two companies, such that no issues of co-employment are allowed to exist, will go a long way toward establishing this relationship. This requires both good leadership and good management to be successful. It is the effective use of people skills, coupled with flexibility and diversity of resources available, that reduces costs, not the mere "brokering" of labor.

The term "Contract Maintenance," or "Maintenance by Contract," actually began in Canada in 1952, at what is now the Shell Oil Refin-ery in Sarnia, Ontario. The (unionized) concept eventually spread into the United States, primarily in oil refineries. The concept centered around the idea: "Who better to maintain it than the contractor who built it?" The original agreement was called "The General President's Committee Project Agreement for Maintenance by Contract in Can-ada." In the United States, it is called "The Blue Book Agreement" or "The General Presidents' Project Maintenance Agreement." The con-cept, under this agreement, has spread over the years from oil refin-eries and chemical plants to steel and paper mills. The advantage of this agreement over single trade agreements is shown in Figs. 11.1 and 11.2. Single trade union agreements generally form an annex to the main collective agreement and in fact, now bear many similarities to the terms, conditions, and wording of The General Presidents Agreement.

In fact, there are locals specifically established to provide multi-trades under one collective agreement, with terms, conditions, and rates specific to a jobsite. Therefore union contractors, literally, can have a shopping list from which to create an agreement, specific to the needs of an owner, not necessarily bound to the local ICI agree-ment. These types of agreements facilitate work in office towers, gov-ernment facilities, school boards, military bases, laboratories, etc.

The non-union environment, though not bound to collective agree-ments, as such, usually has agreements specifying working conditions and local wage rates. The prime advantage of nonunion over union is not base rate, but the ability to multi skill or cross trade. Some union agreements contain emergency work clauses which permit multitasking/skilling. It is union benefits and jurisdiction that gen-erally create the competitive disadvantage because they are tied to the ICI sector main body agreements; the typical non-union agree-ment may be 10 to 12 percent less (total package) than the union agreement. However, only if management skills, human resources and

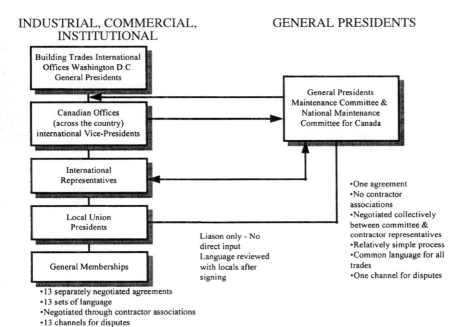

INDUSTRIAL, COMMERCIAL, INSTITUTIONAL

GENERAL PRESIDENTS

Building Trades International
Offices Washington D.C.
General Presidents

Canadian Offices
(across the country)
international Vice-Presidents

International
Representatives

Local Union
Presidents

General Memberships

General Presidents
Maintenance Committee &
National Maintenance
Committee for Canada

•One agreement
•No contractor
associations
•Negotiated collectively
between committee &
contractor representatives
•Relatively simple process
•Common language for all
trades
•One channel for disputes

Liason only - No
direct input
Language reviewed
with locals after
signing

•13 separately negotiated agreements
•13 sets of language
•Negotiated through contractor associations
•13 channels for disputes

Figure 11.1 General negotiating structure: ICI vs GP agreements.

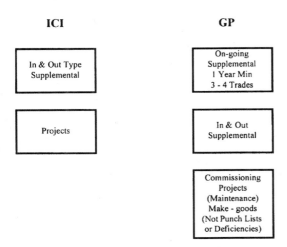

ICI

In & Out Type
Supplemental

Projects

GP

On-going
Supplemental
1 Year Min
3 - 4 Trades

In & Out
Supplemental

Commissioning
Projects
(Maintenance)
Make - goods
(Not Punch Lists
or Deficiencies)

Figure 11.2 Maintenance agreement vs ICI agreements.

work planning/scheduling approach are equal, is the nonunion approach far more competitive. Therefore, *do not delete the union contractor from your bid selection* process, until you have evaluated all the factors required.

Maintenance by Contract is the "leasing" or provision of skilled craftsmen and/or technically skilled professional personnel and expertise, on an as required basis, for temporary or seasonal level changes in activity, such as shutdowns, modifications, miscellaneous and minor construction, outages or ongoing routine or specialty type plant maintenance requirements. It can provide an economically effective alternative to the initial costs of setting up the maintenance organizational structure and the ongoing costs of employing full-time maintenance personnel, as these costs must be considered in relation to the overall production unit costs.

The generally accepted advantages of contract (outsourced) maintenance are shown in Table 11.1.

TABLE 11.1   Advantages of Contract Maintenance

Ability to fluctuate the workforce to suit operational requirements.	Ensures employment and retention of the most productive workers.
Secure source of competent and skilled craftspersons and supervision.	More productive on undesirable and repetitive type work.
Ability to have a two-tiered labor rate system depending on the type of labor agreement used.	Hours of work can be tailored to the owners' needs.
Potential to move personnel to different sites under the same agreement.	Reduces equipment costs, capital investment, annual taxes, repair costs and idle time costs.
Emergency work clause supports flex crafting in a union environment.	Reduced involvement in labor relations issues.
Greatly reduced administrative and cost burdens.	Better results on gearing up for overtime assignments.
Access to proven safety and labor relations expertise.	Access to tools and specialized equipment.
Ability to quickly fluctuate technical support staffing requirements.	Establishes competitive atmosphere with in-house forces.
Access to contractor's maintenance costing and management data.	Resource for variety of support services: management, labor, engineering, special skills and equipment, CMMS and QA/QC.
Provision of government approved welding and QA procedures.	Pay for time worked only.
Reduced involvement in labor relations issues.	Reduced "make work" expenditures.

As described throughout this chapter, outsourcing can be obtained from a large variety of resources from the single contractor, to a third party professional services company who provide contracted management support and trade labor, through independent contractor agreements with the professional services company, in a totally nonunion environment. Such companies are gaining in popularity with a large data base of personnel available to the owner and, in some cases, more flexibility than a contractor, to respond quickly to needs.

11.2 Costs of Maintenance

Many companies have little or no concept of the overhead costs associated with employing full time workers. Few have an adequate understanding of how to properly use an automated maintenance management system to collect the costs associated with operating and maintaining their equipment. Most companies simply consider the base rate plus benefits to be the deciding factor.

Walter Munroe, of Ogden Allied Industrial, has suggested the following costs as a means of assessing the financial feasibility of handing over the management of part, or all, of your maintenance function to a contract services company:[1]

11.2.1 In-house costs

1. *Direct*
   (a) Payroll
   (b) Benefits
   (c) Savings/pension
   (d) Cost of money
                                              Subtotal
2. *Legal / Labor*
   (a) Pay equity/employment litigation
   (b) WHIMIS
   (c) Workers' compensation
   (d) Labor negotiations
   (e) Arbitrations
   (f) Wrongful dismissals
   (g) Hiring/layoffs/firings
   (h) Labor relations litigation

---

[1]Walter Munroe, "You versus Us: Resolving The Key Contracting Problems," pp. 18–22, Plant Services, November 1990.

Subtotal
3. *Administrative*
   (a) Payroll
   (b) Taxes
   (c) Benefits/insurance
   (d) Personnel
   (e) Recruitment
   (f) Training
   (g) Recordkeeping

Subtotal
4. *Technical staff support*
   (a) Projects
   (b) New products
   (c) Safety
   (d) Risk management
   (e) Liability insurance

Subtotal
5. *Miscellaneous*
   (a) Volume discounts
   (b) Systematic operating programs

Subtotal
Grand total

Assuming you know what your maintenance costs are for repair and overhaul of equipment, outsourcing a particular function, on a trial basis, with a performance driven contract, will not only promote improved productivity of your in-house force but improve the opportunity to increase current equipment availability and reliability. Performance driven contracts need about 12–18 months to provide proper results.

Other costs of maintenance, not particularly addressed by Munroe, might be:

- CMMS software
- Tools and equipment
- Shops
- Equipment availability (related to lost orders)
- Environmental quality (air, water)
- Quality of utilities (gas, oil, electricity)

11.3   Value-Added Support to the Maintenance/Facilities Manager

The generic maintenance organization, Fig. 11.3 provides an insight into other areas suitable for outsourcing and long used as such by the

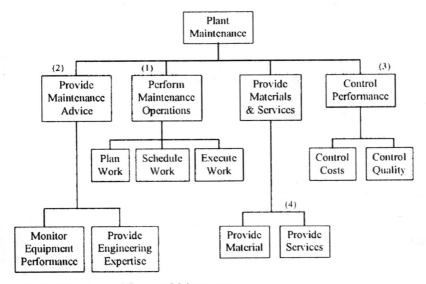

(1) Traditional use of Contract Maintenance.
(2) - (4) New opportunities not fully utilized on a team or resource shared basis.

Figure 11.3   The generic maintenance organization.

petrochemical industry. One can, literally, consider outsourcing just about every facet of the maintenance/facilities organization with the exception of key management and support personnel. Contract maintenance is not limited to just a supplemental labor or "brokered" type force. It can include all of the functions within the maintenance organization and, to a degree, the other functions working in support of maintenance.

- Failure analysis coupled with statistical process control
- Supervisory and craft training
- Development and maintenance of operations and maintenance manuals
- Organization and staffing
- Performance reports
- Daily and weekly scheduling
- Work order system
- Modern unique tools, equipment, and methods
- Planning, estimating, scheduling
- Preventive/predictive maintenance programs

- Spare parts and material control
- Workforce utilization controls
- Work performance assessment parameters
- Development of commissioning procedures and management controls
- Warranty management
- Purchasing

The primary decision to outsource any or all of your operations/maintenance functions should begin with a clear understanding of the nature of your business and where your prime business efforts are to be devoted.

To "spot outsource" is not always the best answer if your contractor cannot provide you with other "value-added" benefits as part of the contract. You want your contractor to develop a potentially long term relationship with your organization. You want to talk with as few people and organizations as possible. Consultants may tell you what to do, but remember that they walk away leaving someone else responsible for field implementation. If your contractor is capable of providing you with consultative and full management support, while doing the work, chances are good that your long term cost will be less and implementation time shorter. Table 11.2 will assist you in your selection process.

Although the owner/client must be a good corporate citizen with respect to utilization of local goods and services, there are several good points to consider for limiting the number of contractors on site.

11.3.1   Advantages of the use of one contractor

- Overall potential for reduced costs and fees:
  - Smaller contractors generally require larger profits.
  - Single provision of trailers, tool rooms, facilities at varying prices and conditions.
  - Consistent approach to supply and purchase of tools, materials, equipment and to loss control of tools/material.
- There is much better and more consistent control of safety with the use of one contractor. An improved standard of work habits is also evident.
- With more than one contractor on site, there is greater administrative burden to the owner by way of owner supervision, engineering and payroll audit involvement.

TABLE 11.2 Typical Selection Criteria

Contractor	
Management capability and organizational approach.	management attitude
Financial viability	ease of contacting
Use of a cost-effective labor agreement	response to calls
Approach to and depth of skills in labor relations, including assessment of history of disputes resolution	"good feeling" with key management personnel
Safety record in relation to the industry and expression of a safety policy	level of knowledge with such issues as TPM, TQM
Does it appear dedicated to saving you, the client, money?	resourcing capabilities
Goal setting capabilities and examples of strategic planning	synergy of organizations (contractor and in-house)
Depth of supervisory skills available	financial strength, bonding capability and insurance exposure
Services it can offer	clearly stated corporate vision/mission statement
Familiarity with ISO, TQM, TPM, HAZOP, etc.	a division devoted to maintenance technology
Depth of training capability—supervisory and trades skills upgrading	evidence of teamwork
How does it monitor productivity?	knowledge of CMMS skills in areas such as: —balancing —alignment —predictive techniques —attitude to training
Use of other procedures and techniques	knowledge and experience in union vs. non-union environment
Adaptability of invoicing requirements to meet the needs of the owner	fabrication facilities or capabilities
Use of a quality driven process?	knowledge and support in failure analysis techniques
Use of CMMS	how they can support you in managing maintenance
Existing client base and level of satisfaction	willingness to cooperate in labor cost reduction efforts —maintenance agreements —open book policy on overhead recovery and fees —productivity based fee
Length of time in the business	ability to supply support personnel and technology through single sourcing
Type of contract proposed or alternates	focusing on training and safety
Experience with long-term partnering arrangements	commitment expressed by senior management

- There is much better coordination of effort with one contractor on site; multiple contractors often have to "work around" each other.

- With one contractor having a common collective agreement, there is the advantage of having all employees in and working during outside labor disruptions.

- Effectiveness is increased through the potential for better communications with only one general contractor who communicates directly with a few specialty subcontractors such as refractory, and instruments depending on the owner's contract system and depth of in-house expertise.

- With only one contractor there is a greater potential for uniformity of supervision and work standards.

- The use of a single contractor employing a consistent approach reduces potential jurisdictional disputes.

11.4   A Strategy of Contracting Maintenance Services

With terms such as *Partnering, Total Quality, Value Added,* making their way into tendering documents, a study of industries in which the author has been involved has led to a thought process one should be conscious of when contracting your needs. They key word here, of course, is "needs." Let us examine some of the key issues involved and apply these to a strategy for contracting of maintenance services.

Maintenance management principles—overview. The common thread to introduction of a quality-driven maintenance program is a solid management structured base, with clearly defined objectives and strategy—a proactive rather than reactive approach.

Framework for maintenance management—objectives

- Document strategy for control of cost and manning.
- Outline directional improvements for management and the maintenance organization
- Provide basis for support
  - Link business information with improvement activities
  - Dynamic goal setting
  - Comparison of budget, actual, and long-term objectives
  - Objectives for the changing plant organization
  - A sound history environment for front end base material complaints (QA/QC)

— Assistance in product costing/scheduling

Framework for maintenance management—strategy

1. Selection of personnel involved in the program.

2. Establishment of objectives of the maintenance department. Maintenance philosophy/objectives and overall direction of activities.

3. Establishment of organization charts.

4. Maintenance budget costs and manning by unit/area: own labor/contract/materials/equipment/shop. All contractor count and construction, all chargeouts.

5. Organization/maintenance guidelines—planning, scheduling, resource allocation, control, authority, responsibility. Includes supervisory levels replacement, hiring, and training plans.

6. Stewardship concept costs and manning by unit/area. Include varied performance indicators.

7. Feedback of information budgetary controls and forecasting. The vehicle for improving maintenance standards and practices.

If we look at the framework for maintenance management, a certain amount of structure can begin to take shape to your business plan. Why a business plan? Your plant manager is following a business plan, developed either at the corporate level, *or* he or she has perhaps developed his/her own plan to guide the business with approval of corporate management. In this business plan, customers, the buying public or business marketplace was considered. In maintenance, however, you have probably been included on their P & L in an area titled "cost of doing business" or "cost of goods sold." As maintenance manager, consider preparation of your own business plan, with your market, or customers, being the production department.

Through the development of a business or strategic plan you will force yourself into analyzing the weakness in your current maintenance operation. This plan becomes the directional improvement you will want to take. In order to improve cost control (not collection), you will probably want to work with accounting in developing a better breakdown of accounts by unit or area, machine or piece of equipment and subcomponents of the line or system. Spare parts use can then be better monitored. If you have not got a good handle on inventory turnover, your cost system will become more focused to this need. You will also likely look at a preventive maintenance (PM) and work-order-based CMMS package which will allow you to better analyze failures and trends in maintenance related vs. operational related costs. Your system should also include a method of categorizing and monitoring reason for delays on production equipment. In this way, you can begin to get a handle on costs that are truly within your control and those costs that should really be borne by operations management.

Your CMMS package, which will include a 52-week scheduler, will allow you the ability to begin to monitor progress and effectiveness of your P.M. program and, more importantly, the ability of your existing crewing to respond to the levels required. Your equipment history data collection and monitoring system will then enable you to better analyze failures and develop repair/replacement/overhaul policies and procedures, which can become a more meaningful addition to your scheduling routine.

There appears to be almost a universal mindset that contract maintenance is simply the contracting of labor. Since labor has traditionally been looked at as the highest cost content of maintenance, industry has generally awarded contracts on the basis of lowest price alone. Here is some food for thought on this issue: Maintenance is made up of a number of items, all brought together at a point in time.

- Mechanic
- Tools
- Equipment
- Material
- Work order package indicating problem location
- Open time on machine
- Other sources

If all these are not coordinated through effective planning and scheduling, the lower rate accomplishes two things:

- Masks inherent system weaknesses
- Possible low skill level may mean longer repair time equating to a cost which will include the even higher cost of lost product.

Therefore, the labor rate alone should not be construed to provide the cheapest price. It will, with all things being equal; however, ninety-nine percent of the time it will not.

The same holds true for using price as the determinant factor in buying tools, equipment, materials and services. Contract maintenance, encompasses the contracting of all and necessary labor and services to meet the needs of the maintenance department, defined in your business plan.

Figures 11.4 and 11.5 indicate functions of maintenance management and what I refer to as TIPM or Total Integrated Plant Manage-

The Functions of Maintenance. . .

Figure 11.4 Functions of maintenance.

ment. Your maintenance needs can be defined with reference to these two concepts. Practically all maintenance facility managers are continually inundated on a daily basis with words such as:

- Maintenance cost management
- Maintenance quality

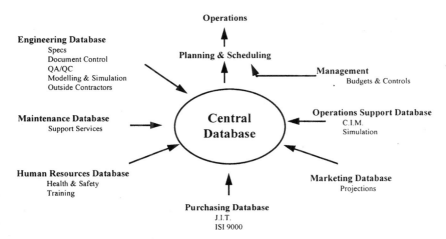

Figure 11.5 Total integrated plant management.

- Selling maintenance to management
- Resource management
- Staffing
- Quality enhancement
- Downsizing
- Organization
- Skills
- Forecasting
- Stores
- Work orders, planning, materials, scheduling
- Designing effective maintenance functions
- Predictive, preventive maintenance
- Labor/equipment justification

You will note that these are all contained within the context of Figs. 11.4 and 11.5. They all deal with labor and services and all can be included in the phrase "contract maintenance."

Your dilemma as maintenance manager or plant engineer is, "What do I contract and on what basis do I justify an award?"

Trends in maintenance.   Figure 11.6 outlines the current trends in the approach that industry has been taking toward maintenance. Indeed, the successful contractors are focusing their approach to respond to

Systems / Methods
   •Preventive / Predictive vs Breakdown / Corrective
   •Historical Equipment Data
   •Spare Parts Inventory Control

Focus
   •Introduction of ISO 9000 to Maintenance Work
   •Equipment Reliability / Availability
   •Automated Control and Operation
   •Increase output with Fewer People
   •Move toward Strategic partnering to Achieve Required Skill Sets

Training Needs
   •Increased Multi-Craft Flexibility
   •Improved QC through statistical process control
   •Tightening Government Regulation with respect to Safety and
      Hazardous waste Disposal

Figure 11.6   Trends in maintenance.

these trends. Where the industry is introducing some sort of quality program and, most importantly, where senior management is dedicated to implementation, maintenance is taking an increasingly more active role. In such plants, maintenance is being allowed to carry out improvements they have always recognized the need for, but seldom been able to do because of "the short-term cost factor."

11.4.1 How to Select Your Contractor Depends on Your Specific Requirements

The methodology used to select your contractor relates directly to the way you interface with the contractor in your organization. If it is only to supplement your crew for "task at hand" type work, you will likely select on the basis of least total wage rate. Below are a series of guidelines on how to utilize your contractor and interface with him to provide true cost effectiveness.

- Is the contractor reporting to a central coordinating figure, knowledgeable of field work and contract maintenance and who has decision-making authority—usually the *maintenance manager, superintendent* or *plant manager* (common titles for the same position)?

- Is the contractor a part of the organization through a steering committee approach?

- Insist and clearly state in the commercial agreement that the contractor acts as an independent employer.

- Ensure sufficient delegation of responsibility (and accountability) and authority to the contractor site management as part of your team.

- Establish mutual cost cutting goals and required joint participation toward achievement. Allow him to form an active part of your continuous improvement or TQM initiatives.

- Set standard practices for liaison between owner/client and contractor, including uniform policies and practices, contract compliance and client support group interface.

- Ensure that the contractor is rewarded with greater responsibilities, just as you would within your own organization infrastructure.

Remember:

Contractor Strengths + Owner / Client Strengths = Effective Resource Utilization

The selection process is typically done through a committee of direct users, using a matrix approach to scoring. The direct users are those persons (both supervisory and tradespersons) who will be working and interfacing directly with the contractor's personnel on a daily basis, typically from operations, maintenance, engineering, safety, finance. The questions you will prepare are oriented around your needs in specific areas of safety, labor relations, process knowledge, engineering support, supervision, understanding of partnering, additional support personnel, home office support, support to in-house quality program, attitudes of senior management, etc. Who the contractor brings to the meetings will reflect on its sincerity to work with you.

You should not waste time on interviewing a large number of contractors simply for political purposes. A preselection process would include a list of questions such as:

- Sales volume
- Number of employees
- Profile of senior management
- Profile on clients with references
- Profile on types of projects
- Financial reports
- Bonding and insurance capabilities with references
- Request to visit their facilities.

The answers to these questions and the professionalism of response will provide sufficient information to effectively short list three to five candidate contractors.

The key indicators in preparing your short list will be:

- Financial strength (10)
- Insurance/bonding capabilities (10)
- Client references (10)
- Dedication to maintenance within the organization (10)
- Professionalism of presentation (15)

Select those scoring between 30 and 55. Your selection committee will score independently with the results based on the average of each factor.

Your final selection matrix, based on an actual example, might look like the following:

1.0 Management Committment to Quality-details of processes currently in place to assure and maintain quality of services to its customers.

1.1 Management & Organizational Structure
    1.1.1 Organization Chart with Names
    1.1.2 Resume of Senior Management
    1.1.3 Length of Time in Job Function
    1.1.4 Staff Turnover
    1.1.5 TQM (or ISO) implementation levels reached in organization
1.2 Employee Training Programs
    1.2.1 Status of implementation of SPC techniques
    1.2.2 Skills/Training
    1.2.3 Supervisory Training
    1.2.4 Safety Training
    1.2.5 Dollar percent of operating cost spent on training
1.3 Management Systems
Information on what systems company currently has in place to monitor.
    1.3.1 Continuous Process Improvements Through Application of SPC
    1.3.2 Industrial injuries
    1.3.3 Payroll

2.0 Total Costs

Provide information on how costs are allocated and tracked with a percentage breakdown of overheads and submit a total cost model.

3.0 Project Management Capability

Provide details on PM capabilities
    3.1 Number of employees with project management training.
    3.2 Experience & qualifications
    3.3 Types of program used.

4.0 Labor Relations

    4.1 History of labor relations performance
    4.2 Types of labor agreement used.

5.0 Safety

Provide most recent 2 years of workers compensation performance reports indicating safety statistics.

6.0 Financial Information

Provide most recent annual report or financial statement signed by a company officer together with key information covering the past four (4) years.

7.0 General Information
- A. Background of the company's history and a statement of it's mission or business principles.
- B. Information on three current contracts, one of which should be a maintenance contract.
  - 7.2.1 Customer name, contact, telephone number
  - 7.2.2 Type of business
  - 7.2.3 Type of contract
  - 7.2.4 Value and duration of contract
  - 7.2.5 Average annual person hours

8.0 Proposals
- 8.1 Brief outline of how you would manage a maintenance project at our plant with emphasis on:
  - 8.1.1 Total quality management/continuous improvement
  - 8.1.2 Team work
  - 8.1.3 Training
  - 8.1.4 Safety
- 8.2 Describe priorities when setting up an organization to meet the challenges of partnering.
- 8.3 List in priority, five (5) total cost reduction initiatives which you would propose for our plant.

This request (cost plus contract) was aimed at specific requirements needed by in-house management. Each individual plant will have its own special requirements. No handbook or text book can provide you with the "specific" questions to ask. You know your plant and your requirements best. You are looking for a contractor committed to supporting you, as a partner, in achieving your objectives. Figures 11.7 to 11.11 suggest how this might be achieved.

The effectiveness of your own in-house maintenance/facilities organization is directly related to it's ability to function within the total organization infrastructures of the company, division or plant. What portion you contract or "outsource," is entirely up to you. There are no rules except guidelines relating to employer/employee relationship and safety. A structured format for selection of your contractor, with performance based contractual wording will go a long way to ensuring true and measurable value for your money.

The performance based contract will provide a section relating to performance assessment factors upon which you would assess your own maintenance department. These factors are monitored by the contractor and owner together. There is generally either a 10 percent holdback from each invoice or an amount of money, decided by the steering committee, in advance, that is open to loss should the con-

Literally turns traditional style upside-down to promote new approaches

```
                          ┌──────────────────────┐
                          │  Partnering Agreement │
                          └──────────────────────┘
┌────────────────────────────────┐        ┌──────────────────────────────┐
│ Client Management Committee     │        │ Contractor Steering Committee│
└────────────────────────────────┘        └──────────────────────────────┘
                    ┌──────────────────────┐
                    │ Executive Committee   │
                    └──────────────────────┘
                                          ┌──────────────┐
                                          │  Building     │
                                          │ Trades Councils│
                                          └──────────────┘
                    ┌──────────────────────┐
                    │ Steering Committee    │
                    └──────────────────────┘
┌──────────────────┐  ┌──────────────────┐  ┌──────────────────────┐
│ Client Resources │  │ Site Combined     │  │ Contractor Resources │
└──────────────────┘  │ Management Group  │  └──────────────────────┘
                      └──────────────────┘
          ┌──────────────────────────────────┐
          │ Continuous Improvement Team       │
          │ Subcommittees      Subcommittees  │
          └──────────────────────────────────┘
┌──────────────────┐                  ┌──────────────────────────┐
│ Client Maintenance│                 │ Contractor Maintenance    │
│ Workforce         │                 │ Workforce                 │
└──────────────────┘                  └──────────────────────────┘
```

Figure 11.7 How partnering works in a TQM/CIP environment.

tractor be unable to meet the agreed upon assessment factors. These factors will refer to such things as:

- Speed of response
- Reduction of emergency type work
- Reduction of stores inventory
- Increased availability/reliability of equipment

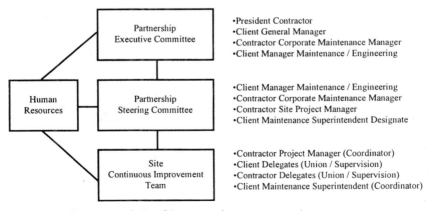

Figure 11.8 Corporate relationship-partnering arrangement.

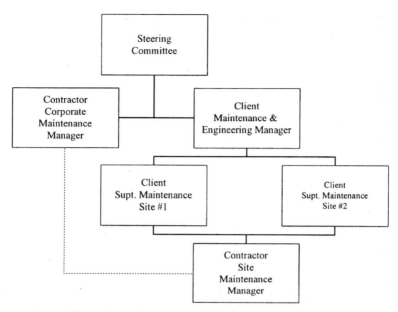

Figure 11.9 Plant relationship-partnering arrangement.

■ Reduced utility costs

The contractor, through the steering committee, can assist in developing the factors, should there not be a good base line of information from which to begin. A contractor, with good knowledge of ISO and/or SPC is well suited to support this long term management initiative. Such contracts require 12–18 months to formulate and verify cost savings.

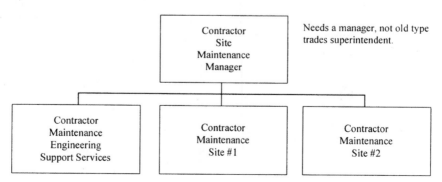

Figure 11.10 Plant support relationship-partnering arrangement.

- Steering committee meets monthly
- Executive committee meets quarterly
- Steering committee will create a framework for management through goals
- Project manager resolves problems through committees and personal contact, training, and problem / causal definition, employing maximum use of (trades)persons trained in problem recognition, definition, and resolution.

Figure 11.11   Problem resolution.

Before deciding to "contract out," or outsource, be sure there is good control with respect to workload planning and scheduling, allowing you the ability to forecast requirements over and above the base support crew. Be sure that your cost system allows you the means to not only control costs but also measure the effectiveness of those costs with respect to equipment and plant reliability. If you are unable to do this, a cost plus, performance driven contract will provide the ability to reduce person power, reduce costs and increase availability of plant equipment.

There are a number of different types of contracts to respond to generally in the form of:

- Lump sum
- Unit price
- Cost plus

The owners, when creating terms and conditions are advised to put themselves in the place of the contractor when preparing contracts. Many owners request a lump sum bid to a prepared schedule and bill of material. Generally the owner is not skilled at being a general contractor or a contracts manager. The good contractors are skilled at responding to specific types of contracts and the owner suffers with "contract extras" or the traditional contractor/owner conflagrations. Maintenance is not construction, it is not "green field" oriented. There are many unknowns.

Owners, generally, no longer have a complete in-house project engineering/management capability. Select your contractor, as described and let him work with you in preparation of the schedule and bill of material. You know the end date for requirement, the good contractor knows how to meet or exceed it. Generally, for piping, plant and equipment repairs, overhaul, and installation replacement, the cost plus format works the best. It is extremely difficult in a maintenance environment to respond to the lump sum contract format which includes parts, supplies and materials.

Your own forces are cost plus, why not the contractors' as well—the contractor is really only an extension of your own workforce and management team. Therefore, if the management aspect of your organization is not in place, you will probably not save anything in contracting out. The labor differential you originally (incorrectly) based your need and selection on, will get you eventually. Remember that if you can not manage within your own structure, nor can the reputable contractor, unless you are prepared to work with the contractor to make changes in how the work is scheduled and resources managed. In the end, you will only be "brokering" labor and will lose out on the ability to ensure a good supply of skilled and knowledgeable labor and supervision through a managed workforce.

If a proven in-house management capability exists, you may want to look at just contracting labor swings with the contractor home office used to provide management support as necessary (on a fixed rate hourly basis).

Contractors having a business unit devoted to contract maintenance have the proper long term focus with which to support their client from a maintenance management perspective. Those contractors obviously using "maintenance" to supplement construction downturns, *I suggest*, are not committed to nor understand the concept of maintenance.

Open and closed shops can be equally as competitive where there is a strong team-driven spirit toward labor relations, rather than the traditional confrontational approach. The owner usually fears relinquishing control to the maintenance contractor. The good manager will always use the talent at his or her disposal to effect what has to be done.

11.4.2 Partnering

The concept of partnering with your contractor is becoming more prevalent in "world class" facilities. It is based on a long term relationship between the owner/client and the contractor, aimed at allowing each party to fulfill its own corporate mission while doing those things that it does best, to the mutual financial benefit of each party to the agreement. Such an arrangement can not be implemented merely to overcome a short-term market-driven condition, affecting either or both parties. The combined management strength of both organizations will determine the merits and success of such an arrangement, with respect to long-term contract maintenance, as it involves the joint sharing of defined goals and objectives between contractor and owner through regular meetings and strategic planning sessions. Attributes of partnering include the following two components:

Requirements for Standard of Excellence:

- Commitment
- Co-operation
- Communication
- Confidence and Trust
- Long-term in scope
- Team building
- Defined benefits
- Flexibility
- Method of response from both parties

Elements of the Partnering Plan:

- Definition of partnering
- Goals
- Benefits to both parties
- Common goals/objectives
- Work scope of each party
- Needs analysis
- Evaluation and selection plan
- Relationships
- Problem resolution
- Communication
- Contract terms
- Sharing of risks/liabilities
- Organization
- Team structure and relationship vertical/lateral
- Training
- Quality management
- Continuous performance improvement

■ Risk reduction and loss control

As the concepts of a quality-driven economy develop, due to survival in a global marketplace, facilities professionals are assessing how they contract-out work. With downsizing, there is less support staff to assist the maintenance manager in assessing all opportunities for cost reduction.

The advent of total quality management into industry is resulting in a reassessment of strengths and weaknesses defining what the business really is. The contracting philosophy, resulting from such analysis, is creating a trend toward single source long-term, or partnering relationships, away from bottom line price selection criteria. It is creating a selection process where purchasing and maintenance, together with operations, effect implementation of the relationship.

Your contractor selection and management strategy should include a joint in-depth analysis, by both maintenance and operations, of how you are now doing business. You should be focusing more on contractor reliability and team work, than on the bottom line pricing. If you focus your strategy on *needs,* you will improve effectiveness of your maintenance function through achievement of maintenance oriented targets which translate directly into production and quality improvements, targeted in parallel with operations. Your contractor, through a well-developed long-term partnering relationship becomes a key team player in achieving those targets as depicted in Fig. 11.12.

Partnering is not something that can be implemented to overcome a short term, market driven condition affecting either or both parties. The amenable labor pool available to both parties is essentially the same. Therefore, the owner/client has to assess its management strength from the highest level in the company, plant or division, in order to assess the merits of long term contract maintenance.

Partnering involves the joint sharing of defined goals and objectives, between contractor and owner/client, through daily and regular meetings and strategic planning sessions. It may involve, literally, opening up all but the most confidential of records to each party, in order to see how each can benefit from the other's strengths and, in some cases, weaknesses.

A partnering arrangement may involve financial output, from either party, to support a long-term planning process involving such areas as implementation of new technology, training, or process. It is intended to provide a quality product to the owner/client from the supplier (contractor), just the same as long-term supply arrangements are made within the automotive industry. To maintain the relationship, even to enter into the relationship, a quality process (or program) acceptable to the owner, must be evident. There must also be an ongoing effort towards improvement, measurable in quantitative terms.

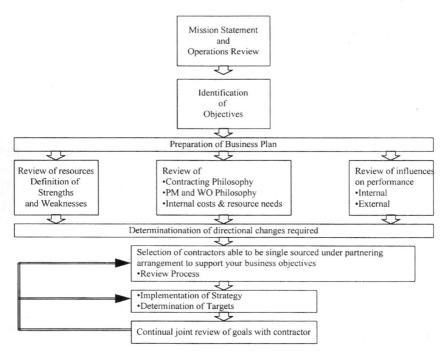

Figure 11.12 Strategic planning is needed for effective contracting of maintenance services.

The decision to partner must come from the highest level within the organization and acceptance must be evident from all functions within the organization, since the effectiveness of maintenance is truly affected by each function within the organization.

Selection of the contractor should be by committee, chaired by the highest level in the organization and administration of the agreement should be by a steering committee representing all functions of the organization in a team approach with the contractor (Figs. 11.7 to 11.10).

There have been projects where contract maintenance was used without full understanding and/or support from the entire organization. The results were a continuous struggle between contractor and owner with neither party achieving full benefit. The contractor is dissatisfied because he or she can't perform to potential and correct problems inhibiting performance. The owner/maintenance manager is concerned that his or her authority or knowledge is continually being questioned. There never is a real assessment of cost savings other than manpower, but at what price? Inefficiencies still exist that are never addressed. According to Paul Inglis, "A partnership means benefits to both parties...this is not a program for those with meager resources and poor internal communications...the selection...must be based on an analysis of current performance and future potential."[1]

Leadership from the highest level will provide the guidelines and goals. Contractor and owner management, together, left alone with clear guidelines and mandates, will produce the results required.

Look at your internal strengths and weaknesses, perhaps select a contractor that you know and trust and ask him to perform a parallel review at his cost. This becomes the basis for your arranagement — it may be in process or maintenance engineering, or QA/QC, or planning and scheduling, or even development of cost controls, or monitoring and administration of failure analysis and equipment utilization improvement studies. The contractor will become an integral part of your organization, responsible for adherence to goals in the same fashion as your own organization.

11.4.3 Summary

The field of maintenance management is taking on new dimensions with the external forces imposed by global competition. We in maintenance are now being recognized as capable of generating profits rather than costs. We have always had this capability; for some reason it has never been recognized. As senior operating management begins to change attitude and welcome us into the field of true management status, we can now begin to use all the tools available to us. Continue to be hands-on practical managers, assess our needs, and closely examine all the functions of maintenance management. The contractor/supplier group are the best resource suited to help you, through long term, value added partnering arrangements. They can be part of your team. After a careful selection process, make them responsible and accountable within the organization.

Let your contractor become part of your productivity improvement efforts. Support and encourage his or her input. They have valuable experience from exposure to other plants.

11.5 Warranty Considerations

The owner should never forget to include in the commercial contract, an article relating to requirements for the contractor to provide warranty protection to the owner. This is true at both the construction and maintenance levels.

The warranty control process model, Fig. 11.13, provides a series of items the owner should include in the contract to ensure compliance with OEM warranty requirements. The key to a successful warranty program is firstly to ensure that all items under warranty are, in fact, included in the program. This begins at the commissioning stage. (See

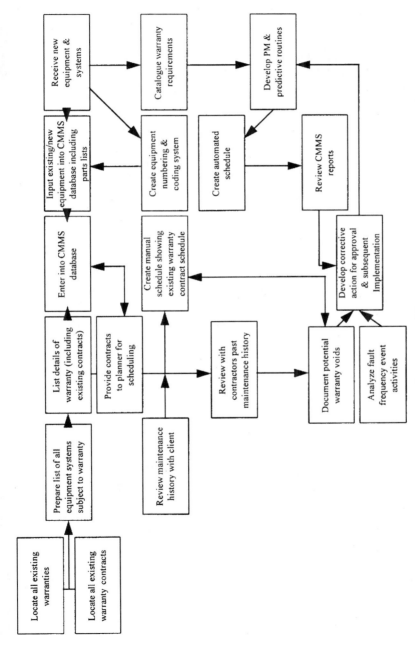

Figure 11.13 Maintenance program warranty control-process model.

121

Section B: Commissioning) A CMMS, properly structured, will contain an auditable trail for the OEM and owner/client of not only the PM routines required but also equipment operating history related to failures and equipment operating conditions.

This model was extracted from a project where outsourcing was replacing in-house maintenance and warranty of equipment was of particular concern to the owner. The owner wanted to know how the contractor was going to be able to assist in monitoring and control of costs associated with warrantys. One of your selection matrix criteria might include a specific requirement related to your own need for controlling costs associated with warranty protection. If you are an ISO company, your question might require an answer related to the contractor quality process specifically on warranty management. Your contractor, providing miscellaneous minor construction services as well as maintenance, as previously described, will provide document control in accordance with a quality process such that the warranty program is considered (Figs. 11.14 and 11.15).

Document	CM	PM	COM	SC	MM	O	OEM
Construction field quality control test reports	P	X	X				X
System summary lists	P		X	X	X		X
Maintenance deficiency lists	X	X	P	X	P		X
Transfer to commissioning reports	X	X	P	X	X		X
Safety Inspection Test Reports	X	X	P	P	X		X
Maintenance equipment test report			P		X		X
Transfer to on-load report	X	X	P	X	X		X
Completion test-run request	P	X	X	X	X	X	
Completion test report	X	X	P		X	X	X

Note: Documentation sets the stage for a controllable warranty program with OEM involvement.

CM	=	Construction Manager
PM	=	Project Manager
COM	=	Commissioning Manager
SC	=	Safety Coordinator
MM	=	Maintenance Manager
O	=	Owner
OEM	=	Vendors
X	=	Distribution
P	=	Prepares Original

Figure 11.14 Commissioning document control distribution.

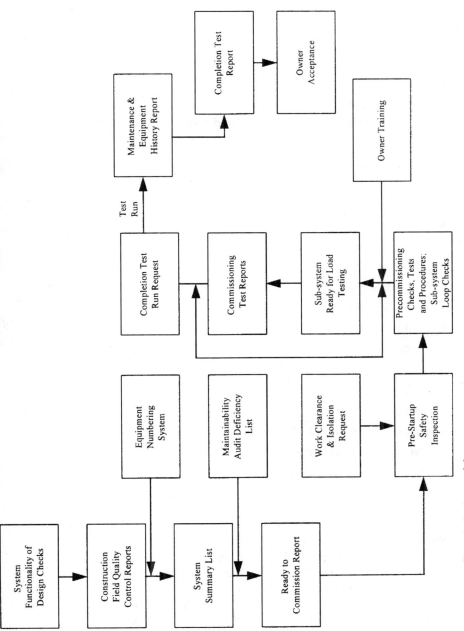

Figure 11.15 Commissioning model.

11.6   What to Outsource/the Business Planning Process

Typically, outsourced maintenance/facility related work is:

- HVAC
- Piping
- Roofing repair/inspections
- Exchanger cleaning
- Lubricating oil analysis
- Vibration analysis-establishing and monitoring program
- Motor rewinding
- Steam traps
- Motor repair
- Lifting equipment certification and maintenance
- Sewers
- Road repairs
- Custodial
- Groundskeeping
- Millwrighting
- Plumbing
- Carpentry
- Security
- UPS generators
- Instrumentation
- Electrical (480 V, 47 kW)
- Warehouse organization
- CMMS selection/implementation
- Steam trap program
- Thermography

Such programs intuitively can be done in a long-term partnership with your supplier/contractor. Your supplier is quite capable of using contractor developed management programs and processes to do so. All these items can be done with a minimum contracted force giving you the ability to promote familiarity, ownership, quick response, quick tool up, common planning, and scheduling at minimum cost. You are left with a base PM or emergency crew, able to work on problem solving and cause reduction. Where company branch plants are located in a geographical area, use a single contractor rotating crew with a co-ordinated approach to effective manpower utilization.

Just about anything can be outsourced as long as it fits within your business plan—The business plan developed by the maintenance/facilities department.

The purpose of your business plan is twofold:

1. Provide you with an organized plan on which to guide your management thought process and a focus to achieve your defined objectives;
2. Assist, through an organized approach, in development of justification for acquisitions to senior management and strengthen your position, by using your plan to support your company's overall plan.

Before you can develop a plan, you have to know where you want to be (goals). Before you can do that, you have to know where you are now (baseline of datum signature). Your plan will describe how you are going to achieve your objectives and where you are now. I assume your company has some type of mission statement. You, too, should have one for your department. This statement will be the overall guiding principle under which you and all your people, operate. An example might be:

> The maintenance department will keep the plant and facility able to produce a quality product at maximum design capacity, in a cost effective manner, while respecting the safety and integrity of its employees, suppliers and contractors.

We will accomplish this by optimizing maintenance activities, through an integrated approach with all departmental organizations, application of the latest technology and techniques available, operating in an environment of total continuous improvement.

Now you can begin to create a typical business plan which will contain, roughly, the following elements:

1. What your product is.
2. What your market is.
3. Size of your market.
4. What factors, internal and external, affect your performance. One of these will be the financial and management reports now being used.

5. What your organization is, how it functions, why it is needed and how it interrelates within itself and the entire organization.

6. Your contracting philosophy, your PM philosophy, your work order philosophy.

7. Desired achievement, strategy and monitoring activities (your strategic plan).

8. Budget showing assumptions and limiting factors. Include project dollars in your budget, as you, ultimately, will be responsible for maintaining someone else's errors. Including projects in your budget, will allow you to have control over the installation and to be able to schedule manpower at least cost to the projects. (yours and in-house contract workforce.) The budget is not the traditional escalated previous year actuals, but dollars based on required improvements and PM's plus an emergency work allowance target.

This step by step plan will force you to be very introspective of your department, and its interface within the entire organization. This is the getting organized phase for which you should allow 3–4 months. You may want to get your people involved with you to avoid bias and promote team effort. Companies entering a quality process are doing these things.

During this review process, you will be able to arrive at solutions, not excuses, for such things as inventory control, use of contractors, work order effectiveness, organization changes, equipment downtime and delay assessment. During the process you may be making minor changes within the system. Record these changes to show a cause/effect relationship.

With respect to contractors, keep a log or spreadsheet, tracking, generally, the following areas.

- Name of contractor
- Equipment or systems worked on
- Dates
- Costs
- Work orders assigned
- Whether work was done in combination with own forces
- Supervision — contractor or assigned in-house
- Speed of response
- Speed of completion
- MTBF of equipment worked on
- Failure patterns with the equipment

- Trend maintenance costs
- Cooperation of the contractor with your systems

If you have a CMMS, properly structured, you can track all this information by judicious use of work orders, using a well thought out coding system.

Remember, if your work order (accounting) system divides equipment down far enough, you will be able to spot trends. For example, in a steel mill, contractors may be assigned to a group of rolls, a table, a piping system, etc. In a refinery, it may include a unit or subunit, broken down by pieces of equipment.

Begin work with your contractors to see in what other areas they can provide assistance; get them involved in the planning and scheduling of the work they will be doing, in concert with your own people.

Test their management and support skills. Try to promote a sense of ownership for the work for which they are responsible. If you do not have facilities, or knowledge to analyze failure, can they assist you rather than hiring yet another party? What suggestions do they have for your work order system, PM program etc? Do not let pride stand in your way—your job is at stake!

Your contractor works under many management styles, stores systems, work order and PM systems. With this knowledge, he is bound to be able to assist you. Only *you* can lose control of your position—it was not the contractor's to begin with.

Remember that you contract for work to be done. To do that work effectively, requires preparation of a responsibility matrix as suggested in Figs. 11.16 and 11.17.

Depending on the volume of work to be done, you cannot expect the contractor to provide exactly the same people all the time. Therefore, what you are really contracting is assistance in managing the work. This concept is confusing and will be explained as we proceed.

Your plan, when completed, will provide reasoning for what services and how much you will contract out. It will, hopefully, enlighten you on the extreme importance of a good planning/scheduling and equipment performance monitoring system plus the reason why maintenance must be totally integrated with the rest of the organization.

11.6.1 Planning and the 52 week scheduler

If your strategic plan is working, you will have your mechanics totally involved in the effectiveness improvement program, working through co-ordinators. They will be working on such things as:

- Equipment data collection and verification
- Equipment numbering

	Maintenance	*Operators*	*Contractor*
the equipment to be available when agreed to		X	
nature of the work to be clearly defined by a work order package	X	X	
a workforce and equipment, tools, etc., to do the work	X		X
management of the work (notice I did not say workforce)	X		X

Figure 11.16   Responsibility matrix.

	Operations	Maintenance	Contractor	Purchasing
Preparation of Work Order	X	X	X	
Description of W.O. Completion	X	X	X	
Define Equipment Availability	X			
Define Work Force Requirement		X	X	
Define Materials Requirement		X	X	X
Review Work List	X	X	X	
Define Tooling Requirement		X	X	
Execution of PM's	X	X	X	
Define Corporate Mission	X		X	
Define Maintenance Mission		X	X	
Preparation of Management Reports	X	X	X	
Analysis of Management Reports	X	X	X	
Develop Skills Survey	X	X	X	
Develop / Implement Training Program	X	X	X	
Failure Analysis Requirements		X	X	
Planning and Scheduling		X	X	
Define Purchasing Requirements		X	X	X
Define Equipment Requirements		X	X	X

Figure 11.17   Work derivance responsibility matrix.

- Verifying, preparing and updating PM routines
- Suggesting changes, if any, to your work order system
- Performing activity analysis to spot holdups and implementing corrective action
- Verifying inventory, numbering and creating an organized stores and warehouse

Your contractors may even be able to provide assistance.

How large or small a section of the plant you do initially, is up to you. Whether it is done manually or by use of a computer (only speeds up process for you—it does not replace your own knowledge), enter all your project work for the year, PM routines by day, week, month, quarter, etc.; routine and emergency work orders as they come into the system. After you have scheduled all your planned work and resources (tools, materials, equipment and labor) you will see what work you can do and what you have to contract out. Your records review will assist you in improvements that have to be made and where your skills weakness are. Contracting training by osmosis, and maximum use of single contractor resourcing can assist you by reducing administrative costs. *This continuous scheduling function is your most important tool.*

The importance of planning cannot be overstated. Measurement ratios, such as work orders worked over work orders scheduled, are a beginning point for performance assessment. A TQM/TPM process doesn't look solely at these because trending creates no corrective action. We now prefer to look additionally at MTTR, MTBF, and interprocess product quality. Planning is the root cause of poor ratios, not scheduling. Such tools as activity analysis that you have done during your business plan process, as well as listening to your people, will highlight poorly planned work orders such as "fix steam leak" in steam pipe at pump P-404." This creates delays while people find where it is, how big it is, and what tools and people are needed. If it is not fully planned, it cannot be properly scheduled with production. The maintenance contractor, can get very frustrated because of poorly planned work orders impacting on their performance. If it frustrates them, think what it does to your own people with respect to morale and productivity. Think of what it does to cost of repair. We have to know *what, where, why,* and *how* The *when* is scheduling. We like to see what we call a full work order package with drawings, contacts, tools, parts, etc., as do your own people.

11.6.2 The maintenance organization

In assessing your organization, during development of The Business Plan, refer to the functions depicted in Fig. 11.3. How you assign accountability is your choice. The contractor can relate to this functional organization to get the information he needs.

Traditionally, the owner has contracted labor to execute the work. Some contractors have been involved, to a small degree, in planning and scheduling the work.

Few organizations look to the single-sourcing capability of their contractor in the establishment of a long-term partnering relationship.

It is advantageous to have as few relationships existing as possible, in project and ongoing maintenance work, so that familiarity and ownership of the equipment can be established by both your contractor and your in-house crew. It has the added impetus of creating the accountability in performance that you require of your contractor. Remember that *you must maintain control,* but your contractor can be a partner with you in the control process.

11.6.3 Contracting philosophy

The business plan you have prepared jointly with operations and spent considerable time in preparing, will define what your primary business is and where you, as plant people, should be expending your efforts. Your maintenance organization will respond to Fig. 11.3. Your maintenance contractor can easily become a fully participating part of that organization.

You will have selected targets for improvement and your focus must be in these areas.

These targets might take the form of

- Reduce inventory requirements by 2 percent
- Increase bearing life on "Y" machine by 50 percent (indicated by analyzing MTBF)
- Reduce lubrication costs by 5 percent (by standardizing lubricants used and analysis of PM program)
- Reduce overtime by 5 percent

These targets are derived from preparation of your strategic or business plan, and translate directly into dollars.

Quality programs are assessing each stage of the manufacturing process, which should facilitate the management aspect of maintenance. Your job as manager of maintenance costs and equipment reliability is to keep the equipment capable of operating at design capacity while producing a quality product with a minimum loss of product. You want to accomplish this, at the least cost possible. One of the first things you want to do is to determine the true out of pocket cost of each maintenance and support employee, and facility on site. Next, you want to, using your 52-week schedule, examine what work you have time to do with your existing work force. You want to highlight those delays for which you are responsible and isolate those charged to you, yet caused by product or operations problems/causes. This is a key item in being able to justify your maintenance costs.

Prepare a check list of contractors you have dealt with and what their specialties and strengths are. Do not prejudge union vs. nonunion, nor the ability of union contractors to work in your, perhaps, nonunion environment. You need help in managing the work to be done. Your job is not to see that the work gets done—that is the contractor's job. Be specific with him or her on defining your expectations and needs. If you lock him into a lump sum job, you'll end up, 90 percent of the time, paying more in the long run.

This concept of contractor responsibility appears to be, perhaps, the most difficult for plants to understand. If you prepare the responsibility matrix with your contractor beforehand and your contractor has the ability to perform unimpeded, you will be able to truly judge his or her performance.

Remember that maintenance is not construction. The philosophical approach is different. Your contractors should be selected based upon their experience in maintenance, providing service and management assistance. You do not want to have to talk to a myriad of people to get things done. You want to look at a contractor who understands such things as:

- Support
- Responsibility/accountability
- Commitment
- Cooperation
- Team work
- Planning
- Ownership
- Scheduling

His or her performance and therefore longevity of contract, should be measured the same way as you measure your own performance.

During your contractor selection process, ensure that he or she understands accountability and management, through a series of interviews and discussions with his or her client base. Exclude pricing from these initial discussions. In the final analysis, their bottom line consists of the same factors as your own; true labor cost, and overhead. Is he or she willing to open his or her books to you for audit to ensure all profit isn't derived from your plant? Is he or she consistent with application of profit? With your contractor, before the work is to be done, perform a responsibility matrix, similar to that shown in Fig. 11.17.

11.6.4   Summary

As the concepts of a quality-driven economy develop, due to survival in a global marketplace, plants are assessing how they contract out work to be done. With downsizing, there is less support staff to assist the maintenance manager in assessing all opportunities for cost reduction.

The advent of Total Quality Management into industry is resulting in a reassessment of the strengths and weaknesses defining what the business actually is. The contracting philosophy, resulting from such analysis, is creating a trend toward single-source, long-term, or partnering relationships away from bottom line price selection criteria. It is creating a selection process where purchasing and maintenance, together with operations, effect implementation of the relationship.

Your contractor selection and management strategy should include a joint in-depth analysis, by both maintenance and operations, of how you are now doing business. You should be focusing more on contractor reliability and teamwork, than on bottom line pricing.

If you focus your strategy on needs, you will improve effectiveness of your maintenance function through achievement of maintenance orientated targets which translate directly into production and quality improvements targeted, in parallel, by operations. Your contractor, through a well-developed, long-term partnering relationship, becomes a key team player in achieving those targets.

11.7   The Typical HVAC Contract

The rage in the '90s has been to establish performance based contracting, based on energy savings, with commitment beginning at the end of construction and continuing throughout the energy guarantee period. Such contracts have proven very successful for the contractor but what happens at the end of 3–5 years? Without proper maintenance and management control, statistics will show that costs begin to return to those existing prior to the retrofit.

The typical HVAC contract is a lump sum calling for delivery of a preventive maintenance program. The contractor bids, based upon past history of repair parts to sometimes poorly defined specific PM spec, provided by the owner. Emergency calls and call-ins are extra, as are parts beyond a certain value. There are no factors with which the contractor's performance will be held accountable. The contractor wins the bid based on labor rate and absolute adherence to the PM required. The owner provides little or no interface with the contractor other than invoice approval.

Generally, a number of different contracts are let involving chillers, piping, EMS, water treatment, cooling towers, fans, blowers, and controls. This adds to the administrative nightmare.

From an investigative point of view the PM contractor has generally, nothing to do with providing maintenance reports to the owner. It is generally left to the owner to enter data to a manual maintenance management or automated system. The contractor is not accountable. The contractor, at times, seldom interfaces with the owner unless it is specified, or unless the contractor is interested in doing so. Cost control is generally left to the owner unless clearly defined parameters are provided in the contract. A number of different HVAC system contracts adds administration to the owner.

11.8   References

1. John Brewster, "Contract Maintenance a Partnership in Cost Cutting," *IIE 8th Annual International Maintenance Conference Proceedings,* pp. 279–290, November 1991.
2. John Brewster, "The Maintenance Operations Review," *1st Annual Canadian Maintenance Management Conference Proceedings,* Toronto, June 1989.
3. J. D. Brewster, "Simplifying the Myths about Contract Maintenance," *AIPE Facilities,* March/April 1995, pp. 43–47.
4. J. D. Brewster, "Doing More With Less, A Strategy For Contracting of Maintenance Services," *AIPE Facilities America '95,* Portland, Oregon, October 17, 1995.
5. Walter Munroe, "You versus Us: Resolving The Key Contracting Problems," pp. 18–22, *Plant Services,* November 1990.
6. Paul Inglis, "Preparing for Vendor Partnership," pp. 31–33, *Purchasing Management,* May 1990.

11.9   Index of Abbreviations

Page	Symbol	Meaning
96	ICI	Industrial, Commercial and Institutional
	GP	General Presidents
98	QA	Quality Assurance
	QC	Quality Control
	CMMS	Computerized Maintenance Management System
99	WHMIS	Workplace Hazardous Material Information System

Page	Symbol	Meaning
103	CMMS	Computerized Maintenance Management System
	ISO	International Standards Organization System Of Quality Measurement
	TQM	Total Quality Management
	TPM	Total Preventive Maintenance (Japanese Philosophy and System)
	HAZOP	Hazardous Operability Review
	CMMS	Computerized Maintenance Management System
110	TQM	Total Quality Management
	ISO	International Standards Organization System Of Quality Measurement
111	SPC	Statistical Process Control
	PM	Preventive Maintenance
112	CIP	Continuous Improvement Process
119	QA/QC	Quality Assurance/Quality Control
120	OEM	Original Equipment Manufacturer
124	PM	Preventive Maintenance
128	TPM	Total Preventive Maintenance (Japanese Philosophy and System)
	TQM	Total Quality Management
	MTTR	Mean Time To Repair
	MTBF	Mean Time Between Failure
131	EMS	Energy Management System

Computerized Maintenance

George M. Player, CPE
Assistant Director-Engineering

Howard J. McKew, P.E., CPE
Vice President-Engineering
William A. Berry & Son, Inc.
Danvers, Massachusetts

12.1 Introduction

Preventive maintenance (PM) is planned maintenance. For the educated building owners, this PM process is a proactive initiative that will usually be balanced by an equally planned operating budget. It is strategic investing today for tomorrow, an intangible insurance policy that promises equipment reliability, extended equipment life, and efficient use of energy.

Managing a facility should be considered maintaining an asset versus putting capital into a cost center. Much has been written about this subject of facility management and the quest to do a better job at maintaining this asset has come under intense corporate scrutiny in recent years. An example of proactively maintaining this building asset has been through the introduction of computer technology as the new cornerstone of the facility management process. Applying today's Computer Maintenance Management Software (CMMS) technology can transform this cornerstone into a foundation of facility management success.

For a health care facility, the patient care environment and system reliability are two strategic issues that should be an integral part of the facility management mission. For an educational institution, customer comfort and system reliability becomes part of the facility strategic plan. Whether it is health care or academia, all facility management entities share a third common agenda and that is to maintain cost-effective operations. The CMMS system can be the tool to organizing this process and achieving these strategic requirements.

Those who invest in planned maintenances are people who recognize and appreciate the need to routinely care for the building systems. You don't get nothing for nothing! Analogous to the upkeep of your automobile, PM is something that you would rather not do but know proper care is essential if your car is going to provide you with a reliable means of travel. If the car does not start; it runs but doesn't get really good gas mileage, or the engine overheats, then investing in a PM tune-up is something the owners can relate to. They know they will receive an immediate return on their investment.

Routinely changing the oil, oil filter, and air filter have less obvious returns on their investment. The automobile owners do not experience the same immediate feeling of satisfaction from this investment as they do with the tune-up of the engine. They quite often miss that "money well spent" experience.

It is this empty feeing that raises the question "When do you change the oil in your refrigerant system?" The car manufacturer may recommend the oil be changed every three thousand miles and the refrigeration equipment manufacturer may recommend changing oil every year. The end user is not provided statistical data to support these directives so most people will not strictly adhere to the manufacturer's recommendations. Instead they will select PM "tasking" based on their experience and/or what they can afford within their operating budget. *Planned maintenance is not an exact science but rather an educational experience.*

Once building owners recognize that they must invest in a PM program, the next issue will be to recognize the scope of work, the cost to start and the cost to continually manage this program. Today's technology is inherently driving the planned maintenance process via the use of CMMS systems. Manual PM systems are antiquated and CMMS systems are the processes for today that comes in good, better and best software. *What may be good for a small simplistic application may not be good for a large multibuilding complex.* A more sophisticated CMMS system may be the best solution for the large campus settings and yet, as a first step the process of choice frequently decrees "learn to walk before your can run."

12.2 Applying Quality Management Tools

Recognizing the need for a CMMS system is best solved through the Quality Process of "Plan, Do, Check, and Act." Planning entails data collection based on what is essential for this program. It is this first analytical step that will help the potential CMMS candidate with the pertinent information necessary for the process to be a long term success.

Using a matrix analysis, we have established a menu of needs and have applied a scoring system for (3) different building applications (Table 12.1).

Using a matrix analytically to estimate the need for a proactive PM system, the building owner can mathematically determine if they should invest in a CMMS system. Based on Table 12.1, the hospital would be a strong candidate for this system with its score of 24 points out of a possible score of 28 points (86 percent). The college and industrial facilities scored more than 50 percent "need" with approximate scores of 64 percent and 71 percent, respectfully. Based on these two scores, further analysis may be required to better define the CMMS application. Building "owner-specific" categories may need to be added and scored, such as, the institution's long-term plans to remain at this site; the economic situation of this business and/or will the facility being considering outsourcing the maintenance management?

Having introduced an analytical means to determine if a CMMS system is applicable for your needs, the next step is to commit to purchasing this system. The Quality Process requirements continue the "Do" step with a plan to invest in a CMMS system. Herein we will discuss two such plans. The first case study will be a new CMMS system at a site where there was no previous planned maintenance.

TABLE 12.1 Needs analysis matrix

Criteria	Hospital	College	Industry
Agency compliance	4	2	2
Life safety	4	4	4
Indoor air quality	4	4	4
Owner paid utilities	4	4	4
Extended life cycle	4	4	2
System reliability	4	2	2
Computer-literate staff	0	0	2
Total	24	18	20
Maximum possible score	28	28	28

Scoring: 4 = essential   2 = preferred   0 = Not Needed

The second case study will be an example of introducing a new CMMS system to a facility that had already had a system in place. In the second example, the client had failed to carefully Plan, Do, Check, and then "Act" on the process. As a result, their existing CMMS system failed to satisfy the goals of facility management staff.

It has been noted by many sources that successful CMMS systems are in the minority and that most planned maintenance efforts fail because there is no "buy-in" by the people who will be responsible for its success: the workforce. The process requires close management and monitoring of the implementation phase. This commitment to continually check the "Act" sends a signal to the workforce that this CMMS system needs to work, be continually improved and continually measured. In the case of the existing system that failed, the process lacked day-by-day management necessary for it to succeed. Instead of routinely analyzing the data received, the CMMS system never established itself sufficiently to displace the unscheduled work order process. Our discussion will focus on how to start a CMMS system, manage it and continually improve the plan. Eventually, each of these case studies will contribute operating cost savings and extend long-term replacement needs.

12.3 Process Requirements

Before selecting a new CMMS, facility managers should evaluate their needs and set priorities on their requirements and goals (refer to Table 12.2, CMMS Customer Needs Matrix). There are many customer needs, and this matrix is just one customer's priority list. In general, the questions are:

- Will this system be used for PM, as a repair work order system, materials management (spare part ordering system), a combination of some or all of the modules available?

- Most CMMS systems will meet your requirements, but can they be simultaneously implemented?

Experienced facility managers will tell you that implementing more than a couple of modules at the same time can be counter productive and time consuming for the staff involved. Based on our experience, customers who set out to use a CMMS system (the "Act" phase of the Quality Process) will frequently fall short of their anticipated goal. In turn, the system will be ineffective, resulting in the software not being used to its full potential and staff disappointed with the results. Attempting to do-it-all, at one time has become the Achilles' heel of CMMS success. Strategically planing a phased implementation of the CMMS system must be part of the customers' priority matrix.

TABLE 12.2 CMMS customer-needs matrix

CUSTOMER ISSUES	PRIORITY	COMMENTS
CMMS has several options	PM work orders	Phase II - Expanded
Compatible with ...	IBM/Apple	Required Capacity ___ RAM
"Windows" driven	Yes	
Purchase or lease	Lease	Budget for $
On-site training	2 days	Unlimited telephone support
Automatic upgraded	Preferred	
Awareness letter	Preferred	
On-line support	Required	24 Hour preferred
Customized reports	Required	"Word Perfect" preferred
Graphics software	Required	"Powerpoint" preferred
Data retrieval assignment	Out-source labor	Preferred
Bar code adaptable	Preferred	Budget for $____
Automatic scheduling	Required	Automatic load-leveling
Estimated data retrieval hours	$____budget	____hours, co-op
Data base software	$____budget	To benchmark
Data input hours	$____budget	____hours, secretary
Scheduling hours	$____budget	____hours, project manager
Customized reports	$____budget	____hours, dir. of engineering
Customized graphics	$____budget	____hours, coordinator
Upgrade #1, Team tasking	$____budget	____hours, coordinator
Upgrade #2, JCAHO compliance	$____budget	____hours, coordinator
Upgrade #2, code compliance	$____budget	____hours, coordinator
CMMS Administrator	Required	$____, salary
Annual training	Required	____hours, coordinator
CMMS updates	Required	____hours, coordinator

A Powerful Tool:
The Customer-needs matrix is a tool that can be modified by the user to outline costs and hours associated with the implementation of a CMMS. It should also be used to establish the time line and milestones for implementation

Referring to the CMMS Customer-needs matrix, Table 12.2, the customer should also document the available resources that the management group can cost-effectively commit to the process. In our experience, having all the available modules does not necessarily mean this product offers any significant advantage over less comprehensive software because there may not be the resources available to apply these additional modules successfully. This is particularly important in the first year of use. Referring back to the decree that you "learn to walk before you can run," setting up a CMMS program requires a carefully planned progression of successful steps. Purchasing a simple, and possible less costly CMMS software may be a very effective first step, while allowing for growth of the system as needed.

Another important factor that must be considered when selecting a CMMS for use at the facility is the compatibility of the data in the CMMS. Can this information be transferred and coordinated with other computer software systems currently in place at the facility. If there is a current work order system in place and a new planned maintenance system is introduced, the man-hours tracked by both systems will need to be paralleled until everyone is in agreement that all the tasking data was successfully transferred to the new CMMS system.

It was recently noted that industry standards have recorded that 15 percent of the work orders have been planned PM and that 85 percent have been unscheduled work orders. Our position is that as properly conducted and measured PM work orders increase, unscheduled work orders will decrease. Planned maintenance work should become the driving force for the department.

12.4   Reviewing Your Options

Moving from the "To Do" to the "Check" phase of selecting a CMMS program that fits the previously discussed needs analysis, potential buyers can pick up almost any facility trade publication and find advertisements and information for numerous CMMS systems. The American Institute of Facilities Engineers (AIFE) journal includes periodically a detailed guide to several CMMS products. The latest listing at publication time is shown at the end of this chapter, but the reader should verify later listings with AIFE.

This lists more than 75 systems that address such issues as operating characteristics, data languages, price, and performance. This listing, in parallel with the previous Table 12.2 matrix can be an effective and analytical means to efficiently document all the needs to be met with the purchase of the CMMS software. It would also be beneficial to contact other facility managers in your industry to get their perspectives and comments on CMMS they are currently using.

Step 3, "Act" should not overlook the time necessary to successfully complete the entire process. In evaluating CMMS system purchase and implementation, it will require establishing a time line and milestones for the successful implementation of the system.

The listing notes several options. Some options may not be immediately used, but consideration should be given to future growth of the system's use. For example. a user may elect to implement the planned maintenance module immediately, and later make use of the corrective work order system, the material management module or time card module as needs dictate.

12.5 Implementing the Process for the First Time

When starting the CMMS maintenance program, the customer must be prepared to "think the process through." The user needs to outline all the steps required for successful implementation, write it down and then methodically follow-through with the process. This concept requires a total commitment to applying these initial steps and then continue to review the process. Once these initial steps have been reached and the staff has bought into the process, then additional CMMS modules can be added at the appropriate intervals. Implementing a CMMS system can become a well-documented Total Quality Management process to ensure that all groups involved have assisted in its development and successful implementation.

In one recent case, a college campus, the customer's requirements began with the need to simply initiate a CMMS program. This client had restricted proactive PM of their building systems and instead had adopted a "pay-as-you-go" approach. Through an array of service contracts, the college would contact the respective vendor when a repair or maintenance was needed. As a result, the facility management was trapped in a reactive versus proactive mode of operation and relied on vendor work slips for documentation of maintenance.

The college recognized the need to change this process. With our new facility management team, the customer purchased a CMMS program for the first time. With thirty buildings spread over its campus, a CMMS system allowed the college to now proactively manage its facility as an asset. Additionally the CMMS system had to be able to track the scheduled maintenance performed by its service vendors, which accounted for a large portion of the facility's maintenance activity. As the college moved from a nine-month academic schedule to a twelve-month schedule, the need to schedule the required maintenance at the proper time became a function that the subcontract service vendors could not rely on to perform. Management incentives would be scheduled in the system to prompt the facility manager to call and schedule the required vendor.

Once the benefits have been determined, and the decision has been made to bring the planned maintenance system into the computer age with a new CMMS, the most important assignment with the implementing of a new CMMS is determining who will be in charge of the day-to-day process. Our experience indicates that there must be one person controlling and evaluating the process to achieve the expected goals and outcome of implementing a CMMS program. When initiating a CMMS software system, controlling the data retrieval, maintaining the project will be on-schedule, and managing the personnel assigned to this start-up job are key to its success. The project manager should schedule job progress meetings, held on a regular basis, to monitor the progress of the project and to focus efforts in areas where additional needs are required.

To perform the task of equipment data retrieval at the college, we chose to use engineering students from two different technical colleges who were participating in their school's co-op training program. Although the students possessed an engineering background and had some computer skills, training was required for the data retrieval and equipment tasking process. In order to ensure all the required equipment information would be recorded properly and consistently, a data retrieval form was developed and tested by the co-op students. A revision to the form was completed, and this document became the current data retrieval form being used with all CMMS installation projects (refer to Mechanical-Electrical Data Retrieval, Figs. 12.1a and 12.1b).

Construction and renovation projects are ongoing and always changing day-to-day activities for the facility staff. In order to ensure this newly installed equipment would be captured by the PM system, the data retrieval form was issued to the construction manager, and their subcontractors would record the equipment information. The form would then be sent to the facility maintenance department and input into the CMMS system. Construction and renovation also means removal of equipment. In order to delete replaced equipment in the CMMS, the data retrieval form would be utilized. It would be marked as deleted equipment and only the identification number and location of the equipment would be required for this process.

12.6 Step 4—Implementation/Replacement of an Old CMMS System

At one such health care facility, a CMMS software was put on-line and failed almost immediately at being a proactive solution to pro-

MECHANICAL - ELECTRICAL DATA RETRIEVAL		
FACILITY:	COST CENTER:	EQUIPMENT CATEGORY:
EQUIPMENT DESCRIPTION:		I.D.#:
SERVICE:		PRIORITY:

LOCATION	BUILDING/FLOOR:	ROOM:
	DESCRIPTION:	

EQUIPMENT	MAKE:
	MODEL NUMBER
	SERIAL NUMBER:
	CAPACITY:

DATE INSTALLED: / /	VENDOR/MFG:	WARRANTY:

COMMENTS:

COMPONENTS
(Fuses, Filters, Belts, Etc.)

	INV. CODE #	DESCRIPTION	QUANTITY
1.			
2.			
3.			
4.			

Special Instructions:

MSDS Information:

Figure 12.1a Mechanical-electrical data retrieval form (front).

grammed maintenance because of an overly ambitious implementation plan. In hindsight, the tactical planning associated with this CMMS start-up was deficient. The management staff had not completely thought-out and prioritized the critical steps to accomplishing their plan. Beginning with the necessary commitment of time and labor, the process lacked follow-through by management.

We quickly identified these process shortcomings with the hospital's initial implementation plan and took action to correct a process that

Motor			
Vendor/Mfg.			
Make:		Model #:	
Serial #:		Frame:	
HP	RPM	Type	
Cycle	Phase	Design	Serv. Fac
Voltage	Amperage	Code	NEMA Desg.
Dr. End Bear.		Opp End Bear.	
Ambient Temp:		Rating:	
Pressure Vessel Specification			
NB#	Certified By:		MAWP
Year Built	CRA#		Receivers
Head	Shell		
Pressure Specifications			
PSI	MAX WP	FT To Head	
Max Work Press Shell		Max Working Pressure Tubes	
Compressor Air Max WP			
HVAC			
Refrigerant		LBS	OZS
PUMP			
GPM	RPM	Pump Head	
Max WP			
Miscellaneous			
Belt	Motor Sheave	Fan Sheave	Controller
Relief Valve	Regulator		

Figure (1) Back

Figure 12.1b Mechanical-electrical data retrieval form (back).

had been to ambitiously implemented and subsequently abandoned. We evaluated where the process was and established a project schedule to complete data retrieval and data input. As a result, milestones were established for project completion.

A major feature of CMMS programs, and an essential customer requirement for hospital facility managers, has been the ability to record accurate maintenance records. By using CMMS, maintenance record-keeping becomes a by-product of successful implementation and use of the system. At this health care site the strategic planning had not been thorough and the subsequent record-keeping became a missed opportunity that our management of the process later regained.

At the hospital, we reviewed the existing system and analyzed its capabilities. We recognized that the hospital's requirements were not being met because the existing software was very limited in its system options and its capacity. It was found that new equipment that was being entered into the CMMS was saved while original equipment data was lost due to lack of system capacity. As a result, the work order process was flawed and equipment was not being maintained on a scheduled basis. In addition, the required documentation for the Joint Commission for Accreditation of Hospital Organizations (JCAHO) requirement for routine inspections and testing was missing. This missed opportunity within the existing software, which was an essential customer requirement for the facility management mission, would later become a part of the renewed strategic plan.

From the Quality Process of analytical analysis and data review, it was determined that the present system be replaced and a renewed commitment to the CMMS process be made by the entire staff. To avoid the pitfalls of the existing system, while not overreacting to its shortcomings, we elected to use a simple but effective software to begin their assignment. They recognized that it did not have the sophistication of other more expensive and more elaborate CMMS systems but the hospital needed to "learn to walk before it could run." The focus was to be on a practical and effective process, performed in a timely manner.

When replacing the hospital's existing system, we inherited an existing data base that was not easily down-loaded into the new system. Keeping the new CMMS application independent of the existing system would be prudent while the new system was being installed. Operating both systems in parallel for a predetermined period would be an effective way to start the new system. Paralleling would allow the synchronization of the two systems databases, whereas, combining the two from the start would have resulted in transferring existing "bad data" problems into the new program. Planned maintenance could continue as equipment data was being loaded into the new CMMS. Associated work tasks, scheduling, and work "load leveling" could be properly distributed. Additional administrative support was brought in to complete this step in a time-efficient manner.

At both facilities, the customers' requirements include receiving a return on their investment and a means to measure the success of the CMMS process. By doing the maintenance on a scheduled basis and inspecting systems regularly, a reduction in unscheduled repairs and in equipment down time could be tracked and measured. At each of these complexes, we assigned a Project Manager who was responsible for keeping the process on schedule and measuring the results at each milestone. This Project Manager also met with each trade shop on numerous occasions to ensure their concerns and questions were answered. In turn, we were able to achieve the facilities's CMMS goals.

12.7   Data Retrieval Form

An integral tool in any data retrieval process is the application of standardized data forms that interface with the CMMS. We developed our own Mechanical-Electrical Data Retrieval Form. In order to record all pertinent information from the equipment needed for the CMMS program, the Data Retrieval form was modified to meet the systems requirements. This form should be used as a tool for the programs implementation and should be customized as needed. The form provided consistency when assigning several different personnel for data retrieval. It also proved its value later when providing construction management services at both the college and the hospital. As the Construction Manager, we would request the design engineer to insert this form into the contract specification. In turn, the equipment manufacturer would then be obligated to complete the data retrieval during the shop drawing phase of the construction project. This thought process applied the Quality Process rule "do it right the first time" by having the data retrieval completed during the shop drawing phase of construction.

During the data retrieval period, our Project Manager required that the facility mechanic install a bar code identification number on each piece of equipment while verifying data retrieval information. An alternative to this method of bar coding can be installing a "fabricated" ID number and, when the piece of equipment PM is printed for the first scheduled maintenance, the data retrieval information can be verified and the ID bar code sticker installed.

At the hospital, the original data retrieval process did not initially follow a predetermined implementation schedule or procedure. Emphasis on updating the existing CMMS system was sporadic, and when initiated, some trade shops completed data retrieval on only their specialty equipment (i.e., HVAC Shop completing inventory of HVAC equipment), while other shops were taking inventory on all equipment in a given equipment room. In addition, data entry was random with no specific individuals assigned the task of loading data into the computer program. This reinforced the need for a tactical plan to be developed by us based on our experience with CMMS systems. The previous CMMS implementation lacked sufficient direction and follow up. Our managed approach was to require a project schedule, assignment of tasks and associated due dates, and regularly scheduled meetings designed to ensure the timely success of the process.

While both the hospital and the college had a Project Manager, at the hospital they also chose members of the existing staff to complete the survey of existing equipment. The size of the hospital and the ability to access all areas, including sensitive areas, made the decision to use on-site personnel to complete data retrieval proved to be very effective. At the college, the introduction of CMMS system was an entirely new process. The project schedule required weekly meetings to update the progress and make adjustments as needed. The process of hiring two co-op students, training them, completing inventory of 30 buildings, loading the equipment into the CMMS program, tasking the equipment, scheduling for programmed maintenance, and finally issuing work orders proved to be equally successful. Accurate data retrieval was critical to the success of the project and future acceptance of the system by the personnel involved in the planned maintenance process. At the hospital, the staff was skeptical of introducing another CMMS system based on their own experience with the existing system that was already in place. Similar skepticism arose at the college because the school had functioned without such a system. We believed it to be essential that all the needed information and the location of each piece of equipment be accurate if the technicians were to accept the new system. Bar coding was an integral part of both CMMS projects.

It should be noted that during the retrieval process it was not unusual to not be able to locate a piece of existing equipment. On these occasions, it was determined that this equipment had been removed during a modernization or energy retrofit project but existing record drawings failed to reflect these changes. This proved to be a very important point to stress to facility personnel when discussing Continuous Quality Initiative (CQI) and the need to maintain accurate CMMS system database.

Special operating instructions or special shutdown procedures were also noted on the data retrieval form and included on the PM work order. For example, every three years, health care facilities must pass a JCAHO inspection. At this hospital, JCAHO requirements were noted in a customized file provided within the new software program. The customized file allowed the facility manager to identify all equipment that was related to patient care with a designated "Utility Management File" established by the user. Additional custom files for OSHA, NFPA, Life Safety and Refrigerant Management were also established.

Start-up of these custom files allowed the user to sort through all the equipment listings and note only those that were assigned to a specific custom file (i.e., Utility Management File or JCAHO designated equipment). When sorting through over 5,000 pieces of equipment and needing only those that are associated with the JCAHO utility management requirement, this custom file sort proved to be very efficient use of process management time. This day-to-day JCAHO compliance was a culture change for the hospital group traditionally experienced to what we called, "a JCAHO Fire Drill" every three years. Now the staff was providing scheduled compliance as needed each day, week, and month of the year depending on the task. Special instructions also included who to contact when a mechanical and/or electrical system shut down was required or to a piece of equipment serving a sensitive area, such as, PM work orders for isolation rooms. Included on the standardized work order was a "note section," which was used to list any special tools required to complete the PM tasks or any required personal protective equipment that may be required by OSHA guidelines.

At the hospital, where data entry was completed by the existing staff (and not by co-op students) this provided an accessible line of communication when additional information was reneeded. However, other daily functions distracted both the facility technicians and the support staff from concentrating on the project at hand. At the school, the data retrieval was completed by co-op students with very little communication with the occupants. In addition, data input was completed by off-site support from our home office. Although this arrangement provided for better data retrieval and input control and allowed facility staff to concentrate on their daily work schedules, the co-op's did not have the experience to ask any pertinent questions to the occupants.

12.8   Equipment Tasking & Scheduling

Equipment tasking was the next critical step of the CMMS implementation process. Equipment tasks were developed from equipment operation and maintenance manuals, the American Hospital Association Maintenance Management for Health Care Facilities manual, as well as the existing CMMS database. The existing database at the hospital proved to contain a large amount of duplicate tasks, as well as tasking that was assigned to equipment no longer at the facility that needed to be removed from the system. Once the equipment task database was refined and all tasking was assigned to the proper category of equipment, this database became a standard and was transferred to the CMMS being installed at the college facility. Although the frequency of maintenance was higher at the hospital than required at the college, the tasks could be transferred and the frequency of main-tenance adjusted. It was found that many of the PM software came without standardized tasks. As a result, we had to establish our own, referencing the resources noted above and making adjustments as nec-essary during scheduled maintenance.

Anticipating the application of "differed" maintenance, the database would later be revised to reflect this change in philosophy. It was de-cided that the facility would consider differed maintenance (i.e., all electric motors below 25HP [18 kw]). It was agreed to perform only the routine maintenance, (i.e., grease bearings) and not record voltage, amperage, or additional maintenance performed by a highly skilled mechanic. The cost of performing the additional maintenance was compared to the cost of replacing the motor at the end of its useful life. This was an effective solution to containing maintenance within the operating budget constraints and maintaining staff at its current level. Spare motors were inventoried and stocked and arrangements were made to access motors from vendors 24 hours a day.

When scheduling the hospital equipment for planned maintenance, the existing CMMS system had a database with existing, predeter-mined frequencies for maintenance. What was lacking in the process and may be lacking in other CMMS programs was "load leveling." Scheduling work orders requires a sensitivity toward load leveling of maintenance tasking to ensure an even flow of work from one week to the next. For example, some software will require selected equip-ment to receive PM quarterly. Without load levels, all the quarterly tasked equipment will come out the first week of the quarter. This work probably cannot be all completed during that specific week do to the numerous work orders, and a back log will be unavoidable.

Software with load leveling will automatically spread these quar-terly work orders over a few weeks at the beginning of the quarter. With the system selected, we knew no load leveling option existed. Instead, the Project Manager had to systematically spread the work-load manually for all equipment entered in the database. This manual process avoided an unnecessary back log of PM work orders while ensuring all equipment would be maintained, tours would be com-pleted in compliance with required agencies, OSHA, NFPA and JCAHO, and team tasking would be performed. Value-added assess-ment of load leveling frequency also included an assurance that pa-tient care equipment would be routinely maintained for operational efficiency. Other pieces of equipment servicing a common space (i.e., garage area) had its tasking frequency reduced to meet available work force. Equipment that could be replaced for the cost of performing routine maintenance was categorized under the "differed" maintenance. This equipment did not affect patient care at the hospital or student services at the college.

In parallel with the load-leveling exercise, it was found that the majority of existing facility maintenance staff worked the first shift, which is also the busiest period for unscheduled work orders and daily emergencies. To focus on completing planned maintenance in a timely manner with a minimum of rescheduled PMs, we reorganized the staffing so that planned maintenance could be completed during "off-hours" second and third shifts. For the staff to be able to do all required maintenance during these other two shifts, personnel were selected who possessed skills in all require disciplines.

In addition to the tasking for mechanical equipment, management incentives such as monthly reports, safety committee reports, and monthly training were scheduled through the CMMS system. This allowed the facility manager at the hospital to track all JCAHO required training and documentation. The hospital staff was now being cross-trained as a department, and the excuse "it's not my job" was becoming a thing of the past. All engineering personnel were receiving training in all mechanical electrical systems at the facility which gave them the ability to answer questions from users with some basic knowledge of the system.

At the college campus, mandatory and mechanical system training was now being documented and the staff was being held responsible to put into action the training they had received. With a smaller on-site mechanical staff, training gave them the basic knowledge of all mechanical and electrical systems while increasing their scope of work and improving job satisfaction.

With every facility person being introduced to the CMMS system, "cross-training" of current staff became an inherent result of CMMS education. Traditionally, staffing had included mechanics that specialize solely in one area of facility maintenance, such as, steam fitting, refrigeration, or electrical. To be cost-effective and to increase job satisfaction as well as security, it had become necessary to possess multiskill capabilities. Neither the hospital nor the college could afford to continue "business as usual." With outside pressures to bring in cross-trained technicians, our strategic plan was to cross-train the existing staff using the CMMS system as the foundation to the curriculum. Tactically, this made a lot of sense at the hospital because the existing personnel were already very familiar with the two and one half million square foot complex. For the staff it became imperative that they all possess a willingness to learn multiskills through training and be capable to perform more than one function. The by-product of this plan would be to increase productivity and a reduction in specialized trade shops. By training the mechanics to do multifunctional tasks, their customers had on-hand mechanics who could respond quicker to a variety of issues.

12.9 CMMS On-Line

As the new CMMS systems came on-line at both the hospital and the college, the change was met with some reservation and doubt. Change is seldom embraced when a team of people have been doing the same thing for many years. For many of the facility mechanics at the hospital this was the third CMMS system. For the college, this was the first ever CMMS on site and the growing pains would be expected with implementing the process.

Start-up of a CMMS did not end with the completion of data retrieval or with equipment work orders scheduled and loaded leveled. Our own CQI process required that the CMMS system be continuously monitored, measured, and updated. Including new equipment into the system and deleting equipment removed was as important as gathering all the existing equipment data.

This new information, along with the existing database, would be measured against the man-hours needed to do planned maintenance with "estimate versus actual" man-hours. This means-of-measurement would be routinely updated and tracked with an eye toward next year's manpower needs analysis for staffing levels.

Another part in the strategic plan was to allocate the necessary time to fine-tune the application of the program (i.e., equipment scheduled only annual maintenance.) For this equipment it would take a full year's cycle before all adjustments to tasking or scheduling could be assessed. In addition, the more software modules implemented in the Year One phase of the process, the more difficult and challenging it will be to achieve complete success. Beginning in Year Two, we would provide annual CMMS training, activated via a certification work order similar to the reminders that specific agency approvals are needed. Training would include all personnel within each shop, as well as the office staff. A policy and procedure was also needed that documented the standardized CMMS system plan. This too would be part of the annual training and include a time line for completion of work orders [refer to CMMS Planned Maintenance (PM) Flow Chart, Fig. 12.2] that established assigned responsibilities, work flow, close out of completed work, and reporting process.

Quality control required a means of measurements for the process to be a continued success along with maintaining status of all work

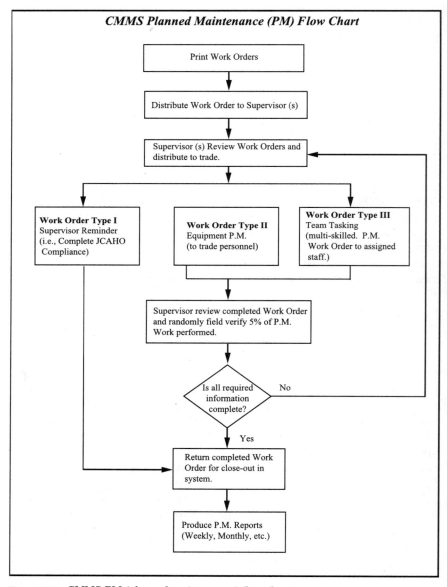

Figure 12.2 CMMS PM (planned maintenance) flow chart.

orders and manpower. Benchmarks and monitoring the success and posting the results would document these accomplishments and encourage a team approach to improvement. The ultimate goal at both the hospital and the college would be a reduction in unscheduled work orders and a reduction in equipment down time.

CQI measurements through posting of performance graphs clearly showed everyone improvements were being made. These measurement tools proved to be the heart and road map to the CMMS success. Measuring the results, the progress, and the deficiencies were all integral to this and any CMMS system implementation. Those systems that failed did so because management and the workforce failed to take the time to complete the Plan-Do-Check-Act rules. After investing thousands of labor hours to collect equipment data, issue work orders, and complete these work orders, it is unfortunate that so many of these planned maintenance systems flounder and eventually fail.

It is imperative that goals be set with the objectives clearly written to identify why these goals are important to the process. Milestones need to be tabulated into the goal's time line, so that progress can be clearly tracked. In addition, goals should focus in on system deficiencies, completion of a specific task, and/or improvements when compared to past operating periods.

Posting the results keeps everyone informed and involved (refer to measurement graph, Fig. 12.3, PM Work Orders—Scheduled vs Closed). This "means-to-measure" in not only good for employee morale, it keeps others within the building complex informed of the facilities' proactive maintenance activities.

12.10   An On-Going Process

Completion of the PM work order and inputting this information back into the program is as important as issuing the work order on time. Timely input of this information will allow the facility management team to monitor workload, staffing levels, and plan/budget overtime when a backlog of work persists.

As a quality measure, the issued workorders were measured against those completed for a given month. A goal of 80 percent completion within 24 hours (refer to CQI graph, Fig. 12.4) and a 90 percent completion for the month was expected. As each month went by and the same measurement was graphed, the backlog was also tracked to ensure all issued work was being completed. By tracking the PM percent complete for each week of the year it allowed the facility management to know all scheduled maintenance had been completed.

Through the use of the graphs that tracked work order completion, manhours, etc., we were able to review and monitor the success of the

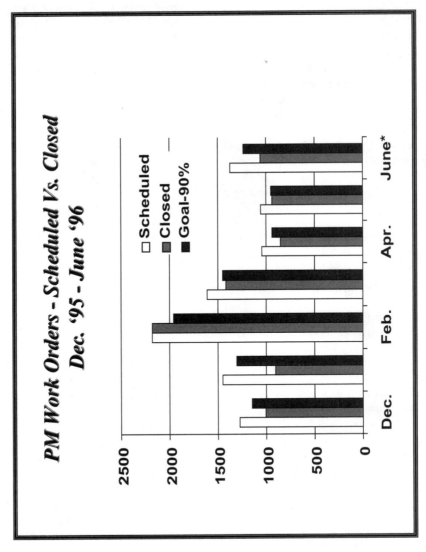

Figure 12.3 PM work orders-scheduled vs. closed.

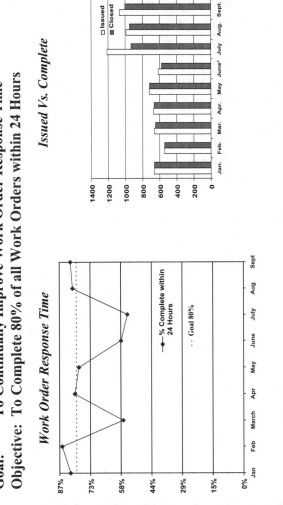

Figure 12.4 Monthly work orders.

program. Analyzing this information allowed for the adjusting of frequency and staffing to meet the needs of the facility. Equipment that required JCAHO inspection needed to be maintained on a higher frequency than a piece of equipment that serves a common area. Only through constant monitoring and ongoing review of the data could the correct amount of manhours and staff be determined.

Revising the database was also an ongoing process that would reinforce the continued success of the CMMS system. The total number of pieces of equipment that were in the system would change as construction and renovation projects were conducted. Changes in the database would include modifying equipment categories, as well as tasks that were assigned to each piece of equipment. Changes to equipment tasks could be as simple as including lock out/tag out information or personnel protective equipment required.

Another on-going process was the use of the database for the development of custom files. The custom files allowed for sorting through the entire database of equipment and searching out one specific detail (i.e., equipment that should be in a JCAHO Utility Management equipment file).

Predictive Maintenance such as vibration analysis, infra-red thermoscanning, eddy current testing, and other nondestructive tests should be incorporated into the CMMS system at the appropriate time. This type of testing performed on an annual or other recommended basis allows the facility manager to detect a problem that may not be uncovered through routine PMs. This knowledge allows the facility management team to plan a shutdown and budget for repairs while avoiding a potential failure of a piece of mechanical equipment.

Once the Planned Maintenance module had been up and running, staff trained on all revisions handled in a timely manner, it will now be the opportune time to introduce additional modules and/or functions of the CMMS system. Module, such as spare parts management, material management, or time cards can move a function once completed on paper form into the computer age with no additional cost of a software program. Most CMMS systems are now "window-based" programs which have replaced the DOS applications. The software programs now being used allow for much expansion of the system as required by the customer and, at the same time, are cost-effective enough that a smaller facility can take advantage of the same technology without a large capital budget.

12.11   Summary

30 years ago, operation and maintenance of mechanical and electrical systems was simpler. Then, when you walked into a large boiler or

chiller plant, the equipment was big with ample redundancy. A Facilities Engineer would have 50 percent (if not 100 percent), boiler and chiller back-up capacity sitting there ready to come on-line. When it was time to pull maintenance on one of these big units, there was another unit ready to go.

The operating strategy back then was also quite simple: on-off control. Energy and labor were relatively inexpensive, water was clean and plentiful, and environmental guidelines were not very stringent. The Facility Engineer title was probably Chief Engineer or Plant Engineer. Certainly, the individual was not perceived as "Manager." If anything, the entire operation and maintenance staff was considered a necessary evil, an overhead cost of doing business.

Maintenance procedures, like system operations, were also relatively simple then. Preventive maintenance consisted on a file cabinet full of neatly typed, alphabetized equipment record cards. Each card noted what maintenance needed to be done and when this work should be done. No practice measures, predictive maintenance or run-time action agenda. Simple and to the point, like the systems themselves.

Things have changed with the introduction and advancement of computers. In addition, the cost of energy and the cost of doing business have had a dramatic impact on facility management. Electronics have slowly been replacing pneumatic automatic controls, and with these new direct digital controls, computer software offer the facility manager numerous opportunities to automate planned maintenance. Enter the computer! The panacea of building management and the "cure-all" to operation and maintenance, this exciting tool dramatically changed the way everyone did business. For the Facility Manager however, it has become a double-edged sword. It has placed maintenance on the "cutting edge" of its industry and the "bleeding edge" of performance. Because the computer could inherently offer so much from within such a small box, Facility Managers were expected to use its software options, as soon as possible. Maintenance was now going hi-tech, and the people doing the work were expected to act accordingly.

It has been documented often that planned maintenance traditionally takes a back seat to "fire fighting" unscheduled work orders. Purchasing a CMMS system will not put out these fires unless there is a mission, a plan, and a commitment to methodically change how maintenance is done. We have seen how operation and maintenance have changed significantly over the past 30 years and we know how that has adversely affected the skill levels of these crews. Planned maintenance can't be achieved unless facility management staff "makes the time" and commits themselves to a quality process that will keep them current with computer-age technology. In addition, they need to be patient and plan their venture into CMMS, one step at a time.

When discussing operation and maintenance performance with Vice Presidents of Support Services who have succeeded today, they will point out three priorities; first, train the staff to be the best they can be; second, reverse the trend of unscheduled work over planned maintenance; and third, reduce operating cost. Nowhere do they say CMMS is essential to successful facility management, but it can be used as a tool to improved performance.

Getting started with CMMS parallels these same three priorities. Training begins with a "reintroduction to one's job description." Responsibilities have changed over the last 30 years, and personnel must keep abreast of computer technology. Those who don't embrace this mandate will find someone else doing their job. With more qualified people capable of doing a wider range of services proficiently, a proactive approach to maintenance can be the next step. Enter the CMMS system, and a focus on planned maintenance only! A commitment to apply the preventive maintenance module of a CMMS software and follow-through on completing the work orders will be a positive first step to accomplishing the next priority, reversing the trend of unscheduled works.

It has been found that implementing a strong, planned maintenance program will initially increase the number of unscheduled work orders. The reason being that the workforce, proactively doing maintenance, will find other tasks that needed to be done. In time, the unscheduled work decreased because systems were being maintained and equipment wasn't breaking down. Reliable system operation equates to reduced operating cost, energy costs, and no product disruptions, the third priority.

With that much work ahead, how can Facility Managers consider applying all those other CMMS computer software modules too? Keeping up with computer-age technology and getting started with CMMS requires careful planning and a commitment to take one step at a time. After all, it took us 30 years to get to this version of facility management.

Following is a listing of available computer maintenance management systems, reprinted with the permission of the Association of Facilities Engineering (AFE), formerly the American Institute of Plant Engineering (AIPE). This is the latest now available, and the reader should contact AFE to verify the latest version.

COMPUTERIZED MAINTENANCE MANAGEMENT SYSTEMS

Company	AD/C Solutions 1445 Grant Rd. Los Altos, CA 94024 (415) 969-3979 Fax: (415) 967-0126	AEC Data Systems Inc. 75501H-10 West, Ste. 400 San Antonio, TX 78229 (800) 659-9001 Fax: (210) 308-9015	Angus Consulting Management Ltd. 1129 Leslie St. Don Mills, Ontario, Canada, M3C 2K5 (416) 443-8304 Fax: (416) 443-8323
System name	Work Request & Asset Management	Facility Management System (FM)	Angus Maintenance Management System
COMPUTER SYSTEM IBM=IBM PCs and Compatibles OTH=Other PCs and Micros MIN=Minis MAI=Mainframes	IBM OTH	IBM OTH MIN	IBM OTH MAI
Operating System(s)	Windows, Apple Macintosh	Windows, UNIX	DOS, Windows, OS/2, UNIX, VMS
FEATURES			
On-line work request/approval	■	■	■
Work order origination	■	■	■
Employee competence, skills, training, etc.	■	■	■
Customize outputs	■	■	■
Customize field names, labels, screen, etc.	■	■	■
Ad hoc reporting/on-line inquiries	■	■	■
Display report data in graphical format		■	■
Priority origin., management, states tracking	■	■	■
On-line stds./procedures/engineering data, oper. specs	■	■	■
Auto. P.O. request, generation, P. O. tracking	■	■	■
EDI interface with suppliers			
Inventory management, analysis, control	■	■	■
Long/short interval scheduling of labor	■	■	■
Bar code reading/printing	■	■	
On-line OSHA, MSDS, EPA data			
Failure, root-cause mgmt, analysis		■	■
Remote on-line support via modem		■	■
Established links	Predictive Maintenance	CADD Systems, Construction Estimating Systems	
Communication systems, file service		TCP/IP	
Data language supported		Informix, Oracle, Sybase, DB212, Watcom	C, C++, FoxPro
Export data format	Most	ASCII, dBase 2, dBase 3, Excel, SYLK	ASCII
★ PRICE: Single-user package	$3,795	$2,500	$6,000-$12,000
Network (2-25)	$3,995-$8,795	$20,500-$82,500	$12,000-$13,000
Network (25-50)	$8,795-$13,795	$82,500-$145,000	$13,000-$16,000
Large Package (Unlimited)		Negotiable	
Cost for maint. contract	$250/yr.	20% of software license plus services	$700
Updates included	Yes	Yes	Yes
CMMS INSTALLATION POPULATION No. of systems sold per year	50	15	
CMMS installed/sold to: Government		60	100
Manufacturing			100
Process Manufacturing			10
Facilities		15	300
Other		12, Universities & School Districts	150
No. of application licenses			
CMMS PROVIDER STAFF No. of CMMS license/customer rep.		6-10	
Help Desk Rep.			2
Programmer/systems-data language analyst		14	8
Installation Assistance		Yes	Yes
Operation Assistance		Yes	Yes
Special benefits and features		Work order management, preventive maintenance, estimating, inventory control, scheduling, labor and material use tracking.	

* Prices as of 1995.

COMPUTERIZED MAINTENANCE MANAGEMENT SYSTEMS

Applied Computer Technologies Corp. 113 E. Main St., P.O. Box 1032, Coats, NC 27521, (800) 488-4128, Fax: (800) 455-3326	Argos Software 3949 Sparks Dr., SE, Grand Rapids, MI 49546, (616) 949-6400, Fax: (616) 949-5577	Arsenault Associates 610 Jackson Rd., Atco, NJ 08004, (800) 525-5256, FAX: (609) 768-8289	Astar Inc. 1275 Minhinette Dr., Ste. 200, Roswell, GA 30075, (404) 641-1486, Fax: (404) 641-3044	Benchmate Systems Inc. P.O. Box 5663, Bellevue, WA 98006, (206) 391-2757, Fax: (206) 391-3184
ACT 1000 Maintenance Management System	Abecas™	Maintenance Dossier	Mapcon	Benchmate for Windows
IBM	IBM	DOS	IBM	IBM
DOS, Windows, WIN '95	DOS, Windows, OS/2		DOS, Windows	DOS, Windows
■	■		■	■
■	■		■	■
■	■		■	■
■	■	■	■	■
■	■		■	■
■	■		■	■
■			■	
■	■		■	■
■	■	■	■	■
■			■	
■	■	■	■	■
■	■	■	■	■
■		■	■	
■	■		■	■
■	■		■	■
	■	■	■	■
MRO, Predictive Maintenance, CADD, Construction Estimating, OEM, Project Management, Facilities Planning & Facility Accounting Systems	Construction Estimating & Facility Accounting Systems		CADD Systems	CADD Systems
	Networks, eg. Novell, NT	ASCII	Advanced Revelation & Open Insight	CTREE
C, XBase, Access, VB, ODBC ASCII	ASCII		ASCII, dBase, Lotus	dbf
$495-Custom quotes	$2,000-$8,000 $3,000-10,000 $5,000-$14,000 $10,000-$20,000	$595-3,495 $1,395-4,895	$4,790-13,750 $7,750-17,000 $17,000-29,750 $17,750-29,750	$1,995 $2,995 $3,995
Custom quote		$472/year		
$495 +/training/programming Yes	2% per month Yes	Yes	15% of purchase Yes	$350+ Yes
95-130	30-40	300	50-70	50
		25 1,400 Vehicle Fleets	70% 20% 10%	
750, Universities & School Districts	Transportaion 700		700+	400+
92	13	200 3 Yes	10-20 9	
Yes Yes	Yes Yes	Yes	Yes Yes	Yes Yes
Designed exclusively for educational facilities management with complimenting service and training options and owners meeting.	Integrates with activity-based costing system plus full accounting, if desired. Modular system with options configurable to user.	Vehicle and equipment maintenance,repair order histories and full report features.	Flexibility, user-friendly, proven, variable size records, designed by maintenance professionals.	

COMPUTERIZED MAINTENANCE MANAGEMENT SYSTEMS

Company	Bender Engineering Co. 3585 Farguhar Ave. Los Alamitos, CA 90720 (800) 255-5675 Fax: (310) 596-7143	Black & McDonald Ltd.. 101 Parliament St. Toronto, Ontario, Canada M5A 2Y7 (416) 366-2541 Fax: (416) 361-5918	Bonner & Moore 2727 Allen Parkway Houston, TX 77019 (713) 522-6800 Fax: (713) 522-1134
System name	MainStar	CMM	Compass 7
COMPUTER SYSTEM IBM=IBM PCs and Compatibles OTH=Other PCs and Micros MIN=Minis MAI=Mainframes	IBM MIN MAI	IBM	IBM MIN OTH
Operating System(s)	DOS, Windows, OS/2, Macintosh	DOS	Windows, Windows NT Advanced Server
FEATURES On-line work request/approval	■	■	■
Work order origination	■	■	■
Employee competence, skills, training, etc.	■	■	■
Customize outputs	■	■	■
Customize field names, labels, screen, etc.	■		■
Ad hoc reporting/on-line inquiries	■	■	■
Display report data in graphical format	■	■	■
Priority origin., management, states tracking	■	■	■
On-line stds./procedures/engineering data, oper. specs	■	■	■
Auto. P.O. request, generation, P.O. tracking	■	■	■
EDI interface with suppliers	■	■	■
Inventory management, analysis, control	■	■	■
Long/short interval scheduling of labor	■	■	■
Bar code reading/printing	■		■
On-line OSHA, MSDS, EPA data	■		■
Failure, root-cause mgmt, analysis	■	■	■
Remote on-line support via modem	■		■
Established links	MRO, Pred. Maint., CADD, Constr. Estim., OEM, Proj. Mgmt., Fac. Plan., Fac. Acct.	Predictive Maintenance, Construction Estimating Systems, Project Management Systems, Facilities Planning Systems & Facility Accounting Systems	MRO, Pred. Maint., CADD, Constr. Estim., OEM, Proj. Mgmt., Fac. Plan, Fac. Acct.
Communication systems, file service Data language supported	Power Builder, Watcom, Oracle, etc.	XBase	SyBase, SQL Server, Oracle 7
Export data format	ASCII Dense	Word, Excel, TDF	ASCII, Others
★ PRICE: Single-user package Network (2-25) Network (25-50) Large Package (Unlimited)	$2,000-$2,500 $2,500-$10,000 $5,000-$15,000 $10,000-$18,000	$6,000-$8,000 $8,000-$14,000 $8,000-$14,000 $8,000-$14,000	$26,000-$91,000 $91,000-$156,000
Cost for maint. contract Updates included	15% of software cost Yes	$720-$1,500 Yes	15% of license/year Yes
CMMS INSTALLATION POPULATION No. of systems sold per year	300	20	
CMMS installed/sold to: Government Manufacturing Process Manufacturing Facilities Other	100 300 100 700 100	5 2 2 8 6/Airlines, Schools	9 42 80 42
No. of application licenses		70	
CMMS PROVIDER STAFF No. of CMMS license/customer rep.	200	10	
Help Desk Rep.	6	1	
Programmer/systems-data language analyst		3	
Installation Assistance	Yes	Yes	Yes
Operation Assistance	Yes	Yes	Yes
Special benefits and features	Ease of use is main feaure. Can be customized by the user—no programming knowledge is required.	Easy lookups, fast setup, best ad-hoc report writer, great filters, short learning curve and new billing module for cost recovery.	Developed using modern object-oriented programming techniques. Written in Visual C++ and operates within the Windows NT server environment.

* Prices as of 1995.

COMPUTERIZED MAINTENANCE MANAGEMENT SYSTEMS

CK Systems Inc. 772 Airport Blvd. Ann Arbor, MI 48108 (313) 665-1780 Fax: (313) 665-6074	CanaTech Consulting Int'l. Ltd. 1157 Landsdowne Dr. Coquitlam, BC, Canada V3B 4VI (604) 944-7285	Candlewood Computer Services 4 Oakwood Drive New Fairfield, CT 06812 (203) 746-1181	Clayton Systems Associates Inc. 8420 Delmar Blvd., Ste. 207 St. Louis, MO 63124 (314) 993-8685 Fax: (314) 993-0588	Creative Management Systems 400 Riverside Ave. Jacksonville, FL 32202 (800) 874-5554 Fax: (904) 366-2690
Maintimizer	ProMaintainer for Windows (Version 1.0)	PMaint	Quick Start V, Advanced Maintenance Management	CAMS-S2L Maintenance Management Systems
IBM OTH	IBM OTH	IBM	IBM MIN	IBM
DOS, Windows, OS/2	DOS, Windows	DOS	Windows	DOS
■	■		■	■
■	■		■	■
■	■	■	■	■
■			■	■
			■	
■	■		■	■
■	■		■	■
■	■		■	■
			■	
■	■		■	■
■		■	■	■
■	■		■	■
■	■		■	
	■			
CADD Systems, MRO Supplier Order Writing Systems			MRO, Pred. Maint., CADD, Constr. Estim., OEM, Proj. Mgmt., Fac. Plan, Fac. Acct.	
BTrieve ASCII	Visual Objects (Computer Assoc.) XBase, dbf	dBase ASCII	All Windows Oracle, SQL Server, Access Multiple	SQL ASCII
$3,995 $7,995 $7,995 $7,995 $750 Yes	$1,800	$495	$8,000-$10,000 $10,000-$35,000 $35,000-$80,000 $80,000 + Varies Yes	$4,900-$10,000 $10,700,-$25,000 $25,000-$50,000 $980-$3,500 Yes
	Yes			
200 600 200 200		20 10 80 30	20-30 40% 30% 30% 50+	130 30 50 50
				25 50
Yes Yes Yes		Yes Yes	Yes Yes	Yes Yes
Ease of use and customer support.	Can tie in spare parts to assets; auto printing of PM work orders and purchase orders.	Price and ease of use.	Intelligent database learns as it is used. Improves estimating, planning, materials management.	Offers a variety of strategic planning and evaluation services. The first step, the on-site survey, maps out a successful implementation plan.

COMPUTERIZED MAINTENANCE MANAGEMENT SYSTEMS

Company	DFM Systems Inc. 1119 High St. Des Moines, IA 50309 (515) 244-6114 FAX: (515) 244-4918	DP Solutions 4249 Piedmont Pkwy., Ste. 105 Greensboro, NC 27410 (910) 854-7700 FAX: (910) 854-7715	Datastream Systems Inc. 1200 Woodruff Rd., Ste. C-40 Greenville, SC 29607 (800) 955-6775 FAX: (803) 627-7227
System name	Mapcon 95	PMC for Windows	MP2 for Windows
COMPUTER SYSTEM IBM=IBM PCs and Compatibles OTH=Other PCs and Micros MIN=Minis MAI=Mainframes	IBM	IBM	IBM
Operating System(s)	DOS, Windows	Windows, OS/2	Windows
FEATURES			
On-line work request/approval	■	■	■
Work order origination	■	■	■
Employee competence, skills, training, etc.	■	■	■
Customize outputs	■	■	■
Customize field names, labels, screen, etc.	■	■	■
Ad hoc reporting/on-line inquiries	■	■	■
Display report data in graphical format	■	■	■
Priority origin., management, states tracking	■	■	■
On-line stds./procedures/engineering data, oper. specs	■	■	■
Auto. P.O. request, generation, P. O. tracking	■	■	■
EDI interface with suppliers	■		■
Inventory management, analysis, control	■	■	■
Long/short interval scheduling of labor	■	■	■
Bar code reading/printing	■		■
On-line OSHA, MSDS, EPA data		■	■
Failure, root-cause mgmt, analysis		■	■
Remote on-line support via modem	■	■	■
Established links	MRO, CADD Systems, Construction Estimating Systems & Project Management Systems	Predictive Maintenance, CADD Systems, & Project Management Systems	MRO Supplier Order Writing Systems & Predictive Maintenance Systems
Communication systems, file service Data language supported Export data format	LAN Advanced Revelation/Open Insight ASCII 0IFFTIFF	ODBC Microsoft Access dbf, WKS, TXT, ASC, Paradox, BTrieve, ODBC	Paradox for Windows ASCII, Quattro, Lotus, Excel, Paradox
★ PRICE: Single-user package Network (2-25) Network (25-50) Large Package (Unlimited)	$4,750 $7,450 $15,750 $19,750	$4,995 $7,495 $7,495 $7,495	$2,995 $4,995-$7,995 $10,995 $10,995+
Cost for maint. contract Updates included	15% Yes	Yes	Yes
CMMS INSTALLATION POPULATION No. of systems sold per year CMMS installed/sold to: Government Manufacturing Process Manufacturing Facilities Other No. of application licenses	100 10% 50% 20% 20%	300 5 300 50 75 3,800+	2,000+ 240 660 925 295
CMMS PROVIDER STAFF No. of CMMS license/customer rep. Help Desk Rep. Programmer/systems-data language analyst Installation Assistance Operation Assistance	20 10 Yes Yes	380 18 Yes No	500 Yes Yes
Special benefits and features	Mapcon 95 is available in DOS/Windows or both on same server/network.	User-changeability of fields, screens, reports and graphs. Implementation support using proprietary MAP/CMMS project management services Multi-language versions.	Includes interface to *Grainger Electronic Catalog.* Voted 1994 Software Product of the Year by *Plant Engineering* magazine.

* Prices as of 1995.

COMPUTERIZED MAINTENANCE MANAGEMENT SYSTEMS

Decision Dynamics Inc. 696 McVey Ave. Lake Oswego, OR 97034 (800) 872-0061 FAX: (503) 636-1562	Dima Litvak Corporation 400 Hillside Ave. Needham, MA 02194-1226 (617) 444-1031 FAX: (617) 444-2079	EFAX Corporation 444 North York Rd. Elmhurst, IL 60126 (708) 279-9292 FAX: (708) 530-0521	EQ2 Inc.. 209 Battery St. Burlington, VT 05401 (802) 865-0920 FAX: (802) 865-0928	Eagle Technology Inc. 10500 N. Port Washington Rd. Mequon, WI 53092 (800) 388-3268 FAX: (414) 241-5248
DynaStar 2000	Grand-PM™	PROBE III	HEMS/AMEP	Expert Maintenance Management for Windows
IBM	IBM	IBM OTH MIN MAI	IBM OTH	IBM OTH MIN MAI
Windows	DOS, Windows	DOS, Windows, OS/2, UNIX, Macintosh, etc.	DOS, Windows, OS/2	Windows, OS/2, Windows NT
■	■	■	■	■
■	■	■	■	■
■	■	■	■	■
■	■	■	■	■
■	■	■	■	■
■	■	■	■	■
■	■	■	■	■
		■		■
■	■	■	■	■
■	■	■	■	■
■	■	■	■	■
	■	■		■
■	■	■	■	■
CADD Systems		MRO, Predictive Maintenance, CADD, Project Management, Facilities Planning & Facility Accounting Systems	MRO, CADD, Construction Estimating, OEM Technical Support, Project Management, Facilities Planning & Facility Accounting Systems	MRO, Predictive Maintenance, CADD, Construction Estimating, OEM Technical Support, Project Management, Facilities Planning & Facility Accounting Systems
Fax Watcom, ODBC Database ASCII, Spreadsheet		Native to operating system ASCII, Excel, Access, etc.	FoxPro ASCII, Lotus, Quattro, Excel, Text etc.	Microsoft ODBC Databases, C++ Microsoft ODBC ASCII, ODBC
$3,295 $4,295+ $4,995+	$2,700-S3,400 $5,400	$5,000-$10,000 $10,000-$20,000 $20,000-$40,000 $50,000-S100,000	$3,500-S6,900 $7,900-S17,000 $23,695-S38,695 S58,685	$2,950 $4,000-S15,000 $4,000-S15,000 $4,000-S15,000
15% Yes	$350 Yes	15-18% Yes	$1,200 Yes	S595 per year Yes
100 5 50 15 25 5/Vessels & Vehicles 700		20 5 50 30 20	40 20 2 505	200 30 500 120 350 40 1,040
200 3 10 Yes Yes	Yes Yes	5-10 Yes Yes	105 105 4 Yes Yes	250 250 Yes Yes
Radio frequency bar code, image storage and retrieval and easy-to-use graphical interface.	Easy to use.	Can locate any work order, part or equipment without knowing corresponding identification number. Has fast response, flexibility and is modular.	Depreciation—capital equipment acquisition modules and asset life cycle management are key features. Service and warranty provided.	Client/server version. Spanish, German and other languages offered. Fully integrated and user modifiable.

COMPUTERIZED MAINTENANCE MANAGEMENT SYSTEMS

Company	Ecta Corporation 321 Norristown Rd., Ste. 110 Ambler, PA 19002 (215) 540-0250 FAX: (215) 540-0847	Epix Inc. 1851 Sherbooka St. East, Ste. 1100 Montreal, Quebec, CN H2K 4LS (514) 522-3749 FAX: (514) 522-1656	Equipac Systems 1089 Fairington Dr. Sidney, OH 45365 (513) 498-7080 FAX: (523) 498-2180
System name	M-PET—Maintenance Productivity Enhancement Tool	WISE	Equipac for Windows
COMPUTER SYSTEM			
IBM=IBM PCs and Compatibles	IBM	IBM	IBM
OTH=Other PCs and Micros			OTH
MIN=Minis	MIN	MIN – AS 400 IBM	
MAI=Mainframes			
Operating System(s)	Windows, OS/2, UNIX	DOS, OS/400 for IBM, Windows ('96)	DOS, Windows, NT, WIN '95
FEATURES			
On-line work request/approval	■	■	■
Work order origination	■	■	■
Employee competence, skills, training, etc.		■	■
Customize outputs	■	■	■
Customize field names, labels, screen, etc.		■	■
Ad hoc reporting/on-line inquiries	■	■	■
Display report data in graphical format	■		
Priority origin., management, states tracking	■	■	■
On-line stds./procedures/engineering data, oper. specs	■	■	
Auto. P.O. request, generation, P. O. tracking	■	■	■
EDI interface with suppliers			
Inventory management, analysis, control	■	■	■
Long/short interval scheduling of labor	■	■	■
Bar code reading/printing	■	■	■
On-line OSHA, MSDS, EPA data			
Failure, root-cause mgmt, analysis	■	■	■
Remote on-line support via modem	■	■	■
Established links	Predictive Maintenance Systems, Project Management Systems & Facility Accounting Systems	CADD Systems, Facilities Planning Systems, Facility Accounting Systems	CADD Systems
Communication systems, file service			Novell, WIN NT, WIN '95
Data language supported	Watcom, Oracle, Any ODBC		Access, Oracle SQL, Microsoft SQL
Export data format	ASCII & others	ASCII	ASCII
* PRICE: Single-user package	$10,000-$15,000	$5,000-$10,000	$1,995-$2,500
Network (2-25)	$15,000-$30,000		$1,995-$3,000
Network (25-50)	$30,000-$50,000		$1,995-$3,000
Large Package (Unlimited)	$50,000-$100,000	$12,000+	$1,995-$3,000
Cost for maint. contract	Percentage of purchase	12% of software value	1st year free/$500 second year
Updates included	Yes		Yes
CMMS INSTALLATION POPULATION			
No. of systems sold per year	50		200+
CMMS installed/sold to: Government	20		5
Manufacturing	100		150
Process Manufacturing	50		
Facilities	100		10
Other			
No. of application licenses			200+
CMMS PROVIDER STAFF			
No. of CMMS license/customer rep.	50		50
Help Desk Rep.	4		4
Programmer/systems-data language analyst	4		4
Installation Assistance	Yes	Yes	Yes
Operation Assistance	Yes	Yes	Yes
Special benefits and features	M-PET Scheduler allows one-screen execution of all critical work order functions from planing to posting.	Maintenance specialist, full turnkey service vector and raster files, fax, B coding, work and time standards, WHMIs and ADDHOC reporting.	Windows-based, full imaging, barcode, network ready, SQL capable, CAD viewer and complete scheduling capability.

* Prices as of 1995.

COMPUTERIZED MAINTENANCE MANAGEMENT SYSTEMS

Facility Management Technology Inc. 13005 16th Ave. North, Ste. 500 Plymouth, MN 55441 (612) 557-6749 FAX: (612) 557-6929	Fleming Systems Corp. 6291-7 Dorman Rd. Mississauga, Ontario, Canada L4V 1H2 (800) 361-9630 FAX: (905) 673-0058	Fluor Daniel 100 Fluor Daniel Dr. Greenville, SC 29607-2762 (803) 281-5840 FAX: (803) 676-7677	Four Rivers Software Systems Inc. 1001 Ardmore Blvd., Ste. 202 Pittsburgh, PA 15221 (412) 243-5600 FAX: (412) 243-5799	GBS Associates Inc. 240 Kent Rd. Warminster, PA 18974 (215) 674-3949 FAX: (215) 674-2808
Maintenance/View-Win	4Site for Windows	Orion	TMS—Total Maintenance System	Profit Oriented Maintenance Manager
IBM OTH	IBM MIN MAI	IBM OTH MIN MAI	IBM OTH MIN MAI	IBM
Windows	Windows Clients, Most Servers	Windows, OS/2, UNIX, Open VMS	DOS, Windows, OS/2, Unix, VMS	DOS, Windows
■	■	■	■	■
■	■	■	■	■
	■	■	■	■
	■	■	■	■
■	■	■	■	■
■	■	■	■	■
■	■	■	■	■
■	■	■	■	■
	■	■	■	■
	■	■	■	■
■	■	■	■	
■	■	■	■	■
■	■	■	■	■
■	■	■	■	■
	■	■	■	
■	■	■	■	■
MRO Supplier, Order Writing Systems, Predictive Maintenance Systems & Facility Accounting Systems	MRO, Predicitve Maintenance, CADD Systems, Project Management Systems, Facilities Planning Systems & Facility Accounting Systems	MRO, Predictive Maintenance, CADD, Construction Estimating, OEM, Project Management, Facilities Planning & Facility Accounting Systems	Pred. Maint., CADD Systems, Constr. Estim., Proj. Mgmt., Facilities Planning SYstems	
LAN-Novell, Windows NT	Most	Novell, Pathworks, TCP/IP	ODBC	Various
Paradox, C++	SQL Server, Sybase, Oracle	Sybase, Oracle		FoxPro, CC, DDER
ASCII, DDE, NETDDE	ASCII, Any SQL format	ASCII	ASCII, Excel	ASCII
$5,500-$5,500 $5,500-$20,000 Site License Site License	$10,000-$19,000 $12,000-$72,000 $46,000-$120,000 $140,000-$210,000		$10,000-$15,000	$2,300+ Up to $9,500
$250/hr.-$350/hr. on-line Yes	1st year free/15% Yes	Yes	15% Yes	$300-$450 Yes
10 2 2 6/ Utilities 50	20 2 20 50 4	30-40 10 100 100 20 20	 1,600	
75 3 Yes No	30 Yes Yes	 Yes Yes	150 Yes Yes	
One application of an integrated application guide, including operator interface and data management/reporting.	Windows client server for medium to large organizations commitment, activity-based accounting. Repairable spares tracking.	Multi-lingual, document linking, graphics, open integration, highly intuitive, user defined screens, word processing and client server.	Ease of use, ability to interface with external programs and devices.	A maintenance control system, not just software.

COMPUTERIZED MAINTENANCE MANAGEMENT SYSTEMS

Company	GP Solutions Inc. 10400 Little Patuxent Pkwy., Ste. 480 Columbia, MD 21044 (410) 730-9661 FAX: (410) 730-9679	General Energy Technologies 100 Via Quito Newport Beach, CA 92663 (714) 673-4222 FAX: (714) 673-4122	Hilco Technologies Inc. 3300 Rider Trail South, Ste. 300 Earth City, MO 63045 (314) 298-9100 FAX: (314) 298-1729
System name	GP Mate	Predictive Mainatence Control System	Monitrol℠
COMPUTER SYSTEM IBM=IBM PCs and Compatibles OTH=Other PCs and Micros MIN=Minis MAI=Mainframes	IBM	IBM MIN MAI OTH	MIN – HP, IBM, DEC
Operating System(s)	DOS, Windows	DOS	UNIX
FEATURES On-line work request/approval	■	■	
Work order origination	■	■	
Employee competence, skills, training, etc.	■	■	
Customize outputs	■	■	
Customize field names, labels, screen, etc.	■	■	
Ad hoc reporting/on-line inquiries	■	■	■
Display report data in graphical format	■	■	■
Priority origin., management, states tracking	■	■	
On-line stds./procedures/engineering data, oper. specs	■	■	
Auto. P.O. request, generation, P. O. tracking	■		
EDI interface with suppliers	■		
Inventory management, analysis, control	■	■	
Long/short interval scheduling of labor	■		
Bar code reading/printing	■		■
On-line OSHA, MSDS, EPA data	■		
Failure, root-cause mgmt, analysis	■	■	
Remote on-line support via modem	■		
Established links	MRO Supplier Order Writing Systems, Predictive Maintenance, CADD Systems, Project Management Systems, Facility Accounting Systems	Facility Accounting Systems	
Communication systems, file service			
Data language supported		BTrieve	Oracle, Ingres, Progress, Sybase
Export data format	ASCII	ASCII thru Clipboard	
★ PRICE: Single-user package Network (2-25) Network (25-50) Large Package (Unlimited)	$3,995+	$770-$2,770 $3,770 $5,770 $7,770	$4,990-$19,900
Cost for maint. contract	$1,500	10% of original price	15% of software price
Updates included	Yes	Yes	Yes
CMMS INSTALLATION POPULATION No. of systems sold per year	75	65	100+
CMMS installed/sold to: Government	30	23	
Manufacturing	75	87	20
Process Manufacturing	100	16	80
Facilities	30	187	
Other	50/Power plants	96/Misc.	
No. of application licenses			200+
CMMS PROVIDER STAFF No. of CMMS license/customer rep.		22	
Help Desk Rep.			
Programmer/systems-data language analyst		3	
Installation Assistance	Yes	Yes	
Operation Assistance	Yes	Yes	
Special benefits and features	Repairable item and serial number tracking, process safety management, project/account budgeting.	Easy to set up and use. Powerful, modular design. Does what you want, the way you want to do it.	Data collection, presentation, storage and distribution from plant floor systems to planning systems.

* Prices as of 1995.

COMPUTERIZED MAINTENANCE MANAGEMENT SYSTEMS

	Innovative Tech Systems	Inter-Data Systems Inc.	JB Systems	Josalli Inc.	Kakari Systems Ltd.
	1250 Easton Road, Ste. 250	7322 S.W. Frwy., Ste. 1500	21800 Oxnard St. #1000	P.O. Box 460	201 10720-113 St.
	Horsham, PA 19044	Houston, TX 77074	Woodland Hills, CA 91367	Enka, NC 28728	Edmonton, AB, Canada T5H3H8
	(215) 441-5600	(800) 275-6833	(800) 275-5277	(704) 252-9146	(800) 661-9460
	FAX: (215) 441-5989	FAX: (713) 777-3291	FAX: (818) 716-4168	FAX: (704) 252-9146	FAX: (403) 423-1911
	SPAN.FM for Windows	Promax Manufacturing/Dist.	Mainsaver	Preventive Maintenance System	Igor
	IBM OTH	MIN	IBM MIN	IBM	IBM
	DOS, Windows, OS/2, Win NT, Win 95	UNIX, New Era	Windows, OS/400	Windows, DOS	Windows, DOS, OS/2
	■	■	■	■	■
	■	■	■	■	■
	■				
	■		■		■
	■	■	■	■	■
	■	■	■	■	■
		■	■		■
	■	■	■	■	■
	■	■	■		■
	■	■	■		■
	■	■	■		
	■	■	■	■	■
	■		■	■	■
	■	■	■		
	■		■		■
	■	■	■		■
	CADD Systems, Construction Estimating Systems, Project Management Systems, Facilities Planning Systems & Facility Accounting Systems	MRO Supplier Order Writing Systems, Predictive Maintenance, CADD Systems, Project Management Systems, Facility Acct. Systems	CADD Systems, Predictive Maintenance, Project Management Systems, Facilities Planning Systems, Facility Acct. Systems		Predictive Maintenance, CADD Systems, Construction Estimating Systems, Project Management Systems
		TCP/IP	TCP/IP		Novell, WFW, Lantasic, Banyan Vines
	ODBC	Informix 4GL	Oracle, Watcom, Sybase, SQL, Server, Informix	dBase, Paradox	FoxPro
	ASCII, dBase, Excel	ASCII	ASCII	ASCII	ASCII , DBF
	$9,500	$5,000 $7,000-$14,000 $14,000-$35,000	$5,000-$10,000 $8,500-$33,500 $29,500-$54,500 $34,500-$62,500	$595-$900	$5,000-$10,000 $8,000-$25,000 $30,000-$60,000
	15% of contract	15% annual	15% of software price		10% of software
	Yes	Yes	Yes	Yes	Yes
	150 10 40	50 130+	250 100 800 400 400	250 150 75 25	20 5 25 30 5
	10 Yes Yes	5 15-25 Yes Yes	35 19 Yes Yes	Yes Yes	6 4 Yes Yes
	Powerful CMMS application that is easy to implement and easy to use.	Enterprise-wide solution addresses all manufacturing distribution with financial modules, including EDI, bar code and multi-language.	Mainsaver has direct API connections to databases. Dispatch module is a paperless work management system.	Flexible and easy to learn and use system for scheduling of maintenance and inventory activities.	Dynamic, task driven system that offers unparalleled flexibility. Designed for maintenance people, regardless of computer literacy.

COMPUTERIZED MAINTENANCE MANAGEMENT SYSTEMS

Company	M2 Limited 9210 Wightman Rd., Ste. 110 Gaithersburg, MD 20879 (301) 977-4281 FAX: (301) 926-5046	Maintenance Automation Corp. 3107 W. Hallandale Beach Blvd. Hallandale, FL 33009 (305) 962-8800 FAX: (305) 962-9046	Marcam Corporation 95 Wells Avenue Newton, MA 02159 (800) 965-0220 FAX: (617) 965-7273
System name	Mainplan	Chief Advantage™ for Windows	Marcam Maintenance
COMPUTER SYSTEM IBM=IBM PCs and Compatibles OTH=Other PCs and Micros MIN=Minis MAI=Mainframes	OTH	IBM	IBM MIN
Operating System(s)	Apple Macintosh	Windows	Windows, UNIX
FEATURES On-line work request/approval	■	■	■
Work order origination	■	■	■
Employee competence, skills, training, etc.		■	■
Customize outputs	■	■	■
Customize field names, labels, screen, etc.		■	■
Ad hoc reporting/on-line inquiries	■	■	■
Display report data in graphical format	■	■	■
Priority origin., management, states tracking	■	■	■
On-line stds./procedures/engineering data, oper. specs	■	■	■
Auto. P.O. request, generation, P.O. tracking	■	■	■
EDI interface with suppliers		■	■
Inventory management, analysis, control	■	■	■
Long/short interval scheduling of labor	■	■	■
Bar code reading/printing		■	■
On-line OSHA, MSDS, EPA data			■
Failure, root-cause mgmt, analysis		■	■
Remote on-line support via modem	■	■	■
Established links		Predictive Maintenance, Project Management Systems, Facilities, Planning Systems	MRO, Project Management Systems, CADD Systems
Communication systems, file service Data language supported Export data format	4D ASCII	ASCII, Access, Paradox, FoxPro	Oracle, DBL/400 ASCII
* PRICE: Single-user package Network (2-25) Network (25-50) Large Package (Unlimited)	$4,995- $6,995-$36,780	$4,995 $6,995-$11,795 $11,795-$17,795	$100,000-$200,000 $300,000-$500,000 $500,000 +
Cost for maint. contract Updates included	10-20% of total price Yes	12% of software list price Yes	Yes
CMMS INSTALLATION POPULATION No. of systems sold per year CMMS installed/sold to: Government Manufacturing Process Manufacturing Facilities Other No. of application licenses	15 10 5	450+	50 4 89 500 20 500
CMMS PROVIDER STAFF No. of CMMS license/customer rep. Help Desk Rep. Programmer/systems-data language analyst Installation Assistance Operation Assistance	10 Yes Yes		80 Yes Yes No
Special benefits and features	One of the only Macintosh- and Power Macintosh-based CMMS. Based on a totally graphical environment. User-training is cut in half.		Open, object-oriented technology, allowing users to integrate new components easily over time.

* Prices as of 1995.

COMPUTERIZED MAINTENANCE MANAGEMENT SYSTEMS

Microwest Software Systems Inc.	Minneapolis Software	Modern Management Inc.	Nielsen-PM Associates Inc.	OmniComp Inc.
10992 San Diego Mission Road San Diego, CA 92108-2445 (619) 280-0440 FAX: (619) 280-0467	2499 Rice Street Roseville, MN 55113 (612) 484-5684	7421 Carmel Executive Park Charlotte, NC 28226 (704) 542-6546 FAX: (704) 542-1533	19 Lent Avenue LeRoy, NY 14482 (716) 768-2282 FAX: (716) 768-6852	220 Regent Court, Ste. E State College, PA 16801 (800) 726-4181 FAX: (814) 238-4673
Advanced Maintenance Management System	PM Manager	Madcam	PM-MMS	Service Call® Maintenance Management
IBM OTH	IBM OTH	IBM	IBM MIN OTH	IBM
DOS, Windows	Windows	DOS, Windows	Windows, Windows 95, UNIX	DOS, Windows
■	■	■	■	■
■	■	■	■	■
■			■	
■			■	
■	■	■	■	■
■	■		■	■
■	■	■	■	■
■	■	■	■	■
■	■	■	■	■
	■			■
	■			■
■	■	■	■	■
■	■	■	■	■
■	■	■	■	■
		■		
		■	■	
■	■	■		■
Predictive Maintenance, CADD, Project Management Systems, Facility Accounting Systems	Facilities Planning Systems, Facility Accounting Systems	Predictive Maintenance, Project Management Systems, Facility Accounting Systems, Construction Estimating Systems	Predictive Maintenance, Project Management Systems, Facility Accounting Systems	
BTrieve SQL I/F ASCII, Custom ASCII		dBase, BTrieve ASCII, DBF	TCP/IP Informix-SE, On-Line ASCII	dBase III DBF, ASCII, WKS
$3,000-$7,500 $4,500-$10,000 $7,500-$15,000 $10,000-$18,000	$3,295-$4,500 $7,295+ $7,295+ $7,295+	$7,950+ $7,950+ $7,950+ $7,950+	$14,000-$40,000 $40,000-$100,000	$2,950-$6,950 $4,900-$10,900 $4,900-$10,900 $4,900-$10,900
10% of software Yes	$300/year No, $200/yr.	$,1000+ Yes	Yes	No
100+ 10 30 20 20 20 700+	120 15 35 100 530 680	35 5% 60% 25% 10% 275		150-200 78 68 40 765
150 4 Yes Yes	170 4 4 Yes Yes	30 8 Yes Yes	Yes Yes	250 250 Yes Yes Yes
User customization lets you change screen layouts, create reports, forms, ASCIIs on the fly.	Interfaces with any graphics or other database programs for reading data files for import and export purposes.	First CMMS made for PCs. Since its inception in 1981, it has continued to be an industry leader.	Client/server GUI version of PM-MMS. Both PCs and character terminals access same database key tracking module.	User-friendly operation utilizing pull-down menus and point-and-click mouse functionality. Has same-day toll-free technical support and has professional engineering services to help users set up quickly.

COMPUTERIZED MAINTENANCE MANAGEMENT SYSTEMS

Company	Ounce of Prevention Software 1638 Pinehurst Ct. Pittsburgh, PA 15237 (800) 852-8075 FAX: (412) 364-9088	Owen Engineering & Management 5353 W. Dartmouth Ave., Ste. 307 Denver, CO 80277 (303) 969-9393 FAX: (303) 986-8868	PMS Systems Corp. 2800 28th St. Santa Monica, CA 90405 (310) 450-1452 FAX: (310) 450-1311
System name	Ounce of Prevention System	Turbo Maintenance Manager	Smart/MMS
COMPUTER SYSTEM			
IBM=IBM PCs and Compatibles	IBM	IBM	IBM
OTH=Other PCs and Micros			OTH
MIN=Minis	OTH		MIN
MAI=Mainframes			MAI
Operating System(s)	DOS, Windows, OS/2	DOS	Windows, OS/2, UNIX, MVS, VMS
FEATURES			
On-line work request/approval	■		■
Work order origination	■	■	■
Employee competence, skills, training, etc.	■	■	■
Customize outputs	■	■	■
Customize field names, labels, screen, etc.	■		■
Ad hoc reporting/on-line inquiries	■		■
Display report data in graphical format	■		■
Priority origin., management, states tracking	■	■	■
On-line stds./procedures/engineering data, oper. specs	■		■
Auto. P.O. request, generation, P.O. tracking	■		■
EDI interface with suppliers			■
Inventory management, analysis, control	■	■	■
Long/short interval scheduling of labor	■		■
Bar code reading/printing	■		■
On-line OSHA, MSDS, EPA data			■
Failure, root-cause mgmt, analysis	■		■
Remote on-line support via modem	■		■
Established links			
Communication systems, file service			TCP/IP
Data language supported	Knowledgeman	Pascal	SQL
Export data format	ASCII	ASCII	All
* PRICE: Single-user package	$2,995	$895	
Network (2-25)	$3,995-$10,000		
Network (25-50)			
Large Package (Unlimited)			
Cost for maint. contract	$330		
Updates included	Yes	Yes	Yes
CMMS INSTALLATION POPULATION			
No. of systems sold per year	100	20	N/A
CMMS installed/sold to: Government		20	
Manufacturing			
Process Manufacturing			
Facilities		230	
Other			
No. of application licenses			
CMMS PROVIDER STAFF			
No. of CMMS license/customer rep.	100		
Help Desk Rep.			
Programmer/systems-data language analyst	6	1	
Installation Assistance	Yes	Yes	
Operation Assistance	Yes	Yes	
Special benefits and features	User-customizable in work orders, input screens, reports, graphs, look-ups and help screens.	Improves maintenance productivity, browse mode-online. User friendly and easy to master.	Truly scalar architecture runs identical software across mainframe and client server environments.

* Prices as of 1995.

COMPUTERIZED MAINTENANCE MANAGEMENT SYSTEMS

PSDI	Panda Software	Pearl Computer Systems Inc.	Penguin Computer Consultant	Phoenix Data Systems Inc.
20 University Rd.	1907 Bardstown Rd.	705 Birchfield Dr.	P.O. Box 20485, Dept. E.	24293 Telegraph Rd.
Cambridge, MA 02138	Louisville, KY 40205	Mt. Laurel, NJ 08054	San Jose, CA 95160	Southfield, MI 48034
(617) 661-1444	(502) 459-6622	(609) 983-9265	(408) 997-7703	(810) 358-3366
FAX: (617) 661-1642	FAX: (502) 452-9511	FAX: (609) 778-0451	FAX: (408) 927-0570	FAX: (810) 358-3166
Maximo	PM Plus	Ultimaint	Maintenance and Inspection System	AIMS-Asset Information Management System
IBM OTH	IBM	IBM	IBM	IBM
DOS, Windows, OS/2, Unix, NLM	DOS, Windows	DOS, Windows, Apple Macintosh	DOS	DOS, Windows
■	■	■	■	
■	■		■	■
■	■		■	
■	■	■	■	
■	■	■	■	
■	■	■	■	■
■	■		■	■
■	■	■		■
■	■	■		■
■	■	■		■
■	■	■	■	■
■	■	■		■
■		■		■
■				
■		■		
■	■			
MRO, Predictive Maintenance, CADD, Construction Estimating, OEM, Project Management, Facilities Planning & Facility Accounting Systems		Predictive Maintenance, CADD Systems & Facility Accounting Systems		
Novell, IBM, LAN SQL Base, Oracle ASCII, SQL, CSV, DIF, Lotus, dbase	DOS-BTrieve, Windows-Paradox DOS QFast 5, WIndows Delphi ASCII and others	Windows, Novell, Lantastic, Banyan Vines FoxPro 2.6 ASCII, dbf	Custom Link Database ASCII	Novell MDBS ASCII
$40,000+	$2,995-$5,995 $4,495-$7,495 $4,495-$7,495 $4,495-$7,495	$2,000-$4,000 $5,000-$10,000 $6,000-$10,000 $6,000-$10,000	$1,450-$2,500	$995-$6,000 $995-$6,000 $995-$6,000 $5,985-$12,000
15% per year Yes	$995 Yes	$750 per year Yes	$250/per year No	$541/$1,070 Yes
300+	50	35	20	30
65	3	2%	7	2
650	22	45%	150	1
310		10%		
575	220	43%	50	13
60/Education			143/Hospitals, HVAC, Service	28/Clinical, Hospitality, Utility
1,800			350	
	4	10		150
14	1	20	2	4
Yes	Yes	Yes	No	Yes
Yes	Yes	Yes	Yes	Yes
Client/server, scaleable database server, Windows, customizable, easy to use, interface with corporate systems, SQL.	Has hazardous materials section, security and key control. Also features personnel management, policies and procedures.	User-definable fields and reports, and AUTO CAD compatible. PM by calendar or meter and project management.	Unlimited space for work order instructions and work comments.	Fully adaptable to various maintenance management environments. Customization, technical support and management services available.

COMPUTERIZED MAINTENANCE MANAGEMENT SYSTEMS

Company	Plus Delta Performance 1500 Green Bay St. La Crosse, WI 54601 (608) 788-7755 FAX: (608) 788-7700	Precision Maintenance Systems Inc. 133 South Rd. East Harland, CT 06027 (203) 653-9863 FAX: (203) 653-9863	Primavera Systems Inc. Two Bala Plaza Bala Cynwyd, PA 19004 (610) 667-8600 FAX: (610) 667-7894
System name	mPulse	PMS-CMMS	Primavera Project Planner
COMPUTER SYSTEM IBM=IBM PCs and Compatibles OTH=Other PCs and Micros MIN=Minis MAI=Mainframes	IBM	IBM	IBM
Operating System(s)	Windows	DOS	DOS, Windows
FEATURES On-line work request/approval	■	■	
Work order origination	■	■	
Employee competence, skills, training, etc.	■	■	
Customize outputs	■	■	■
Customize field names, labels, screen, etc.	■	■	■
Ad hoc reporting/on-line inquiries	■	■	
Display report data in graphical format	■	■	■
Priority origin., management, states tracking	■	■	■
On-line stds./procedures/engineering data, oper. specs	■	■	
Auto. P.O. request, generation, P. O. tracking		■	
EDI interface with suppliers		■	
Inventory management, analysis, control	■	■	■
Long/short interval scheduling of labor		■	■
Bar code reading/printing		■	
On-line OSHA, MSDS, EPA data			
Failure, root-cause mgmt, analysis		■	
Remote on-line support via modem	■		■
Established links	CADD Systems	Predictive Maintenance, CADD, Construction Estimating, Project Management, Facilities Planning & Facility Accounting Systems	Construction Estimating Systems, Project Management Systems, Facilities Planning Systems & Facility Accounting Systems
Communication systems, file service		PC Anywhere	
Data language supported	Paradox	Focus	BTrieve
Export data format	ASCII	ASCII, Xcel, Lotus 123, dBase	ASCII, Lotus, Excel
★ PRICE: Single-user package Network (2-25) Network (25-50) Large Package (Unlimited)	$795 $1,995-$3,000 $3,000 $3,000	$5,000-$8,500 $10,000-$35,000 $35,000-$40,000 $40,000+	$4,000 $4,000-$40,000 $40,000-$75,000 $75,000
Cost for maint. contract Updates included	90 days free support Yes/$100	$80 per hour Yes	$800/20% purchase price Yes
CMMS INSTALLATION POPULATION No. of systems sold per year	>1,000	2	7,000
CMMS installed/sold to: Government Manufacturing Process Manufacturing Facilities Other		2 1 4 1/Base support contract	6,000 4,000 2,000 2,000 200/Construction, Oil, GLS
No. of application licenses			
CMMS PROVIDER STAFF No. of CMMS license/customer rep. Help Desk Rep. Programmer/systems-data language analyst Installation Assistance Operation Assistance	15 Yes Yes	2 2 Yes Yes	45 50 Yes Yes
Special benefits and features	Full-featured, windows-based package for under $1,000. Includes modules for key/lock management, building systems, fleet management and CMMS modules. Literate in English and Spanish.	Menu driven, totally integrated, customized to client requirements.	Recognized leader in sophisticated, high-end project planning and control software for professional project managers. Offers project scheduling, resource and cost control.

* Prices as of 1995.

COMPUTERIZED MAINTENANCE MANAGEMENT SYSTEMS

Project Services International, Inc. Robinson Plaza #3, Ste. 300 Pittsburgh, PA 15205 (800) 860-0111 FAX: (412) 747-0114	QBIC III Systems Inc. 1005 Old Columbia Rd., Ste. N-160 Columbia, MD 21046 (800) 455-9805 FAX: (301) 854-3930	Qqest Software Systems P.O. Box 171288 Salt Lake City, UT 84117 (801) 272-1697 FAX: (801) 272-1699	RMS Systems Two Scott Plaza Philadelphia, PA 19113 (610) 521-2817 FAX: (610) 521-0112	Rainbow Enterprises 15127 NE 24th St., Ste 152 Redmond, WA 98052 (206) 881-7243 FAX: (206) 881-7243
Totally Integrated Maintenance Management (TIMM)	QBIC III	Maintenance Management	Trimax	Maintenance Management
IBM	IBM	IBM	MIN–AS/400, IBM PCs	IBM
DOS, Windows	DOS, Windows	DOS	Windows, DS-400	DOS, Windows
■	■	■	■	■
■		■	■	■
■			■	■
■	■	■	■	■
■	■		■	■
■	■	■	■	■
■	■		■	■
■	■	■	■	■
■	■		■	■
■	■	■	■	■
	■	■	■	■
	■	■	■	■
	■		■	
			■	
■	■		■	
	■			
Predictive Maintenance	Predictive Maintenance, CADD, Construction Estimating, Project Management, Facilities Planning & Facility Accounting Systems		Predictive Maintenance, CADD Systems, & Facility Accounting Systems	
All Supported by Paradox Paradox All supported by Paradox	C++ Clipper, SQL ASCII, dBase III	All ASCII		XBase (Clipper) ASCII, XBase
$1,750-$8,000	$7,500-$10,000 $12,000-$30,000 $30,000-$40,000 $50,000-$100,000	$595-$995 $995-$1,595	$26,000-$225,000	$750-$1,000 $1,000-$1,250 $1,000-$1,250 $1,000-$1,250
$350-$1,600 Yes	10% Yes	$195 Yes	12% of license fee Yes	No
50 20 150 150 30 30/Utilities	200 100 50 50 600	1,000+	5 100 40 2 175	15 10 15 5 10 5/Military
40 3 Yes Yes	50 Yes Yes	No Yes	35 5 10-15 Yes Yes	15 Yes Yes
Five separate modules using Paradox RDBS gives the ultimate flexibility and unlimited capability in any situation.	Offers flexibility. QBIC III can be modified to mirror any type of operation and integrate with most technologies.	Comprehensive maintenance solution that features preventive maintenance for facilities grounds and equipment.	Keeps you in control of your operations. Flexible, easy to implement and easy to use.	Simple to use, easy to understand and has powerful features.

COMPUTERIZED MAINTENANCE MANAGEMENT SYSTEMS

Company	Ramco Systems Corporation 2201 Walnut Ave. Fremont, CA 94538 (408) 522-8080 FAX: (408) 522-8081	Revere Inc. One Perimeter Park S., Ste. 230-S Birmingham, AL 35243 (205) 967-4905 FAX: (205) 967-4751	S&J Enterprises 3025 Red Wing Ct. Bettendorf, IA 52722-2174 (800) 508-4410 FAX: (319) 332-4252
System name	Ramco Marshal Maintenance Management Systems	Immpower	Maintenance-7
COMPUTER SYSTEM			
IBM=IBM PCs and Compatibles	IBM		IBM
OTH=Other PCs and Micros			OTH
MIN=Minis	MIN	MIN–DEC, HP, IBM	
MAI=Mainframes			
Operating System(s)	DOS, Windows, UNIX, OS/2, MS, NT	DOS, Windows, OS/2, UNIX	Windows, Apple Macintosh
FEATURES			
On-line work request/approval	■	■	■
Work order origination	■	■	■
Employee competence, skills, training, etc.	■	■	
Customize outputs	■		■
Customize field names, labels, screen, etc.	■	■	
Ad hoc reporting/on-line inquiries	■	■	■
Display report data in graphical format	■	■	
Priority origin., management, states tracking	■	■	■
On-line stds./procedures/engineering data, oper. specs	■	■	■
Auto. P.O. request, generation, P.O. tracking	■	■	
EDI interface with suppliers	■	■	
Inventory management, analysis, control	■	■	■
Long/short interval scheduling of labor	■	■	■
Bar code reading/printing	■	■	
On-line OSHA, MSDS, EPA data	■	■	
Failure, root-cause mgmt, analysis	■	■	
Remote on-line support via modem	■	■	■
Established links	Predictive Maintenance, CADD Systems, Project Management Systems & Facility Accounting Systems	Predictive Maintenance, CADD Systems & Project Management Systems	
Communication systems, file service	Netware, TCP/IP, Banyan Vines		Novell, Banyan Vines, Apple Share Omnis 7
Data language supported	Sybases, Oracle, Microsoft SQL		
Export data format		ASCII	ASCII, DIF, SYLK, WKS
★ PRICE: Single-user package			$995-$1,450
Network (2-25)	$40,000-$60,000		$200-$3,500
Network (25-50)			
Large Package (Unlimited)		$50,000-$250,000	
Cost for maint. contract	15% of license cost	15% of user license	$295
Updates included	Yes	Yes	Yes
CMMS INSTALLATION POPULATION			
No. of systems sold per year	20-30	52	100+
CMMS installed/sold to: Government	3		
Manufacturing	9	6	70%
Process Manufacturing	8	46	15%
Facilities			15%
Other	5/Utilities		
No. of application licenses	30		25%
CMMS PROVIDER STAFF			
No. of CMMS license/customer rep.		14	3
Help Desk Rep.			
Programmer/systems-data language analyst	300	12	3
Installation Assistance	Yes	Yes	Yes
Operation Assistance	Yes	Yes	Yes
Special benefits and features	Maintenance management system is part of an integrated solution that includes manufacturing, financial and distribution systems.	Truly an open system. Graphical user interface provides intuitive use of Immpower's holistic design. Complete open CMMS for large-scale process industry.	Ease of operation and cross platform compatibility.

* Prices as of 1995.

COMPUTERIZED MAINTENANCE MANAGEMENT SYSTEMS

SSA Maintenance Management	Selfware Inc.	Specific Designs Inc.	Synergen Associates Inc.	TMA Systems Inc.
1000 Boone Ave. N, Competence Center Minneapolis, MN 55427 (612) 797-8700 FAX: (612) 797-8800	8618 Westwood Ctr. Dr., Ste. 450 Vienna, VA 22182 (703) 506-0400 FAX: (703) 506-0580	21062 Brookhurst St., Ste. 103 Huntington Beach, CA 92646 (714) 965-8988 FAX: (714) 965-8987	2121 N. California Blvd., Ste. 800 Walnut Creek, CA 94596-7304 (510) 935-7670 FAX: (510) 935-9748	5220 E. 69th Place Tulsa, OK 74136-3407 (918) 494-2890 FAX: (918) 494-4892
BPCS Maintenance Management	Opmist	EM/dBS	Synergen Series 7	TMA
MIN	IBM	IBM	IBM OTH MIN MAI	IBM OTH
OS/400	DOS	DOS, Windows, OS/2	DOS, Windows, OS/2, UNIX, Macintosh	Windows, OS/2, Macintosh, Windows NT
■	■	■	■	■
■	■	■	■	■
■	■	■	■	■
■	■	■	■	■
■	■	■	■	■
■	■	■	■	■
■	■	■	■	■
■		■	■	■
■		■	■	■
■		■	■	■
■		■	■	■
■	■	■	■	■
■		■	■	■
■			■	
■			■	
■	■	■	■	■
MRO Supplier Order Writing Systems, Predictive Maintenance, Project Management Systems, & Facility Accounting Systems	Project Management Systems	Project Management Systems, Facilities Planning Systems & Facility Accounting Systems	MRO, Predictive Maintenance, CADD, Construction Estimating, OEM, Project Management, Facilities Planning & Facility Accounting Systems	Pred. Maint., CADD, Constr. Estim., OEM, Proj. Mgmt., Fac. Plan., Fac. Acct.
DB2/400	Novell compatible Foxpro ASCII, XBase	Novell, Lantastic, Banyan Vines dBase ASCII, dBase	Any supported by Oracle Oracle ASCII or Binary	Novell and others Omnis 7, Oracle ASCII, dBase, Lotus, Excel
$1,500-$15,000 $3,000+	$895-$1,495 $1,795-$62,250	$795-$3,295 $1,295-$5,795 $1,295-$5,795 $1,295-$5,795	$50,000 $58,000-$150,000 $150,000-$250,000 $250,000	$4,995-$11,995 $6,995-$22,000 $22,000-$39,000
18% Yes	20% Yes	$395-$595 Yes	15% of license fee Yes	Varies Yes
	50 10 10 30	15 40 20 145 110	4-10 2 3 16 3 6/Utilities, Mining	50 15 2 135
3,000	200	300		
60 25 Yes Yes	25 Yes Yes	Yes Yes	5 7 Yes Yes	50 50 50 Yes Yes
Computerized maintenance management systems for equipment, vehicles and facilities in all industries.	Flexible, extensive customization capabilities and source code options. Integrated project management.	Flexible and easy to use program, moderately priced, interface modules for bar code and mainframes.	WorkFlow Agent™ feature provides the management tool to streamline the workflow process.	Easy to learn CMMS for building maintenance. Detailed inventories and work histories, PM task library and much more.

COMPUTERIZED MAINTENANCE MANAGEMENT SYSTEMS

Company	TMT Software Co. 6320 Quadrangle Dr., Ste. 355 Chapel Hill, NC 27514 (919) 493-4700 FAX: (919) 489-1449	TSW International Inc. 3301 Windy Ridge Pkwy. Atlanta, GA 30339 (800) 868-0497 FAX: (404) 952-2977	Today Systems Inc. 1291 High St., E. Denver, CO 80218-2658 (303) 322-2679 FAX: (303) 322-5346
System name	Transman®	Enterprise MPAC	Operating Contol System
COMPUTER SYSTEM			
IBM=IBM PCs and Compatibles	IBM	IBM	
OTH=Other PCs and Micros	OTH—VAX/Unix		
MIN=Minis	MIN—AS/400	MIN	MIN—Hewlett Packard 9000, IBM RS/6000
MAI=Mainframes			
Operating System(s)	DOS	DOS, Windows, UNIX, WIN '95, NT	UNIX
FEATURES			
On-line work request/approval	■	■	■
Work order origination	■	■	■
Employee competence, skills, training, etc.	■	■	■
Customize outputs	■	■	■
Customize field names, labels, screen, etc.	■	■	■
Ad hoc reporting/on-line inquiries	■	■	■
Display report data in graphical format	■	■	■
Priority origin., management, states tracking	■	■	■
On-line stds./procedures/engineering data, oper. specs	■	■	■
Auto. P.O. request, generation, P. O. tracking	■	■	■
EDI interface with suppliers	■	■	■
Inventory management, analysis, control	■	■	■
Long/short interval scheduling of labor	■	■	■
Bar code reading/printing	■	■	■
On-line OSHA, MSDS, EPA data	■	■	■
Failure, root-cause mgmt, analysis	■	■	■
Remote on-line support via modem	■	■	■
Established links		MRO Supplier Order Writing, Systems, Predictive Maintenance, CADD Systems, Proj. Mgmt. Systems & Facility Accounting Systems	Predictive Maintenance
Communication systems, file service	ASCII	TCP/IP, SPX, IPX	Any
Data language supported	BTrieve	SQL	Oracle, Informix
Export data format	ASCII	ASCII, OLE, DDE	ASCII, Oracle, Informix
★ PRICE: Single-user package	$1,000-$14,000		$29,520-$51,790
Network (2-25)		$40,000-$95,000	$63,190-$86,420
Network (25-50)		$75,000-$250,000	
Large Package (Unlimited)	$10,000-$25,000	$250,000+	Up to $206,995
Cost for maint. contract	15% user price	18% of system price	15% of license fee
Updates included	Yes	Yes	Yes
CMMS INSTALLATION POPULATION			
No. of systems sold per year	100	50-75	
CMMS installed/sold to: Government	15	30	10%
Manufacturing	15	300	20%
Process Manufacturing	25	100	20%
Facilities	10	150	
Other	500/Mobil Equipment	200/Utilities	50%/Mining & Natural Resources
No. of application licenses	450	600+	
CMMS PROVIDER STAFF			
No. of CMMS license/customer rep.	50	16	
Help Desk Rep.	5		
Programmer/systems-data language analyst	3	2	
Installation Assistance	Yes	Yes	Yes
Operation Assistance	Yes	Yes	Yes
Special benefits and features	Operates on PC or AS/400 platforms. Modules include warranty and strong parts management. Custom programming service available.	Multi-mode presentation (advance user interface), stored procedures/triggers (distributed function) and nation DB (PL/SQL) and transact SQL.	Integrated financials, personnel, materials management subsystems, component tracking, transparent 4GL switching of hardware platforms, databases, etc.

* Prices as of 1995.

Reliability Centered Maintenance

Donald A. Morton, P.E., C.P.E.
System Planning Corp., Arlington, Virginia

13.1 Introduction

13.1.1 Background

(A glossary and list of abbreviations/acronyms appears in sections 13.13 and 13.14 at the end of this chapter.)

From approximately 1960 until the late 1980s, preventive maintenance (PM) was the most advanced technique used by progressive facilities maintenance organizations. PM is based upon the principle that equipment failures result largely from age or use. Therefore, PM is time-driven or usage-interval based. It assumes that failure probabilities can be determined statistically and that parts can be replaced before they fail.

The development of new technologies in the late 1980s made it possible to determine the actual condition of equipment, rather than relying upon estimates of when it might fail based upon age or use. Also, it was discovered there are many different equipment failure characteristics, only a small number of which are age- or use-related. This new knowledge has increased the emphasis on condition monitoring, commonly called *Predictive Maintenance* (PdM), and has caused a reduction in reliance upon PM.

In the early 1990s, a new concept that employs the best practices of several different types of maintenance moved into the facilities maintenance mainstream. It is called *Reliability Centered Maintenance,* or RCM. RCM combines the techniques of reactive maintenance (run-to-failure or breakdown maintenance), PM, PdM, and proactive maintenance. RCM applies these four techniques where each is most appropriate based upon the consequences of equipment failure, particularly its impact upon organizational mission. This combination produces optimum reliability at minimum maintenance cost. RCM is a new way of thinking about facilities and equipment maintenance. Its combined benefits far exceed those from using any one maintenance technique.

13.1.2 Reliability centered maintenance (RCM) definition

RCM is the optimum mix of reactive, preventive, predictive, and proactive maintenance, as shown in Fig. 13.1. The principal features of each strategy are shown below their block in Fig. 13.1. These maintenance strategies, rather than being applied independently, can be melded to take advantage of their respective strengths in order to maximize facility and equipment operability and efficiency while minimizing life-cycle costs. These strategies are defined and discussed in paragraph 13.3.

13.2 Reliability Centered Maintenance (RCM) Program

13.2.1 Philosophy

The RCM philosophy employs reactive, preventive, predictive, and proactive maintenance techniques in an integrated manner to increase the probability that a machine or component will function in the required manner over an extended life cycle with a minimum of maintenance. RCM requires that maintenance decisions be based on maintenance requirements supported by sound economic justification.

13.2.2 Analysis

The RCM analysis carefully considers the following questions:

- What does the system or equipment do?
- What functional failures are likely to occur?
- What are the likely consequences of these functional failures?
- What can be done to prevent these functional failures?

Figure 13.2 shows the RCM decision logic tree based upon the answers to these four questions.

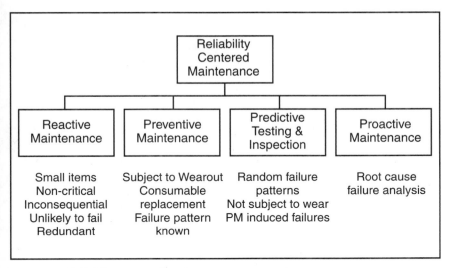

Figure 13.1 Reliability centered maintenance components.

13.2.3 Principles

There are several RCM principles. RCM:

a. *Is Function Oriented.* It seeks to preserve system or equipment function, not just operability for operability's sake. Redundancy of function through multiple equipment improves functional reliability.

b. *Is Reliability Centered.* It treats failure statistics in an actuarial manner. The relationship between operating age and the failures experienced is important. RCM is not overly concerned with simple failure rate; it seeks to know the conditional probability of failure at specific ages (the probability that failure will occur in each given operating age bracket).

c. *Acknowledges Design Limitations.* Its objective is to maintain the inherent reliability of the equipment design, recognizing that changes in inherent reliability are the province of design rather than maintenance. Maintenance can, at best, only achieve and maintain the level of reliability for equipment which is provided by design. However, RCM recognizes that maintenance feedback can improve on the original design.

d. *Is Driven by Safety and Economics.* Safety *must* be ensured before cost-effectiveness becomes the criterion.

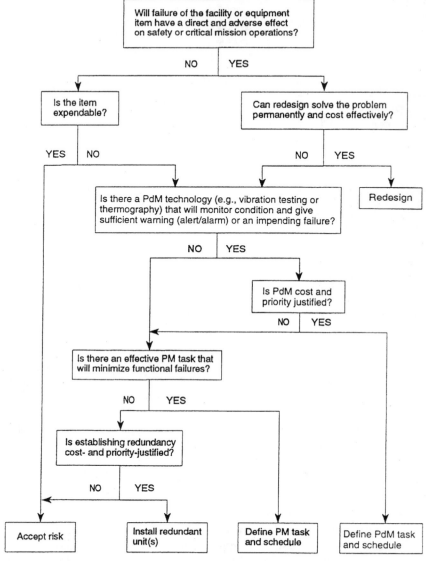

Figure 13.2 Reliability centered maintenance (RCM) decision logic tree.

e. *Defines Failure as Any Unsatisfactory Condition.* Therefore, failure may be either a loss of function (operation ceases) or a loss of acceptable quality (operation continues).

f. *Uses a Logic Tree To Screen Maintenance Tasks.* A logic tree provides a consistent approach to the maintenance of all kinds of equipment.

g. *Tasks Must Be Effective and Applicable.* The tasks must reduce the number of failures or ameliorate secondary damage resulting from failure.

h. *Acknowledges Three Types of Maintenance Tasks.* These tasks are time-directed (PM), condition-directed (PdM), and failure-finding (one of several aspects of Proactive Maintenance). Time-directed tasks are scheduled when appropriate. Condition-directed tasks are performed when conditions indicate they are needed. Failure-finding tasks detect hidden functions which have failed without giving evidence of pending failure.

i. *Is a Living System.* It gathers data from the results achieved and feeds this data back to improve design and future maintenance. This feedback is an important part of the Proactive Maintenance element of the RCM program.

Note that the RCM decision process shown in Fig. 13.2 results in decisions concerning whether a particular piece of equipment should be reactively maintained (*accept risk* and *install redundant unit(s)* blocks), PM'ed (*Define PM task and schedule* block), or predictively maintained (*define PdM task and schedule* block).

A rigorous RCM analysis of each system, subsystem, and component is normally performed only on systems in which failures can have severe, including fatal, consequences such as aircraft flight control systems. Such analyses are expensive and time consuming. The analysis to determine the type of maintenance to be used for most facilities and equipment items is usually much more informal than a detailed, technical RCM analysis.

13.2.3.1 Functional failures. Functional failures are ways in which a system can fail to meet the functional requirements designed into the equipment. A system which is operating in a degraded condition but does not impact any of the requirements addressed in a later paragraph (Systems, System Boundaries, and Facility Envelopes) has not experienced a functional failure.

It is important to determine all the functions of an item that are significant in a given operational context in order that functional failure for that item may be properly defined; for example, aircraft brakes not only stop the aircraft, but also provide differential braking for maneuvering, and both modulated stopping and an antiskid capability.

Prevention of potential failures preempts functional failures. The function of PdM is to identify potential failures.

13.2.3.2 Failure modes. Failure modes are equipment- and component-specific failures that result in functional failure of systems or subsystems. For example, a machinery train composed of a motor and pump can fail catastrophically due to complete failure of the windings, bearings, shaft, impeller, controller, or seals. In addition, a functional failure also occurs if the pump performance degrades such that insufficient discharge pressure or flow exists to meet operating requirements.

Dominant failure modes are those failure modes responsible for a significant proportion of all the failures of the item. They are the most common modes of failure.

Not all failure modes warrant preventive or predictive maintenance because the likelihood of their occurring is remote or their effect is inconsequential.

13.2.3.3 Failure probability. Reliability is the probability that an item will survive a given operating period, under specified operating conditions, without failure. The *conditional probability of failure* means the probability that an item **entering a given age interval** will fail during that interval. If the conditional probability of failure increases with age, the item shows wear-out characteristics. The conditional probability of failure reflects the overall adverse effect of age on reliability. It is not a measure of the change in an individual equipment item.

Failure rate or frequency plays a relatively minor role in maintenance programs because it is too simple a measure. Failure frequency is useful in making cost decisions and determining maintenance intervals, but it tells nothing about which maintenance tasks are appropriate or about the consequences of failure. A maintenance solution should be evaluated in terms of the safety or economic consequences it is intended to prevent. A maintenance task must prevent failures or ameliorate failure consequences in order to be effective.

13.2.3.4 Failure characteristics. Conditional probability of failure (P_{cond}) curves fall into six basic types, as graphed (P_{cond} vs. Time) in Fig. 13.3.[1] The data resulted from a study of the U.S. airline industry conducted by United Airlines for the Department of Defense. It may be considered to fairly represent equipment found in facilities as well.

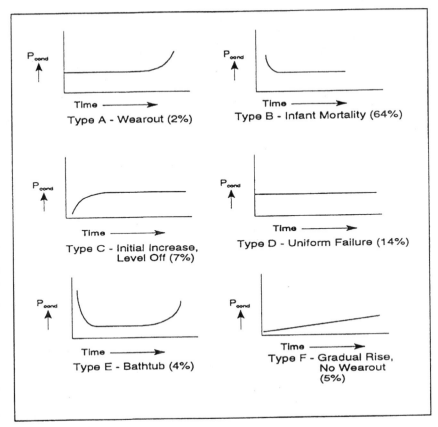

Figure 13.3 Equipment failure characteristics.

The percentage of equipment conforming to each of the six wear patterns is also shown in Fig. 13.3.

Type A — Constant or gradually increasing failure probability, followed by a pronounced wear-out region. An age limit may be desirable. (Typical of reciprocating engines.)

Type B — Infant mortality, followed by a constant or slowly increasing failure probability. (Typical of electronic equipment.)

Type C — Low failure probability when the item is new or just overhauled, followed by a quick increase to a relatively constant level.

Type D — Relatively constant probability of failure at all ages.

Type E — Bathtub curve; i.e., *infant mortality* followed by a constant or gradually increasing failure probability and then a pronounced *wear-out* region. An age limit may be desirable, provided a large number of units survive to that age where wear-out begins.

Type F — Gradually increasing failure probability, but no identifiable wear-out age. Age limit usually not applicable. (Typical of turbine engines.)

Conventional wisdom holds that most types of equipment exhibit bathtub curve characteristics; whereas only about 4 percent of equipments have such characteristics.

Types A and E are typical of single-piece and simple items such as tires, compressor blades, brake pads, and structural members. Most complex items have conditional probability curves of types B, C, D, and F.

The basic difference between the failure patterns of complex and simple items has important implications for maintenance.

Single-piece and simple items frequently demonstrate a direct relationship between reliability and age. This is particularly true where factors such as metal fatigue or mechanical wear are present or where the items are designed as consumables (short or predictable life spans). In these cases an age limit based on operating time or stress cycles may be effective in improving the overall reliability of the complex item of which they are a part.

Complex items frequently demonstrate some infant mortality, after which their failure probability increases gradually or remains constant, and a marked wear-out age is not common. **In many cases scheduled overhaul increases the overall failure rate by introducing a high infant mortality rate into an otherwise stable system.**

13.2.3.5 Preventing failure.   Every equipment item has a characteristic that can be called *resistance to or margin to failure*. Using equipment subjects it to *stress* which can result in failure when the *stress* exceeds the *resistance to failure*. Figure 13.4 depicts this concept graphically. The figure shows that failures may be prevented or item life extended by:

a. Decreasing the amount of stress applied to the item. The life of the item is extended for the period f_0–f_1 by the stress reduction shown.

b. Increasing or restoring the item's resistance to failure. The life of the item is extended for the period f_1–f_2 by the resistance increase shown.

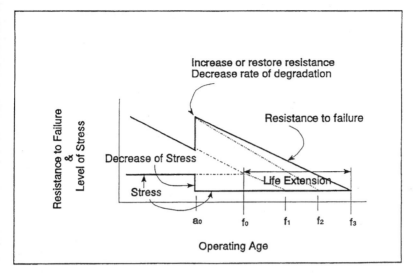

Figure 13.4 Preventing failure.

c. Decreasing the rate of degradation of the item's resistance to or
   margin to failure. The life of the item is extended for the period
   f_2–f_3 by the decreased rate of resistance degradation shown.

Stress is dependent upon use and may be highly variable. It may
increase, decrease, or remain constant with use or time. A review of
the failures of a large number of nominally identical simple items
would disclose that the majority had about the same age at failure,
subject to statistical variation, and that these failures occurred for the
same reason. If one is considering preventive maintenance for some
simple item and can find a way to measure its resistance to failure,
one can use that information to help select a preventive task.

Resistance to failure can be increased by adding excess material
that wears away or is otherwise consumed. Excess strength may be
provided to compensate for loss from corrosion or fatigue. The most
common method of restoring resistance is by replacing the item.

The resistance to failure of simple items decreases with use or time
(age), but a complex unit consists of hundreds of interacting simple
items (parts) and has a considerable number of failure modes. In the
complex case, the mechanisms of failure are the same, but they are
operating on many simple component parts simultaneously and inter-
actively so that failures no longer occur for the same reason at the
same age. For these complex units, it is unlikely that one can design

a maintenance task unless there are a few dominant or critical failure modes.

13.2.4 Systems, system boundaries, and facility envelopes

13.2.4.1 Systems. A system is any user-defined group of components, equipments, or facilities that support an operational requirement. These operational requirements are defined by mission criticality or by environmental, health, safety, regulatory, quality, or other requirements.

13.2.4.2 System boundaries. Most systems can be divided into unique subsystems along user-defined boundaries. The boundaries are selected as a method of dividing a system into subsystems when its complexity makes an analysis by other means difficult.

A system boundary or interface definition contains a description of the inputs and outputs that cross each boundary.

13.2.4.3 Facility envelope. The facility envelope is the physical barrier created by a building, enclosure, or other structure; e.g., a cooling tower or tank.

13.2.5 RCM goals

The RCM goals are to identify for each system and equipment the failure modes and their consequences and to determine the most cost-effective and applicable maintenance technique to minimize the risk and impact of failure. This allows system and equipment functionality to be maintained in the most economical manner.

13.3 RCM Program Components

An RCM program includes reactive, preventive, predictive, and proactive maintenance. RCM application requires an understanding of system boundaries, facility envelopes, functional failures, and failure modes, all of which are critical components of the RCM program. The following paragraphs describe these key RCM components.

13.3.1 Reactive maintenance

Reactive maintenance is also referred to as breakdown, repair, fix-when-fail, or run-to-failure (RTF) maintenance. When applying this maintenance technique, maintenance, equipment repair, or replacement occur only when the deterioration in an equipment's condition causes a functional failure. This type of maintenance assumes that failure is equally likely to occur in any part, component, or system. Thus, this assumption precludes identifying a specific group of repair parts as being more necessary or desirable than others. If the item fails and repair parts are not available, delays ensue while parts are obtained. If certain parts are urgently needed to restore a critical machine or system to operation, a premium for expedited delivery must be paid. Also, there is no ability to influence when the failures occur because no (or minimal) action is taken to control or prevent them. *When this is the sole type of maintenance practiced,* a high percentage of unplanned maintenance activities, high replacement part inventories, and inefficient use of maintenance effort typify this strategy. A purely reactive maintenance program ignores the many opportunities to influence equipment survivability.

Reactive maintenance can be used effectively when it is performed as a conscious decision based on the results of an RCM analysis that compares the risk and cost of failure with the cost of the maintenance required to mitigate that risk and cost of failure.

13.3.2 Preventive maintenance (PM)

PM consists of regularly scheduled inspections, adjustments, cleaning, lubrication, parts replacement, calibration, and repair of components and equipment. PM is also referred to as time-driven or interval-based maintenance. It is performed without regard to equipment condition.

PM schedules periodic inspection and maintenance at predefined intervals in an attempt to reduce equipment failures for susceptible equipment. Depending on the intervals set, PM can result in a significant increase in inspections and routine maintenance; however, it should also reduce the frequency and seriousness of unplanned machine failures for components with defined, age-related wear patterns.

Traditional PM is keyed to failure rates and times between failures. It assumes that these variables can be determined statistically, and therefore that one can replace a part due for failure before it fails. The availability of statistical failure information tends to lead to fixed schedules for the overhaul of equipment or the replacement of parts subject to wear. PM is based on the belief that the overhaul of machinery by disassembly and replacement of worn parts restores the machine to a like-new condition with no harmful effects.

Failure rate or its reciprocal, Mean-Time-Between-Failures (MTBF), is often used as a guide to establishing the interval at which the maintenance tasks should be performed. The major weakness in using these measurements to establish task periodicities is that failure rate data determines only the *average* failure rate. The reality is that failures are equally likely to occur at random times and with a frequency unrelated to the average failure rate. Thus, picking a specific time to conduct periodic maintenance for a randomly failing item is risky.

For some items, while failure is related to age, it is not equally likely to occur throughout the life of the item. In fact, the majority of equipment is not subject to wear-out (sharply increasing conditional probability of failure at a specific operating age). Therefore, timed maintenance can often result in unnecessary maintenance.

In summary, PM can be costly and ineffective when it is the sole type of maintenance practiced.

13.3.3 Predictive maintenance (PdM)

PdM uses advanced technology to assess machinery condition. It replaces arbitrarily timed maintenance tasks with maintenance scheduled only when warranted by equipment condition. Continuing analysis of equipment condition-monitoring data allows planning and scheduling preventive maintenance or repairs in advance of catastrophic and functional failure. Thus, PdM is also referred to as condition-based maintenance.

The results of PdM technology are used in one of several ways to determine the condition of the equipment. The methods of analysis include:

- Trend analysis
- Pattern recognition
- Data comparison
- Tests against limits and ranges
- Correlation of multiple technologies
- Statistical process analysis.

PdM normally is used for facility items that require more sophisticated means than PM to identify maintenance requirements.

PdM does not lend itself to all types of equipment and therefore *should not be the sole type of maintenance practiced.*

13.3.4 Proactive maintenance

Proactive maintenance employs six basic techniques to extend machinery life. These techniques include the following:

- Specifications for new/rebuilt equipment
- Precision rebuild and installation

- Failed-part analysis
- Root-cause failure analysis
- Reliability engineering
- Rebuild certification/verification.

Proactive maintenance improves maintenance through better design, installation, maintenance procedures, workmanship, and scheduling.
The characteristics of proactive maintenance are:

- Using feedback and communications to ensure that changes in design or procedures are rapidly made available to designers and managers.
- Employing a life-cycle view of maintenance and supporting functions.
- Ensuring that nothing affecting maintenance occurs in isolation.
- Employing a continuous improvement process.
- Optimizing and tailoring maintenance techniques and technologies to each application.
- Integrating functions which support maintenance into maintenance program planning.
- Using root-cause failure analysis and predictive analysis to maximize maintenance effectiveness.
- Adopting an ultimate goal of fixing the equipment forever.

A proactive maintenance program is the capstone of the RCM philosophy. The six most commonly recognized proactive techniques, listed at the beginning of this section 13.3.4, are described below:

13.3.4.1 Specifications for new/rebuilt equipment. The design and fabrication of many new types of equipment too often fail to provide for easily obtaining reliable data for verifying equipment condition while the equipment is operating. It has been alleged that this oversight might result from design using CAD/CAM systems that require little practical experience on the part of the designers. Existing standards, often 25 to 30 years old and not reflective of changes in technology, are usually inadequate, typically addressing only general or minimal performance criteria. Additionally, the life cycle costs and failure histories of families of equipment are rarely documented for purchasing and contract personnel who, bound by regulation, procure conforming products based on least cost.

To solve this problem, reliability engineers must write proper specifications, test the equipment of different vendors, and document problems. These specifications should include, as a minimum, vibration, alignment, and balancing criteria. Documenting historical data, so that engineers can write verifiable purchasing and installation specifications for new and rebuilt equipment, is the basis of this proactive technique. Performance testing is then conducted (1) in the factory prior to shipment, (2) as the equipment is installed prior to acceptance, and (3) to establish a performance baseline as the equipment begins operation.

13.3.4.2 Precision rebuild and installation. Equipment requires proper installation to control life-cycle costs and maximize reliability. Poor installation often results in problems routinely faced by both maintenance personnel and operators. Rotor balance and alignment, two common rework items, are often poorly performed or neglected during initial installation. The adoption and enforcement of precision standards can more than double the life of a machine. For example, the contract specification for leveling equipment being installed should include a maximum acceptable slope of the base and the frame; e.g., maximum slope of 0.001 inch per foot (80 μm per m). The specification also should include the type and accuracy of the instrument used for measuring the slope; e.g., a 12-inch (30 cm) machinist's level graduated to 0.0002 in./ft (16 μm per m). After the criteria have been included in the contract specifications, the installation should be checked to ensure that the mechanic has complied with the specification.

13.3.4.3 Failed-part analysis. This proactive process involves visually inspecting failed parts after their removal to identify the root causes of their failures. More detailed technical analysis may be conducted when necessary to determine the root cause of a failure.

Bearings are generally the weakest equipment components. *Only 10 to 20 percent of bearings achieve their design life.* The root causes of bearing failures may relate to improper installation, poor lubrication practices, excessive balance and alignment tolerances, or poor storage and handling techniques. Failed-bearing analysis provides methods to categorize defects such as scoring, color, fretting, and pitting and to relate these findings to the most probable cause of failure.

Over half of all bearing problems result from contamination or improper installation. While indicators of contamination normally appear on the internal surface of bearings, indicators of installation problems generally are evident on both internal and external surfaces.

13.3.4.4 Root-cause failure analysis (RCFA). In some cases, plant equipment fails repeatedly, and the failures are accepted as a normal idiosyncrasy of that equipment. Recurring problems such as short bearing life, frequent seal fracture, and structural cracking are symptoms of more severe problems. However, maintenance personnel often fix only the symptomatic problems and continue with the frequent repairs. Repeated failures result in high costs for parts and labor and in decreased customer goodwill and mission support reliability. Further, unreliable equipment may pose a continuing personnel safety hazard.

While a PdM program can identify most equipment faults at such an early stage that they never result in equipment failure, the program often does not include discovering the underlying reason for the faults. For example, a bearing may fail repeatedly because of excessive bearing loads caused by an underlying misalignment problem. PdM would most likely predict a bearing failure and thus allow the bearing to be replaced before it fails, but if no one recognizes the misalignment and eliminates it, conditions causing the failure will remain and failures will recur and continue to require unnecessary corrective work and downtime.

RCFA proactively seeks the fundamental causes that lead to facility and equipment failure. Its goals are to:

- Find the cause of a problem quickly, efficiently, and economically.
- Correct the *cause* of the problem, not just its effect.
- Provide information that can help prevent the problem from recurring.
- Instill a mentality of "fix forever."

13.3.4.5 Reliability engineering. In combination with other proactive techniques, reliability engineering involves the redesign, modification, or improvement of components or their replacement by superior components. Sometimes a complete redesign of the component is required. In other cases, upgrading the type of component metal or adding a sealant is all that is required. Progressive maintenance organizations have a reliability engineer dedicated to this effort.

13.3.4.6 Rebuild certification/verification. When new or rebuilt equipment is installed, it is essential to verify that it is operating properly. To avoid unsatisfactory operation and early failure, the equipment should be tested against rigid certification and verification standards.

13.4 Predictive Maintenance (PdM) Technologies

This section describes each of the primary PdM technologies in terms of its purpose, techniques, application, effect, equipment required, operators, training available, and cost.

13.4.1 Vibration Analysis

a. *Purpose*

   (1) Vibration analysis is used to detect, identify, and isolate specific component degradation and its causes prior to the component's failure or damage. Vibration monitoring helps to determine the condition of rotating equipment, a system's structural stability, and the sources of airborne noise.

   (2) When equipment is operating properly, its vibration baseline is established by taking operational vibration measurements. Subsequent vibration readings can then be compared to the baseline, and any components causing deviant readings, the magnitude of any problem, and the rate of component deterioration can be determined.

b. *Techniques*

   (1) Frequency and Time Domain Measurement.
   (2) Shock Pulse Analysis.
   (3) Torsional Vibration Monitoring.

c. *Application*

   (1) All rotating and reciprocating equipment; i.e., motors, pumps, turbines, compressors, engines and their bearings, shafts, gears, pulleys, blowers, belts, couplings, etc.

   (2) Induction motors, using frequency analysis to diagnose for broken rotor bars, cracked end rings, high resistance joints, winding faults, casting porosity, and air-gap eccentricities.

   (3) Structural support resonance testing, equipment balancing, faulty steam trap detection, and airborne noise measurements.

d. *Effect*

   (1) Detects equipment component wear, imbalance, misalignment, mechanical looseness, bearing damage, belt flaws, sheave and pulley flaws, gear damage, flow turbulence, cavitation, structural resonance, and fatigue. Vibration analysis can provide several weeks or months warning of impending failure.

   (2) When measurements of both amplitude and frequency are available, diagnostic methods (spectrum analysis) are used to

determine both the magnitude of any problem and its probable cause.

e. *Equipment Required.* Vibration analysis systems include microprocessor data collectors, vibration transducers, equipment-mounted sound discs, and a host personal computer with software for analyzing trends, establishing alert and alarm points, and assisting in diagnostics.

f. *Operators*
   (1) Requires personnel who have the ability to understand the basics of vibration theory and who have a basic knowledge of machinery and failure modes.
   (2) Manning requirements are site specific.

g. *Training Available*
   (1) Training is provided by equipment vendors and the Vibration Institute.
   (2) The Vibration Institute has published certification guidelines for vibration analysis. Passing a written examination is required for certification.

h. *Cost (1993 dollars).* $20,000 to $70,000 for system, software, and primary training. The high-priced systems are multichannel, installed systems.

13.4.2 Thermography

a. *Purpose*
   (1) Temperature monitoring, using both noncontact- and contact-type devices, detects temperature variances in machines, electrical systems, heat transfer surfaces, and structures and determines the relative severity of those temperature variances. Severe temperature changes often precede equipment failure.
   (2) Temperature monitoring, infrared thermography in particular, is a reliable technique for finding the moisture-induced temperature effects that characterize roof leaks and for determining the thermal efficiency of heat exchangers, boilers, building envelopes, etc.
   (3) Deep-probe temperature analysis can detect buried pipe energy loss and leakage by examining the temperature of the surrounding soil. This technique can be used to quantify energy losses and their cost.
   (4) Temperature monitoring can be used as a damage control tool to locate mishaps such as fires and leaks.

(5) Coupled with other technologies such as vibration and oil analysis, temperature monitoring aids in identifying the equipment condition and the root cause of failures.

b. *Techniques*

(1) Infrared thermography (noncontact).

(2) Contact devices (thermometers, resistance temperature detectors, thermocouplers, decals, crayons).

(3) Deep probe temperature analysis.

c. *Application.* Heat exchangers, electrical distribution and control systems, roofing, building envelopes, direct-buried pipes carrying steam or hot water, bearings, conveyors, piping, valves, steam systems, air handlers, and boiler insulation, casing, and tubes.

d. *Effect*

(1) Temperature monitoring techniques are used to locate hot spots due to loose, corroded, or dirty connections; friction; damaged or missing insulation; thermal cavities; leaks; and blockages.

(2) Infrared thermography localizes the moisture indicative of roof leaks and is used in energy conservation programs for locating sources of heating and air-conditioning losses through building envelopes.

(3) Deep probes for measuring soil temperatures near buried pipes detect insulation system failures and leaks. With knowledge of soil properties, these losses can then be estimated. This technique requires knowledge of piping locations.

(4) Noncontact heat measurement can be done from a distance and will accurately measure temperatures on items that are hard to reach, such as power lines or inaccessible equipment.

e. *Equipment Required*

(1) Equipment ranges from simple, contact devices such as thermometers and crayons to full color imaging, computer-based systems that can store, recall, and print thermal images.

(2) The deep-probe temperature technique requires temperature probes, analysis software, and equipment to determine the location of piping systems.

f. *Operators*

(1) Operators and mechanics can perform temperature measurements and analysis using contact-type devices with minimal training on how and where to take the temperature readings.

(2) Because thermographic images are complex and difficult to measure and analyze, training is required to obtain and inter-

pret accurate and repeatable thermal data and to interpret the data. With adequate training (Level I and Level II) and certification, this technique can be performed by electrical/mechanical technicians and/or engineers.

(3) Maintenance personnel can apply deep-probe temperature monitoring after being trained, although this service is often contracted.

g. *Training Available*

(1) Training is available through infrared imaging system manufacturers and vendors.

(2) The American Society of Nondestructive Testing (ASNT) has established guidelines for nondestructive testing (NDT) thermographer certification.[2] These guidelines, intended for use in nondestructive testing, may be useful for thermography in PdM if appropriately adjusted. General background, work experience, thermographic experience, and thermographic training are all considerations for certification. The thermographer's employer normally pays for training and for making the decision to certify the thermographer. The trainer provides course completion certificates and recommends certification criteria, but the ultimate certification decision is the employer's.

h. *Cost (1993 dollars)*

(1) Noncontact infrared thermometers cost less than $1,000. Full color microprocessor imaging systems with data storage and print capability range from about $25,000 to $70,000.

(2) Average thermographic system rental is approximately $1,500 per week.

(3) Thermographic contractor services cost approximately $1,000 per day. Contract services for deep-probe temperature analysis cost from $1,500 to $2,000 per day, with $5,000 to $6,000 for the first day.

(4) Operator training costs approximately $1,250 per week.

13.4.3 Passive (airborne) ultrasonics

a. *Purpose.* Passive ultrasonic techniques measure the sound or vibration resulting from cavitation, flow turbulence, or influx (in the case of vacuum systems) or escape of gas or fluid.

b. *Techniques*

(1) Acoustic detectors measure frequencies within the range of human hearing.

(2) Airborne ultrasonics measure frequencies above the range of human hearing.

c. *Application.* Ultrasonics are used for piping and process systems, compressed gas and vacuum systems, boiler and heat exchanger tubes, steam traps, refrigeration systems, electrical switchgear, and rotating machinery.

d. *Effect*
  (1) Leak detection techniques are used to detect gas, fluid and vacuum leaks, locate areas of turbulent or restricted flow, and measure corrosion and erosion in piping and vessels.
  (2) In addition to detecting leaks, ultrasonic technology can also be used to detect electrical switchgear malfunctions, gear noise, faulty rolling element bearings, and other harmful friction in plant equipment. Ultrasonic frequencies range between 20 kHz and 100 kHz.

e. *Equipment Required*
  (1) Ultrasonic monitoring scanner for airborne sound or ultrasonic detector for contact mode through metal rod.
  (2) Vibration monitoring equipment (see "vibration analysis," part e above).

f. *Operators.* Maintenance technicians and engineers.

g. *Training Available.* Minimal training required.

h. *Cost (1993 dollars).* Scanners and accessories range from $1,000 to about $8,000.

13.4.4 Lubricant and wear particle analysis

13.4.4.1 Lubricant analysis

a. *Purpose*
  (1) Lubricant analysis is used to determine the condition of oil, fuel, or grease by testing for viscosity; particle, fuel, and water contaminants; acidity/alkalinity (pH); breakdown of additives; and oxidation.
  (2) Coupled with other technologies such as vibration and temperature measurements, lubricant analysis aids in identifying the equipment condition and the root cause of failures.

b. *Techniques*
  (1) Physical analysis.
  (2) Infrared spectrography.

c. *Application*

    (1) Engines, compressors, turbines, transmissions, gear boxes, sumps, transformers, and storage tanks.

    (2) Receipt inspection of incoming lubricating and fuel oil and grease supplies for condition, viscosity, and contamination.

    (3) Spot-checking new, rebuilt, or repaired equipment as part of the acceptance process.

d. *Effect*

    (1) Monitoring the condition of lubricants determines whether they are suitable for continued use or should be changed.

    (2) Analysis of both the quantity and type of metal particle contamination in a sample can identify a specific component experiencing wear.

    (3) Maintaining exceedingly clean lubricating fluids extends the life of bearings and other components. Maintaining proper acidity/alkalinity and composition of additives controls lubricant corrosiveness.

    (4) Lubricant monitoring protects equipment warranties which would not otherwise be honored by a manufacturer based on the manufacturer's claim that the equipment operated with contaminated oil.

    (5) Use of oil analysis as part of the quality control equipment acceptance test indicates if all lubrication or hydraulic systems are properly installed, cleaned, flushed, and filled with the appropriate lubricant.

    (6) Long-term trending of oil analysis data can identify poor maintenance or repair practices that contribute to high maintenance costs, downtime, and reduced machine life.

e. *Equipment Required.* Extensive, costly laboratory equipment is required for detailed analysis; thus, in-plant analysis is usually not justified. However, portable, standalone analyzers are available for prescreening samples on site to determine if a more thorough or specific analysis is warranted.

f. *Operators.* One person should be trained in lubricant analysis and should in turn train equipment operators and maintenance personnel in proper sample-taking techniques.

g. *Training Available.* Training is available from equipment vendors and independent laboratories that perform oil analysis.

h. *Cost (1993 dollars)*

    (1) Free to approximately $150 per sample, depending on the type of analysis desired, disposal fees, and the level of service provided by the vendor.

(2) $13,000 to $20,000 for equipment (on-site, standalone analyzer for prescreening) and training.

13.4.4.2 Wear particle analysis

a. *Purpose*
   (1) Wear particle analysis is an analysis technique that determines the condition of a machine or machine components by examining particles contained in a lubricating oil sample. Wear particles are separated and subject to ferrographic and microscopic analysis.
   (2) Coupled with other technologies such as vibration and temperature measurements, wear particle analysis aids in identifying the equipment condition and the root cause of failures.

b. *Techniques*
   (1) Direct-reading ferrography.
   (2) Analytical ferrography.
   (3) Magnetic chip/particle counters.
   (4) Graded filtration/micropatch.

c. *Application*. Engines, compressors, turbines, transmissions, and hydraulic system gear boxes.

d. *Effect*
   (1) Analysis of both the quantity and type of metal particle contamination in a sample can identify a specific component experiencing wear, the magnitude of the wear, and the type of wear being experienced.
   (2) Particle count indicates the effectiveness of filtration and measures overall system cleanliness.
   (3) Long-term trending of oil analysis data can identify poor maintenance or repair practices that contribute to high maintenance costs, downtime, and reduced machine life.

e. *Equipment Required*. Extensive, costly laboratory equipment is required for detailed analysis; for this reason, in-plant analysis is rarely justified. However, portable, direct-reading contamination monitors and analyzers are available for prescreening samples on site to determine if a more thorough or specific analysis is warranted.

f. *Operators*. One person should be trained in wear particle analysis. That person should then train equipment operators and maintenance personnel on sampling techniques.

g. *Training Available*. Training is available from equipment vendors and independent laboratories that perform oil analysis.

h. *Cost (1993)*

    (1) Free to approximately $150 per sample, depending on the type of analysis desired, disposal fees, and the level of service provided by the vendor.

    (2) $10,000 to $20,000 for equipment (on-site, standalone analyzer for prescreening) and tribology training.

13.4.5 Electrical condition monitoring

a. *Purpose*

    (1) Electrical testing is used to measure the complex impedance and insulation resistance of electrical conductors, starters, and motors. It detects faults such as broken windings, broken motor rotor bars, voltage imbalances, and cable faults.

    (2) Current, voltage, and power factor are monitored to determine power quality and to form a basis for reducing energy costs.

    (3) Coupled with other technologies such as temperature monitoring and ultrasound, electrical testing aids in identifying the equipment condition and the root cause of failures.

b. *Techniques*

    (1) Megohmmeter testing.

    (2) High potential (high-pot) testing.

    (3) Surge testing.

    (4) Conductor complex impedance.

    (5) Time Domain Reflectometry (TDR).

    (6) Motor current spectrum analysis.

    (7) Radio Frequency (RF) monitoring.

    (8) Power factor and harmonic distortion.

    (9) Starting current and time.

    (10) Motor Circuit Analysis (MCA)™

(*Note.* High-pot and surge testing should be performed with caution. The high voltage applied during these tests may induce premature failure of the units being tested. For that reason these tests normally are performed only for acceptance testing, not for condition monitoring.)

c. *Application.* Electrical distribution and control systems, motor controllers, cabling, transformers, motors, generators, and circuit breakers.

d. *Effect*

    (1) Electrical testing is used to monitor the condition or test the remaining life expectancy of electrical insulation; motor and

generator components such as windings, rotor bars, and connections; and conductor integrity.

(2) Electrical testing is used as a quality control tool during acceptance tests of electrical systems or plant property equipment such as new or rewound motors.

(3) During equipment start-up, electrical testing is used to check proper motor starting sequencing and power consumption.

(4) Electrical testing is used to monitor power factor so that improvements can be made to reduce electrical power consumption.

e. *Equipment Required.* A full electrical testing program includes the following equipment: multimeters/volt-ohmmeters, current clamps, time domain reflectrometers, motor current spectrum analysis software, and integrated motor circuit analysis testers.

f. *Operators.* Electricians, electrical technicians, and engineers should be trained in electrical PdM techniques such as motor current signature analysis, motor circuit analysis, complex phase impedance, and insulation resistance readings and analysis.

g. *Training Available.* Equipment manufacturers and RCM consultants specializing in electrical testing techniques provide classroom training and seminars to teach these techniques.

h. *Cost (1993 dollars)*

(1) Equipment costs vary from $20 for a simple multimeter to approximately $40,000 for integrated motor-current analysis (MCA) testers. A full inventory of electrical testing equipment should cost from about $30,000 to $50,000.

(2) Training averages between $750 and $1,000 per week per person trained.

13.5   RCM Program Benefits

13.5.1   Reliability

RCM places great emphasis on improving equipment reliability, principally through the feedback of maintenance experience and equipment condition data to facility planners, designers, maintenance managers, craftsmen, and manufacturers. This information is instrumental in continually upgrading the specifications for equipment to provide increased reliability.

The increased reliability that comes from RCM leads to fewer equipment failures and therefore greater availability for mission support. This increase in reliability lowers maintenance costs.

13.5.2 Cost

Due to the initial investment required to obtain the technological tools, training, and equipment condition baselines, a new RCM Program typically results in a short-term increase in maintenance costs. This increase is relatively short lived. As shown in Fig. 13.5, the cost of reactive maintenance decreases as failures are prevented and preventive maintenance tasks are replaced by condition monitoring. The net effect is a reduction of both reactive maintenance and a reduction in total maintenance cost. Often energy savings are realized from the use of PdM techniques.

13.5.3 Scheduling

The ability of a condition monitoring program to forecast maintenance provides time for planning, obtaining replacement parts, and arranging environmental and operating conditions before the maintenance is done. PdM reduces the unnecessary maintenance performed by a time-scheduled maintenance program which tends to be driven by the minimum "safe" intervals between maintenance tasks.

A principal advantage of RCM is that it obtains the maximum use from equipment. With RCM, equipment replacement is based upon

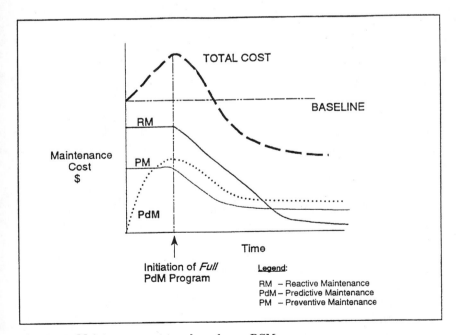

Figure 13.5 Maintenance cost trends under an RCM program.

equipment condition, not on the calendar. This condition-based approach to maintenance thereby extends the life of the facility and its equipment.

13.5.4 Efficiency/productivity

Safety is the primary concern of RCM. The second most important concern is cost effectiveness. Cost effectiveness takes into consideration the priority or mission criticality and then matches a level of cost appropriate to that priority. The flexibility of the RCM approach to maintenance ensures that the proper type of maintenance is performed on equipment when it is needed. However, maintenance which is not cost effective is identified and not performed.

RCM's multifaceted approach promotes the most efficient use of resources. The equipment is maintained as required by its characteristics and the consequences of its failure.

13.6 Impact of RCM on the Facilities Life Cycle

To achieve maximum effectiveness, RCM must be a consideration throughout the facilities life cycle and integrated into the life cycle at an early stage. The life cycle is typically divided into four phases:

- Planning
- Design
- Construction
- Maintenance and operations. (M&O)

The facilities life cycle is often divided into two broad stages: (1) acquisition (planning, design, and construction) and (2) operations. RCM affects all phases of the acquisition and operations stages to some degree, as shown in Table 13.1.

Decisions made early in the acquisition cycle profoundly affect the life-cycle cost of a facility. Even though expenditures for plant and equipment may occur later during the acquisition process, their cost is committed at an early stage. As shown conceptually in Fig. 13.6, planning (including conceptual design) fixes two-thirds of the facility's overall life-cycle costs. The subsequent design phases determine an additional 29 percent of the life-cycle cost, leaving only about 5 percent of the life-cycle cost that can be impacted by construction and maintenance and operations.

Thus, the decision to include a facility in an RCM program, including PdM and condition monitoring, which will have a major impact on

TABLE 13.1   RCM Implications in Facilities Life Cycle

Life-Cycle Phase	Acquisition Implications	Operations Implications

Planning	Requirements
	Contract Strategy
	RCM Implementation Policy
	Funding Estimates
	   Construction
	   Equipment
	   Labor
	   Training
	   Operations
	A&E Scope of Work
		Requirements Development
		   Modifications
		   Alterations
		   Upgrades
		A&E Scope of Work
		Funding Estimates

Design	A&E Selection
	Drawings
	Specifications
	Acceptance Testing
	   Requirements
		A&E Selection
		Drawings
		Specifications
		Acceptance Testing
		   Requirements

Construction	Contractor Selection Construction Acceptance Testing	Contractor Selection Construction Acceptance Testing

Maintenance and Operations	Not Applicable	RCM Operations
		   Training
		   Certification

D042

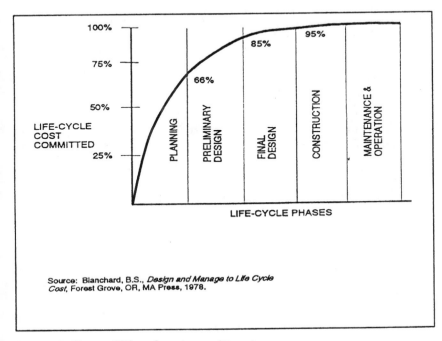

Source: Blanchard, B.S., *Design and Manage to Life Cycle Cost*, Forest Grove, OR, MA Press, 1978.

Figure 13.6   Stages of life-cycle cost committment.

its life-cycle cost, is best made during the planning phase. As RCM decisions are made later in the life cycle, it becomes more difficult to achieve the maximum possible benefit from the RCM program.

Even though maintenance is a relatively small portion of the overall life-cycle cost, being typically 3 to 5 percent of a facility's operating cost, RCM is still capable of introducing significant savings during the M&O phase of the facility's life. Savings of 30 to 50 percent in the annual maintenance budget are often obtained through the introduction of a balanced RCM program.

13.7   RCM Requirements During Facilities Acquisition

This chapter discusses RCM requirements during the acquisition stage of the facilities life cycle; i.e., during facilities planning, design, and construction.

13.7.1   Planning

Early in the planning of a new facility, consideration must be given to the extent that RCM will be used to maintain the facility and its equipment. The fundamental determination is the amount of built-in condition monitoring, data transfer, and sensor connections to be used. It is more economical to install this monitoring equipment and connecting cabling during construction than later. Planning, designing, and building in the condition monitoring capability ensures that it will be available for the units to be monitored. Continuously monitored equipment tied into performance analyzers permits controlling functions and monitoring degradation. In the future, online monitors will become increasingly capable and important. Installed systems also reduce manpower requirements as compared with collecting sensor data with a portable data collector. However, for many uses, portable condition monitoring equipment does provide the advantages of lower cost and flexibility of application as compared with post-acquisition installed systems.

13.7.2   Design

The following should be considered during the design phase:

a. *Maintainability and Ease of Monitoring.* In recent years great strides have been made in designing new equipment to ensure a high level of reliability. By extending this approach to the maintainability and ease of monitoring of equipment during design, one further enhances reliability and ensures improved maintainability for the life of the equipment.

(1) Maintainability factors that are within the control of the designer are:

    (a) *Access.* Equipment, its components, and facilities should be accessible for maintenance.

    (b) *Material.* Choose materials for durability, ease of maintenance, availability, and value (optimal cost vs. special requirements tradeoffs).

    (c) *Standardization.* Minimize use of special or one-of-a-kind materials, fittings, or fixtures. Maximize commonality of equipment component parts. Choose standard equipment for multiple uses where feasible.

    (d) *Quantitative Maintenance Goals.* Use quantitative measures of maintenance (i.e., mean-time-between-maintenance (MTBM), maintenance downtime (MDT), etc.) to set goals for maintainability which will influence design.

(2) Ease of monitoring factors within the control of the designer are:

    (a) *Access.* Provide clear access to collect equipment condition data with portable data loggers or fluid sample bottles.

    (b) *Online Data Collection.* Installed data collection sensors and links (wire, fiber optics, or radio frequency (RF) links are possible) may be justified for high-priority, high-cost equipment or inaccessible equipment.

    (c) *Management Indicators.* Management (or performance) indicators are discussed in paragraph 13.11. For an RCM program, the management indicators and the analysis method are incorporated into the system design. Often the performance parameters monitored for equipment or system control may be used to monitor equipment condition.

    (d) *Performance Measures.* RCM performance measures such as operating time or equipment loading are directly equipment related. The data to be used and the collection method are incorporated into the system design.

b. *Technology Review*

(1) Conducting a PdM technology review at an early stage of the design is necessary to establish which technologies are to be used in the RCM program.

(2) A continuing review of other maintenance programs for new and emergent predictive technologies assists in keeping the program current by incorporating the latest technological developments.

c. *Feedback.* Update or improve the design based on feedback, prior experience, and lessons learned.

d. *Scope of Work.* Clearly establish in the A&E contract scope of work that RCM maintainability and ease of monitoring requirements must be met.

e. *Qualification.* Contractor qualifications should provide for a demonstration of familiarity and/or understanding of the PdM technologies planned for the RCM program.

f. *Specifications and Drawings*

(1) The designer has a major impact on the life-cycle cost of the equipment by incorporating the RCM lessons learned from the history of similar or identical equipment.

(2) The design specifications should address the use of condition-based maintenance in lieu of interval-based maintenance and its effect on manufacturer's warranties.

(3) Specifications for equipment procurement should incorporate maintainability and ease of monitoring requirements. Maintainability requirements should be specified in terms of an appropriate combination of measurements. The attainment of these requirements is an important aspect of equipment performance.

(4) Construction contracts should specify the type of acceptance testing to be performed on building materials and equipment prior to shipment to the building site. The designer also should specify the PdM technologies to be used and the related acceptance criteria as part of the post-installation acceptance testing for all collateral equipment and structures and for determining if a building is ready for occupancy.

13.7.3 Construction

During the construction phase, one major concern is to monitor the progress and quality of construction to ensure that the planning and design work from earlier phases is effectively implemented. This includes monitoring for conformance to specifications, drawings, bills of material, and installation procedures. The following important steps should be taken during construction:

a. The training of maintenance personnel in the use of PdM technologies and equipment should start during construction.

b. It is during this phase the RCM analyses are done, the maintenance program tasks should be chosen, and maintenance procedures and instructions should be written.

c. Operations and maintenance personnel should take advantage of the construction phase to become familiar with the details of construction which will no longer be easily accessible after the facility is completed.

d. The construction inspectors should assure that the construction contractor properly installs, aligns, and checks the equipment in accordance with the contract specifications and the equipment manufacturer's recommendations. Incorrect installation can void the manufacturer's warranty and cause early equipment failure.

e. When equipment is specified by performance rather than proprietary name (this is preferable), the owner should require that the contractor submit catalog descriptions of the equipment he intends to provide for approval by the A&E/owner. The equipment should be approved by the owner (including the RCM technologists) to ensure that it satisfies the requirements of the contract specifications. If the contract specifications require that the contractor establish a vibration and thermographic baseline for the equipment, the owner should ensure that this is done and that the baseline information is properly documented and turned over to the owner with the equipment technical manuals and other information required from the contractor.

13.7.4   Maintenance and operations (M&O)

How the facility and its equipment will be operated and maintained must be considered during the planning, design, and construction phases. During these phases, maintenance and operations needs are best served by carefully and realistically identifying and defining the PdM requirements. Although the performance of maintenance and operations tasks occurs during the operations stage of the life cycle of the equipment or facility, some preparatory activities may be carried out during the latter part of the acquisition stage. These activities include personnel selection, PdM training, procedure preparation, review of specifications, and collection of nameplate data.

13.8   RCM Considerations During Facilities Operation

This chapter discusses RCM requirements during the operational stage of the facilities life cycle (maintenance and operations).

13.8.1   Maintenance data

The need to maintain an adequate level of funding is a continuing requirement during the life cycle of a facility. As a result, there is a recurring need for RCM program data as an input to an Annual Work Plan (AWP) and other budgetary planning documents. To support these periodic requirements, the collection of data to provide justification should not be left to chance or to the last minute.

It is important that RCM maintenance costs, costs avoided, and program savings be documented as they occur. A good Computerized Maintenance Management System (CMMS) has the capability to track this information. Direct maintenance program savings are usually apparent, but maintenance personnel must become attuned to recognizing *and documenting* the less obvious "costs avoided."

One use of maintenance data is in tracking the "maintenance burden"; i.e., the man-hours and/or labor costs expended on all types of maintenance on a particular equipment. By identifying the equipment on which maintenance effort is spent and categorizing the type of maintenance (i.e., RM, PM or PdM), the maintenance burden may be allocated to end use. With labeled maintenance man-hour expenditures, both the high burden items and the distribution of the burden within the facility can be tracked over time. The use of Pareto analysis, or some other method of prioritization of maintenance resource assignment, will permit easy identification of areas which demand attention from the standpoint of maintenance effectiveness or the need for modification to improve reliability. As the intensive PdM efforts take hold, the effects on maintenance burden can be analyzed to determine their costs and relative values, especially as increased availability and improved quality impacts each activity.

The task of maintaining and updating drawings and specifications to keep facility documentation abreast of the modifications and alterations should not be neglected. The effectiveness of the facility and the RCM program will reflect the degree to which this updating is accomplished.

13.8.2   Maintenance feedback

During the life of the facility and its equipment, a continuing design effort is necessary for modifications and alterations as operators and maintenance personnel gain experience with the facility. Sometimes these changes result from an expanded or modified facility mission. As the facility matures, new methods and tools become available to accomplish the mission and tasks in a more efficient or cost-effective manner. These needs for changes and modifications must be documented and returned to the designers as feedback.

An important function of feedback during the life cycle of any facility is to improve and optimize the performance of the equipment. Design plays a crucial role in this improvement process. As shown in Fig. 13.7, design improvements from one generation of equipment to the next

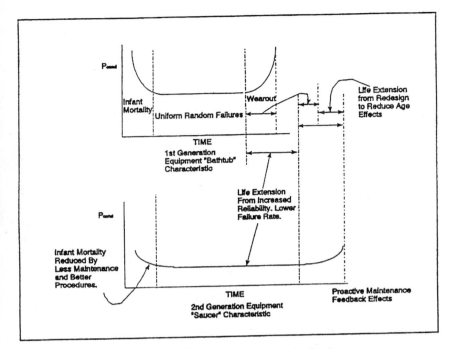

Figure 13.7   Design improvement through maintenance feedback.

can have a significant effect on reliability. The feedback which the designer integrates into each successive generation yields lower overall failure rates and extended lives. These improved characteristics reduce maintenance costs as well.

13.8.3   Maintenance and operation (M&O)

The following are RCM considerations during the M&O phase:

a. *Labor Force.* The RCM program requires a labor force with the special skills necessary to accomplish the program's reactive, preventive, and PdM tasks successfully. The quality of the maintenance accomplished is reflected in the equipment post-maintenance *infant mortality* rate. The quality of maintenance is influenced by the skill of the maintenance technicians, their workmanship, the quality of supporting documentation and procedures, and the technologies in use.

b. *System Experts.* The responsibility for condition monitoring, predictive analysis, and maintenance planning should be assigned to someone not subject to random or frequent reassignment. To maintain continuity of effort, gain experience in a single individual or group, and ensure a consistent technical approach, a single person should be assigned maintenance oversight responsibility for a given equipment or group of similar equipment.

The involvement of a single person in all maintenance actions for the equipment or system(s) for which he or she is responsible is the heart of the *system expert* concept and is an excellent way to provide continuity of experience and expertise with regard to the equipment and systems. The system expert is the person responsible for monitoring and analyzing the maintenance data for an assigned equipment or system. The system expert should receive all the related data, work requests, and test documents for review, analysis, and filing. He or she is designated to maintain expertise and provide continuing oversight, analysis, and continuity for designated facility systems or equipment. The *system expert* concept results in a significant contribution to better overall management of facility maintenance and provides the critical elements of corporate memory and analysis.

The system expert is not intended to replace or supplant the technical expertise normally found within the design organization of the original equipment manufacturer (OEM) or the facility. The system expert is expected to be a member of the facility maintenance team who is conversant with the maintenance history of his or her assigned equipment and systems. The system expert is also involved with integrating maintenance support from facility resources outside of the normal maintenance organization.

c. *Training*. Training plays a major role in reaching and maintaining the required RCM workforce skill level. The training is both technology/equipment specific and of a more general nature. Management and supervisory personnel benefit from training which presents an overview of the RCM process, its goals, and methods. Technician training includes the training on specific equipment and technologies, RCM analysis, and PdM methods.

Table 13.2 summarizes the overall maintenance training requirements:

d. *Equipment*. During the operation of the facility and its equipment, a continuing improvement program should be in place. The need for equipment modifications and alterations is documented by feedback from the operators and maintenance personnel.

An important function of feedback during the life of the facility is to improve the performance of the facility and its equipment. The M&O-to-designer linkage is crucial in this improvement process. As shown in Fig. 13.7, design improvements from one generation

TABLE 13.2   Maintenance Training

Position/Title	Maintenance Overview	RCM/PdM	Root Cause Predictive Analysis	Technology/ Equipment
Managers/Supervisors	X	X	X	
System Experts	X	X	X	X
Maintenance Technicians	X	X	X	X
Support Personnel (Logistics)	X			

D042

of equipment to the next have a significant effect on operability, maintainability and reliability.

e. *Maintenance History.* An important function of the CMMS is to collect, organize, display, and disseminate maintenance data. The CMMS provides the facility's maintenance history, which is critical to the success of the RCM program because:

(1) It contains information to establish whether an age-reliability relationship (wear-out) exists for equipment, and if so, what is the critical failure-onset age.

(2) It contains the data analyzed to trend and forecast equipment failure.

(3) It forms the basis for system expert analysis of long-term equipment and system performance trends.

(4) It provides the statistics used to determine the failure rates, which influence spare part stock levels.

(5) It contains test results, performance data, and feedback information used to improve equipment performance.

f. *Procedural Documentation.* For a maintenance organization to be truly world class, it must perform reactive maintenance expeditiously and correctly. Complex repair tasks should be planned in advance. The use of maintenance planners to generate maintenance work orders, which include detailed procedures, repair parts and tools, maintenance personnel training, and specific test procedures, is part of a good work order system for a complicated or critical repair. This detailed work planning has been proven to be vital by quality maintenance organizations. Work preparations ensure smooth repair with few surprises and greatly reduce the incidence of "infant mortality" which often accompanies the difficult repairs. The lessons learned from each repair are included in the equipment history for future reference. As a maintenance organization gains experience, it builds a file of work orders that may be reused repeatedly.

The use of standard procedures in PdM ensures that data is collected in a consistent and comparable manner without being dependent on the experience of the collecting technician. Improvements in data collection techniques can be applied to all similar collection efforts by modifying the standard procedures.

13.9   Criteria for RCM Priorities Based on Mission Criticality

Table 13.3 suggests the criteria to be used in determining RCM priorities.

13.9.1   Reactive maintenance criteria

Table 13.4 suggests the criteria to be used in determining the priority for repairing or replacing the failed equipment in the reactive maintenance program.

13.9.2   Preventive maintenance criteria

Preventive maintenance criteria reflect the age-reliability characteristics of the equipment based upon the equipment history. They are not necessarily related to mission criticality. The selection of maintenance tasks is made using the process shown in Fig. 13.2. The selection process guides the determination of the type of task which will be done, but is less helpful in establishing task frequency or periodicity.

TABLE 13.3   Maintenance Priority Levels

Priority		Application
Number	Description	
1	Emergency	Safety; mission impact.
2	Urgent	Continuous operation of facility at risk.
3	Priority	Mission support/project deadlines.
4	Routine	Accomplish on "first come, first served" basis.
5	Discretionary	Desirable, but not essential.
6	Deferred	Needed but unable to accomplish until more resources available.

D042

TABLE 13.4   Reactive Maintenance Priorities

Priority		Criteria Based on Consequences of Equipment/System Failure
Number	Description	
1	Emergency	Safety of life or property threatened. Immediate serious impact on mission.
2	Urgent	Continuous facility operation threatened. Impending serious impact on mission.
3	Priority	Degrades quality of mission support. Significant and adverse effect on project.
4	Routine	Redundancy available. Impact on mission insignificant.
5	Discretionary	Impact on mission negligible. Resources available.
6	Deferred	Impact on mission negligible. Resources unavailable.

D042

13.9.2.1   Determining PM task and monitoring periodicity. This section offers suggestions for selecting equipment monitoring periodicities.

a. *PM Tasks.* Although numerous ways have been proposed for determining the correct periodicity of preventive maintenance tasks, none are valid unless the in-service age-reliability characteristics of the system or equipment affected by the desired task are known. This information is not normally available and must always be collected for new systems and equipment. PdM techniques should be used as an aid in determining equipment condition vs. age.

Careful analysis of similar kinds of hardware in industry has shown that, overall, more than 90 percent of the hardware analyzed showed no adverse age-reliability relationship. This does not mean that individual parts do not wear; they do. It means that the ages at failure are distributed in such a way that there is no value in imposing a preventive maintenance task. In fact, in a large number of cases, imposing an arbitrary preventive task increases the average failure rate through "infant mortality."

The Mean Time Between Failures (MTBF) is often used as the initial basis for determining PM interval. This approach is incorrect in that it does not provide any information about the effect of increasing age on reliability. It provides only the *average* age at which failure occurs, not the most likely age. In many cases a Weibull distribution, as used by the bearing industry to specify bearing life, will provide more accurate information on the distribution of failures.

The best thing that can be done if good information on the effect of age on reliability is lacking, is to monitor the equipment condition.

b. *Equipment Monitoring*. The factors above still apply, but with several important modifications. The aim in monitoring equipment is to (1) determine equipment condition and (2) develop a trend with which to forecast future equipment condition. For trending purposes, a minimum of three monitoring points before failure may reasonably be expected are recommended. Using three data points provides two to establish the trend and the third to provide confirmation. The following techniques are recommended for setting initial periodicity:

(1) *Anticipating Failure from Experience*. For some equipment, failure history and personal experience provide an intuitive feel as to when to expect equipment failure. In these cases, failure is time related. The monitoring periodicity should be selected such that there are at least three monitoring intervals before the anticipated onset of failures. It is prudent in most cases to shorten the monitoring interval as the wear-out age is approached.

(2) *Failure Distribution Statistics*. In using statistics to determine the basis for selecting periodicities, the distribution and probability of failure should be known. Weibull distributions can provide information on the probability of an equipment exceeding some life. For example, bearings are normally specified by their B10 life; i.e., the number of revolutions that will be exceeded by 90 percent of the bearings. Depending on the criticality of the equipment, an initial periodicity is recommended which allows a *minimum* of three monitoring samples prior to the B10 life or, in less severe cases, prior to the MTBF point. In more severe cases a B2 life can be calculated, and the monitoring interval can be adjusted accordingly.

(3) *Lack of Information or Conservative Approach*. The most common practice in industry is to monitor the equipment biweekly or monthly due to a lack of information and poor monitoring techniques. This often results in excessive monitoring. In these cases, significant increases in the monitoring interval may be made without adverse impact on equipment reliability.

When indications of impending failure become apparent through trending or other predictive analysis methods, the monitoring interval should be reduced and additional analysis should be performed in order to gain more detailed information on the condition of the equipment.

13.9.2.2 Alerts and alarms. Alerts and alarms are set to meet specific user requirements. Common methods used to determine alert and alarm values are:

a. Set an arbitrary value for change from a baseline value. In this situation an increase in the reading by a predetermined amount over the initial reading is used to determine maintenance requirements. This approach was widely used by the U.S. Navy submarine force until the middle 1980s at which time it was abandoned in favor of a more statistical approach as described below.

b. Set an alert value to indicate a statistically significant deviation (usually 2σ) from the mean as a warning or alert level. An alarm value is established at 3σ. This approach normally allows sufficient time between the alert and alarm levels and failure to schedule repairs.

c. Refine the alert and alarm values by performing failure analysis of the parts and correlating the as-found condition to the PdM data. For these, it is important to document trends, the values at which failures occur, and the operating environment for future reference. Changing the alert and alarm values are the means used to fine tune the predictive maintenance process and should be based on the best information available.

13.9.2.3 Failure-finding task frequency. To establish the periodicity of a failure-finding task, one simplifies the calculation by making some assumptions about the failure distribution of the affected hidden or infrequent function.

The following is a suggested trial solution:

$$T = -M \log_e (2A - 1)$$

where: T = periodicity
       M = no-task MTBF
       A = the desired/expected availability fraction

Example: No-task MTBF   = 50 hr (estimate)
           Availability   = 0.95
           Task Periodicity (T) = 5 hr (approximately)

Note. This solution assumes exponential survival and no adverse impact on tests of future reliability. If testing actually degrades reliability, then the testing interval should be extended.

13.9.3 PdM criteria

13.9.3.1 Baselines. Baseline data is that condition monitoring data representative of an equipment in a new and/or properly operating condition. The baseline data are the foundation of the predictive trending analysis required to forecast equipment conditions. It is important that this baseline data is established as early as possible in the life of the equipment. The baseline readings and periodic monitoring data should be taken under the same conditions (or as close as can be achieved). These conditions should be recorded. It is only under identical conditions that the relative comparison of data is valid. Significant changes in conditions often affect the data in unquantifiable ways because of unknown or complex relationships. Baseline readings should be reestablished each time equipment undergoes major maintenance.

13.9.3.2 Vibration monitoring. The development of specific vibration criteria for every machine at all possible operating speeds in all applications and in every mounting arrangement is not possible. Vibration amplitude varies with operating speed, load, and mounting arrangement, so developing criteria based solely on amplitude can be misleading. The frequency content of the spectrum is often as important as the amplitude. For example, the presence of a tone associated with an inner race defect on a new bearing is unacceptable regardless of the amplitude. Furthermore, one must consider the difference between the vibration amplitude and frequency content of a reciprocating machine as compared to a centrifugal machine.

The vibration specifications provided in this guide are based on International Standards Organization (ISO), American Petroleum Institute (API), American Gear Manufacturers Association (AGMA), American National Standards Institute (ANSI), MIL-STD-167-1, MIL-STD-740-2,[5–10] and field data acquired on a variety of machinery.

a. *Developing Vibration Criteria.* Specific vibration criteria are provided in this chapter where possible. Where specific criteria are not provided the following procedure is recommended for the guide user for use in developing the vibration criteria:

   (1) Obtain nameplate data.
   (2) Obtain vibration spectra on similar machines. Differences in baseplate stiffness and mass will affect the vibration signature.

(3) Calculate all forcing frequencies, i.e., imbalance, misalignment, bearing defect, impeller and/or vane, electrical, gear, belt, etc.

(4) Construct a mean vibration signature for the similar machines.

(5) Compare this mean vibration signature to the specifications and guidelines provided in this guide.

(6) Note any deviations from the guidelines and determine if the unknown frequencies are system related; e.g., a resonance frequency from piping supports.

(7) Collect vibration data on the new component at the recommended positions.

(8) Compare the vibration spectrum with the mean spectrum determined in step (5) above as well as with the criteria and guidelines provided in this guide.

(9) Any new piece of equipment should have a vibration spectrum which is no worse than a similar unit of equipment which is operating satisfactorily.

b. *Vibration Analysis of New Equipment.* For all large or critical pieces of equipment assembled and run at the factory prior to shipment, a narrowband vibration spectrum should be acquired while the equipment is undergoing this factory performance testing. A baseline or reference spectrum should be retained for comparison with the post-installation vibration check. Equipment failing the vibration criteria should be rejected by the owner prior to shipment.

Vibration tests are recommended under the following situations if the equipment fails the initial test and/or if problems are encountered following installation:

(1) Motor cold and uncoupled.

(2) Motor hot and uncoupled.

(3) Motor and machine coupled, unloaded and cold.

(4) Motor and machine coupled, unloaded and hot.

(5) Motor and machine coupled, loaded and cold.

(6) Motor and machine coupled, loaded and hot.

A significant change in the vibration signature could indicate a problem with thermal distortion and/or bearing overloading due to failure of one of the bearings to float.

c. *Vibration Criteria for Electric Motors*

(1) *General.* All motor vibration spectra should be analyzed at the following forcing frequencies:

(a) One times running speed (1x) for imbalance.

(b) Two times running speed (2x) for misalignment.

(c) Multiples of running speed (Nx) for looseness, resonance, plain bearing defects.

(d) Electric line frequency and harmonics (60 or 120 Hz for AC motors) for stator and rotor problems.

(e) The following is a list of rolling element bearing frequencies:
- outer race defect
- inner race defect
- ball defect (ball spin frequency)
- fundamental train frequency.

(f) Plain or journal bearings indicate faults at harmonics of running speed and at the frequency corresponding to 0.4–0.5 of running speed.

(g) Other sources of vibration in motors are dependent on the number of motor rotor bars and stator slots, the number of cooling fan blades, the number of commutator bars and brushes, and on the firing frequencies for variable speed motors.

(h) Broken rotor bars will often produce sidebands spaced at two times the slip frequency. The presence of broken rotor bars can be confirmed through the use of electrical testing.

(2) *Balance.* The vibration criteria listed in Table 13.5 are for the vibration amplitude at the fundamental rotational frequency or one times running speed (1x). This is a narrowband limit. An overall reading is not acceptable.

(3) *Additional Vibration Criteria.* All testing should be conducted at normal operating speed under full load conditions. Suggested motor vibration criteria are provided in Table 13.6.

d. *Rewound Electric Motors.* Due to the potential of both rotor and/or stator damage incurred during the motor rewinding process (usually resulting from the bake-out of the old insulation and subsequent distortion of the pole pieces) a rewound electrical motor should be checked both electrically and mechanically. The mechanical check consists of post-overhaul vibration measurements at the same location as for new motors. The vibration level at each mea-

TABLE 13.5 Motor Balance Specifications

Motor Speed (RPM)	Maximum Vibration (in/sec, Peak)	Maximum Displacement (mils, Peak-to-Peak)
900	0.02	0.425
1200	0.026	0.425
1800	0.04	0.425
3600	0.04	0.212

TABLE 13.6 Motor Vibration Criteria

Frequency (X RPM) Motor Component	Maximum Amplitude (in/sec Peak)
0.4–0.5	Not detectable
1X	See Motor Balance Specifications
2X	0.02
Harmonics (NX)	Not detectable
Roller Element Bearings	Not detectable
Side Bands	Not detectable
Rotor Bar/Stator Slot	Not detectable
Line Frequency (60 Hz)	Not detectable
2X Line Frequency (120 Hz)	0.02

D042

surement point should not exceed the reference spectrum for that motor by more than 10 percent. In addition, vibration amplitudes associated with electrical faults such as slip, rotor bar, and stator slot should be noted for any deviation from the reference spectrum.

Note. Rewinding a motor will not correct problems associated with thermal distortion of the iron.

e. *General Equipment Vibration Standards*

(1) If rolling element bearings are utilized in either the driver or driven component of a unit of equipment (e.g., a pump/motor combination), no discrete bearing frequencies should be detectable. If a discrete bearing frequency is detected, the equipment should be deemed unacceptable.

(2) For belt-driven equipment, belt rotational frequency and harmonics should be undetectable. If belt rotation and/or harmonics are detectable, the equipment should be deemed unacceptable.

(3) If no specific criteria are available, the ISO 3945 acceptance Class A guidelines[4] should be combined with the motor criteria contained in Table 13.6 and used as the acceptance specifications for procurement and overhaul.

f. *Specific Equipment.* Use the criteria shown in Table 13.7 on boiler feedwater, split case, and progressive cavity pumps:

g. *Belt-Driven Fans.* Use the criteria in Table 13.8 for belt-driven fans:

h. *Vibration Guidelines* (ISO). Table 13.9 is based on International Standards ISO 3945[5] and should be used as a guideline (*not* as an absolute limit) for determining the acceptability of a machine for

TABLE 13.7 Pump Vibration Limits

Frequency Band	Maximum Vibration Amplitude (in/sec Peak)
Overall (10–1000 Hz)	0.06
1X RPM	0.05
2X RPM	0.02
Harmonics	0.01
Bearing Defect	Not detectable

D042

service. The vibration acceptance classes and ISO 3945 machine classes are shown in Tables 13.10 and 13.11, respectively. Note that the ISO amplitude values are *overall measurements* in inches/second RMS while the recommended specifications for electric motors are *narrowband measurements* in inches/second Peak.

13.9.3.3 Lubricant and wear particle analysis. Lubricant analysis monitors the actual condition of the oil. Parameters measured include viscosity, moisture content, flash point, pH (acidity or alkalinity), and the presence of contaminants such as fuel, solids, and water. In addition, the levels of additives in lubricants can be determined. Tracking the acid/alkaline nature of the lubricant permits the identification of an undesirable degree of oxidation, gauging the ability of the lubricant to neutralize contaminants, and aids in the verification of the use of the correct lubricant after a lubricant change. Viscosity (resistance to flow) provides a key to the lubricating qualities of the lubricant. These qualities may be adversely affected by contamination with water, fuel, or solvents and by thermal breakdown or oxidation. The presence of water reduces the ability of the lubricant to effectively lubricate, promotes oxidation of additives, and encourages rust and corrosion of metal parts. In performing a spectrometric analysis, one burns a small

TABLE 13.8 Belt-Driven Fan Vibration Limits

Frequency Band	Maximum Vibration Amplitude (in/sec Peak)
Overall (10–1000 Hz)	0.15
1X RPM	0.10
2X RPM	0.04
Harmonics	0.03
Belt Frequency	Not detectable
Bearing Defect	Not detectable

D042

TABLE 13.9 ISO 3945 Vibration Severity

Ranges of Radial Vibration Severity			Quality Judgement for Separate Machine Classes			
	RMS Velocity in 10–1000 Hz at the Range Limits					
Range	mm/sec	in/sec	Class I	Class II	Class III	Class IV
0.28	0.28	0.011	A	A	A	A
0.45	0.45	0.018	A	A	A	A
0.71	0.71	0.028	A	A	A	A.
1.12	1.12	0.044	B	A	A	A
1.80	1.80	0.071	B	B	A	A
2.80	2.80	0.110	C	B	B	A
4.50	4.50	0.180	C	C	B	B
7.10	7.10	0.280	D	C	C	B
11.20	11.20	0.440	D	D	C	C
18.00	18.00	0.710	D	D	D	C
28.00	28.00	1.10	D	D	D	D
71.00	71.00	2.80	D	D	D	D

D042

amount of the fluid sample and analyzes the resulting light frequencies and intensities to determine the type and amount of compounds present based on their absorption of characteristic light frequencies. Sample results are compared to the characteristics of the same new lubricant to measure changes reflecting the lubricant's reduced ability to protect the machine from the effects of friction. The continued presence of desirable lubricant additives is another key indicator of lubricant quality. Infrared spectrometry is also capable of detecting and measuring the presence of organic compounds such as fuel or soot in the lubricant sample.

In wear particle analysis, analysts examine the amount, makeup, shape, size, and other characteristics of wear particles and solid contaminants in the lubricants as indicators of internal machine condi-

TABLE 13.10 Vibration Acceptance Classes

Class	Condition
A	GOOD
B	SATISFACTORY
C	UNSATISFACTORY
D	UNACCEPTABLE

D042

TABLE 13.11   Machine Classifications

Machine Classes for ISO 3945	
Class I	Small size machines to 20 HP
Class II	Medium size machines (20–100 HP)
Class III	Large machines (600–12,000 RPM) 400 HP and Greater Rigid mounting
Class IV	Large machines (600–12,000 RPM) 400 HP and Greater Flexible mounting

tion. With experience and historical information, one can project degradation rates and estimate the time until machine failure. Wear particle analysis includes ferrography, which is a technique used to analyze metal wear products and other particulates. Elemental spectrographic analysis is used to identify the composition of small wear particles and provide information regarding wear sources. Analyzing and trending the amount, size, and type of wear particles in a machine's lubrication system can pinpoint how much and where degradation is occurring. Figure 13.8 illustrates the relationship between wear particle size, concentration and equipment condition.

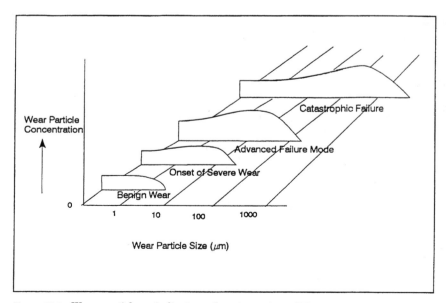

Figure 13.8 Wear particls as indicators of equipment condition.

13.9.3.4 Thermography. There are two basic criteria for evaluating temperature conditions. They are temperature difference (ΔT) and absolute temperature. Each is described below.

a. *Temperature Difference (ΔT).* Temperature difference criteria are simple, easy to apply in the field, and provide an adequate qualitative screening system to identify thermal exceptions and problems. The ΔT criteria compares component temperature to the ambient temperature and may be used for electrical equipment. ΔT may also be used for mechanical components.

The typical ΔT criteria, which may be modified easily based on experience, are:

<20°F (11°C) ΔT:	No immediate corrective action required; maintenance can be postponed until a scheduled maintenance period.
20°–40°F (11°–22°C) ΔT:	Corrective action should be taken at the *next* scheduled maintenance period.
40°–80°F (22°–44°C) ΔT:	Corrective action should be taken at the *first available* opportunity.
>80°F (44°C) ΔT:	Corrective action is required *immediately.*

b. *Absolute Temperature.* Absolute temperature criteria are generally specific to an equipment model, type of equipment, class of insulation, service use, or any of many other salient characteristics. As a result, absolute temperatures are more suited to quantitative infrared thermography and critical temperature applications. The mechanical temperature specifications come primarily from manufacturer's manuals. Electrical temperature specifications are set by three principal electrical standards organizations:

(1) National Electrical Manufacturers Association, (NEMA) .[11]
(2) International Electrical and Electronic Engineers, (IEEE).[12]
(3) American National Standards Institute, (ANSI).[8]

A very useful summary of temperature criteria is found in the *Guideline for Infrared Inspection of Electrical and Mechanical Systems,* published by the Infraspection Institute of Shelburne, Vermont.[13]

13.9.3.5 Airborne ultrasonics. The use of a passive ultrasonic instrument as a leak detector; i.e., listening for the ultrasonic noise characteristic of a pressure/vacuum leak, is qualitative. There are no numerical thresholds.

Many common passive ultrasonic devices operate on a relative, rather than a calibrated, absolute scale. However, by using relative changes in intensity from baseline readings, the degradation process may be trended and tracked.

13.9.3.6 Motor circuit analysis. Motor circuit analysis measures natural electrical motor circuit characteristics, such as:

- Individual phase resistance from bus disconnect through the motor windings (milli-ohms)
- Inductance of the motor coils (millihenries)
- Capacitance of each phase to ground (picofarads)

During the same test series, one can measure resistance to ground of each phase (megohms) through the use of low voltages (both AC & DC) and low currents, which are not harmful to the motor or motor circuits.

The following procedure is recommended for use when performing motor current analysis (MCA):™

- Deenergize the circuit (Comply with safety instructions)
- Eliminate stray currents/voltages
- Do not disconnect/exclude components except:
  — Power factor correction capacitors
  — Solid state controllers
- Place as much of the circuit under test as possible, including
  — Disconnects
  — Motor controller contactors
  — Circuit breaker(s).

A motor circuit analysis test set can provide indications of circuit problems directly in electrical terms that can be used by maintenance personnel to pinpoint and correct faults. The test set may then be used to perform post-repair tests.

The following are MCA™ resistance imbalance guidelines:

MCA™ Conductor Path Resistance Imbalance Guidelines

$<2\%$ expected and acceptable when new
$>5\%$ plan troubleshooting to locate cause of increased resistance when convenient
$>10\%$ schedule effort to locate and eliminate problem in order to preserve motor life

$$\% \text{ imbalance} = \frac{R_{\text{high}} - R_{\text{avg}}}{R_{\text{avg}}} \times 100$$

where:
$$R_{\text{avg}} = \frac{(R_1 + R_2 + \cdots R_n)}{n}$$

MCA™ Inductive Imbalance Guidelines

<10%	acceptable from OEM or rewind shop
10–15%	acceptable in service
>15%	isolate cause(s), increase monitoring frequency
>25%	begin planning for motor repair or replacement
>40%	be prepared for failure, within weeks

$$\% \text{ imbalance} = \frac{X_{\text{high}} - X_{\text{avg}}}{X_{\text{avg}}}$$

where
$$X_{\text{avg}} = \frac{X_1 + X_2 + \cdots X_n}{n}$$

The following are MCA™ capacitance imbalance guidelines:

MCA™ Capacitance Imbalance Guidelines

Capacitance is reflective of moisture and dirt in the vicinity of motor circuit components.

Analysis is performed by trending and by making relative comparisons as follows:

- Take initial reading
- Compare readings for similar motor circuits
- Compare follow-on readings, watch for upward trends
- Identify significant differences and schedule inspection to resolve differences

13.9.3.7 Motor current spectrum analysis. Mechanical/electrical interactions associated with magnetic forces in and around the rotating element of a motor are "reflected" in and around AC power supply cables. They are readily identifiable and display repetitious flux field characteristics centered at the electrical power line frequency (F_L) of 60 Hz (3600 CPM). Through analysis and interpretation of the relationship between the AC power line frequency and its motor generated sidebands, motor current spectrum analysis may be used to detect:

- Broken rotor bars or high resistance joints (braze, crack)
- Defective rotor shorting rings (alignment, porosity, integrity)
- Rotor and stator (air gap) eccentricity (dynamic and static)
- Unbalanced magnetic pull.

As the difference between the amplitude of the power line frequency and the pole pass sideband frequencies decrease in magnitude, the greater the concern should be for the condition of the rotor. Results indicate:

- A slight decrease may trigger increased monitoring.
- A moderate decrease may indicate increasing resistance between the rotor bars and the end ring or that a crack is developing in either.
- Further decreasing values indicate rotor bar breaks.

13.9.3.8   Insulation resistance. There are several forms of insulation resistance measurement:

a. *Megohmmeter Testing.* Used to measure phase-to-phase and phase-to-ground resistance. It gives an overall indication of insulation condition, however, reliable trending requires temperature correction. Even the best insulation will show unacceptably low insulation resistance readings just following an oven bake or dry-out.

b. *Dielectric Absorption Ratio and Polarization Index.* Insulation resistance ratios are frequently used to evaluate insulation-to-ground conditions in order to avoid having to compensate for temperature. A single DC voltage, usually slightly higher than the motor rated voltage, is impressed continuously on a winding. The current induced by the DC voltage has three components:

   (1) Capacitive charging of the circuit, which fades quickly.
   (2) Leakage current to ground, a constant value.
   (3) Current, which polarizes the molecules of the insulation surrounding the motor circuit conductor path and fades slowly from its initial value.

The ratio of readings taken at two different times indicates the condition of the insulation, as follows:

$$\frac{\text{Megohm reading at 1 min}}{\text{Megohm reading at 30 sec}} = \text{Dielectric absorption ratio}$$

$$>1.5 \text{ OK}$$
$$<1.25 \text{ Danger}$$

$$\frac{\text{Megohm reading at 10 min}}{\text{Megohm reading at 1 min}} = \text{Polarization index}$$

>2 up to ~4 OK
<2 insulation weak
>5 insulation possibly too dry and brittle

c. *Leakage Current.* An overall indicator of insulation condition and cleanliness of the equipment. The accumulation of dirt and moisture provides a path for leakage current. There are no absolute values by which to gauge the deterioration of insulation using leakage current. The comparison of successive values permits the comparison and trending of leakage current.

13.9.3.9 Surge testing. A surge test uses high voltage, high energy discharge pulses that are inserted into two windings of a polyphase motor simultaneously. These pulses cycle between motor windings and the test set. Current waveform analysis may indicate problems to an experienced and trained analyst. The test also indicates the voltage level at which the insulation breaks down.

Surge testing, in the view of the Electric Power Research Institute, accomplishes the following:

- The surge test could be destructive, inducing failure in weakened turn insulation. For this reason, test impulse voltage levels and rise times should be carefully selected.

- This is a go/no-go proof test and, as such, does not provide information that will allow making an assessment of remaining life.

- Experience is required to perform the test and to interpret the results, especially if complete windings rather than individual coils are tested.

13.9.3.10 Start-up tests. There are two types of start-up tests as described below:

a. *Coast-Down Time.* A simple and often overlooked test. In this test the time for the motor to coast to a stop after the removal of power is recorded and trended. This data tracks the mechanical condition of the motor bearings over the life of the machine.

TABLE 13.12   RCM Quality Assurance Planning Considerations

Planning QA Checklist	✔
Are the RCM requirements for the facility determined?	
Is built-in monitoring planned where cost effective?	
Has the use of performance data for condition monitoring been considered and planned?	
Is the collection of cost, cost avoidance, and cost savings incorporated?	
Does the CMMS require a PdM module?	
Is there a mechanism provided for maintenance feedback?	
Have PdM technologies that are appropriate for the equipment been selected?	

D042

b. *Peak Starting Current.* Another simple test which involves periodically recording the peak starting current of a motor. This test also provides an indication of the mechanical condition of the motor.

13.10   Checklists for RCM Quality Assurance

13.10.1   Planning

Table 13.12 contains some factors to be considered in the planning phases of facilities acquisition.

13.10.2   Design

Table 13.13 contains some factors to be considered for quality assurance during the design phase of facilities acquisition.

TABLE 13.13   RCM Quality Assurance Design Considerations

Design QA Checklist		✔
Are maintainability factors considered?	Access	
	Material	
	Standardization	
	Quantitative goals set	
Are ease of monitoring factors considered	Access	
	Online data collection	
	Performance indicators	
Are predictive technologies specified and incorporated?		
Do the contractor qualifications match the RCM requirements?		
Has a predictive analysis capability been provided?		
Has the distribution of raw data to those who could use it been incorporated?		
Are test and maintenance results distributed to users?		
Have design modifications based on maintenance feedback been incorporated?		

D042

TABLE 13.14 RCM Quality Assurance Construction Considerations

Construction QA Checklist	✓
Are acceptance testing requirements established?	
Do the contractor's qualifications indicate an adequate understanding of RCM?	
Is the contractor conforming to specifications and drawings?	
Is the contractor conforming to bills of material?	
Is the contractor conforming to installation procedures?	
Has the training of maintenance personnel been initiated?	
Is the selection of maintenance tasks in progress?	
Has the writing of maintenance procedures and instructions been initiated?	
Are baseline condition and performance data recorded and made available for equipment as it is installed?	
Does the contractor need specialized subcontractors?	
Have contractor equipment submittals been approved?	

D042

13.10.3 Construction

Table 13.14 contains some factors to be considered during the construction phase of facilities acquisition.

13.10.4 Equipment procurement

Table 13.15 contains some factors to be considered during the equipment procurement phase of facilities acquisition.

13.10.5 Maintenance and operations

Table 13.16 contains some factors to be considered during the maintenance and operations phase of a facility life cycle.

13.11 Management Indicators

13.11.1 General

There are a number of management indicators used to measure the effectiveness of an RCM program. The most useful indicators are nu-

TABLE 13.15 RCM Quality Assurance Equipment Procurement Considerations

Equipment Procurement QA Checklist	✓
Are specifications determined to meet RCM requirements?	
Is acceptance testing specified?	
Are contractor qualifications matched to RCM requirements?	
Is a feedback system in place for continuous equipment improvement?	
Has provision been made for equipment condition monitoring, if applicable?	
Are baseline data required?	
Are equipment life cycle costs required/provided?	
Are embedded (online) sensors required/provided?	

D042

TABLE 13.16   RCM Quality Assurance Maintenance and Operations Considerations

Maintenance and Operations QA Checklist	✓
Is the inventory of skills to support the RCM program available?	
Is training planned to fill skill and technical shortcomings?	
Does the training support the development of predictive analytical skills?	
Does the training support RCM management and supervisory skills?	
Are the documentation, procedures and work practices capable of supporting RCM?	
Are the responsibilities for systems and equipment defined and assigned?	
Are the maintenance history data and results distributed to proper users?	
Is there a feedback system in place for continuous maintenance program improvement?	
Is root-cause failure analysis in use and effective?	
Are failed components subject to post-failure examination and results recorded?	
Are predictive forecasts tracked and methods modified based on experience?	
Are PM task and CM monitoring periodicities adjusted based on experience?	
Does the CMMS fully support the maintenance program?	
Are maintenance cost, cost avoidance and cost savings data collected, analyzed and disseminated?	
Are baseline condition and performance data updated following equipment major repair or replacement?	
Are appropriate measures of maintenance performance (metrics) in use?	

D042

merical. The numerical, or quantified, indicators, often referred to as "metrics," can be expressed as goals and objectives, measured and displayed in several ways for the purpose of analysis and management decision making.

13.11.2   Metrics definition

Management (or performance) indicators are relationships used for indicating the effectiveness of an operation and for comparing performance with goals and objectives. There are two types of management indicators: words and numbers (or metrics). Metrics are preferable because they are quantitative, objective, precise, and more easily trended than words. Metrics consist of a descriptor and a benchmark.

13.11.2.1   Descriptor.   A descriptor is a word or group of words describing the units, the function, or the process to be measured. Examples are the number of corrective actions developed by the PdM program, the number of hours of equipment run time, and the equipment availability expressed as a ratio of equipment hours available to equipment hours required.

13.11.2.2   Benchmark.   A benchmark is a numerical expression of a goal or objective to be achieved. It is that against which one measures one's performance. It can be an absolute number or a range. For example, the benchmark for equipment availability might be 90%. The metric (descriptor and benchmark) would therefore be:

$$\frac{\text{Equipment available (hrs)}}{\text{Equipment required (hrs)}} = 90\%$$

13.11.3 Sample metrics

The following are sample metrics one might choose to measure the effectiveness of an RCM program. The benchmarks suggested are averages taken from approximately 50 major corporations worldwide surveyed in the early 1990s:

Metric   **Benchmark**

a. *Equipment Availability*   96%

$$\% = \frac{Hrs\ each\ unit\ of\ equipment\ is\ available\ to\ run\ at\ capacity}{Total\ hrs\ during\ the\ reporting\ time\ period}$$

b. *Maintenance Overtime Percentage*   5% or less

$$\% = \frac{Total\ maintenance\ overtime\ hrs\ during\ period}{Total\ regular\ maintenance\ hrs\ during\ period}$$

c. *Emergency Percentage*   10% or less

$$\% = \frac{Total\ hrs\ worked\ on\ emergency\ jobs}{Total\ hrs\ worked}$$

d. *Percentage of Candidate Equipment Covered by PdM*   100%

$$\% = \frac{Number\ of\ equipment\ items\ in\ PdM\ program}{Total\ equipment\ candidates\ for\ PdM}$$

e. *Percentage of Emergency Work to PdM and PM Work*   20% or less

$$\% = \frac{Total\ emergency\ hrs}{Total\ PdM\ and\ preventive\ maintenance\ hrs}$$

f. *Percent of Faults Found in Thermographic Survey*   3% or less

$$\% = \frac{Number\ of\ faults\ found}{Number\ of\ devices\ surveyed}$$

g. *Percent of Faults Found in Stream Trap Survey*   10% or less

$$\% = \frac{Number\ of\ defective\ steam\ traps\ found}{Number\ of\ steam\ traps\ surveyed}$$

h. *Ratio of PM / PdM Work to Reactive Maintenance Work*

A = 70% PM/PdM

B = 30% Reactive Maintenance

$$A\% = \frac{Manhrs\ of\ PM\ /\ PdM\ work}{Manhrs\ of\ reactive\ maintenance\ +\ PM\ /\ PdM\ work}$$

$$B\% = \frac{Manhours\ of\ reactive\ work}{Manhours\ of\ reactive\ maintenance\ +\ PM\ /\ PdM\ work}$$

A% + B% = 100%

13.11.4 Trending indicators

The following indicators also are recommended for consideration for trending as maintenance program management tools:

a. Equipment (by classification) percentage out-of-service time for repair maintenance.

b. Mean time between equipment overhauls and replacement.

c. Number of vibration-related problems found and corrected per month.

d. Number of vibration-related work orders open at the end of the month.

e. Number of vibration-related work orders over 3 months old.

f. Number of problems found by other PdM techniques (i.e., infrared thermograpy, ultrasonics, lube oil analysis, etc.) and corrected per month, work orders open at the end of the month, and work orders over 3 months old.

g. A monthly record of the accumulated economic benefits or cost avoidance for the various PdM techniques.

h. Number of spare parts eliminated from inventory as the result of the PdM program.

i. Number of overdue PM work orders at the end of the month. (Total number of PM actions should decrease.)

j. Aggregate vibration alert and alarm levels (trending down).

TABLE 13.17   Plant Survey

Indicator	Best	Worst
Number of Hourly Maintenance Workers per Total Number of Plant Employees	5%	49%
Number of Maintenance Supervisors per Number of Hourly Maintenance Employees	3%	10%
Plant Replacement Value per Each Hourly Maintenance Employee	$6.5M	$1.1M
Annual Plant Maintenance Cost per Plant Replacement Value	0.9%	4.5%
Stores Inventory per Plant Replacement Value	0.5%	1.9%

13.11.5   Best and worst industry examples

Finally, the Table 13.17 tabulations are representative of a cross section of the industrial organizations surveyed. The "best" and "worst" columns refer to the two extremes of the best and worst evaluated plants. For example, for the first indicator shown below the best plant has a ratio of the number of wage-grade maintenance workers to the total number of plant employees of 5%, while the worst plant has a ratio of 49%.

13.12   References

1. Nowlan, F. Stanley and Heap, Howard F., *Reliability-Centered Maintenance,* Dolby Access Press, 1978.
2. American Society of Non-Destructive Testing, *Recommended Practice Number SNT-TC-1A,* Columbus, OH, 1992.
3. American Society for Testing and Materials, Philadelphia, PA.

    ASTM D877   *Dielectric Breakdown Voltage of Insulating Liquids Using Disk Electrodes,* April 1987.

    ASTM D924   *Power Factor,* June 1993.

    ASTM D974   *Acid Neutralization by Color Indicator,* March 1994.

    ASTM D2285   *Interfacial Tension of Electrical Insulating Oils of Petroleum Origin Against Water by the Drop-Weight Method,* November 1985.

    ASTM 1289   *Test Method for Density, Relative Density (Specific Gravity or API Gravity) of Crude Petroleum and Liquid Petroleum Products by Hydrometer Methods,* February 1991.

    ASTM D92   *Flash and Fire Point by the Cleveland Open Cup,* May 1991.

    ASTM D97   *Pour Point of Petroleum Oils,* February 1994.

    ASTM D1533   *Water and Insulating Liquids,* November 1989.

    ASTM D445   *Kinematic Viscosity of Transparent and Opaque Liquids,* April 1989.

    ASTM C1060-90   *Thermographic Inspection of Insulation in Envelope Cavities in Wood Frame Buildings,* December 1990.

    ASTM C1153-90   *Standard Practice for the Location of Wet Insulation in Roofing Systems Using Infrared Imaging,* September 1990.

    ASTM E1186-87   *Standard Practices for Air Leakage Site Detection in Buildings,* May 1993.

4. National Fluid Power Association, NFPA T2.9.1-1972, *Method for Extracting Fluid Samples from the Lines of an Operating Hydraulic Fluid Power System for Particulate Particle Contamination Analysis,* 1992.
5. International Standards Organization, American National Standards Institute, 1430 Broadway, New York, 10018.

ISO 6781   *Thermal Insulation-Qualitative Detection of Thermal Irregulari-ties in Building Envelopes-Infrared Method, 1983.*

ISO 3945   *Mechanical Vibration of Large Rotating Machines with Speed Range from 10–200 rev / s — Measurement and Evaluation of Vibration Severity in Situ, 1985.*

6. American Petroleum Institute, API 670, *Vibration, Axial-Position Bearing-Temperature Monitoring System,* New York, NY, November 1993.
7. Hewlett Packard Application Note 243-1, *Effective Machinery Measurement Using Dynamic Signal Analyzers, 1990.*
8. *Temperature Specifications* American National Standards Institute, New York.
9. MIL-STD-167-1, *Mechanical Vibrations of Shipboard Equipment,* May 1974.
10. MIL-STD-740-2, *Structureborne Vibratory Acceleration Measurements and Acceptance Criteria of Shipboard Equipment,* December 1986.
11. *Temperature Specifications* National Electrical Manufacturers Association, Washington, DC.
12. *Temperature Specifications* International Electrical and Electronic Engineers, New York.
13. Infraspection Institute, *Guidelines for Infrared Inspection of Electrical and Mechanical Systems,* Shelbourne, VT, 1993.
14. Hewlett Packard Application Note 243, *The Fundamentals of Signal Analysis,* 1991.
15. Jackson, Charles, *The Practical Vibration Primer,* Gulf Publishing Company, Houston, TX, 1979.
16. Mitchell, John S., *Machinery Analysis and Monitoring,* Penn Well Books, Tulsa, OK, 1981.
17. Moubray, John, *Reliability-Centered Maintenance,* Butterworth-Heinemann Ltd., Oxford, England, 1991.
18. Harris, Tedric A., *Rolling Bearing Analysis,* John Wiley & Sons, New York, 1984.
19. Proceedings of the 44th Meeting of the Mechanical Failures Prevention Group, *Current Practices and Trends in Mechanical Failure Prevention,* Vibration Institute, 1990.
20. *Oil Analysis Seminar Coursebook,* PdMA Corporation, Tampa, FL, 1990.

OTHER SOURCE MATERIAL

Johnson Space Center Reliability Centered Maintenance Guide, prepared by PdMa Corp. and System Planning Corporation, Arlington, VA, June 1994.

13.13   Glossary

Applicable PM Task — A PM task which is capable of improving equipment reliability by modifying its failure behavior (how or when it fails).

Computerized Maintenance Management System (CMMS) — Computer software programs designed to reduce manual administrative and technical workloads and paper files while assisting management and operational personnel in making decisions for efficient and effective facility and equipment maintenance. They provide historical data, report writing capabilities, job analyses, and more. The data describe the equipment, parts, jobs, crafts, costs, step-by-step instructions, and other information involved in the maintenance effort. This information may be stored, viewed, analyzed, reproduced, and updated with just a few keystrokes.

Critical Failure — A failure involving a loss of function or secondary damage that could have a direct adverse effect on operating safety, on mission, or have significant economic impact.

Critical Failure Mode — A failure mode that has significant mission, safety or maintenance effects that warrant the selection of maintenance tasks to prevent the critical failure mode from occurring.

Dominant Failure Modes — The failure modes that are most likely to occur during the lifetime of the item, component, or equipment.

Effective PM Task — The characteristic of a preventive maintenance task when it is capable of improving equipment reliability to a given level under specific constraints (i.e., cost effective).

Failure Mode and Effects Analysis (FMEA) — Analysis used to determine what parts fail, why they usually fail, and what effect their failure has on the systems in total. An element of Reliability Centered Maintenance (RCM).

Infrared Thermography — A predictive technique that uses infrared imaging to identify defects in electrical and electromechanical devices such as fuse boxes, circuit breakers, and switchgear. It also can be used effectively in a nonpredictive manner to detect thermal cavities and leaks in walls, ceilings, and rooftops, the correction of which can result in sizeable reductions in heating and air conditioning expenses. Thermal imaging is extremely sensitive, and since it evaluates the heat an object emits, emittance and reflective factors of the viewed object and its environment must be considered.

Maintainability — A design objective which provides for easy, accurate, safe, and economical performance of maintenance functions.

Motor Circuit Analysis (MCA) — A predictive technique whereby the static characteristics (i.e.; impedance, capacitance to ground, inductance) of a motor or generator are measured as indicators of equipment condition.

Motor Current Spectrum Analysis (MCSA) — A predictive technique whereby motor current signatures provide information on the electromechanical condition of AC induction motors. It detects faults such as broken rotor bars, high resistance joints, and cracked rotor end rings by collecting motor current spectrums with clamp-on sensors and analyzing the data.

Performance Standards — Those standards which an item is required to meet in order to maintain its required function. The performance standard defines functional failure for the item.

Predictive Maintenance (PdM) — (1) Those testing and inspection activities for facility items that generally require more sophisticated means to identify maintenance requirements than those of preventive maintenance. (2) Sometimes referred to as "Condition-Based Maintenance" and "Predictive Maintenance." Use of advanced technology to assess machinery condition. Replaces maintenance scheduled at time intervals with maintenance scheduled only when the condition of the equipment requires it. The PdM data obtained allows for planning and scheduling preventive maintenance or repairs in advance of failure.

Preventive Maintenance (PM) — The planned, scheduled periodic inspection, adjustment, cleaning, lubrication, parts replacement, and minor repair (no larger than trouble call scope) of equipment and systems for which a specific operator is not assigned. Preventive Maintenance (PM) consists of many checkpoint activities on items that, if disabled, would interfere with essential Center operation, endanger life or property, or involve high cost or long lead time for replacement. Also called "time-based maintenance" or "interval-based maintenance." Depending on the intervals set, PM can result in a significant increase in inspections and routine maintenance; however, it should also reduce the frequency and seriousness of machine failures for components with defined, age-related wear patterns.

Proactive Maintenance — Application of predictive maintenance technologies toward extending machinery life. It seeks to eliminate the need for maintenance through better design, better installation, precision balance and alignment, and root-cause failure analysis.

Reactive Maintenance — Often called "breakdown maintenance," "reactive maintenance," or "run to failure (RTF)." Maintenance or equipment repairs are performed only when the deterioration in a machine's condition causes a functional failure. A high percentage of unplanned maintenance work, high replacement part inventories, and the inefficient use of maintenance personnel typify this strategy.

Reliability — The ability of an item to perform a required function under stated conditions for a given time interval (usually expressed as a probability).

Reliability Centered Maintenance (RCM)—Also called "reliability-based maintenance." A maintenance strategy that logically incorporates I into a maintenance program the proper mix of reactive, preventive, predictive, and proactive maintenance practices. Rather than being used independently, the respective strengths of these four maintenance practices are combined to maximize facility and equipment operability and efficiency while minimizing required maintenance time, materials, and consequently, costs. For example, a small pump might be run to failure, a gasoline engine might be placed on a 1000-hr PM program, and a critical turbine might be monitored with online diagnostic sensors. This strategy often includes performing a statistical analysis of historical data related to failures to determine the optimal investment of maintenance resources and risk assessment methods, called "Failure Mode and Effects Analysis (FMEA)," to identify those processes or systems that statistically exhibit the greatest chance of catastrophic failure. The equipment is then modified or replaced accordingly. Thus, the result is a shift in maintenance resources to areas of greatest mission consequence.

Repair—That facility work required to restore a facility or component thereof, including collateral equipment, to a condition substantially equivalent to its originally intended and designed capacity, efficiency, or capability. It includes the substantially equivalent replacements of utility systems and collateral equipment necessitated by incipient or actual breakdown.

Ultrasonic Analysis—A predictive technique incorporating the use of airborne and pulse-echo, ultra-high-frequency sound. Commonly used to pinpoint steam leaks and steam trap failures. Additionally, ultrasonic thickness gauges are used routinely to measure wall thinning in vessels and pipes and in other similar applications.

Vibration Analysis—The dominant technique used in predictive maintenance. Uses noise or vibration created by mechanical equipment to determine the equipment's actual condition. Uses transducers to translate a vibration amplitude and frequency into electronic signals. When measurements of both amplitude and frequency are available, diagnostic methods can be used to determine both the magnitude of a problem and its probable cause. Vibration techniques most often used include broadband trending (looks at the overall machine condition), narrowband trending (looks at the condition of a specific component), and signature analysis (visual comparison of current versus normal condition). Vibration analysis most often reveals problems in machines involving mechanical imbalance, electrical imbalance, misalignment, looseness, and degenerative problems.

13.14   Abbreviations/Acronyms

AC	Alternating Current
A&E	Architect & Engineer
AGMA	American Gear Manufacturer's Association
API	American Petroleum Institute
ANSI	American National Standards Institute
ASNT	American Society for Non-destructive Testing
ASTM	American Society for Testing and Materials
AWP	Annual Work Plan
CAD/CAM	Computer-Aided Design/Computer-Aided Manufacturing
CMMS	Computerized Maintenance Management System
CPM	Cycles per Minute
DC	Direct Current
FMEA	Failure Mode and Effects Analysis
HP	Horsepower

HIGH-POT	High Potential Testing
Hz	Hertz; Cycles per Second
IEEE	International Electrical and Electronic Engineers
ISO	International Standards Organization
M&O	Maintenance and Operations
MCA™	Motor Circuit Analysis
MCE™	Motor Circuit Evaluation Test Set
MCSA	Motor Current Spectrum Analysis
MDT	Mean Down Time
MTBF	Mean Time Between Failures
MTBM	Mean Time Between Maintenance
NEMA	National Electrical Manufacturers Association
NDT	Non-Destructive Testing
NFPA	National Fluid Power Association
NLGI	National Lubrication and Grease Institute
OEM	Original Equipment Manufacturer
PI	Polarization Index
PdM	Predictive Maintenance
pH	Hydrogen Ion Content
PM	Preventive Maintenance
RCFA	Root-Cause Failure Analysis
RCM	Reliability Centered Maintenance
RF	Radio Frequency
RFP	Request for Proposal
RFQ	Request for Quotation
RM	Reactive Maintenance
RMS	Root Mean Square
RTF	Run to Failure
SAE	Society of Automotive Engineers
SUS	Saybolt Universal Seconds
TDR	Time Domain Reflectometry
TQFM	Total Quality Facility Maintenance

Chapter 14

Stores and Parts Management

Ron Moore
The RM Group
Knoxville, Tennessee

14.1 Introduction

Stores should be run like a store—clean, efficient, everything in its place, not too much or too little, run by a manager with a focus on customer (facility maintenance) needs. More over, stores should be viewed as an asset, not a liability or cost. Maintaining a good, high-quality stores operation is in fact the low-cost approach to operating a facility. Yet many operations treat their stores function as if it were a necessary evil, a burdensome cost, a non-valve adding function. This is not an enlightened approach. If properly managed, stores will help assure a high quality, low cost operation. If not, stores will continue to be a "nonvalue adding" and expensive "necessary evil."

*Reliability and stores management go hand-in-hand. Reduce inventory, **and** increase reliability.*

14.2 What are Stores?

In most facilities, stores are typically viewed as spare parts. However, a broader and more accurate perspective is that all items not consumed directly or indirectly in production are included under the heading of stores. Stores can generally be classified into five groups:[1]

- *Hardware and Supplies,* e.g., bolts, small tools, belts, pipe, valves.
- *Materials,* e.g., paint, lubricants, cement, refractory.

- *Spare Parts,* e.g., bearings, gears, circuit boards, specific components.

- *Spare Equipment,* e.g., complete assemblies and machines.

- *Special Items,* e.g., lubricants, catalyst, steel banding, construction surplus, pilot.

The stores function is to provide high-quality spare parts and other material as needed and where needed, primarily supporting the maintenance function. Ideally, stores would always have exactly what was needed, when it was needed, and be able to place it where it was needed at a moment's notice, *and* it would have no more than what was needed—only that material and spare parts that are needed immediately, and only for a minimum period in stores. Unfortunately, few of us live in this ideal world, and more often than not, stock-outs are frequent, reorders are often needed, delivery of spares is sporadic, etc. Many companies however, have found a "solution" to these problems—they carry lots of spare parts in considerably larger quantities than would ordinarily be needed on a day-to-day basis. This results in excess inventory, poor cash flow management, and often, sloppy management practices. Yes, there *is* a better way.

14.3   The "Cost" of Stores

Spare parts and stores expenditures typically run 25 to 50 percent of a given maintenance budget. Further, annual carrying costs (labor, space, insurance, taxes, shrinkage, utilities, etc.), typically run 30 percent ± of the value of stores. Perhaps more importantly, if the parts are not available, or are of poor quality because of poor stores practices, then facility function can be lost for extended periods of time. The loss associated with this is often larger than the carrying costs. The cost of stores can be characterized as losses associated with the following:[1]

- *Working capital 'losses'*—overstocked, underused material, parts, etc., sitting in stores

- *Carrying cost 'losses'*—for maintaining the stores operation

- *Facility capability 'losses'*—due to lack of timely availability of parts

- *Maintenance inefficiency 'losses'*—due to poor management—wait time, transit time, etc.

- *Expediting cost 'losses'*—due to poor planning

- *Shrinkage 'losses'*—due to poor control, i.e., theft, deterioration, obsolescence, etc.

At the same time, if we could eliminate or at least minimize these losses, then the "value" of the store's function would be readily apparent. More enlightened managers understand the need to minimize these losses, and therefore put in place practices to assure a so-called world class operation—minimal losses, maximum support capability.

How do we set up a good stores operation?

14.4 What Stores are Needed—Kind and Quantity

Most people already have a stores operation, but often it is not functioning like a modern store to routinely meet kind, quantity, and availability requirements. Suppose, for example, you went into a local department store in your home town, and found it to be dirty, lacking the items you were seeking, managed by people who didn't seem to care about the store, or you! Would you ever go back? In most facilities, we only have one store, and that's the one we must use. Unfortunately, all too often it is not being managed like a modern store. To do so, that is, to build a good stores operation, requires an in-depth understanding of the customer's needs—of facility maintenance. Therefore, kind, quantity, and availability needs must be driven by an keen understanding of maintenance needs. Further, to facilitate good communication and understanding, maintenance must have a good maintenance management system in place for: equipment identification, work orders, planning and scheduling system, the link to stores, kitting and delivery of parts for given work orders, and a cost accounting link for effective management. So, where do we start?

14.5 Bill of Material and Catalog

If you don't have a good bill of material for every critical item in your facility, then develop one in cooperation with maintenance and their definition of critical equipment and understanding of spares needs. Use this bill of material in conjunction with other needs to develop a catalog of all stores and spare parts required in your facility. This is a dynamic document and must be updated 2–3 times per year, or as needed, e.g., for a major change in suppliers, for major equipment additions, etc. This catalog should include unique identifier codes, generally numerical, for each catalog item, and should provide for a logical grouping of material for use in Pareto analyses, component use frequencies, equipment history analyses, etc. It should also include a standard nomenclature and descriptors for each item. All this (catalog, standard nomenclature, grouping, descriptors) may not be as simple as it sounds, and as noted, should be developed in cooperation with others whom it may affect, e.g., maintenance, design, suppliers, construction and capital projects. This will ensure buy-in, acceptance, common understanding, etc., and will ensure greater probability of success in the effort to run a world class stores organization. Finally, while hard copies are routine, the catalog should also be "on-line" and staff must be trained in using the catalog, both manual and "on-line."

14.6 Management Methods

Simple policies and procedures should be developed in cooperation with maintenance and engineering to assure consistency of process and quality of operations. This should include a policy and/or procedure for:[1]

Development, use, and modification of the catalog system

Inventory classification, including process for obsolescence

Vendor stocking and consignment programs

Economic order practice—quantity, point, level, and dynamic adjustment

Consolidation of parts, suppliers

Repair/replace policy, including managing reconditioned (vs. new) equipment

Alternate sourcing (qualification, detailed drawings availability, etc.)

Communication to maintenance, engineering, purchasing of key policies

An audit process

Whether or not you develop a policy and a separate procedure in these areas is dependent on your current practices and philosophy, the level of skill of your staff, the resources available, etc.

Finally, a quick tip: Many companies have a repair/replace policy for their motors which essentially says that if the cost of the repair/ rewind is less than, say 50 percent, of a new motor, then they will repair. This practice ignores the efficiency loss of about 3 percent which results from a rewind, which translates to increased power consumption and shorter life for the re-wound motor. The cost of increased power consumption, and reduced motor life must be considered and put into a policy for repair/replacement. You are encouraged follow up on this and even develop other examples.

14.7   Partnerships with Suppliers

Suppliers should be consolidated and standardized as much as possible. When selecting suppliers, key performance indicators should be used, in conjunction with key performance objectives. Supplier alliances should include:

- Partnership agreements detailing the basis for the agreement
- Supplier stocking and consignment terms and methods for reducing physical inventory
- Blanket orders wherever possible
- Maintainability and reliability requirements
- Feedback process for resolving problem areas
- Use of electronic communications and order entry wherever possible
- Measurement (and minimization) of emergency or spot orders
- Measurement of stock types

Further, suppliers are routinely asked to provide information regarding spare parts recommendations and frequency of preventive maintenance (PM) requirements. A typical response is a list of spares to be used for annual PM. Rather than simply accept the recommendations at face value, which may be related more to next year's sales plan than to good maintenance practices, the supplier should be required to provide the statistical basis for their recommendations, including the basis related to the application of the equipment and its working environment. It's important to remember that overhaul PMs essentially presume that all your equipment is average, a highly unlikely situation. It may also be that your application is very different from a typical application, and that you do not have consistency in your operation and maintenance practices.

Suggestions for developing a partnership are provided in Table 14.1.

14.8   Standardization

Supplier partnerships will facilitate standardization, but these partnerships must also be combined with seeking opportunities for further standardization. For example, methods for standardization should include: 1) using materials which fulfill common if not identical requirements, 2) reviewing new equipment and purchases for standardization opportunities; assuring consistency with corporate supplier opportunities. The standardization process must include the design and

TABLE 14.1   Outline for Creating Strategic Supplier Partnership

- Analyze operating results—equipment life, process performance, initial and operating and maintenance costs.
- Baseline current operation re: key performance indicators for a partnership.
- Set increases in business levels with partner tied to specific performance criteria, including costs reduction targets—total cost, including operation and maintenance.
- Establish basis for PM and spare parts, including statistical analysis; specific applications and environment for the equipment; and maintainability requirements for ease of maintenance.
- Commit a minimum percent of the business to your partner based on meeting these criteria.
- Set dates for achievement of improved performance standards from the baseline data.
- Agree upon the measurement criteria and the techniques for their measurement.
- Provide for warranty of performance, e.g., better than any of competitor products; better than average of three prior years; free of defects; minimal performance criteria.
- Define basis for communication, frequency, report formats, etc.
- Define basis for QA/QC, and failure analysis process.
- Define basis for resolving disputes.
- Define lead times, forecasts, quantities, and overall business relationship.

capital projects staff, as well as purchasing. All appropriate parties must be trained in the standards which have been developed, and a process must be put in place which defines how the standards are changed and/or waived.

14.9   The Store

The store should be well managed. There are a number of techniques to assist in managing the store itself:[2]

Layout

Work environment and practices

Technology/methods

People

14.9.1   Layout

Routine work should be reviewed from a "time/motion" perspective. For example, ask yourself: Are material and spare parts conveniently located for minimizing the amount of time required to pick up and deliver to the counter? Is the material received in a way to minimize the time and effort for stocking? Are material and parts conveniently located, especially for frequently needed items? Is the issue counter near the receiving counter? Is management near the hub of the activity—issue and receipt? Is a delivery system in place to provide the parts at the job location? The layout of the stores should be reviewed with minimizing the time required to provide the needed deliveries to maintenance, all things considered.

14.9.2 Procedures

Are processes computerized—is bar coding in routine use? Is receipt inspection routinely practiced? Are automated reduction and inventory control points highlighted? Are receipts recorded quickly and inspected? Have you considered contracting miscellaneous material purchases, such as shoes, towels, jackets, coveralls, etc?

14.9.3 Work environment and practices

You should answer the following questions and use these answers to make a judgment about whether or not you are using best practices in your stores facility.

Are the floors and walls clean and painted (floors with epoxy, non-skid for reducing dust)? Is the stores area clean, comfortable, and well lighted? Is it access controlled and/or managed, e.g., limited to users of a swipe card? Is there an air conditioned, nonstatic area for PC boards and other electronic equipment? Are you using high density storage for appropriate items? Are bearings and gaskets protected, sealed, stored to minimize damage or deterioration? Are motors covered to minimize deterioration? Do large, critical motors have heaters to eliminate moisture accumulation in the windings? Are the shafts of rotating machinery rotated regularly to avoid false brinneling? Are shafts to critical rotors stored vertically, and in a nitrogen-sealed, pressurized enclosure? Is the carbon steel equipment, which may corrode, coated with a thin, protective film?

Has the catalog been completed and input into a stores management system, which is linked to the maintenance management system and to accounting? Is bar coding used to minimize clerical requirements? Are carousels in use and controlled by a controller pad linked to the stores management system? Will the carousel bring the material to the counter, and charge the withdrawal to the customer, and mark the withdrawal against quantities for order point determination? Is this process tied to maintenance planning and control? Is material and/or parts kitted by work order, and as required, delivered to the job location? Is a process in place for cycle counting (verifying inventory quantities on a periodic basis)? Is an electronic data entry, and order entry process in place, especially for key suppliers? Is there a process for managing equipment to be repaired or overhauled and restocked, e.g., separate area for "to be repaired and restocked," cost accounting procedure, etc.? Are the following in place relative to suppliers:

Supplier stocking and consignment terms and methods for reducing physical inventory

Blanket orders wherever possible

Feedback process for resolving problem areas

Measurement (and minimization) of emergency or spot orders

14.9.4 People

People represent perhaps the easiest, and simultaneously the most difficult, issue in the stores management function. Most people want to do a good job, given the proper training, tools, systems, procedures, and encouragement. In many stores operations, the stores person has been assigned to the function without much training in store management, other than what might have been garnered through on the job training, and the practices which have been handed down over the years. Yet the cost and value considerations for effectively managing a store described above dictate much more than a casual and historical approach to stores management.

Having answered the questions posed above regarding stores environment and practices, you should now develop an organization with the right people, in the right mix and quantity, and provide the tools, training, and encouragement to assure superior performance. A matrix should be created which outlines the training which your people need in order to accomplish the tasks identified. Further, you should routinely keep measurements of the effectiveness of the stores operations. Several are listed below. And, you should routinely perform customer satisfaction surveys of the maintenance function, and seek to develop a strong supportive relationship with maintenance, as well as purchasing, operations, and engineering.

14.9.5 Training

Training should be formulated based upon a series of strategic questions:

What are my key objectives (reference key performance indicators)?

What are the skills required to achieve those objectives? In what quantities?

What are the skills of my people today?

What new equipment and/or systems are coming into use in the future?

What are may workplace demographics — age, ability, etc.?

What are my training requirements in light of my answers?

Use the answers to these questions to develop your strategic training plan, one which assures first that the skills are put in place on a prioritized basis to minimize losses and add value to the organization.

14.9.6 Contracting the stores function

While the author does not share many other's enthusiasm for contracting maintenance (reduces ownership and core competencies, increases loss of equipment and process knowledge, increases risk of down time, etc.), it may have its place in the stores function. You may review the characteristics and prospective value of a good stores function, and conclude that you would be very hard pressed to achieve this level of competence in a short period of time. If so, you may decide that all things considered a good high quality stores management function could be put in place using contractors. This is what was concluded in reference 2, but in that situation there were also additional considerations related to the intransigence of the union, its work rules, etc. Apparently, the union was unwilling to work with management to improve productivity and performance and was replaced by a contractor, who is reported to be doing a very good job. Further, others have found that certain items may be better handled by a "roving trailer" type contractor who handles safety shoes, uniforms, coveralls, etc. All in all, all unions must recognize that they are in fact competing for the same jobs as contractors. All things equal, companies will normally stay with its employees. If major differences are demonstrable, or even appear to be compelling, then contracting must be considered.

Finally, though not specifically a contracting issue, in one area a supplier park was created through a cooperative effort of the purchasing managers of several large manufacturing operations. Essentially the park was built on 'spec,' and space was leased to a number of suppliers, e.g., routine bearings, lubricants, hose, piping, o-rings, and belts. These suppliers used a common same-day and next-day delivery system, electronic ordering, routine review of use and repair histories (from their records), and an integrated relationship with their suppliers to achieve a superior level of performance and lower inventories than what was otherwise achievable within an individual stores operation at the plants.

14.10   Key Performance Indicators

The following are suggested as key performance indicators for your consideration. Above all, performance indicators should provide an indication of the success of the organizational objectives which have been established.[1,2]

	Best	Typical
Stores values—% of facility replacement value	.25–.50%	1–2%
Service level—stockouts	<1%	2–4%
Inventory turns (see discussion below)	2–3	1
Line items processed per employee per hour	10–12	4–5
Stores value per store employee	$1.0M–1.5M	$0.5M–1M
Stores disbursements per store employee	$1.5M–2M	$0.5M–1M

Further, measures should also be set up for the following:

Utilization of catalog items in stock—percent, quantities

Utilization of preferred suppliers—percent, quantities

Carrying costs—Total and percent of store's value

Receipt and issue backlog—percent and delay days

If you establish all the processes described above, and begin to measure your performance, you should be well on your way to improved performance. However, it may be instructive to review a case history on how one company improved its inventory turns and managed its stores operation more effectively. This case history is provided below. Please note, however, that this company already had most of the systems described above in place, so that it could manage its stores operation.

14.11   Minimizing Stock Levels— A Cast Study[3]

Most companies are continuously seeking to reduce inventory levels. Certainly they should, since inventory generally represents capital which is not generating a return. Indeed, it costs money to store and maintain inventory. It could be compared to stuffing money under a mattress—which you're renting in order to store your money.

The drive to reduce inventory is often intense. Many companies will issue decrees—"A world class level for inventory turns of spare parts is 3. Therefore, we will be at three turns on our inventory in two years!," or some other equally arbitrary objective. Middle management is then left with the goal to reduce inventory, usually with limited guidance from senior management about the strategy of how to achieve this objective, or whether or not this objective can reasonably be accomplished. None the less, most will make a good faith effort to do so. However, most will be caught between the proverbial rock and hard place. If they simply reduce inventory across the board, they could jeopardize their ability to quickly repair failed equipment in a timely manner, due to lack of parts which resulted from reducing inventory, risking unplanned down time, or incurring extra costs. Further, many times inventory will be "reduced" only to find its way into a desk drawer, filing cabinet, storage closet, etc. for future needs. This is especially true in an organization which is highly reactive in its maintenance function, e.g., lots of breakdown maintenance, emergency work orders, run-to-failure, etc. So, what should we do?

Inventory should be driven by reliability and capacity objectives, and a systematic strategy, not by arbitrary decrees. The first step in establishing a basis for spares inventory management is to segregate inventory into categories which can be managed, such as:

- Obsolete—to be disposed as economically as possible
- Surplus—quantity > economic order point; to be managed in cooperation with suppliers
- Project—excess of which is to be returned at the end of the project
- High volume/use items—most to be put on blanket order and delivered as needed
- Low volume/use critical spares—most to be stored in house using specific procedures
- Low volume/use noncritical spares—most to be ordered when needed

You may have other, better categories, but this should start the thinking process for better inventory management, while still assure maximum equipment reliability. Integral to this, consideration should be given to other issues, such as:

- Machine failure histories
- Parts use histories
- Lead times

- Supplier reliability (responsiveness, quality, service)
- Stock-out objectives
- Inventory turns objective

And finally, all this should be integrated with the following:

- Strategic objectives as to reliable production capacity
- Application of a reliability based strategy for knowing machinery condition

There are other issues which may come into play at any given manufacturing facility, but these points should illustrate the principles involved.

14.12   A Hypothetical Example

ABC company, a heavy industrial manufacturer, has been directed by senior management to reduce spares inventory, such that turnovers are 2.0 by year end. On assessing their current position, they find the following:

- Current inventory level: $10M
- Current annual parts expenditure: $10M

They are presently turning their inventory at 1.0 times per year, which essentially means that inventory must be cut in half—a major challenge.

Senior management had further determined that unplanned down time, and maintenance costs were excessive as compared to industry benchmarks. Reactive, run-to-failure maintenance was resulting in substantial incremental costs due to ancillary damage, overtime, the unavailability of spares (in spite of high inventory levels), and most importantly lost production capacity from the down time. A team of the staff determined that application of a reliability based strategy would assure a reduction in down time and maintenance costs. At the heart of this strategy was the application of machinery condition monitoring technologies (vibration, infrared, oil, motor current, etc.) to facilitate knowing machinery condition and therefore:

- Avoid emergency maintenance and unplanned down time
- Optimize PMs—do PMs only when necessary, since few machines fail at precisely MTBF

- Optimize stores—plan spares requirements based on machine condition, move to JIT
- Assist in root cause failure analysis to eliminate failures, extend machinery life
- Commission equipment to ensure its "like-new" condition at installation, extending its life
- Systematically plan overhaul work requirements, based on machine condition
- Foster teamwork among production, engineering, and maintenance.

In doing so, the team felt they could extend machinery life, lowering spares requirements, and planning spares requirements more effectively for both routine and overhaul needs.

On reviewing their inventory, they find that the $10M in current inventory can be broken down as follows:

Obsolete	$1.0M
Surplus	$2.0M
Project	$1.0M
High volume/use	$4.0M
Low volume critical	$2.0M
Low volume non-critical	$1.0M

Note the total adds to more than $10M, because of overlapping categories. We'll ignore this for the moment for ease of demonstrating the principles, but this could be handled through a matrix for multiple categories. Their next step was to do the following:

Obsolete equipment was disposed as economically as possible, hopefully some value could be obtained through a broker, or other means. An inventory control process was established to dispose of inventory as it became obsolete, not to continue to hold it in storage. The company took a $800,000 charge on inventory disposal, a painful expense, but one which shouldn't occur again in a lump sum such as this with the new process in place. Inventory level $9M; turns at 1.11— progress.

Next, project inventory was reviewed with the following results. Half would be consumed by the project before year end. Another 25 percent was returned to the supplier with a 20 percent return penalty. This was accomplished only after considerable pressure was applied to the supplier. In the future they will have a policy built into major capital projects for returns, and will make this policy a part of the

contract with their suppliers. The final 25 percent was decided to be necessary as critical or important spares for future use. Result: $0.75M reduction in inventory. However, note that a $50K charge was incurred. Note also that in the future project inventory and the timing of its use/disposal/return must be considered when developing inventory turn goals. Inventory turns is now at 1.21 — more progress. When combined with a commissioning process to test and verify proper installation based on standards for vibration, IR, oil, motor current, and other process parameters, they felt they could substantially improve machinery life and avoid rework. Their experience had been that half of
their machinery failures occurred within two weeks of initial installation. They also applied these same standards to their suppliers.

Next, surplus inventory was reviewed, considering each of the categories shown. Economic order points were reviewed and it was determined that a total of $200K of the surplus could be used before year end. More stringent policies were put into place regarding order points, considering stock-out objectives. More importantly, the condition monitoring program allowed the company to reduce the quantity of its order points. They now planed to provide their suppliers with at least 5 days' notice on certain spare parts needs. With good support and ease of shipping to the plant, inventory could be reduced even further in the future. For now though, inventory turns are at 1.24 — more progress.

Next, a detailed review was performed of critical equipment (equipment whose failure leads to lost production). The equipment was detailed and listed down to the component level which was kept in inventory. Lead times were reviewed. A zero 0 percent stock-outs policy was established for this equipment. After a team from production, purchasing, maintenance, and engineering reviewed the listing, lead times, and inventory in stock, the conclusion was reached that excess inventory amounted to $500K, but that fortunately most of that would be used during an outage planned later in the year. All things considered, they expected to reduce critical low turning inventory by $300K by year end. Over the long term, they met with other plants concerning their inventory, identified common equipment and spares, and anticipate that some $500K in spares will be defined as critical/common, meaning several plants will use this as part of their slow moving, critical inventory. This inventory, while stored at this plant, was placed under the control of the division manager. Inventory turns now at 1.29.

Next, high volume, high use rate inventory was reviewed. This inventory typically turned at two or more times per year. Concurrently the inventory was reviewed against machinery histories, identifying critical machinery overlaps. The inventory was categorized by vendor $, and reviewed as to lead times in major categories. Reorder points were reviewed, and in some cases trimmed. Condition monitoring was used as a basis for trending equipment condition, and determining and planning needed spares. Suppliers were contacted and asked to maintain, under blanket order and/or consignment agreement, appropriate quantities of spares for delivery within specified time periods. A stock-out objective not to exceed 5 percent was determined to be acceptable for most routine spares. All things considered, it was expected that high use spares could be reduced by $1M by year end, and even more over the next 2–3 years. Inventory turns now at 1.48.

Next, low volume, non-critical spares were reviewed, again, in light of lead times, machinery history, use history, stock-out objectives, etc. Suppliers were contacted concerning maintaining guaranteed spares under a blanket order and/or consignment agreement, with minimum quantities, etc. All told, it was felt that half of the low turning, non-critical spares could be eliminated by year end, with most of the balance eliminated by the end of the following year. Inventory turns is now at 1.6, substantially below the decree of 2.0, but substantially above the historical level of 1.0. The company has positioned itself to increase available cash by $3.75M by year end. A reduction of an additional 1.25M or more (considered well within their grasp) over the next 1–2 years would yield the objective of inventory turns of 2.

They were now ready to present their findings to management. After their presentation to management, the objective of two turns on inventory was considered not to be achievable before year-end, but that in light of capacity objectives, a good plan had been established to achieve that objective within 18 months. The plan was approved, with the proviso that the team would report quarterly on its progress, including any updated plans for additional improvement.

Consistent with this, the company now put in place a strategic plan for reducing inventory which included the approach described above and had the following characteristics:

1. Production targets considered achievable were:
   - 95 percent production availability, including 4 percent planned and 1 percent unplanned down time
   - 90 percent production capacity, including 5 percent downtime for lack of market demand
2. Inventory targets were:
   - 0 percent—stock-outs for critical spares;
   - 5 percent—fast turning noncritical;
   - 10 percent—slow turning noncritical

3. Supplier agreements under blanket order were effected for inventory storage and JIT shipment.

4. Interplant sharing of common critical spares, placed under control of the Division Manager.

5. Systematic application of condition monitoring to:
   - Baseline (commission) newly installed equipment
   - Trend equipment condition to anticipate and plan spare parts needs
   - Comprehensively review equipment condition prior to planned overhauls
   - Engage in root cause failure analysis to eliminate failures

The company was well on its way to improved facilities reliability, and concurrently reduced costs and inventory.

14.13 Summary

Good stores management first requires that management recognize the value-adding capability of a good stores function — increased working capital, increased facilities capability, reduced carrying costs, reduced shrinkage, and improved maintenance efficiency. The losses associated with poor stores management practices should be intolerable to senior management.

In setting up a quality stores management function, a number of issues must be addressed:

- Development of a comprehensive catalog for stores requirements

- Development of policies and procedures which facilitate effective management of stores

- Establishment of procedures and practices which facilitate maintenance excellence

- Establishment of supplier partnerships, blanket orders, consignment, etc.

- Establishment of a stores layout which assure efficient operation

- Comprehensive training of all appropriate staff

- Consideration of contracting the stores function to accomplish these tasks

- Comprehensive management of stores and inventory turns by classification

- Comprehensive performance measurements for ensuring a superior stores capability

Doing all this in a comprehensive, integrated way will ensure world class stores, and therefore maintenance and facilities operations — your path to financially successful performance.

14.14 References

1. Edwin K. Jones, *Maintenance Best Practices Training,* PE-Inc., Newark, DE 1974.
2. G. Hartoonian, Maintenance Stores and Parts Management, *Maintenance Journal,* 8(1), Jan/Feb 1995, Mornington, Australia.
3. R. Moore, Establishing an Inventory Management Program, *Plant Engineering,* March 1996, Chicago, IL.

Maintenance and Repair Technology

15.1

Piping

The Mechanical Contracting Foundation
Rockville, Maryland

15.1.1 System and Component Guidelines

Piping systems interconnect equipment and piping components of heating, ventilating and air conditioning (HVAC) systems in order to transmit liquids and gases to the equipment for heat transfer processes. Pipe failures in the HVAC system are remote when properly installed; however, the loss of media from the pipe at joints or system components does occur when connections are not properly made by the installer.

Very minor leaks at joints will not affect system operation because most hydronic systems have automatic make-up of media. Systems without automatic make-up will have loss of efficiency and equipment "lock out" when leaks occur; therefore, repair procedures should be initiated promptly.

When selecting piping configurations and equipment locations, proper piping practices must be followed. An improperly—or poorly—installed piping system can later prove to be a major service problem. In addition, it is possible to have a piping configuration that is functionally correct, but it is configured in such a manner that servicing the equipment is difficult, and in some instances impossible.

Never install lines in a position where they will block equipment access panels. Never install piping systems containing liquids where condensation or leaks could occur directly over electrical panels or electrical equipment without proper protection. Electrical access panels must have at least 36 in (92 cm) of clear access.

Piping should be installed to provide clear access for servicing equipment. Piping systems should be properly cleaned and flushed to avoid original-installation debris from accumulating in the system or in connected equipment.

The position of the equipment to be installed should be given initial consideration. The position selected should make the installation less complicated, while creating a safe and workable atmosphere for future service and maintenance work. To address the serviceability of piping, it is necessary to address the types of equipment and pipe accessories that are available.

Maintenance features should address pipe surface temperatures, expansion and contraction, corrosion, vibrations, system pressures, pipe joints, and connections to system components. Maintenance is necessary because of the broad range of operating temperature, pressures, and many different connected components.

Piping systems discussed are: hot water—low, medium, and high temperature, chilled water, condensing water, steam, condensate, refrigerant, natural gas, and oil.

A. *Hot water piping systems* are normally constructed of copper or black steel pipe. Systems are insulated to retain heat and protect individuals from surface temperatures. Corrosion at joints is an indication of a system leak; the joint should be repaired. Medium and high temperature systems are subjected to considerable expansion and contraction of the pipe materials. Attention should be given to expansion joints, anchors, and pipe guides, which are installed to control the movement of pipe as a result of expansion. Excessive pipe movement can cause damage to insulation, piping components, pipe joints, or surrounding equipment.

B. *Chilled water piping systems* are normally constructed of copper or black steel pipe. Insulation of these systems must maintain external surface temperature above the surrounding dew point and maintain a sealed vapor barrier. Condensation will collect on pipe surface if insulation is defective. This condition shall cause the pipe to rust and components to corrode. Defective insulation must be repaired immediately.

Hot and chilled water pipe systems must be totally filled with system fluid because a shortage of fluid will permit air to be retained in the system. This condition will restrict flow to some system components. The system must be vented to obtain full operating conditions. If air is retained in the system, internal corrosion can occur.

C. *Condensing water piping systems* are normally constructed of galvanized, black steel, copper, or plastic pipe. Internal corrosion is a major maintenance problem because the system is open to the atmosphere. Periodic inspections should be regularly scheduled to control this condition. Chemical treatment and cleaning are necessary to maintain clean piping and system components. System bleed off must be performed to reduce the quantity of solids that could collect in the water system and to prevent excessive corrosion. Closed circuit systems should be maintained as other water systems.

D. *Steam and condensate pipe systems* are normally constructed of black steel pipe. Fiberglass reinforced plastic (FRP) is frequently used for condensate systems. Maintenance conditions are the same as those for the hot water piping systems.

E. *Refrigerant pipe systems* are constructed of copper pipe for freons and black steel for ammonia. Internal pipe surfaces must be clean and free of debris to prevent damage to system components; therefore, system piping maintenance procedures must be followed. Oil slicks at joints are an indication of leakage and should be repaired. Vibration of piping as a result of compressor operating should be controlled by use of vibration eliminators and pipe anchoring.

F. *Natural gas piping systems* are normally constructed of black steel or plastic pipe. These systems are usually maintenance-free when properly installed.

G. *Oil piping systems* are normally constructed of copper, black steel, or plastic pipe. Oil slicks at joints are an indication of leaks and must be repaired.

15.1.2 Installation Guidelines

15.1.2.1 Preassembly procedures

Piping should be cut accurately to establish measurements and then neatly installed either parallel to or at right angles to walls or floors. Materials should be worked into place without springing or forcing. Sufficient head-room should be provided to clear lighting fixtures, ductwork, sprinklers, aisles, passageways, windows, doors, and other openings. Materials should not interfere with access to other equipment.

Materials should be clean (free of cuttings and foreign matter inside) and exposed ends of piping should be covered during site storage and installation. Split, bent, flattened, contaminated, or damaged pipe or tubing should not be used.

Sufficient clearance should be provided from walls, ceilings, and floors to permit welding, soldering, or connecting of joints and valves. A minimum 6- to 10-inch (15–25 cm) clearance should be provided.

Installation of material over electrical equipment such as switchboards should be avoided. Piping systems should not interfere with safety valves or safety relief valves.

15.1.2.2 Assembly procedures

A means of draining the piping systems should be provided. A ½- or ¾-inch (12 mm–18 mm) hose bib (provided with a threaded end) should be placed at the lowest point of the water piping system for draining. Constant grades should be maintained for proper drainage, and piping systems should be free of pockets or traps due to changes in elevation.

Unions should be installed in an accessible location in the piping system to permit dismantling of piping or removal of equipment components. Clearance should be provided for using wrenches of sufficient size to break unions after many years of service. Union joints should be clean of system debris prior to installation to permit proper sealing.

Flanges should be installed in an accessible location in the piping system to permit dismantling of piping or removal of equipment components. Clearance should be provided for using wrenches of sufficient sizes to break flange bolts after many years of service.

Mechanical joints, when used for dismantling or removal of piping components, should be installed in an accessible location in the piping system. Clearance should be available for using wrenches of sufficient size.

When pipe threads are cut too long, it allows the pipe to enter too far into the fitting or system component. For example, on a valve installation, the pipe end would hit against the valve seat area distorting the seat and body, and this could prevent a tight shutoff when the valve is closed (See Fig. 15.1.1).

When a system component is installed on threaded pipe, it is important to use a wrench on the end of the component nearest the joint. Never use a wrench on the component end opposite from the end being joined to the pipe as this could distort the component body or threaded end of the pipe (See Fig. 15.1.2).

When pipe joint compound is applied to internal (female) threads, the compound may be forced into a valve seating area, contaminating the fluid in the pipeline. Pipe joint compound must be applied to external (male) threads only. The same rule applies when using Teflon® tape.

Cracked flanges are usually the result of unequal stress due to improper makeup or poor bolt-tightening procedures. Flanges must be properly aligned before bolts are tightened (See Fig. 15.1.3). There should be no gaps between the flanges. Flange bolts should not be used to correct alignment problems.

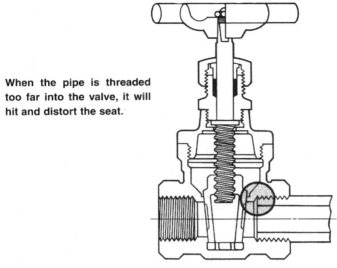

When the pipe is threaded
too far into the valve, it will
hit and distort the seat.

Figure 15.1.1   Threading pipe into a valve. (Courtesy Grinnell Corp.)

There are at least three stages of bolt tightening that shall be followed. First, properly align the flanges and gaskets, install the bolts
and hand tighten the nuts. Second, snug up the bolts *but to less than
the specified torque.* Depending on the specific requirements of the
component being installed, more than one intermediate level of torque

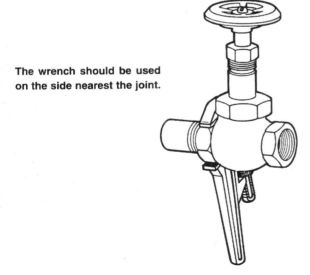

The wrench should be used
on the side nearest the joint.

Figure 15.1.2   Using a wrench on a valve. (Courtesy Grinnell Corp.)

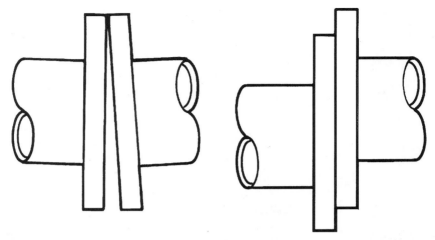

Figure 15.1.3 Examples of improper flange alignment and fit-up. (Courtesy Grinnell Corp.)

may be prudent before applying the final/specified torque. The bolt-tightening sequence demonstrated in Fig. 15.1.4 shall *always* be followed. Third, apply final/specified torque.

Flange gasket selection is also very important. Most flanges have a raised face or a flat face that is the mating surface where the flanges fit together. Ring gaskets should be used with raised-face flanges, and full-face gaskets should be used with flat-face flanges.

The most common gasket material of the wide variety available is red rubber. Several asbestos substitutes are being developed for applications where asbestos is no longer approved.

The art of soldering joints is well detailed in other manuals, thus it is not covered here. In general, the most common problem with solder joint connections is when excessive heat is applied while soldering the

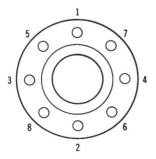

Figure 15.1.4 Flange bolts tightening sequence. (Courtesy Grinnell Corp.)

joint. Resilient seats become scorched, and metal seat rings become distorted.

Valves should be fully open during the soldering process. In addition, most of the heat should be applied to the tubing, not the component.

Debris in piping lines is a major cause of contamination. Installation of new piping, in particular, must always be cleared of any foreign material that could penetrate piping.

Pipe lines must be flushed or blown out (1) to eliminate weld spatter, or small pieces of metal that are created when butt welds are made in piping, (2) to eliminate pipe chips, the metal chips created when cutting pipe threads and drilling holes in piping, (3) to eliminate pipe scale, the scale-like chips in steel pipe caused by severe rusting, (4) to eliminate any internal debris.

Refrigerant piping systems should be purged with dry nitrogen when brazing to eliminate internal debris.

A strainer should be installed at unfiltered water sources, pump intakes, control and metering devices to flush debris from the piping system.

15.1.2.3 Storage procedures

The proper storage of material requires a reasonably clean and preferably dry storage area for piping system components, an orderly placement of the various material categories, and adequate access for material removal and/or added deliveries.

Material that is improperly stored is usually subject to exposure by foreign substances, such as mud, dirt, sand, or other debris. During system erection, most of these elements find a way to stay within the system so that during system start-up operations, costly and time-consuming problems generally develop. Most problems can be avoided by using proper storage techniques.

Piping should be stored on racks or dunnage, off of the ground, and protected from the elements. Where practical, pipe fittings should be placed in bins. Larger sizes should be stored on pallets to ensure that clean, dry and easy handling conditions prevail.

Valves and specialities should be stored and protected in a similar manner. In addition, in almost all cases, they should at least be protected from the elements by waterproof tarpaulins or an enclosed storage area. Gasket materials must always be protected from exposure to ultraviolet light and ozone.

When purchasing piping materials, it is recommended that they be ordered for delivery in such a manner that materials and sized parts are delivered in a prepackaged separate condition [i.e., ½-inch (12-mm) copper elbows separate from ½-inch (12-mm) copper tees]. This will facilitate the establishment of an orderly warehouse arrangement where materials are binned in separate compartments or containers.

The orderly arrangement of material by size and type permits the easy location of needed material during system erection, as well as recognition of potential depletion of stocked material prior to running out of an important system component.

The ability to get stored material is vital for an efficient storeroom operation. Larger fittings and valves (where mechanical apparatus is needed for proper handling) should have adequate aisle space for equipment (e.g., forklift). Smaller material that can be manually handled requires only personnel access space.

15.1.3   System Components

Many components are common to various pipe systems. Components requiring maintenance are valves, metering devices, relief valves, strainers, and firestops. Other components, such as unions, hangers, and fittings, should be inspected visually.

15.1.3.1   Valves—common usage types (See
also Chapter 17.2, "Valves").

Proper installation is necessary if valves are to perform as expected. Some of the more common but sometimes overlooked installation and service problems are described below. Valves should be installed in a position that allows servicing of units. Consideration should be given to the removal of bonnets and of internal operating parts, and to the ability to pack units.

Isolation valves should be installed in a position that will permit the removal of the piping system or equipment. Isolation valves should be rated at a pressure differential high enough to accommodate full shut-off, with one side open to atmospheric pressure.

Drain valves should be installed at the bottom of the piping system risers in a practical and serviceable area.

The correct valve position for gate, globe, and check valves is an upright position whenever possible. *Exception: A valve in chilled water systems should have its operating stem in a horizontal position to prevent condensation from damaging insulation.* Gate, globe, and check valves are designed to be installed upright. Positioning valves otherwise invites problems because sediment in the line will settle in the valve's working parts and bonnet cavities. Swing check valves will not function properly in the upside-down position under any circumstances.

If valves do not seat properly when first used, flush foreign matter off of the seat by opening a valve slightly, then closing it, and repeating this procedure as necessary. Since the valve seat may have debris preventing normal closure, it is a mistake to apply excessive force on the handwheel or handle. Abnormal force imposed on valve stems, seats, etc., will distort or break them. Cheater bars or any devices that put more-than-normal force on valve handles should never be used.

Most valve leakage problems fall into one or more of these three categories: *through-the-valve, outside-the-valve, and outside-to-inside-the-valve.*

1. Through-the-valve leakage occurs when flow penetrates beyond the closing member of the downstream side of the valve.

2. Outside-the-valve leakage is characterized by flow escaping through valve seals, or packing, to outside the valve.

3. Outside-to-inside leakage occurs in vacuum service where air is drawn in through worn or damaged packing or seals.

As previously stated, when a valve fails to shut off completely, do not use extra force to close it. Debris on the seat area is often the problem and is handled by opening the valve partially. The resulting turbulent flow will usually flush the debris.

Always open and close one-quarter-turn valves slowly. Operating them too rapidly may result in water hammer and damage to the valve or piping.

15.1.3.2 Balancing valves

These valves are used in hydronic systems to adjust and proportion the flow through various branches and mains to various portions of the system. If balancing valves are omitted from the piping system, the flow will take the path of least resistance, which may result in some sections of the system not receiving any flow. As an example, the use of a balancing valve on a warm-water heating coil will limit the flow through that coil, limiting the amount of energy used to heat the air passing across the coil.

The three most common balancing valve configurations are: (1) calibrated manual balancing valve, (2) flow-measuring venturi or orifice plate with manual ball or butterfly valve, and (3) automatic flow-limiting (or automatic flow control) valve.

15.1.3.3 Steam traps (See also Chapter 22, "Condensate Control").

Soon after steam leaves a boiler, it begins to lose some of its heat to any surface with a lower temperature, and it begins to condense into water. Clearly, it is desirable to separate and remove the accumulated water (condensate) from the steam supply system by sending it back to the boiler through a piped-vacuum or gravity-return system, while retaining the heat or energy that produces steam within the supply steam. This energy-saving function is fulfilled by means of steam traps.

The entire steam distribution system must remain free of air and condensate. Failure to maintain the steam distribution system free of air and condensate often leads to water hammer and slugs of condensate that could damage equipment, system control valves, and steam traps.

Common to all steam distribution piping systems is the requirement for strategically located drip legs. Drip legs permit condensate to escape by gravity from rapidly moving steam within the system. In addition, drip legs store the condensate until it can be discharged by the pressure differential through the steam trap.

The following installation considerations should be considered when installing steam traps.

1. Service, repair, and replacement are simplified by making steam trap configurations identical for each given size and type.

2. Isolation valves are recommended in order to permit the removal of steam traps without shutting down the operating system.

3. When only one union is used, it should be placed on the discharge side of the steam trap. Avoid in-line (horizontal or vertical) installation of more than one union. It is best to install two unions at right angles to each other (one in horizontal run, one in vertical run).

Dirt legs are installed upstream of a trap assembly to prevent scale and dirt from entering the steam trap. Dirt legs should be cleaned periodically by using a blowdown valve or by removing the end cap and free blowing until clear.

Install a strainer upstream of a steam trap, especially where pipe line dirt conditions are present. Some steam traps are made with built-in strainers.

15.1.3.4 Strainers/filters-driers

Strainers should be installed in the piping system at locations that will allow removal of the screens for cleaning. Strainers should be used on the inlet side of system components to protect critical operating parts from foreign particles and debris.

The strainer screen must be clean to permit full-flow. The strainer must be cleaned after piping system is repaired; flush strainers after the system is activated as a result of repair procedures.

Strainer screens are available with various opening sizes and materials. The screen specification must be used to maintain a clean system and to resist corrosion from system media. A fine-mesh type screen may be used for system cleaning after repair; however, the specified screen should be used during normal system operations.

Refrigerant systems normaly use filter-driers that function as a strainer and a moisture collector. A three-valve bypass assembly permits ease of removal and replacement of a clogged unit.

15.1.3.5 Flexible connections

The installation of these connections in the piping system prevents the transmission of vibrations from equipment to piping in other parts of the building.

Piping system components should be supported independently on each side of a flexible connection so that the connector can be easily removed when a piping system is disconnected.

15.1.3.6 Air vents

These vents should be installed in the high points of any piping system where air could accumulate, and they should be installed wherever the water stream reduces velocity, changes direction, or is heated. Any such areas should have either a manual or automatic air vent installed. Automatic air vents should *not* be installed on the suction side of pumps.

If the vent port of an automatic air vent discharges in an area where there is danger of water damage, it should be piped to a drain. Automatic air vents should be installed with a manual isolation valve so that the air vent can be replaced or repaired. If automatic air vents are to be left open to the system, it is necessary to install an automatic make-up water system to compensate for the leakage at the air vent. The air vent should be installed in an accessible area for use and service.

15.1.3.7 Thermometers (or thermometer wells)

These should be installed on the inlet and outlet of each heat transfer device. Consideration should be given to locations that will permit easy reading of the thermometer. Thermometers should be installed in the piping system at points that protect them from damage during normal service activities around the equipment.

15.1.3.8 Gauges and gauge ports with shut-off valves

These gauges and gauge ports should be installed on the inlet and outlet of each piece of equipment that reflects pressure differential in a piping system. One water pressure gauge can be used for connection between the inlet and outlet of a component by manifolding so an accurate pressure-differential reading can be obtained. Pressure gauges permanently installed in a piping system should have a pigtail, or where applicable, a snubbing device, between the gauge and the system shut-off valve. The shut-off valve can be used as a snubber while reading system pressure, and it should be shut off when not reading system pressure.

15.1.3.9 Flow switches

When pressure-differential flow switches are required, they should be installed in the piping system with a minimum of 10 pipe diameters of straight pipe on each side of the pressure-sensing point. Pressure-differential flow switches have proven to be a service problem due to the small pressure drop across the system components.

If the piping-system line size is less than 1-inch (25-mm), a piston-type flow switch should be used. If the piping-system line size is greater than 1-inch (25-mm), a paddle-type flow switch should be used. Paddle-type flow switches require at least a 1-inch (25-mm) pipe diameter spacing and operate best with 10 gpm (38 liters/min) or more of water flow.

15.1.3.10 Anchors and guides

These anchors and guides are installed in piping systems that are subjected to a wide temperature range or pressure shock. Expansion joints or expansion loops are installed to permit system expansion and contraction. Pipe alignment guides and anchors are installed to control direction of movement. Components should be inspected occasionally to verify design performance.

15.1.3.11 Pipe sleeves and firestops

The inspection of pipe sleeves is necessary because damage to the packing materials may eliminate fire protection qualities.

Sleeve packing materials are installed to prevent the intrusion of dirt, water, air, rodents, and foreign objects. Packing materials add air tightness, cathodic protection, corrosion resistance, anchorage support, and shock (sound and vibration) absorption. Common packing materials are lead-and-oakum joints, mastics, casing boots, and link seals.

Project specifications determine the types of materials required to seal the annular space between the pipe and sleeve.

To penetrate fire-rated floors and walls with items such as pipes, a thru-penetration firestop is required. This firestop consists of specific field-installed materials designed to prevent the spread of fire through fire-rated openings.

The firestop codes require that any penetration through a fire-rated wall or floor must be firestopped to a fire rating that is equal to that of the wall or floor before the opening was created. In some cases, state law may require that a building meet NFPA Life Safety Code, Section 101, in addition to local requirements.

The following listings present the materials and hardware associated with firestop installations.

1. *Firestop dam-forming material.* Some form of damming material, such as mineral wool, backer rods, or insulation must be placed in the annular space to help support the fire-stopping caulk or sealant.

2. *Firestop caulk or sealant.* The caulk or sealant in a firestop is the material that actually stops the fire. Firestopping caulk or sealant is available in four chemical compositions to satisfy different applications.
   - Intumescent caulk or sealant expands when exposed to heat.
   - Silicone caulk or sealant is very flexible and allows movement of the pipe.
   - Endothermic caulk or sealant expels moisture in a fire to protect the seal.
   - A water-based material is the fourth kind of caulk or sealant.

3. *Firestop retaining hardware.* Retaining collars, clips or wire mesh may be required in some firestop systems to support the firestop system, or to direct the expansion of intumescent materials.

4. *Firestop intumescent wrap strips, firestop kits / devices.* Strips of intumescent wrap (which may expand more than sealants when exposed to heat) are typically used around combustible penetrants. These materials may be used in conjunction with retaining hardware, fasteners, etc.

15.1.4   Safety Considerations

Whenever a system must be serviced or repaired, or when a piece of pipe must be removed, it is necessary to plan the service or repair procedure carefully to avoid personal injury and damage to valves or piping. As a minimum, the piping system must always be depressurized before attempting the removal or repair of any piping component.

Special caution must be taken to relieve pressure from any piping system before service or removal. The operator is responsible for knowing the type of fluid, gas, or other medium in the piping system. The process for disposal or recovery of the medium also must be known. Relieving the system pressure must be accomplished in accordance with requirements for the specific medium involved.

The depressurization of any refrigerant system must be accomplished through the normal established recovery process. Depressurizaton should be handled in accordance with Environmental Protection Agency (EPA) requirements.

Unknowingly, pressure can remain in a piping system even though all of the valves are shut off and the system has supposedly been drained. Even after a section of piping has been sealed by isolating valves, the medium within the pipe can remain under enough pressure or temperature to cause injury. Gaskets may have been sealed to both flange faces by heat, allowing some flange joints to hold pressure, even though the bolts are removed. Ball and plug valves can trap fluid inside and retain it there, even after removal from the piping.

When inspecting valves, always hold them in a manner that prevents fluid from being spilled onto personnel after removal. When dealing with hazardous or corrosive materials, special care must be taken to prevent personal injury and damage to surrounding materials and surfaces.

Opening a drain valve does not always make a piping system safe to work on as leaking valves can often result in unexpected pressure buildup.

When a flanged connection is being dismantled, the safe practice is to make it leak to relieve pressure before removing all the bolts. Leaking a flanged connection to relieve pressure should be accomplished with any system that is known to be or is suspected to be under pressure; do not depend solely on a pressure gauge to assure a safe condition. *Never trust a pressure gauge; double check line pressure whenever possible!*

15.1.5   System Pressure Testing

Piping systems are usually tested at 1.5 times the working pressure of the system. The system should not be buried, concealed, or insulated until it has been inspected, tested, and approved. All defective joints should be repaired, and all defective materials should be replaced. The following procedures briefly summarize a typical pressure test.

1. Examine the system to ensure proper isolation of equipment and system parts that cannot withstand the test pressures. Examine the test equipment to ensure that it is leak-free and that low pressure filling lines are disconnected.

2. Provide temporary restraints for those expansion joints that cannot sustain reactions resulting from test pressure. If temporary restraints are not practical, isolate expansion joints from testing.

3. Isolate equipment that is not to be subjected to the piping system test pressure. If a valve is used to isolate the equipment, its closure shall be capable of sealing against the test pressure without damage to the valve. Flanged joints, at which blinds are inserted to isolate equipment, need not be tested.

15.1.5.1 Water and steam systems

The following procedures summarize a typical pressure test:

1. Flush with clean water, clean all strainers. Use ambient temperature water as the testing medium, except where there is a risk of damage due to freezing. Other liquids may be used if safe for workers and compatible with the piping system components. The engineer should be consulted if the system is tested with a liquid other than water to ensure compatibility with other system components.

2. Use the manual air vents that have been installed at high points in the system to release trapped air while filling the system.

3. Subject the piping system to a hydrostatic test pressure, which at every point in the system is not less than 1.5 times the design pressure (high-temperature water piping shall not be tested at more than 500 psig [3500 kPa]). The test pressure *shall not exceed* the maximum pressure for any vessel, pump, valve, or other component in the system under test.

4. After the hydrostatic test pressure has been applied for at least four hours, examine the system for leakage. Eliminate leaks by tightening, repairing, or replacing components as appropriate, and repeat the hydrostatic test until there are no leaks.

15.1.5.2 Refrigerant systems

Mechanics must be proficient in testing, evacuation, and dehydration techniques of refrigerant systems to ensure a tight and dry system.

1. Pressurize the complete system with dry nitrogen to 50 psig (350 kPa), and observe over a period of time for any loss in pressure. Joints, fittings, etc., can be leak tested using an approved bubble solution.

2. After the system has been proven leak-tight with dry nitrogen, pressurize with a mixture of dry nitrogen and HCFC-22 (as a leak test gas) to near system-operating pressure or per the applicable local building code. Then test joints, fittings, etc. with an electronic leak detector or other suitable leak detection device.

   Caution: Enough time should be allowed for the refrigerant to mix with the dry nitrogen.

3. Evacuation and dehydration. After determining that there are no refrigerant leaks when the system is pressurized, the system must be evacuated and dehydrated to remove moisture and noncondensables. Evacuation of the system should be accomplished with a vacuum pump to lower the absolute pressure of the system to 500 microns or less. This procedure will reduce the internal pressure of the system below the boiling point of water. External heat may be required to vaporize the water in the system.
   A standing vacuum test should then be performed to ensure that the system can hold a vacuum, and that the rate of rise is less than 50 microns per hour. If this test fails, and the system equalizes at the vapor pressure of the ambient temperature, then moisture is still present, and further dehydration must be performed. If the system pressure continues to rise, it will be necessary to check for leaks again, using the procedures listed above. It is advisable to hold the vacuum for 24 hours.
   When pulling a vacuum on a system, care should be taken to avoid evacuating the system so rapidly that moisture in the system freezes. Moisture in the form of ice can remain in the system and cause internal damage. The vacuum on the system should be broken with the refrigerant specified for the system.

15.1.6 Servicing Piping Systems

The most common defects, cause, and repair/maintenance procedures of piping systems or components are listed in Table 15.1.1. A visual inspection of the systems periodically (quarterly) should reveal de-

TABLE 15.1.1   Piping System Troubleshooting Chart

COMPONENT	DEFECT	CAUSE	REPAIR & MAINTENANCE PROCEDURE
AIR VENTS	Leak	Dirt	Remove, flush, or replace (Depress stem to verify shut-off)
	Plugged	Sediment	
DIELECTRIC UNIONS	Leak	Gasket failure resulting from expansion/ contraction	Clean joint – replace gasket. (Consider anchors and guides to control movement of pipe.)
		Overtightening at time of installation damaged component	Replace fitting
GAUGES	False reading	Dirt in gauge line	Clean line and fittings
		Vibration or pulsation	Use snubber valve and shut-off valve. Open shut-off valve only when reading.
HANGERS	Loose	Vibration/Movement	Adjust rod length and lock nuts. Add units at equipment to support pipe free of components.
	Noise	Vibration	Consider use of spring or fiberglass hanger isolator.
INSULATION	Deterioration	Condensation	Replace – Apply proper vapor barrier
			Metal jacket may be necessary for protection
			Saddles at Supports
PVC PIPE	Leak at fittings	Cracked ends of pipe	Cut off defective ends before makeup at joints.
PIPE CLAMPS/ HANGERS	Loose or missing	Vibration	Tighten – use lock nuts
PIPE STEEL	Internal pitting	Air (oxygen) in system	Bleed system; Check air control systems; Check compression tank air level

fects. Manual valves should be turned a quarter to a half turn to provide freedom of movement of parts. Blow down valves should be opened to flush out any dirt accumulation. Relief valves should be manually activated on steam and water systems to verify proper operation.

TABLE 15.1.1 (*Continued*)

COMPONENT	DEFECT	CAUSE	REPAIR & MAINTENANCE PROCEDURE
PRESSURE REDUCING VALVES	Erratic flow	Pilot line clogged	Clean and flush line
	Fails to shut-off	Contamination	Replace or rebuild valve
REFRIGERATION FILTER DRIERS	Excess pressure drop	Dirt	Replace
	Excess moisture	Contamination	Replace
REFRIGERANT RUPTURE DISC	Leaking refrigerant	Excess temperature or pressure	Replace disc after system repair – comply with refrigerant procedures
RELIEF VALVES	Leaks	Weak diaphragm; Dirt on seat valve spring tension	Flush or clean seat
	Fails to open	Corrosion	Replace valve, if necessary – check pipe for cleanliness. Test at system design.
STRAINERS	Clogged	Dirt in system	Back flush
			Remove screen – clean and replace
			Verify mesh size of screen system cleaning
THERMOMETERS	Separation	Excessive vibration or temperature	Proper location of device
	False Reading	Poor heat transfer	Use compound in well to transmit temperature
		Wrong scale device	Replace with device with scale of media range

The loss of media from hydronic systems may result in internal scaling and oxygen corrosion of pipe. This condition will cause clogged strainers, valve seat defects, flow restriction in heat transfer units, and low system efficiency. The system should be cleaned, flushed, and pressure-tested after repair. Hydronic systems should be filled and have air vented through manual vents to obtain full circulation through all components. The media should be analyzed to determine the need for chemical treatment.

TABLE 15.1.1 (*Continued*)

COMPONENT	DEFECT	CAUSE	REPAIR & MAINTENANCE PROCEDURE
UNIONS	Leak	Dirt	Clean joining surfaces
		Misaligning pipe ends	Align pipe ends – tighten
	Flow turbulence	Installation direction	Install union in proper flow direction
VALVES	Will not turn	Stem corroded	Clean and lubricate stem (Do not paint stems)
	Will not shut-off	Dirt in seats	Clean seats, discs, gates; Replace as required
WATER HAMMER	Lack of flow	Air in system	Install manual air vents at high points to vent system
	Noise	Condensate in steam flow	Install drip leg and steam trap to collect condensate
VIBRATION ELIMINATORS	Leaking failure	Pipe weight or alignment	Support pipe independently; Align pipe

For control valve maintenance, see "System Control Equipment" (Chapter 21, also "Valves," Chapter 17.2.)

As previously stated, piping systems when properly installed should require very limited maintenance. Visual inspections may reveal the need for repairs; however, only skilled mechanics qualified for the specific piping system and understanding the media in that system should attempt any repair procedures.

The piping components of HVAC systems have a much longer life cycle than heat transfer components; therefore, repair is more common than replacement. A highly corrosive atmosphere or internal contamination could result in pipe pitting, scaling, and eventually failing if not maintained. Periodic (annual) inspection and chemical treatment of media should provide good operating conditions for the life of the system.

15.1.7 Bibliography

1. "Guideline for Quality Piping Installation" 1996, Mechanical Contracting Foundation, Rockville, MD.

Ductwork

Michael S. Palazzolo
President, Safety King, Inc.
Utica, Michigan

15.2.1 A Practical Definition

Ductwork is a system of passageways through which air is delivered to, and/or extracted from, the interior spaces in buildings. The image called to mind by the word "ductwork" is usually the long, enclosed, aluminum passageways, rectangular in cross section, through which heated or cooled air is supplied to the rooms of homes or offices by the furnace or air conditioner fan units. But ductwork is not always long, or enclosed, or aluminum, or rectangular in cross section, and does not always convey heated or cooled air and may not even be connected to a furnace or air conditioning. In practice, ductwork must be understood to connect all the air handling components of the heating, ventilating, and air conditioning (HVAC) system (Fig. 15.2.1).

The HVAC system of a building includes return air grilles, return air ducts to the air handling unit, or AHU, (except ceiling plenums, those spaces above a suspended ceiling from which return air may be extracted), interior surfaces of the AHU, air wash systems, filters, filter housings, fan housings and blades, mixing boxes, coil compartments, condensate drain pans, spray eliminators, supply air ducts, humidifiers and dehumidifiers, turning vanes, reheat coils, and supply diffusers. However, the HVAC system of a building sometimes is not considered to include the air heating or cooling equipment (furnaces and air conditioners) themselves.

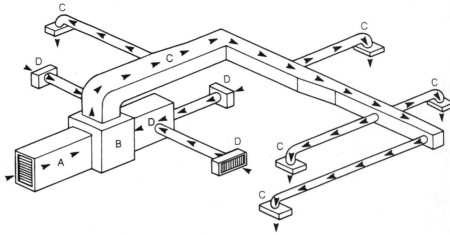

A. Fresh air intake (not always present)
B. Central heating/cooling equipment
C. Supply duct system
D. Return duct system

Figure 15.2.1   Typical HVAC system schematic. [Reproduced by permission of the North American Insulation Manufacturers Association (NAIMI).]

15.2.2   Tasks of the HVAC System

The HVAC System of a building has three primary tasks: to deliver air, to return air, and to make up air.

To fulfill the air delivery task, the HVAC system conveys to occupied building zones or conditioned spaces, air that has been pressurized, heated, cooled, humidified, dehumidified, washed, or filtered. In its return-air role, the HVAC system extracts air from occupied building zones or conditioned spaces in order to pressurize, heat, cool, humidify, dehumidify, wash or filter it. To accomplish the make-up air task, the HVAC system carries air to occupied building zones or conditioned spaces to replace air being removed by equipment such as range hoods or spray booths.

15.2.3   Maintenance Requirements of the HVAC System

Properly maintaining the HVAC system of a building requires several ongoing activities:

15.2.3.1   Inspection checklist, schedule, and log

In all but the smallest buildings, a written inspection checklist (in hard copy or on line) should be drawn up (example, Fig. 15.2.2-four parts). This inspection checklist should be sufficiently detailed to serve unaided as documentation to guide an inspector unfamiliar with both the building and its HVAC system through the entire inspection program. The procedure should be updated at least annually, or more often as changes may dictate, to insure that it reflects the most recent additions to or alterations in the components of the HVAC system. The inspection procedure, each time it is updated, should be dated or coded in such a way that its versions may be controlled. Some care should be exercised to limit the number of copies that are made to

HVAC Checklist - Short Form

Building Name: _____   Address: _____

Completed by: _____ Date: _____ File Number: _____

MECHANICAL ROOM

■ Clean and dry? _____ Stored refuse or chemicals? _____

■ Describe items in need of attention _____

MAJOR MECHANICAL EQUIPMENT

■ Preventive maintenance (PM) plan in use? _____

Control System

■ Type _____

■ System operation _____

■ Date of last calibration _____

Boiler

■ Rated Btu input _____ Condition _____

■ Combustion air: is there at least one square inch free area per 2,000 Btu input? _____

■ Fuel or combustion odors _____

Cooling Tower

■ Clean? no leaks or overflow? _____ Slime or algae growth? _____

■ Eliminator performance _____   _____

■ Biocide treatment working? (list type of biocide) _____

■ Spill containment plan implemented? _____ Dirt separator working? _____

Chillers

■ Refrigerant leaks? _____

■ Evidence of condensation problems? _____   _____

■ Waste oil and refrigerant properly stored and disposed of? _____

Figure 15.2.2 HVAC checklist [part 1].

HVAC Checklist - Short Form

Building Name: _____ Address: _____

Completed by: _____ Date: _____ File Number: _____

AIR HANDLING UNIT

■ Unit identification _____ Area served _____

Outdoor Air Intake, Mixing Plenum, and Dampers

■ Outdoor air intake location _____

■ Nearby contaminant sources? (describe)_____

■ Bird screen in place and unobstructed? _____

■ Design total cfm _____ outdoor air (O.A.) cfm _____ date last tested and balanced _____

■ Minimum % O.A. (damper setting) _____   Minimum cfm O.A. $\frac{\text{(total cfm x minimum \% O.A.)}}{100}$ = _____

■ Current O.A. damper setting (date, time, and HVAC operating mode) _____

■ Damper control sequence (describe) _____

■ Condition of dampers and controls (note date) _____

Fans

■ Control sequence _____

■ Condition (note date) _____

■ Indicated temperatures   supply air _____ mixed air _____ return air _____ outdoor air _____

■ Actual temperatures   supply air _____ mixed air _____ return air _____ outdoor air _____

Coils

■ Heating fluid discharge temperature _____ ΔT _____ cooling fluid discharge temperature _____ ΔT _____

■ Controls (describe)   _____

■ Condition (note date)   _____

Humidifier

■ Type _____ If biocide is used, note type _____

■ Condition (no overflow, drains trapped, all nozzles working?) _____

■ No slime, visible growth, or mineral deposits? _____

Figure 15.2.2 [part 2]

guard against inadvertent use of an expired version of the procedure. Whenever a new version is created, all copies of the previous version should be discarded. A copy of the current version should be under the control of the chief building engineer, or of a person in an equivalent position, whose responsibility it is to see that orderly and regular inspections are carried out as called for in the inspection procedure. This person should be prepared to produce this copy for examination by the building engineer audit team or auditor, should such an audit (or any other similar audit) be common practice in the building.

HVAC Checklist - Short Form

Building Name: _____ Address: _____

Completed by: _____ Date: _____ File Number: _____

DISTRIBUTION SYSTEM

Zone/ Room	System Type	Supply Air		Return Air		Power Exhaust		
		ducted/ unducted	cfm	ducted/ unducted	cfm	cfm	control	serves (e.g. toilet)

Condition of distribution system and terminal equipment (note locations of problems)

■ Adequate access for maintenance? _____

■ Ducts and coils clean and obstructed? _____

■ Air paths unobstructed?   supply _____ return _____ transfer _____ exhaust _____ make-up _____

■ Note locations of blocked air paths, diffusers, or grilles _____

■ Any unintentional openings into plenums? _____

■ Controls operating properly? _____

■ Air volume correct? _____

■ Drain pans clean? Any visible growth or odors? _____

Filters

Location	Type/Rating	Size	Date Last Changed	Condition (give date)

Figure 15.2.2 [part 3]

A schedule for inspections, matching the requirements of the inspection procedure, should be created and maintained in a place accessible to all personnel directly concerned with inspection and maintenance of the HVAC system. The schedule should specify dates and times for inspections to be carried out and identify the engineering personnel responsible for carrying out each one.

A log of inspections should also be created. This log, as it is completed during and following each inspection, becomes the official and

---

HVAC Checklist - Short Form

Building Name: _____ Address: _____

Completed by: _____ Date: _____ File Number: _____

OCCUPIED SPACE

Thermostat types _____

Zone/ Room	Thermostat Location	What Does Thermostat Control? (e.g., radiator, AHU-3)	Setpoints		Measured Temperature	Day/ Time
			Summer	Winter		

Humidistat/Dehumidistat types _____

Zone/ Room	Humidistat/ Dehumidistat Location	What Does It Control?	Setpoints (%RH)	Measured Temperature	Day/ Time

■ Potential problems (note location) _____

■ Thermal comfort or air circulation problems (drafts, obstructed airflow, stagnant air, overcrowding, poor thermostat location)

_____

_____

■ Malfunctioning equipment _____

■ Major sources of odors or contaminants (e.g., poor sanitation, incompatible uses of space)

_____

Figure 15.2.2 [part 4]

sometimes legal record of the maintenance of the building HVAC system. As such, it should be a first priority of building engineering to insure the log is completed for each inspection and that the log is maintained in a place accessible to all personnel directly concerned with inspection and maintenance of the HVAC system. The log should show the date each inspection is carried out, the person or persons conducting the inspection, the findings of the inspection, and the ac-

tion plan (with completion dates shown) for all unscheduled maintenance activity generated as a result of the inspection.

Since all three of these documents, the procedure, the schedule, and
the log, are of such singular importance to the maintenance of the
building HVAC system, building engineering should consider maintaining all three in a loose leaf binder kept in a single location.

15.2.3.2 Balancing of air flow

Proper maintenance of the HVAC system of a building is impossible
if proper balance of air flow is not insured in all occupied building
zones or conditioned spaces. If the air flow in any part of the HVAC
System is allowed to become unbalanced with respect to zone requirements, either too much or too little air flow, the occupants of the
affected zone are likely to take action that can only result in premature or excessive collection of particulate matter and/or serious damage to the equipment itself. These actions include taping over air supply grilles, removal and loss of supply grilles, abnormal thermostat
settings causing excessive running of heating or cooling apparatus,
overly frequent thermostat setting changes causing excessive start/
stop cycling of heating or cooling apparatus, disabling of thermostats
thought to be responsible for perceived lack of air flow balance, and
blocking off of parts of the building creating new zones not allowed
for in HVAC system design. These actions, taken by occupants in order
to secure their own comfort, are usually the result of inadequate attention to air flow balance. All these actions will mean more frequent
and costly cleaning and repair of affected HVAC system components.

Two things are needed to avoid these difficulties and insure proper
balance of air flow in all occupied building zones or conditioned spaces:
First, systematic monitoring of occupants and/or spaces must be carried out. And second, at least one person per shift on the building
engineering staff must be properly and completely trained in the specific air flow balancing techniques and equipment applicable to the
HVAC system components of the building.

In smaller buildings, monitoring of occupants and spaces can be a
simple matter of walking the various spaces of the building, noting
temperature and other air quality characteristics of each space, and
asking occupants, if any, to comment on their perceived level of comfort. In large buildings with many occupants, it may be necessary to
adopt a more formal process, utilizing regular measurements of, logging of, and corrective action on temperature, humidity, particulate
matter, or other air quality measurables connected to HVAC system
air flow balance.

Formal outside training may be necessary to assure one staff person
on each shift capable of adjusting air flow balance throughout the

building. This training can sometimes be obtained from the manufacturer of the HVAC system components affecting air flow balance. Some community colleges also provide courses which include training in specific air flow balance techniques and equipment. More frequently, such information is obtained by engineering staff members by careful study of equipment, blueprints, and engineering manuals supplied with HVAC system components at the time of installation. It is critical that this information, once reliably established, be documented and stored. If air flow balance techniques and equipment use are written down, training of new engineering staff members is greatly simplified and tends to produce a more uniform approach to air flow balance over time. Attention paid to air flow balance throughout the building HVAC system will pay for itself many times over in reduced cleaning and repair costs.

15.2.3.3 Replacement or repair of components—choosing a contractor

Even with best-in-class inspection procedures and consistent attention to balancing of air flow throughout the HVAC system, components still break down or wear out and must be repaired or replaced. Some of these tasks are very simple and can be carried out in-house, but most (particularly major failures) will require the services of one or more outside professional mechanical, heating, cooling, or plumbing contractors. This is particularly true when replacement or repair involves the use of expensive or specialized equipment you do not have in-house. Even when an HVAC system component repair or replacement can be handled in-house, you may often choose to bring in a contractor. In some cases, there may be service contracts in place. In effect, a service contract means that the services of the contractor have already been paid for so there is every reason to make use of the contractor's services. In other cases, the use of a contractor to address a replacement or repair of an HVAC system component may mean faster results. This could be important, especially in occupied buildings in which the component failure deprives occupants of heated or cooled air in work spaces. It is worth remembering too, that the use of a contractor always means that you and your staff can attend to the many vital building maintenance tasks for which you are responsible, rather than putting everything on hold while you address an HVAC system component repair or replacement.

If you decide to employ a contractor, and no service contract is in place, you will need to choose a contractor. This will generally be a matter of consulting historical files or lists to find out which contractors have performed similar services in your building in the past. If you find you must utilize a contractor unknown to you, invest the time it takes to talk with the contractor in advance and satisfy yourself that he or she appears to be knowledgeable and professional. Make sure you see applicable insurance, licenses, and references and do not contract until you have checked references. A few minutes of checking can eliminate expensive "do-it-again" costs and sometimes even higher legal costs in cases of service failure that can only be resolved by litigation.

15.2.4 Cleaning of HVAC System Components

The HVAC system components in your building will require regular cleaning. Because of health, safety, and regulatory issues, many buildings now have in place an HVAC system component cleaning schedule. If you do not have such a preventive maintenance cleaning schedule for HVAC system components, create one. Consult your inspection, repair, and cleaning records and use these to estimate appropriate intervals between scheduled cleanings. Your inspection procedures should be sufficiently thorough to reveal special cleaning needs that may arise for various reasons between regularly scheduled cleanings.

The cleaning can be carried out by your staff or an outside contractor. Typically, an organization will not have in-house the specialized vacuum equipment and other gear required to do the job thoroughly and will need to bring in a professional to do the work. One word of caution: be sure the contractor you have has the insurance and licenses your state requires. Discovering such a lack after an expensive service failure or a lawsuit is too late.

If you must change contractors for some reason or have no service history to guide you, you may have to hire a contractor with whom you have had no experience. Interview the prospective contractor in person or by telephone before you contract. Here are several suggestions to guide the interview.

Ask to see the contractor's insurance certificates, licenses (be sure you know your state's licensing requirements), and references. Tell the contractor you will call at least one of the listed references and ask a fellow building engineering professional whether he or she would hire this contractor to do the kind of work you need done. Check references and the Better Business Bureau after the interview.

Ask the contractor about other professional credentials he or she may have (formal training, degrees, years of experience, membership in professional associations, etc.) and whether the people actually doing the work have the same or similar credentials.

Inquire into the methods the contractor will use. Ask him or her to describe to you how the work will be done. Ask about equipment that may be used and ductwork access practices, if applicable, employed by the contractor. If the job involves porous and/or nonmetallic surfaces, ask what methods are used to clean these surfaces. If the contractor is going to "clean" such surfaces by spraying shellac on them, you need to know that in advance. In some cases, such an approach may make sense, but discuss it in advance with the contractor and make sure you know what "cleaning" you are buying.

Ask what hazardous waste disposal practices the contractor makes use of, should such materials be encountered during HVAC system component cleaning. HVAC system component cleaning occasionally uncovers asbestos, leaky refrigerant lines, and other wastes that pose potential harm to building occupants, cleaning technicians, or the environment.

If the cleaning will involve liquid treatments of HVAC system component surfaces, for deodorizing or sanitizing, find out what substances will be used and ask for MSDS (Material Safety Data Sheet) documentation on each one.

Price is, of course, an important contracting issue. If the contractor's pricing seems excessively high or low, ask him or her to explain the pricing. Investigate at least one other contractor for comparison.

Here is a list of additional questions you may consider asking a potential HVAC system component cleaning contractor:

1. How long has your company been cleaning HVAC system components?

2. What percentage of your business is dedicated to HVAC system component cleaning?

3. What is your experience in cleaning systems similar to those in my facility?

4. Who will be the on-site supervisor responsible for this project? How many projects of a similar scope has he or she been responsible for?

5. Is your firm a member in good standing of the National Air Duct Cleaners Association (NADCA) and can you provide us with a current membership certificate?

6. Will you use *source removal* techniques in accordance with NADCA Standard 1992-01 when cleaning my system?

7. Do you have a complete understanding of NADCA Standard 1992-01 and will you comply with all of its provisions on this job?

8. Do you have a comprehensive in-house safety program with training for employees?

9. Are you knowledgeable about site preparation issues for a project of this scope?

10. Is your equipment in good repair and proper working order? When was it purchased and how long has it been in use?

Once a new contractor has been chosen, and the cleaning work completed, do a post cleaning inspection and verification. In some special cases, it may be appropriate to hire an independent lab to do particulate and biological sampling in work spaces affected by the HVAC system components that have been cleaned. Some labs will do before-and-after-cleaning sampling to quantify certain kinds of indoor air quality measurables. If you intend to have such sampling done, notify the cleaning contractor before the work is done and give a copy of the results to the contractor, whether favorable or not. In most HVAC system component cleaning, such sampling will not be necessary. A before-and-after visual inspection of the HVAC system components involved will tell you whether the contractor has done the job agreed upon.

15.2.5   Procedures for Duct Cleaning*

There are three generally accepted methods:

- Contact vacuum method
- Air washing method
- Power brushing method

15.2.5.1   Contact vacuum method

Principle of operation.   Conventional vacuum cleaning of interior duct surfaces through openings cut into the ducts is satisfactory so long as reasonable care is exercised. The risk of damaging duct surfaces is minimal. Only HEPA (high efficiency particle arrestor) vacuuming equipment should be used if vacuum equipment will be discharging into occupied space. Conventional vacuuming equipment may discharge extremely fine particulate matter back into the building air space, rather than collecting it.

---

*Reproduced by permission of the North American Insulation Manufacturers Association (NAIMA)

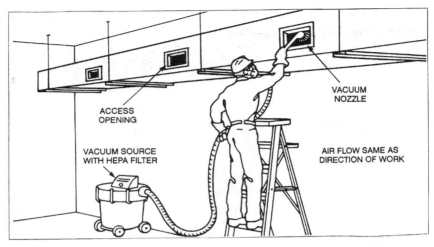

Figure 15.2.3 Contact vacuum cleaning.

This process may leave particulate matter in the duct which may later become airborne and contaminate the occupied space. This may occur because the duct is not under negative pressure during the cleaning operation.

Cleaning procedure. See Fig. 15.2.3.

- Direct vacuuming will usually require larger access openings in the ducts in order for cleaning crews to reach into all parts of the duct. Spacing between openings will depend on the type of vacuum equipment used and the distance from each opening it is able to reach.

- The vacuum cleaner head is introduced into the duct at the opening furthest upstream and the machine turned on. Vacuuming proceeds downstream slowly enough to allow the vacuum to pick up all dirt and dust particles. The larger the duct the longer this will take. Observation of the process is the best way to determine how long it takes before linings are considered sufficiently clean.

- When observation indicates the section of duct has been cleaned sufficiently, the vacuum device is withdrawn from the duct and inserted through the next opening, where the process is repeated.

15.2.5.2 Air washing method

Principle of operation. A vacuum collection device (preferably a powerful truck-mounted vacuum positioned outside the building) is connected to the downstream end of the section being cleaned through a predetermined opening. It is recommended that the isolated area of the duct system being cleaned be subjected to 1″ (25 mm) (minimum) static pressure to ensure proper transport of loosened material (take care not to collapse the duct). Compressed air is introduced into the duct through a hose terminating in a "skipper" nozzle. This nozzle is designed so that the compressed air propels it along inside the duct. This dislodges dirt and debris which, becoming airborne, are drawn downstream through the duct and out of the system by the vacuum collection equipment. (Dirt and dust particles must be dislodged from duct surfaces and become airborne before they can be removed from the duct system.) If the vacuum collector discharges to occupied space, HEPA filtration should be used. The compressed air source should be able to produce between 160 and 200 psi (110 and 140) air pressure, and have a 20-gallon (7 l) receiver tank, for the air washing method to be effective.

This method is most effective in cleaning ductwork no larger than 24″ × 24″ (60 × 60 cm) inside dimensions.

Cleaning procedure.   See Fig. 15.2.4.

- Openings for cleaning purposes often need be no larger than those cut for borescope inspection purposes. These should be drilled through the duct wall and the insulation (if present) at intervals which depend on type of equipment and duct size.

- The vacuum collection equipment is turned on and the proper negative pressure established. The compressed air hose with the skipper nozzle is inserted into the hole farthest upstream.

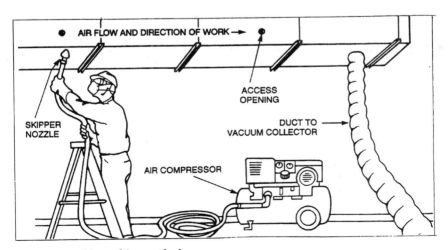

Figure 15.2.4 Air washing method.

- The skipper nozzle is allowed to travel downstream slowly enough to allow the skipping motion to dislodge dirt and dust particles. The larger the duct, the more time this will take; observation of the process in each case is the best way to determine how long. When observation suggests the section of duct has been cleaned sufficiently, the compressed air hose is withdrawn from the duct and inserted in the next downstream hole, where the process is repeated. The condition of the cleaned section may be inspected with a borescope inserted in the hole.

15.2.5.3 Power brushing method

Principle of operation. As with the air washing system, a vacuum collection device (preferably a powerful, truck-mounted vacuum positioned outside the building) is connected to the downstream end of the section being cleaned through a predetermined opening. Pneumatically or electrically powered rotary brushes are used to dislodge dirt and dust particles which become airborne, are drawn downstream through the duct system, and are evacuated by the vacuum collector. Power brushing works satisfactorily with all types of duct and all types of fibrous glass surfaces, provided the bristles are not too stiff nor the brush permitted to remain in one place for such a long time that it may damage them.

If the vacuum collector discharges to occupied space, HEPA filtration should be used.

NOTE: It may prove impossible to use the power brushing method to clean fibrous glass ducts with tie rod reinforcement.

Cleaning procedure. See Fig. 15.2.5.

- Power brushing will usually require larger access openings in the ducts in order for cleaning crews to reach inside and manipulate the equipment. However, fewer openings may be required; some power brushing devices are able to reach up to 20 feet (6 m) in either direction from an opening.

- Once the isolated section of the duct to be cleaned is under negative pressure (as described in the air washing method), the rotary power brushing device is introduced into the duct at the opening furthest upstream. The brushes are worked downstream slowly to allow them to dislodge dirt and dust particles.

- When working in ducts with fibrous glass airstream surfaces, the brushes should be kept in motion so as not to gouge or dig into the surfaces. Again, the larger the duct the more time this will take; observation of the process in each case is the best way to determine

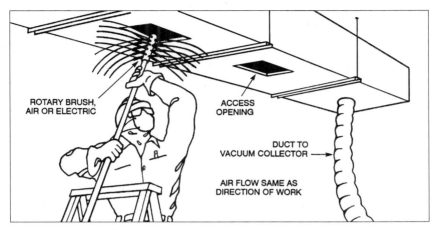

ROTARY BRUSH,
AIR OR ELECTRIC

ACCESS
OPENING

DUCT TO
VACUUM COLLECTOR →

AIR FLOW SAME AS
DIRECTION OF WORK

Figure 15.2.5 Power brushing method.

how long. When observation suggests the section of duct has been cleaned sufficiently, the power brush is withdrawn from the duct and inserted in the next downstream opening, where the process is repeated.

15.2.6 Bibliography

Following is a list of applicable document sources and documents on cleaning of HVAC system components:

Indoor Air Quality Information Clearinghouse
U.S. Environmental Protection Agency
P.O. Box 37133
Washington, DC 20013
1(800) 438-4318/Fax (301) 588-3408

The Inside Story: A guide to Indoor Air Quality

Building Air Quality: A guide for Building Owners and Facility Managers

National Air Duct Cleaners Association
1518 K Street, N.W., Suite 503
Washington, DC 20005
(202) 737-2926/Fax (202) 347-8847

NADCA Standard 1992-01, Mechanical Cleaning of Non-Porous Air Conveyance System Components

Understanding Microbial Contamination in HVAC Systems

Introduction to HVAC System Cleaning Services

National Air Filtration Association
1518 K Street, N.W., Suite 503
Washington, DC 20005
(202) 628-5328/Fax (202) 638-4833

NAFA Guide to Air Filtration
(ISBM 1-884152-00-7)

North American Insulation Manufacturers Association
44 Canal Plaza, Suite 310
Alexandria, VA 22314
(703) 684-0084/Fax (703) 684-0427

Fibrous Glass Duct Construction Standards, 1993

Sheet Metal and Air Conditioning Contractors National Association
(SMACNA)
4201 Lafayette Center Drive
Chantilly, VA 22021
(703) 803-2980

HVAC Duct Construction Standards — Metal and Flexible

U.S. Department of Labor
Occupational Safety & Health Administration
Room N3651
200 Constitution Avenue, N.W.
Washington, DC 20210
(202) 219-6666

All About OSHA (OSHA 2056)

Control of Hazardous Energy (OSHA 3120)

Respiratory Protection (OSHA 3079)

Chemical Hazard Communication (OSHA 3084)

Personal Protective Equipment (OSHA 3077)

American Society of Heating, Refrigerating, and Air Conditioning Engineers
(ASHRAE)
1791 Tullie Circle, NE

Ventilation for Acceptable Indoor Air Quality

16.1

Fans & Blowers

Bayley Fan,* LAU Industries
Lebanon, Indiana

16.1.1 Safety Precautions

Do....

1. Make sure unit is stopped and electrical power is locked out before putting hands into the inlet or outlet opening or near belt drive. We suggest a *LOCK-OUT* and a warning sign on the start switch cautioning not to start the unit.

2. Follow maintenance instructions.

3. Make sure all drive guards are installed at all times fan is in operation. If the inlet or outlet is exposed, a suitable guard should also be installed.

4. Take special care not to open any fan or system access panels while the system is under pressure (negative or positive).

5. Never allow untrained or unauthorized persons to work on equipment.

6. Take special care when working near electricity. Also insure the power is off and can not be turned on while servicing the fan.

7. Keep area near equipment clean.

Caution....

---

*Adapted from Lau Service Manual #0696 C-141.

1. **Do not** put hands near or allow loose or hanging clothing to be near belts or sheaves while the unit is running.

2. **Do not** put hands into inlet or outlet while unit is running. *It is sometimes difficult to tell whether or not it is running* ... be sure it is not running and cannot be operated before doing any inspection or maintenance.

3. **Do not** operate fan with guards removed.

4. **Do not** take chances.

16.1.2 Foundations

A rigid level foundation is a must for every fan. It assures permanent alignment of fan and driving equipment and freedom from excess vibration, minimizing maintenance costs. The foundation must be cast separate from any adjacent floor structure and separated around edges by at least ¾ inch (18 mm) tar felt to prevent transmission of vibration in either direction. The sub-foundation (soil, stone, rock etc.) should be stable enough to prevent uneven settling of fan foundation. Foundation bolt locations are found on the certified drawings. The natural frequencies of the foundation must be sufficiently removed from the rotational frequency of the fan to avoid resonant conditions.

16.1.2.1 Poured concrete foundations (recommended)

Poured Concrete under the fan and all drive components is the best fan foundation. A generally accepted rule of thumb is that the weight of the concrete foundation be at least three (3) times the total weight of the equipment it will support. This weight acts as an inertia block to stabilize the foundation. Where the ground is soft the foundation should be flared or the footing course increased in size to resist settling. The top should extend at least 6 inches (15 mm) outside the outline of the fan base and should be beveled on the edges to prevent chipping.

Anchor bolts in concrete should be L or T-shaped. They should be placed in pipe or sheet metal sleeves approximately 2 inches (7 mm) larger in diameter than the anchor bolts to allow for adjustment in bolt location after the concrete is set. In estimating the length of the bolts, allow for the thickness of the nut and washers, thickness of the fan base, thickness of the shim pack, if required, and extra threads for drawdown.

Seating area for washers and nuts must be clean and threaded area must be clean and lubricated.

16.1.2.2 Equipment-mounted fans

If the fan is mounted on equipment having parts which cause vibration, *it is very important that the fan support is rigid enough to prevent such vibration from being carried to the fan.* The resonant frequency of the support should avoid the fan running speed by at least 20 percent. It may be advisable to use vibration isolators under the fan.

16.1.2.3 Duct-mounted fans

If the fan is mounted in ductwork, horizontally or vertically, indoors or outdoors, the weight of the fan should be sufficiently supported to assure permanent alignment of the fan assembly. Forces imposed on the fan that may cause misalignment must be isolated to prevent serious damage. Horizontally installed fans should be supported with vibration isolators and flexible duct connections. Vertically installed fans should be supported with guy rods and guy wires. The weight of vertical ductwork supported by the fan should not exceed the weight of the fan without additional external duct support.

16.1.2.4 Structural steel foundation

When a structural steel foundation is necessary, it should be sufficiently rigid to ensure permanent alignment. It must be designed to carry, with minimum deflection, the weight of the equipment plus loads imposed by the centrifugal forces set up by the rotating elements. We recommend welded, riveted, or suitably locked structural bolted construction to best resist vibration. In certain applications, it is recommended that vibration isolators, selected specifically for weight and span conditions, be installed.

Fans installed above ground level should be located near to or above a rigid wall or heavy column. An overhead platform or support must be rigidly constructed, level and sturdily braced in all directions.

16.1.3 Assembly and Installation

Most fans and blowers are shipped completely assembled, however, a few are shipped partially knocked-down. Partially knocked-down fans are partially disassembled and the remaining parts match marked for easy field assembly. When installing units allow ample space for removal of impeller, lubrication of bearings, adjustment of motor base and inspection or servicing of complete units.

For Units Shipped Completely Assembled:

- Remove skids and any protective coverings over bearings or housing.
- Move fan to its rigid support. Install vibration isolation and secure fan to the mounting bolts.
- Level unit with spirit level by adjusting vibration isolators or shimming as required.
- Tighten all mounting hardware.

For Units Shipped Partially Assembled:
Fans and blowers shipped partially knocked-down may have parts shipped as follows:

- Fan section with housing, impeller, bearings, and shaft assembled.
- Stack cap with integral gravity back draft damper.
- Motor or V-belt drive separate.
- Variable inlet vanes, screens, outlet damper, etc. separate.

To assemble

- Remove skids and any protective coverings over bearings or housing.
- Move fan to its rigid support. Install vibration isolation and secure fan to the mounting bolts.
- Level unit with spirit level by adjusting vibration isolators or shimming as required.
- Tighten all mounting hardware.
- If fan is V-belt driven and drive is not mounted, assemble sheaves on their shafts and line up V-belts with proper tension. *See V-belt drive section.*
- If fan was shipped separate from a stack cap and roof curb cap, caulk the flanges to be joined with a generous ¼" (6 mm) caulk being careful to caulk completely around all mounting holes. Proper caulking stops rain water leaks. Without proper caulking, leaks will occur because the air passing the flange joints will siphon the rain water into the air stream where it can leak back into the building.
- If variable inlet vane, inlet bell or inlet screen is required, bolt these in position on the fan.
- The fan is now ready to be connected into the system.
- Pay attention to correct direction of rotation.

16.1.4   Alignment

- Level the fan housing on the foundation by shimming where necessary. Use a spirit level on the shaft or on a horizontal portion of the housing. Tighten the nuts on all mounting bolts.

- If fan is received with motor mounted, check tightness of motor hold down bolts. Also check V-belt drive alignment. (Refer to V-belt drive section)

- Spin impeller slowly by hand to see that it clears the fan inlets. Due to transit and handling damage, it may be necessary to loosen the bearing bolts and shift the impeller and the shaft to locate it properly with the fan inlet.

- If fan is on concrete foundation, you are ready to pour to grout. Anchor bolts should be tight. After grout hardens, recheck for final level and alignment of all components.

After motor has been mounted; align and bolt down, wire power supply through a disconnect switch, short-circuit protection and suitable magnetic starter with overload protection. All motors should be connected as shown on nameplate. Install wiring and fusing in accordance with the National Electrical Code and local requirements.

Be sure power supply (voltage, frequency, and current carrying capacity of wires) is in accordance with the motor nameplate.

If fan is on a steel foundation or mounted in duct work, you are ready for operation.

16.1.5   Fiberglass Fans

Fiberglass fans require care similar to steel fans except as noted herein.

- Be sure impeller is not striking the inlet or housing and rotating in the proper direction. Fiberglass may break on impact or fail quickly due to stress caused by improper rotation.

- Do not operate a fiberglass fan in abrasive atmospheres. Abrasives will erode the resin rich surfaces of the FRP material and destroy the corrosion resistance of the fan.

- Do not operate a fiberglass fan in temperatures above 150°F (65°C) without a specific resin and a fan design approved by the fan manufacturer.

- Never attempt to support the fan by one flange. Use mounting brackets, hangers, etc. and appropriate vibration isolators if required. Use *both* flanges if mounted in ductwork.

■ Never allow the fan to operate with a vibration problem. The stress caused by vibration will quickly destroy a fiberglass fan.

16.1.6 Motors

Warning: Always disconnect or shut off electrical power before attempting to service fan and / or motor.

All A-C induction motors will perform satisfactorily with a 10 percent variation in voltage, a 5 percent variation in frequency or a combination voltage-frequency variation of 10 percent. For motors rated 208–220 V, the above limits apply only to 220 V rating. To select control for 208–220 V motors, use same amps for either 208 or 220 V.

Motors are received with bearings lubricated and require no lubrication for some time depending on operating conditions (See maintenance section on motor bearings)

16.1.6.1 To reverse direction of rotation

Single-phase motors

Shaded pole	Rotation cannot be reversed unless motor is constructed so that the shading coil on one half of the stator pole can be shifted to the other half of the stator pole.
Split phase	Interchange connections to supply of either main or auxiliary winding.
Capacitor	All types of capacitor motors are reversed in rotation by interchanging connections to supply of either main or auxiliary winding.
Repulsion	Remove plate on motor end bracket and turn bracket (holding brushes) in direction *opposite* to direction of existing rotation.

Note: It is suggested that rotation change be made for single phase motors by the manufacturer's approved repair shop.

Three-phase motors. To reverse rotation, interchange any two line leads.

Normal operation of motors results in temperature rise. Permitted temperature rise depends on type of motor installation. The total motor operating temperature is the ambient temperature + motor temperature rise. The motor temperature rise includes nameplate temperature rise, service factor allowance, and hot spot allowance.

Important note:

Motors are warranted by the motor manufacturer, the fan manufacturer will assist in locating a local vendor approved repair shop if required.

6.1.7   Motor Lubrication and Maintenance

■ Regrease or lubricate motor bearings according to manufacturer's recommendations. *DO NOT OVER-LUBRICATE.* Motor manufacturer's lubrication recommendations are printed on tags attached to motor. Should these tags be missing, the following will apply:

16.1.7.1   Fractional horsepower ball bearing motors

■ Under normal conditions, ball bearing motors will operate for five years without relubrication. Under continuous operation at higher temperatures (but not to exceed 140°F (60°C) ambient) relubricate after one year.

16.1.7.2   Integral horsepower ball bearing motors

■ Motors having pipe plugs or grease fittings should be relubricated while warm and at a stand still. Replace one pipe plug on each end shield with grease fitting. Remove other plug for grease relief. Use low pressure grease gun and lubricate until grease appears at grease relief. Allow motor to run for 10 minutes to expel excess grease. Replace pipe plugs.

■ Recommended relubrication intervals—this is a general guide only.

■ These ball bearing greases or their equivalents are satisfactory for ambients from −20°F to 200°F (−29°C to 93°C).

TABLE 16.1.1   Recommended Lubrication Intervals

Power Range		Standard Duty	Severe Duty	Extreme Duty—Very Dirty
H.P.*	kW	8 hr/day	24 hr/day dirty, dusty	high ambients
1½–7½	2–10	5 yr	3 yr	9 mo
10–40	13–54	3 yr	1 yr	4 mo
50–150+	67–200+	1 yr	9 mo	4 mo

*Conversion: 1 HP = 746 watts.

Chevron SRIU #2	(Standard Oil of California)
Chevron BRB #2	(Standard Oil of California)
Premium RB	(Texaco, Inc.)
Alvania No. 2	(Shell Oil Company)

- Make certain motor is not overloaded. Check power draw against nameplate.

- *Keep motors dry.* Where motors are idle for a long time, single phase heaters or small space heaters might be necessary to prevent water condensation in windings.

16.1.8 Fan Start-Up

16.1.8.1 Before startup, check

- **Fastenings** — It is recommended that all foundation bolts, impeller hub set screws and bearing set screws be checked for tightness before startup.

- **Access doors** — should be tight and sealed.

- The Fan bearings, whether pillow block or flange mounted, are pre-lubricated and should not require additional grease for startup.

- "Bump" the motor to check for Proper impeller rotation. The motor should be started in accordance with manufacturer's recommendations.

- **Bearings** — check bearing alignment and make certain they are properly locked to shaft.

- **Impeller** — turn over rotating assembly by hand to see that it runs free and does not bind or strike fan housing. If impeller strikes housing, it may have to be moved on the shaft or bearing pillow blocks moved and reshimmed. Check location of impeller in relation to fan inlets. Be sure fan housing is not distorted. See *ALIGNMENT* section.

- **Driver** — check electrical wiring to the motor. See *MOTOR* section.

- **Guards** — make certain all safety guards are installed properly.

- **V-Belt Drive** — must be in alignment; with belts at Proper tension. See *V-BELT DRIVE* section.

- **Duct Connections** — from fan to ductwork must not be distorted. Ducts should never be supported by the fan. Expansion joints between duct connections should be used where expansion is likely to occur or where the fan is mounted on vibration isolators. All joints should be sealed to prevent air leaks and all debris removed from ductwork and fan.

Dampers and VIV's (variable intake ventilators) should operate freely and blades close tightly. All dampers and VIVs should be partially closed during starting periods to reduce power requirements. This is particularly important for a fan designed for high temperature operation being "run in" at room temperature or at appreciably less than design temperature. When air is up to temperature, the damper or VIV may be opened. Completely closing dampers could cause the fan to run rough.

- Fan may now be brought up to speed. Watch for anything unusual such as vibration, overheating of bearings and motors, etc. Multispeed motors should be started at lowest speed and run at high speed only after satisfactory low speed operation. Check fan speed on V-belt driven units and adjust motor sheave to give the desired RPM. Balance system by adjusting damper VIVs.

- **At first indication of trouble or vibration, shut down and check for difficulty.**

16.1.9   Maintenance Pointers

- **Always disconnect or shut off fan before attempting any maintenance.**

- A definite time schedule for inspecting all rotating parts should be established. The frequency of inspection depends on the severity of operation and the location of the equipment.

- Fan bearing alignment should be checked at regular intervals. Misalignment can cause overheating, wear to bearing dust seals, bearing failure and unbalance.

- Fan bearings should be lubricated at regular intervals. Periodic inspection will be necessary. If grease is found to be breaking down, replenish grease by pumping new grease into bearing until all the old grease has been evacuated. See section on BEARING LUBRICATION.

- Bearings on high speed fans tend to run hot, 75° to 105°F (41° to 58°C) above ambient. Do not replace a bearing because it feels too hot to touch. Place a contact thermometer against the bearing pillow block and check the temperature. Before you investigate high temperature, realize that ball or roller bearing pillow blocks can have a total temperature of 225°F (107°C). High temperature bearings may be rated at 425°F (218°C).

- Foundation bolts and all set screws should be inspected for tightness.

- Fans should be inspected for wear and dirt periodically. Any dirt accumulated in housing should be removed. Severely worn spots on components other than the impeller may sometimes be built up with weld material, using care to prevent heat distortion. The impeller may have to be cleaned. A wash down with steam or water jet is usually sufficient, covering bearings so water will not enter the pillow blocks. Impellers having worn blades should be replaced. Impellers require careful rebalancing before being returned to service. Replacement impellers should have the balance checked upon start-up and corrected as required to operate properly in it's specific application.

- Repairing of exterior and interior parts of fans and ducts will extend the service life of the installation. Select a paint which will withstand the operating temperatures. For normal temperatures a good machinery paint may be used. Corrosive fumes require all internal parts to be wire brushed, scraped clean and repainted with an acid resisting paint. Competent advice should be sought when corrosive fumes are present.

- Blow out open type motor windings with low pressure air to remove dust or dirt. Air pressure above 50 psi (345 kPa) should not be used as high pressure may damage insulation and blow dirt under loosened tape. Dust can cause excessive insulation temperatures. **Do not exceed** OSHA air pressure requirements.

- Excessive vibration will shorten the life of any mechanical device. Correct any imbalance situation before returning fan to service.

16.1.10 Lubrication of Antifriction Bearings

Bearings on assembled fans receive their initial lubrication from the bearing manufacturer. Bearings shipped separate from the fan or as a replacement may not be lubricated before shipment. When there is the slightest doubt, the safe practice is to assume that the bearing has not been lubricated. Always turn fan off before lubricating.

For grease lubricated ball or roller bearing pillow block, a good grade of grease free from chemically or mechanically active material should be used. These greases are a mixture of lubricating oil and a soap base to keep the oil in suspension. They have an upper temperature limit where oil and soap base oxidize and thermally decompose into a gummy sludge.

Mixing of different lubricants is not recommended. If it is necessary to change to a different grade, make or type of lubricant, flush bearing thoroughly before changing. Regreasing will vary from 3 months to a year depending on the hours of operation, temperature and surrounding conditions. Special greases may be required for dirty or wet atmosphere (consult your lubricant supplier).

When grease is added, use caution to prevent any dirt from entering the bearing. The pipe plug or grease relief fitting should be open when greasing to allow excess grease to flow out. The pillow block should be about ⅓ full, as excess grease may cause overheating. Use low pressure gun.

- These ball bearing greases or their equivalents are satisfactory for ambients from −20°F to 200°F (−7°C to 93°C).

Chevron SRIU #2	(Standard Oil of California)
Chevron BRB #2	(Standard Oil of California)
Premium RB	(Texaco, Inc.)
Alvania No. 2	(Shell Oil Company)

16.1.10.1 Frequency of lubrication

The bearings are lubricated at predetermined intervals and the condition of the grease established as it is purged out of the seals or by examination of the grease in the housing. An average installation where the environmental conditions are clean and room temperatures prevails may only require bearing lubrication every 3 to 6 months, while operation in a dirty atmosphere at high temperatures will require much more frequent intervals. Base your particular interval on

TABLE 16.1.2 Recommended Frequency of Lubrication—Fan Bearings

Shaft size in	mm	Operating speed (RPM)									
		500	1000	1500	2000	2500	3000	3500	4000	4500	500
		Lubricating frequency (months)									
.50–1.00	12–25	6	6	6	6	6	6	4	4	2	2
1.06–1.44	27–37	6	6	6	6	6	6	4	4	2	1
1.50–1.75	38–45	6	6	6	4	4	2	2	2	1	1
1.88–2.19	49–56	6	6	4	4	2	2	1	1	1	
2.25–2.44	57–62	6	4	4	2	2	1	1	1		
2.50–3.00	64–76	6	4	4	2	1	1	1			
3.06–3.50	77–90	6	4	2	1	1	1				
3.56–4.00	90–102	6	4	2	1	1					

condition of grease after a specific service period. Table 16.1.2 is optimum for various shaft sizes and operating speeds.

16.1.11   V-Belt Drives

Fans shipped completely assembled have had the V-belt drive aligned at the factory. Alignment must be checked before operation.

- Be sure sheaves are locked in position.
- Key should be seated firmly in keyway.
- The motor and fans shafts must be properly aligned, with the centerline of the V-belts at a right angle to the shafts.
- Be sure all safety guards are in place.
- Start the fan. Check for proper rotation of impeller. Run fan at full speed. A slight belt bow should appear on the slack side. Adjust belt tension by adjusting the motor on its adjustable base.
- If belts squeal excessively at startup, they are too loose and should be tightened.
- When belts have had time to seat in the sheave grooves, readjust belt tension. Check belt tension after 8, 24 and 100 hours of operation.

16.1.11.1   V-Belt drive assembly can be mounted as follows

- Clean motor and drive shafts. Be sure they are free from corrosive material. Clean bore of sheaves and coat with heavy oil for ease of shaft entry. Remove oil, grease, rust or burrs from sheaves. Place fan sheave on fan shaft and motor sheave on its shaft. Do not pound on sheave as it may result in damage. Tighten sheaves in place.
- Move motor on base so belts can be placed in grooves of both sheaves without forcing. Do not roll belts or use tool to force belt over grooves.
- Align fan and motor shafts so they are parallel. The belts should be at right angles to the shafts. A straight edge or taut cord placed across the faces of the sheaves will aid in alignment with single grove sheaves. If multiple grooves sheaves are installed, use the centerline of the drive as your alignment point.
- Tighten belts by sliding motor in its base. Correct tension gives the best efficiency. Excessive tension causes undue bearing pressure.

- Be sure all safety guards are in place.

- Start the fan and run at full speed. Adjust belt tension until only a slight bow appears on the slack side of the belts. If slippage occurs, a squeal will be heard at start-up. Eliminate the squeal by tightening the belts.

- Belts require time to become fully seated in the sheave grooves — check belt tension after 8, 24, and 100 hours of operation. Allowing belts to operate with improper tension will shorten belt life substantially.

- If the shafts become scratched or marked, carefully remove the sharp edges and high spots such as burrs with fine emery cloth or a honing stone. Avoid getting emery dust in the bearings.

- Do not apply any belt dressing unless it is recommended by the drive manufacturer. V-belts are designed for frictional contact between the grooves and sides of the belts. Dressing will reduce this friction.

- Minimum belt center distances are available from factory upon request.

- Belt tension on an adjustable pitch drive is obtained by moving the motor — *not by changing the pitch* diameter of the adjustable sheaves.

16.1.12 Summary of Fan Troubles and Corrections

1. Capacity or pressure below rating
   - Total resistance of system higher than anticipated — *system problems.*
   - Speed too low — *adjust drive.*
   - *Dampers or variable inlet vanes improperly adjusted.*
   - Poor fan inlet or outlet conditions — *elbows at or too close to fan.*
   - Air leaks in system — *seal joints — correct damper settings.*
   - Damaged impeller or incorrect direction of rotation — *correct.*

2. Vibration and noise
   - Misalignment of impeller, bearings, couplings or V-belt drive — *loosen, align, tighten.*
   - Unstable foundation — inferior design — *start over.*
   - Foreign material in fan causing unbalance — *remove.*
   - Worn bearings — *replace bearings and shaft.*
   - Damaged impeller or motor — *check and repair.*
   - Broken or loose bolts or set screws — *replace.*

- Bent shaft—*replace.*
- Worn coupling—*replace.*
- Impeller or driver unbalanced—*balance.*
- 60/120 cycle magnetic hum due to electrical input. *Check for high or unbalanced voltage.*
- Fan delivering more than rated capacity—*reduce speed.*
- Loose dampers on VIVs—*adjust and tighten.*
- Speed too high or fan rotating in wrong direction—*correct.*
- Vibration transmitted to fan from some other source—*isolate.*

3. Overheated bearings
   - Too much grease in ball bearings. *Allow run time to purge (24 hours).*
   - Poor alignment—*correct.*
   - Damaged impeller or drive—*inspect, correct or replace.*
   - Bent shaft—*replace.*
   - Abnormal end thrust—*loosen set screws and adjust.*
   - Dirt in bearings—replace bearing—*use filtered grease.*
   - Excessive belt tension—*adjust.*

4. Overloaded motor (draws too many amps)
   - Speed too high—*reduce speed or change HP.*
   - *Discharge over capacity due to existing system resistance being lower than original rating.*
   - Specific gravity or density of gas above design value—*recalculate and correct.*
   - Wrong direction of rotation—*correct.*
   - Poor alignment—*correct.*
   - Impeller wedging or binding on inlet bell—*loosen and adjust.*
   - Bearings improperly lubricated—*see par 10.*
   - Motor improperly wired—*verify and correct.*

5. Motor problems
   - *Check for low or high voltage from power source.*
   - High temperature—*drawing too much current or dirt in windings.*
   - Armature unbalance—*vibration and noise.*
   - Worn bearings—*armature rubs against stator.*
   - *Too much or not enough lubrication in bearings.*
   - *Commutator brushes on d-c motor worn or not seated under proper tension.*
   - Loose hold down bolts—*vibration and noise.*
   - Low insulation resistance due to moisture—*Check resistance with a megohm meter ("Megger") or similar instrument employing a 500 volt d-c potential. Resistance should read at least 1 megohm.*

16.2

Axial Vane Fans

Ben Harstine

Supervisor of Field Service, Joy / Green Fan Division,
New Philadelphia Fan Co.,* New Philadelphia, Ohio
(illustrations by Jon Miller)

16.2.1 Introduction

Axial vane fans may be *direct drive* in which the rotor assembly is mounted on the motor shaft or driven by an external motor with a flexible coupling, or *belt driven* with the motor mounted directly on the fan or base mounted. Axial vane fans may also be fixed-pitch, adjustable-pitch, or controllable-pitch.

We will look at general fan maintenance for axial vane fans, then into specifics for each drive and mounting arrangement.

16.2.2 Safety Procedures

Always lockout and tagout equipment before removing safety screens or guards. *Never attempt to energize a fan while working on the machine.* Always remove all loose material from the fan room or ducted area before energizing fan.

16.2.3 General Fan Maintenance

It is very important to keep filters and coils clean to allow the fan to have access to incoming air flow. Dirty or clogged filters and coils can

---

*Formerly Joy Technologies, Inc.

starve fans for air; this will cause excessive negative pressure on the inlet side and significant rise in power draw to the motors. These conditions can cause aerodynamic stall. If aerodynamic stall continues for a period of time, blades can fail due to fatigue. The result is catastrophic failure. Similar cautions should be observed down stream to be sure all dampers and VAV function properly and sequentially timed.

Axial fans should be lubricated as instructed by the fan manufacturer. If maintenance personnel are not sure of the proper type and quantity of lubricant or at what intervals to lubricate, the maintenance person should contact the fan manufacturer for specific instructions. Incompatible lubricants can break down, loosing the lubricating quality. *Too much lubricant can be as damaging as too little.*

Periodic preventive maintenance programs should be established for all machinery including fans. The frequency of such programs will depend on the environment that the fan is operating in. Fan operating in a clean environment, such as a clean room application where air is prefiltered and where the ambient temperature ranges from 50° to 80°F (11° to 27°C) would only require an annual or biannual cleaning, where as fans ventilating diesel exhaust from a tunnel, for example, would require much more frequent cleaning. Typical HVAC fans should receive an annual preventive maintenance program.

The preventive maintenance program should involve cleaning the blades and the stationary vanes or struts on the discharge side of the fan rotor assembly. Keeping the fan clean will maintain the fan efficiency and save energy. The preventive maintenance program should also include vibration analysis. Vibration signatures should be recorded and stored, then compared with past signatures to establish a history and to warn of any early defects so that repairs can be scheduled rather than cause catastrophic expensive and inconvenient emergency repairs.

Vibration can be measured with a vibration analyzer/balancer or a vibration data collector then stored in a computer for comparison records. Vibration is measured in velocity inches per second (cm/sec) or in mils of displacement. Velocity measures the speed of the movement while mils measures the distance of movement. See the Typical Vibration Severity Chart (Fig. 16.2.1). Indications of one times the operating speed indicates out of balance, two times indicates looseness, three times indicates coupling alignment problems. Other frequencies are bearing frequencies, blade passing (the combination of speed times number of blades and/or the number of stationary vanes). *Never attempt to balance a mechanical problem or a dirty fan.* A fan, or any rotating machine, will not get out of balance unless some mass is added to or taken away from a balanced rotating component.

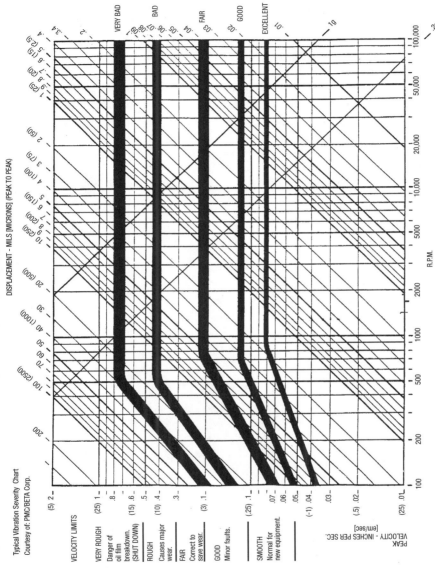

Figure 16.2.1 Vibration severity chart.

311

Vibration signatures should be taken at several locations on the fan. See Fig. 16.2.2. Locations #1 and #3 measure vertical movement, locations #2 and #4 measure horizontal movement and location #5 measures axial movement. When measuring locations #1 through #4, the transducer or accelerometer must be pointed at the center or rotation. These readings will indicate balance, bearing wear and other frequencies such as blade passing. When measuring at location #5, the transducer at the fan flange is placed opposite to, or in the direction of, the air flow. This measures axial movement of the fan and can detect air flow problems such as turbulence or excessive pressure in the system.

Only take and record as much information as is necessary to maintain a history file or to solve a particular problem. Too much or too detailed data may be more confusing than it is helpful. There are many analyzers on the market for maintenance records. A data col-

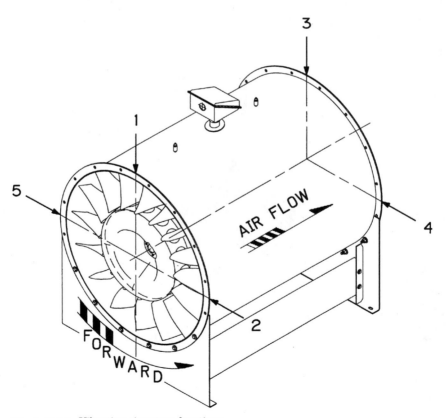

Figure 16.2.2 Vibration signature locations.

lector that gives a printed signature and a log that identifies the frequencies is sufficient for a good preventive maintenance program.

16.2.4   Adjustable- (AP) and Fixed-Pitch (FP) Direct-Drive Fans—Fig. 16.2.3

Adjustable pitch fans are fans which the blade angles may be adjusted manually. Fixed pitch fans are ones which have blades cast or welded to the center hub. These types of fans range in size from a few inches (cm) to several feet (m) in diameter with horsepower ranging from fractional to several hundred. These fans usually have the rotor assembly keyed directly to the motor shaft. This type of fan is usually applied to constant volume systems. When applied to VAV (Variable Air Volume) systems, air flow may be controlled with the use of a variable frequency drive, but never with dampers.

These fans require the least amount of maintenance. An occasional cleaning, proper lubrication and vibration analysis will reward the operator many years of dependable service. Always consult the performance curve before making any blade adjustments and always check motor and amp draw after adjustments are made. See Typical Fan Performance Curve (Fig. 16.2.4).

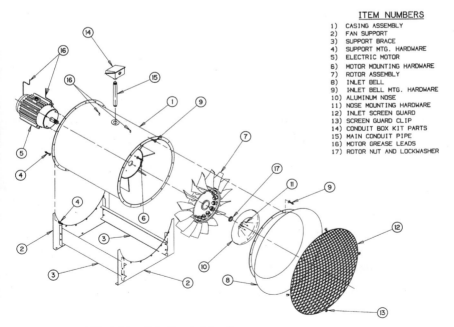

ITEM NUMBERS

1) CASING ASSEMBLY
2) FAN SUPPORT
3) SUPPORT BRACE
4) SUPPORT MTG. HARDWARE
5) ELECTRIC MOTOR
6) MOTOR MOUNTING HARDWARE
7) ROTOR ASSEMBLY
8) INLET BELL
9) INLET BELL MTG. HARDWARE
10) ALUMINUM NOSE
11) NOSE MOUNTING HARDWARE
12) INLET SCREEN GUARD
13) SCREEN GUARD CLIP
14) CONDUIT BOX KIT PARTS
15) MAIN CONDUIT PIPE
16) MOTOR GREASE LEADS
17) ROTOR NUT AND LOCKWASHER

Figure 16.2.3 Adjustable-pitch direct-drive fan.

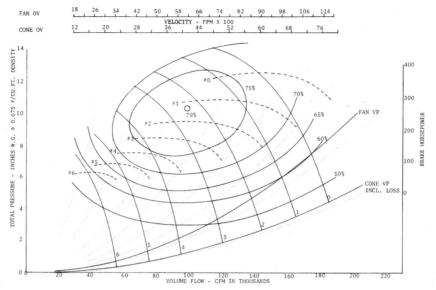

Figure 16.2.4 Typical fan performance curve.

16.2.5 Adjustable Pitch Belt Drive Fans—Fig. 16.2.5

This type of fan is similar to the direct drive fan except, of course, the motor is external; which necessitates the use of a fan shaft, pillow block bearings, belts, and sheaves. Belt-driven fans are used when the motor must be outside the air stream or the owner prefers to use a standard motor.

When replacing bearings in the belt-driven fan, it is very important to install the bearings such that the bearing closest to the fan wheel is fixed to carry the axial thrust generated by the fan and the bearing closest to the sheave be allowed to float and carry the radial load imposed by the belts. If this procedure is not followed, two situations can result, neither of which is desirable.

- If the axial and radial load are both imposed on the sheave end-bearing, it will overload and fail.

- If the fan end bearing does not carry the thrust load or is permitted to run unloaded, it will skid causing excessive heat and be doomed to failure.

Belts should be inspected regularly for signs of wear, cracks, fraying, or scorching due to slippage. Most belt failures are the result of im-

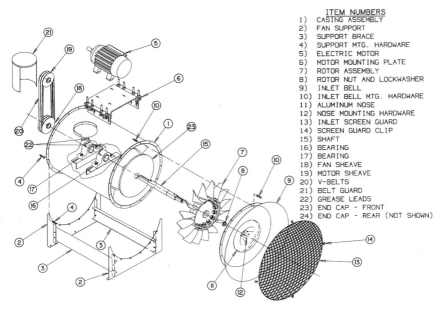

ITEM NUMBERS
1) CASING ASSEMBLY
2) FAN SUPPORT
3) SUPPORT BRACE
4) SUPPORT MTG. HARDWARE
5) ELECTRIC MOTOR
6) MOTOR MOUNTING PLATE
7) ROTOR ASSEMBLY
8) ROTOR NUT AND LOCKWASHER
9) INLET BELL
10) INLET BELL MTG. HARDWARE
11) ALUMINUM NOSE
12) NOSE MOUNTING HARDWARE
13) INLET SCREEN GUARD
14) SCREEN GUARD CLIP
15) SHAFT
16) BEARING
17) BEARING
18) FAN SHEAVE
19) MOTOR SHEAVE
20) V-BELTS
21) BELT GUARD
22) GREASE LEADS
23) END CAP - FRONT
24) END CAP - REAR (NOT SHOWN)

Figure 16.2.5   Adjustable pitch belt drive fan.

proper installation and tensioning. When installing replacement belts, always move the motor base toward the fan so that belts may be easily removed and replaced. *Never pry belts over sheave grooves to install or remove.* This will stretch and damage belt fibers and can cause damage to sheaves. Sheave alignment should be accomplished before tensioning. A string alignment or straight edge can be used to align sheaves. Proper belt tension is the least amount of tension required to operate the fan under full load without slippage. Minimal slippage is acceptable upon start-up. When installing new belts allow the fan to operate for at least eight hours then retention the belts as required. Belt manufacturers do not recommend the use of belt dressings or non-slip chemicals.

16.2.6   Controllable-Pitch (CP) Fans

Controllable pitch fans may be direct or belt driven; these are fans which have the ability to change pitch while the fan is in operation. Blade angle (pitch) is changed by means of an actuator which may be external or internal of the rotating assembly. External actuated controllable pitch fans are considered to have more reliable hysteresis and blade angle positioning than internal actuated fans and are also applied to heavier duty applications.

16.2.7 Externally Actuated Controllable-Pitch Fans—Fig. 16.2.6

Externally actuated CP (controllable-pitch) fans have blade positioning actuators on the outside of the fan casing. Actuators may be pneumatic, electronic, hydraulic, or manual. Others may be a combination of electro pneumatic or electro hydraulic. The actuator is connected to a lever or other push-pull device which is connected to the outer race of a de-spin thrust bearing. The inner race of the bearing is connected by way of a linkage system to each blade.

Maintenance procedures for this type of fan include all of those mentioned for AP and FP (adjustable pitch and fixed pitch) fans. The despin thrust bearing shall receive lubrication which will be similar in quantity as the motor but usually twice the frequency. An annual or bi-annual preventive maintenance program should also include removing the despin thrust bearing, rinsing clean the used lubricant residue, inspecting the bearing, repacking with fresh lubricant and reinstallation. All parts of the rotating assembly should be inspected at this time. Any worn or broken parts should be replaced. Things to check would be worn or sticking rollers, bolts; also check for broken blade clamps. Also check to be sure that the pitch control mechanism

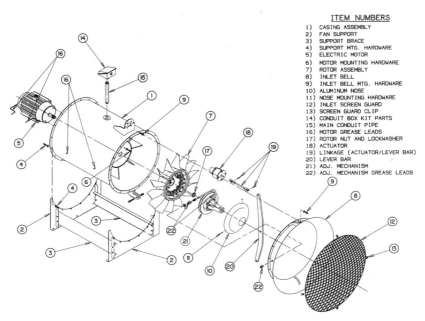

ITEM NUMBERS
1) CASING ASSEMBLY
2) FAN SUPPORT
3) SUPPORT BRACE
4) SUPPORT MTG. HARDWARE
5) ELECTRIC MOTOR
6) MOTOR MOUNTING HARDWARE
7) ROTOR ASSEMBLY
8) INLET BELL
9) INLET BELL MTG. HARDWARE
10) ALUMINUM NOSE
11) NOSE MOUNTING HARDWARE
12) INLET SCREEN GUARD
13) SCREEN GUARD CLIP
14) CONDUIT BOX KIT PARTS
15) MAIN CONDUIT PIPE
16) MOTOR GREASE LEADS
17) ROTOR NUT AND LOCKWASHER
18) ACTUATOR
19) LINKAGE (ACTUATOR/LEVER BAR)
20) LEVER BAR
21) ADJ. MECHANISM
22) ADJ. MECHANISM GREASE LEADS

Figure 16.2.6 Externally actuated controllable-pitch fan.

slides easily on the driven shaft, and check to be sure all blades are set at the same angle of attack.

16.2.8 Internally Actuated Controllable-Pitch Fans—Fig. 16.2.7

Internally actuated fans are equipped with a positioner on the outside of the fan casing. They also have a high-pressure air line supplying air to a diaphragm through a rotating union. Inflating and deflating the diaphragm causes the pitch control mechanism to move. The pitch control mechanism also has a feedback cable to the positioner to tell the positioner how to function. Inflating the diaphragm increases the angle of attack of the blades. A spring apparatus deflates the diaphragm to reduce the blade angle.

Lubrication and cleaning as described in the previous configurations apply to this type of fan except that there is no despin thrust bearing to maintain.

The annual preventive maintenance program should include replacement of the rotating union and feedback cable to positioner. The

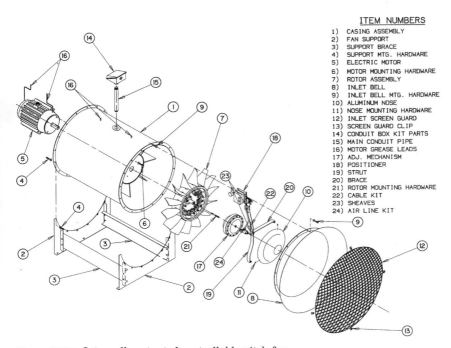

ITEM NUMBERS

1) CASING ASSEMBLY
2) FAN SUPPORT
3) SUPPORT BRACE
4) SUPPORT MTG. HARDWARE
5) ELECTRIC MOTOR
6) MOTOR MOUNTING HARDWARE
7) ROTOR ASSEMBLY
8) INLET BELL
9) INLET BELL MTG. HARDWARE
10) ALUMINUM NOSE
11) NOSE MOUNTING HARDWARE
12) INLET SCREEN GUARD
13) SCREEN GUARD CLIP
14) CONDUIT BOX KIT PARTS
15) MAIN CONDUIT PIPE
16) MOTOR GREASE LEADS
17) ADJ. MECHANISM
18) POSITIONER
19) STRUT
20) BRACE
21) ROTOR MOUNTING HARDWARE
22) CABLE KIT
23) SHEAVES
24) AIR LINE KIT

Figure 16.2.7 Internally actuated controllable-pitch fan.

internal actuator, supply lines, and fittings should be checked for air leaks. Check the positioner calibration and run a vibration analysis. Some fans of this type have radial antifriction thrust bearing where others have teflon rubbing surfaces for each blade. The antifriction bearing design may require complete tear down to clean and relubricate the radial thrust bearings on each blade. As in the previously discussed types of fans, check to see that all blades are at the proper angle of attack.

Before any preventive maintenance program is put into effect, contact the fan manufacturer for specific instructions and assistance to develop a preventive maintenance program for your fans.

George E. Taber
Applications & Field Services Engineer
Taco, Inc., Cranston, Rhode Island

17.1 Overview

17.1.1 Pumps

Most people take pumps for granted until they fail. Whether the application is a small single loop home hot water heating system or a large heating and/or cooling system of a large building, the pump is the heart of the systems. The pump moves the heat transferring fluid from the boiler or chiller to the space conditioning terminal unit or to the atmospheric heat exchanger known as a cooling tower.

The pump pipe sizes could range from 1/2 in. to 18 in. (12 to 46 mm) in diameter, with rated horsepower at 1/25 to 500 (30 W to 375 kW) depending on the pressure and flow produced by the pump.

Pumps used in HVAC applications are *centrifugal pumps*. Small, low flow, low head pumps are called circulators but are still centrifugal pumps. Typically pumps are driven by electric motors.

A centrifugal pump is composed of an impeller, casing, shaft and bearing assembly, and static and dynamic seals. The impeller is comprised of a rotating disk with radial vanes, equally spaced, fixed to the disk. The impeller may or may not have a disk (shroud) with a center opening covering the vanes. The impeller is enclosed and rotates within the casing. The casing is connected to the piping and directs the system fluid to the eye or center of the impeller. The fluid entering the rotating impeller is rapidly accelerated through the passageways formed by the vanes. The fluid speed and mechanical energy content are increased as it is accelerated toward the outer edge of the impeller. As the fluid leaves the impeller, it is collected by the casing or volute. The fluid speed or velocity head is converted to a *pressure head*.

The casing now directs the fluid to the system piping. The magnitude of the velocity head is a function of the speed or the diameter of the impeller. The volume of the fluid pumped is a function of the inlet, outlet, inside casing area, and impeller width.

The pressure head developed by the pump is used to overcome the resistance in the system created by the fluid flowing through the pipes, valves, fittings and heat exchangers. Most HVAC systems are closed loop and the pump has only to overcome system resistance. To drive the fluid to the top of the system, the system is pressurized to equal the elevation. In an open system a portion of the pump head is used to elevate the fluid to the top of the system and the remainder is used for system resistance. A typical application would be a cooling tower loop. Note: For every 1 psi (Pa) of water pressure a column will rise 2.31 ft (10.8 cm).

The casing is the pressure containing envelope, and the rotating shaft has to pass through the envelope to turn the impeller. A dynamic seal is used to maintain the fluid tight integrity. This dynamic seal is one of the maintenance points of a pump.

The rotating shaft is supported by antifriction bearings contained with in a support housing. The antifriction bearings are another point of maintenance and require periodic lubrication.

The motor shaft is also supported by antifriction bearings contained in a housing. These bearings also need maintenance. Close-coupled pumps use the motor shaft to directly drive the impeller. If the motor and pump are separate, they are connected by a flexible coupling which needs to be aligned and depending on the coupler used, could require further maintenance.

Pumps are classified by: (1) How they are connected to the piping, (2) How they are supported, (3) How they are connected to the motor, and (4) How many suction inlets to the impeller. Table 17.1 describes the configurations of the various types of centrifugal pumps and typical applications. Figs. 17.1–17.9 show the different pumps and their construction.

17.1.2   Valves

Valves are a very important part of a pumping system. Valves are needed for automatically or manually controlling the flow, maintaining set pressures, and are depended upon to hold the fluid in the system when a component is to be serviced. Valves, like pumps, must be able to operate from 0°F to 250°F (−18° C to 121°C) at pressure ranges less

TABLE 17.1   Classifications of HVAC Centrifugal Pumps and Uses

Pump type	Pipe size	Applications
Inline	1/2″–3″ (12–76 mm)	Residential hydronic systems
(*Direct-coupled*)		Domestic hot water recirculation
Water-lubricated—Figure 17.1 (no seal, no maintenance)		Primary/secondary hot and chilled water recirculation
Oil-lubricated		Heat pump, closed-loop recirculation
(*flexible-coupled*) Oil and Grease lubricated—Figure 17.2 (mechanical or packed water seal)		
Vertical inline Direct-coupled—Figure 17.3	1 1/4″–12″ (3–30 cm)	Same systems as inlines except larger
Split-coupled—Figure 17.4		Cooling tower loops
Close coupled end suction (Direct-coupled) Figure 17.5	1″–8″ (2.5–20 cm)	Condenser water Chilled and Hot water
Frame-mounted end suction Figure 17.6	1″–8″ (2.5–20 cm)	Primary/secondary loops Condensate return
Horizontal split case (H.S.C.) double-suction impeller Figure 17.7	2″–14″ (5–36 cm)	Boiler feed
Vertically mounted H.S.C. Figure 17.8	2″–14″ (5–36 cm)	
Vertical multi-stage Figure 17.9	1 1/4″–6″ (3.2–15 cm)	High head, low flow

than 0 to 300 psig (kPa). Proper selection of materials and seals are required for valves to operate properly for long periods of time without failure.

There are various valve designs, each with their own function. See further details, section 17.28. A *relief valve* Fig. 17.10*a,b* is typically a right angle globe valve used to protect systems by opening at a preset pressure. A *globe valve* Fig. 17.11, *ball valve* Fig. 17.12, *butterfly valve* Fig. 17.13, or *plug valve* Fig. 17.14 can be used to regulate flow manually or automatically. Typically gate valves Fig. 17.15, are used as service valves and are used wide open or fully closed. A *pressure reducing valve* Fig. 17.16 is typically a globe valve that automatically reduces a higher up stream pressure to a preset lower pressure. A *check valve* is typically a globe valve which automatically allows the flow of a fluid in one direction. A special version of the check valve has a weighted disk, Fig. 17.17, that not only controls the direction of

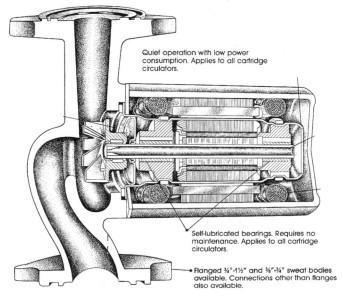

Quiet operation with low power consumption. Applies to all cartridge circulators.

Self-lubricated bearings. Requires no maintenance. Applies to all cartridge circulators.

Flanged ¾″-1½″ and ⅜″-¾″ sweat bodies available. Connections other than flanges also available.

Figure 17.1 Water-lubricated pump.

flow but also the flow created by hotter water rising, and cooler sinking, know as *thermosyphoning*. A lubricated plug valve can be used to control flow and as a *shut off*.

Valves typically have two seals: (1) the dynamic seal around the valve stem, and (2) the internal seal that seals off the internal flow. On some automatic valves there may be a third seal, which would be in the form of a diaphragm that seals the activating fluid. In some automatic valves, the diaphragm takes the place of the stem seal. These seals are the maintenance areas of valves.

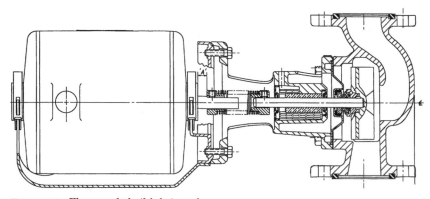

Figure 17.2 Flex-coupled oil lubricated pump.

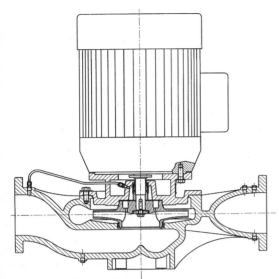

Figure 17.3 Vertical inline pump.

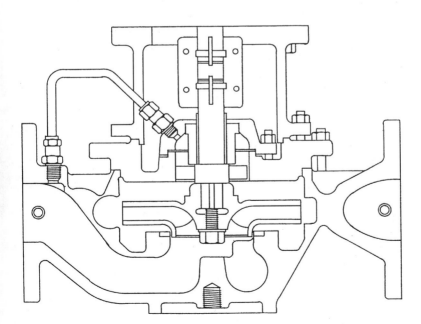

Figure 17.4 Split-coupled vertical inline pump.

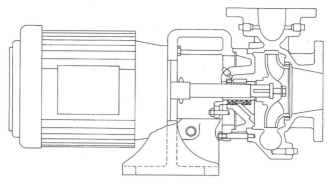

Figure 17.5 Close-coupled end suction pump.

17.1.3 Impact of part failure

Since a pump is made up of components that rely on each other for the pump to function properly, *if one part fails the whole pump fails.* The impeller suction nozzle rotates within a bore with very little clearance. If the bearings should fail or the shaft should snap, the impeller suction nozzle and possibly the casing would be damaged. The dynamic seal depends on the bearings to keep the shaft running true. If the shaft begins to wobble, the dynamic seal is destroyed. You can see from these few examples how important the proper operation of the individual components are to the whole pump. We mentioned earlier that the pump is the device that moves the heat transfer fluid in a system. If the pump fails, *it could cause the whole system to stop functioning as designed.*

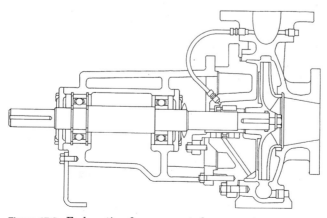

Figure 17.6 End suction frame mounted pump.

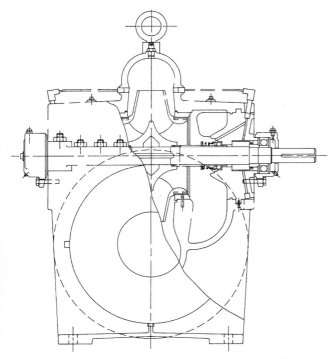

Figure 17.7 Horizontal split-case pump.

Valves do not contain the moving parts as those of a pump, but they do have dynamic seals and static seals that have to function properly. The stem seal keeps the fluid in the system and prevents the fluid from spilling on equipment and furnishings below. The internal seal controls the flow properly, and allows the system to be worked on without fluid loss. (*If the valves do not function properly, system pressures could rise, fluid temperatures could go to extremes, hot or cold, flow rates could differ from design, critical fluids could be lost, and personnel could be injured.*)

17.1.4 The "domino effect"

Because the pump is the device that moves the heat transfer fluid in a system, and valves control the flow of the fluid, a failure of either would cause the whole system to fail. As mentioned before, all components depend on each other for the system to function properly. The failure of one part of a pump would cause other internal parts to fail. Once the pump or valves fail, the living space operating temperature starts to differ from the control set point. This normally leads to dis-

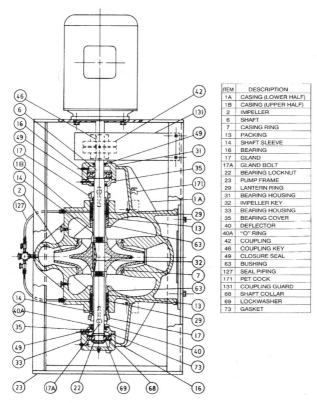

ITEM	DESCRIPTION
1A	CASING (LOWER HALF)
1B	CASING (UPPER HALF)
2	IMPELLER
6	SHAFT
7	CASING RING
13	PACKING
14	SHAFT SLEEVE
16	BEARING
17	GLAND
17A	GLAND BOLT
22	BEARING LOCKNUT
23	PUMP FRAME
29	LANTERN RING
31	BEARING HOUSING
32	IMPELLER KEY
33	BEARING HOUSING
35	BEARING COVER
40	DEFLECTOR
40A	"O" RING
42	COUPLING
46	COUPLING KEY
49	CLOSURE SEAL
63	BUSHING
127	SEAL PIPING
171	PET COCK
131	COUPLING GUARD
68	SHAFT COLLAR
69	LOCKWASHER
73	GASKET

Figure 17.8   Vertical-mounted horizontal split-case pump.

comfort for the occupants. If not corrected, and outside temperatures are below freezing, other components in the system could be damaged due to freezing. If pipes freeze and rupture, the result is property damage due to flooding.

If the relief valve does not operate and the system pressure goes above design, a component in the system could be damaged or rupture, causing property damage due to flooding. A valve that leaks all the time causes new oxygen-laden water to enter the system, causing system components to corrode, which could lead to a very expensive replacement.

If the spring in a check valve fails, a water hammer will be heard whenever a pump shuts down or changes it's duty cycle. The water hammer can be very dangerous as the pressure spikes can be many times normal system pressure. A malfunctioning flowcheck valve can cause a heating loop to heat even when the thermostat is satisfied for that particular zone.

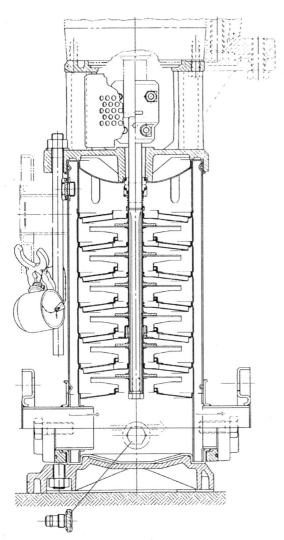

Figure 17.9 Vertical multistage pump.

In large buildings the air is conditioned by alternating chilling and reheating. If the chilled water pumps or valves fail, the chiller could freeze up and shut down, which normally causes a long reset cycle, causing discomfort to the occupants. In the winter time, with outside temperatures below freezing, stoppage of the heating fluid could cause a coil to freeze and cause flooding. If there is a central computer space that requires air conditioning, the computers could shut down due to excessive heat. In the banking business, a loss of the computers means

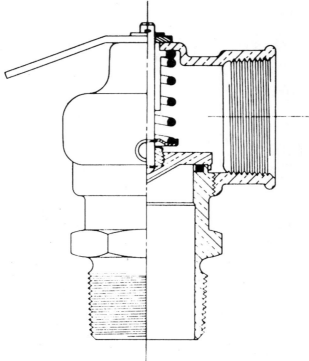

Figure 17.10a   Relief valve (liquid).

the loss of millions of dollars. The lesson is: *maintain the equipment*, repairs will be minimal, and damage loss will be minimal.

17.1.5   Historical failure rate of pumps and valves

Typically pumps and valves, if properly maintained, have a long, trouble-free life. Most pumps don't fail in a catastrophic way. The mechanical and electrical components of a pump *tend to give warning signs that parts are starting to fail.* Periodic checking by maintenance personnel will pick up on these signs, and repairs can be scheduled accordingly.

The *detectable* parts of a pump are generally:

1. Bearings

2. Water seals

3. Couplings

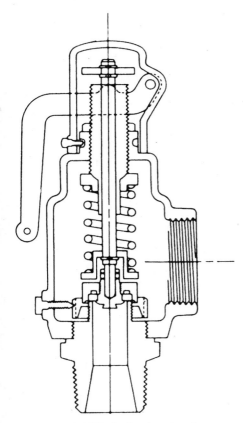

Figure 17.10b   Relief valve (pop type).

4. Motor temperature
5. Impeller wear
6. Cracked or worn through casing
7. Vibration

The *nondetectable* components would be:

1. Pump shaft failure
2. Impeller hub fracture
3. Motor winding insulation breakdown and shorting
4. Elastomer failure of sealing components
5. Spring fatigue and failure
6. Material fatigue and failure

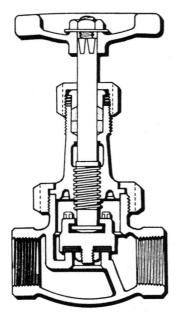

Figure 17.11 Globe valve.

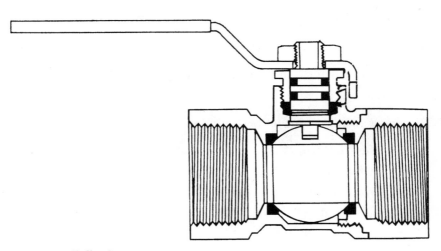

Figure 17.12 Ball valve.

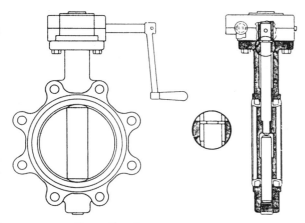

Figure 17.13   Butterfly valve.

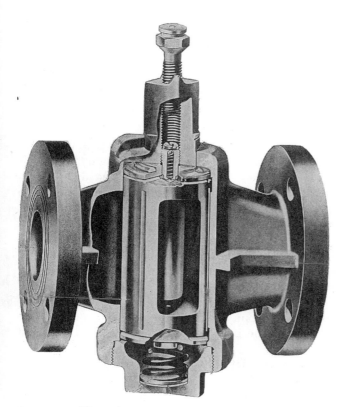

Figure 17.14   Plug valve.

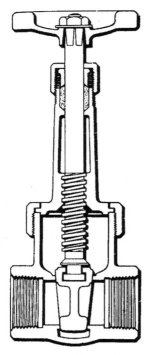

Figure 17.15 Gate valve.

During a repair that requires a teardown of the pump or valve, all parts of the pump and valve can be inspected. Early signs of wear or cracks indicate a future problem that can be corrected during this routine repair and inspection. The various failure modes should be known for each component of a pump or valve to take advantage of the accessibility of the components during a repair.

The two parts of a pump that tend to wear are *water seals* and *bearings*. If the water being pump is clean, properly maintained, and the proper seal materials had been selected the seals can operate for extended periods of time without having to be changed. Bearings, if properly lubricated with the correct lubricant and not exposed to water, heat or abrasives, will last many years. If proper system and pump maintenance are performed, the time between overhauls will also be many years.

Valve failure depends on the activity of a valve. Many valves stay in one position for years on end without being touched, so they don't wear out. On the other hand, relief valves, zone valves, check valves, pressure reducing valves, temperature control valves and active shut

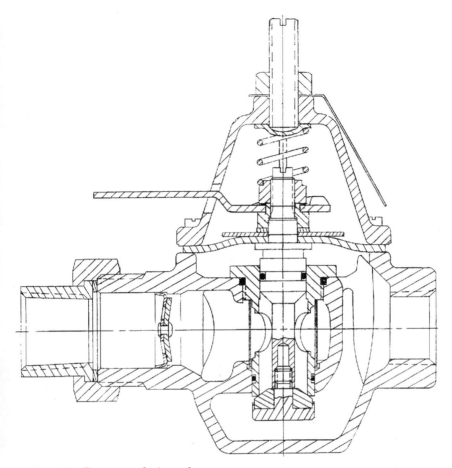

Figure 17.16   Pressure reducing valve.

off valves typically need maintenance on their internal and external seals and controls.

17.2   Preventive Maintenance Procedures

Proper maintenance of valves, pumps, and associated equipment employed in heating and air conditioning is as basic as maintaining equipment and plant cleanliness. Preventive maintenance can also be as sophisticated as performing diagnostic checks on the newest HVAC system's computerized controls.

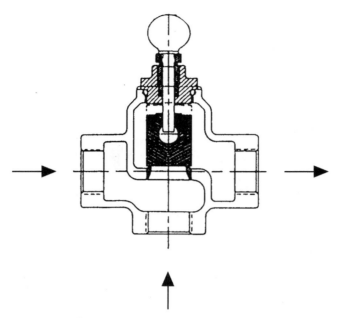

Figure 17.17   Flow check valve.

It is the responsibility of the director of maintenance to establish a preventive maintenance system. An inventory of all equipment, their records, manuals, drawings, and system drawings should be filed in a master file, microfilmed, and duplicates of all these records should be accessible to the maintenance crew as the working records.

Following is an outline of the records that should be set up for a sound maintenance system. It would be advisable to microfilm this material so you will always have a master record (Figs. 17.58, 17.59, 17.60, 17.61, 17.62 at end of the chapter.)

Maintenance Program Requirements

Inventories and Records of Equipment

A. Construction drawings

B. Records of as built drawings

C. Microfilm of as built and specifications — master copy

D. Documentation of system changes, documented on drawings

E. Shop drawings and equipment catalogs

F. Instructions: service and maintenance, trouble-shooting, spare parts

G. Service organizations and spare parts sources

H. Inventory of spare parts

I. Valve charts and system flow diagram

Procedures & Schedules

A. Operating Instructions
   1. Starting and stopping
   2. Adjustments and regulations
   3. Seasonal startup and shut down
   4. Seasonal changeover
   5. Logging and recording — hourly or every 4 hours
      a. Current and voltage
      b. Temperature — air, water: supply and return
      c. Pressure: supply and return
      d. Valves open and amount

B. Inspections
   1. Equipment to be inspected
   2. Points to be inspected
   3. Frequency of inspection
   4. Inspection method and procedure
   5. Evaluation of observations
   6. Recording and reporting

C. Service and repair
   1. Frequency of scheduled service
   2. Scheduled service procedure
   3. Repair procedure
   4. Recording and reporting

D. Monitoring of data
   1. Plot data to predict trends

E. Operating maintenance manuals
   1. Tags identify maintenance to be conducted on a piece of equipment and are attached to the equipment.

Most maintenance checks are fairly easy and quick to perform. The important factors in any preventive maintenance program are *regularity* and *thoroughness*. Many times these factors are overlooked and maintenance personnel start reacting to fires instead of preventing them.

The key is to *prevent* breakdowns that can cause major repair and loss of comfort, production, and loss of other dependent equipment. The best way to avoid this is to establish and maintain a sound preventive and predictive maintenance system. The manufacturer's instructions should be followed as to type and frequency of lubrication, monitoring of temperature ranges, leakage rates, and general wear testing.

When the system is first started and has settled out, performance tests should be run on the pumps. The performance tests should be run at 6 flow rates from 0 to full flow. The data should be recorded on the yearly *Pump Performance Sheet,* Fig. 17.18. This will be the base line from which future performance curves can be compared. Typically,

Pump Performance Data Sheet

Pump #_____   Impeller Dia.___ _____Date_____

Mfg._____Suction Size_____Discharge Size_____

Location_____ Bld._____Floor_____System_____

System Fluid_____Concentration_____Sys. Vol._____

Motor: H.P._____RPM_____F.L.A._____Model #_____

Description	RUN					
	1	2	3	4	5	6
Disch.Press.						
SuctionPress.						
Diff. Pressure						
X 2.31= Ft.						
/ Sp.Gr.						
Head- Ft.						
Velocity Head-Ft						
Total Head- Ft.						
Diff. Press.Flow						
GPM						
RPM						
AMPS.						
VOLTS						
Flow Referance used						

Figure 17.18 Pump performance sheet.

as the wear ring clearance increases more of the flow is bypassed back to the suction side of the pump, Fig. 17.19.

Pumps used in the closed loop HVAC application pump clean water, thus internal abrasion is not generally a problem. If iron oxide is allowed to build up, wear *will* take place, with seals and wear rings deteriorating more rapidly.

Open systems such as cooling towers tend to cause some wear. Wind blown sand can get into the cooling tower water and cycle through the system. Other components in the system will abrade as well. The pump test will be a good indicator of the condition of the rest of the system.

At the annual performance test there will come a time where a decision will have to be made about overhauling the pump. This would be based on unacceptable loss in system flow and/or unacceptable energy cost due to the drop in pump efficiency.

17.2.1 Bearing maintenance

Various types of bearings are used in pumps employed in HVAC systems. These various bearings are associated with various types of pumps.

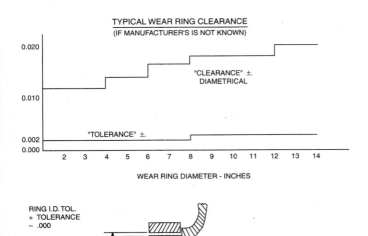

Figure 17.19 Wear ring clearance.

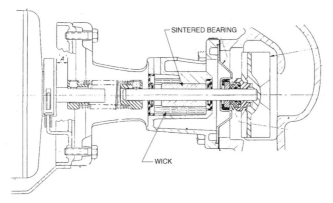

Figure 17.20 Sintered bronze bearing with felt wick.

A *sintered bronze sleeve bearing* is typically used in small *inline* pumps, 1/12–1/3 horsepower (62-250W), Fig. 17.20. A felt wick is wrapped around the bearing and oil is absorbed from the wick, through the porous sintered bronze bearing to the shaft. Because the wick and porous bronze filter the oil, the oil is never changed. Add *10 drops* of *SAE 30 non-detergent oil* at the beginning of *each heating season.* Pumps that run *all year* long should be oiled *every 6 months.*

Plain bronze or babbit bearings are also used in inline circulators. They depend on a wick or slinger to bring the oil from a reservoir to holes in the top of the bearings, Fig. 17.21. With the wick design, the end of the wicks are pushed into the holes in the top of the bearings and contact the shaft. Capillary action in the wick brings the oil to

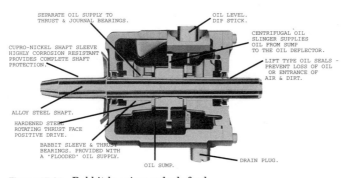

Figure 17.21 Babbit bearing, splash feed.

the bearings. *Add 1 oz. (30 gms) of SAE #20 nondetergent or 10W-30 oil* at the beginning of each heating season. With slinger design, the oil is distributed to the top of the bearing housing where it runs down a rib that directs the oil to the top hole of the bearing. The oil level in the reservoir is checked with a dip stick every 6 months. Add SAE 30 nondetergent oil to the full line on the dip stick. *Do not overfill.* Every 2 years the oil should be changed.

Some fractional horsepower motors use steel backed babbit bearings similar to the pump bearing frames. These bearings are also lubricated using a wicking system. Bell & Gossett, Div. of ITT (B&G) recommends 10 drops every 4 months in each oil cup. Non B&G motors should be lubricated per motor manufacturer's instructions on the nameplate. Emerson recommends 150 drops in each bearing each year for continuous duty, every 2 years for 1/2-year duty. Do not over-oil as excess oil will leak onto the resilient motor mounting rings, causing deterioration, causing the motor to go out of alignment.

17.2.2   Oil bath lubrication

Some pump manufacturers have an optional oil lubrication system for their end suction and horizontal split case pumps. An automatic oiler maintains the oil level just below the centerline of the bottom ball of the bearing, Fig. 17.22. The oil level in the reservoir of the automatic oiler should be kept filled.

17.2.2.1   Oil type and changes.   Under normal working conditions, oil will break down and will have to be replaced at regular intervals. The length of these intervals will depend on the environment and pump operating temperature. Below 200°F (94°C), with clean and dry loca-

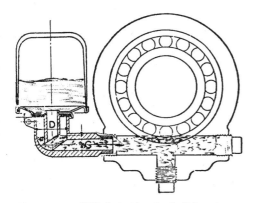

Figure 17.22 Oil lubrication for ball bearings.

tions; change oil once per year. If the temperature exceeds 200°F (94°C) with dirt and moisture present, change the oil every 2–3 months. Typically a SAE 10/20 motor oil or turbine oil is used but check the manufacturer's lubrication instructions for their recommendation.

17.2.2.2 Grease lubrication. Most pumps and motors these days use grease to lubricate their ball bearings. Some of the fractional and low horsepower motors and some of the inline pumps use permanently lubricated bearings. There will be no oil or grease holes. When the bearings start to make noise, change the bearings.

Most large motors and pumps use regreaseable bearings. The regreasing intervals depends on running speed of the unit and the operating temperature:

Pump RPM	Regrease Interval
1750	4250 hr
3450	2000 hr

Most pump manufacturers recommend #2 lithium base petroleum greases. Some recommended greases are:

Manufacturer	Tradename
Exxon	Unisex N2
Mobil	Mobility AW2
Valvoline	Valplex EP
ESL	Tempered W2
Texaco	Premium RBI

17.2.2.3 Relubrication procedure

1. Wipe off the grease fitting.

2. Remove grease outlet plug. Note: Some designs dump the old grease into the inner bearing frame, which will be evident by no visible drain plug.

3. Add recommended amount of grease. Note: Be sure grease to be added is compatible with grease already in bearing frame, as lubrication characteristics could be deteriorated.

4. Start pump and allow excess grease to run out of outlet hole.

5. Replace drain plug. Note: If bearings start to heat when they had run cool before lubrication, there is too much grease. Remove plug and allow excess grease to drain.

TABLE 17.2 Relubrication Intervals for Motors

Hours of service per year	NEMA frame size		
	42 to 215T	254 to 326T	364 to 447T
5000 Hrs.	5 yr	3 yr	1 yr
Continuous normal applications	2 yr	1 yr	9 mos
Seasonal service (motor idle 6 mos or more)	1 yr beginning of season	1 yr beginning of season	1 yr beginning of season
Continuous high ambient, dirty or moist locations, high vibration, or where shaft end is hot (pumps-fans)	6 mos.	6 mos.	6 mos.

Motor Lubrication Recommendations by Balder and U.S. Motors:

Manufacturer	*Tradename*
Chevron	SR1#2
Shell	Dollies R
Texaco	Polestar RB2

These are polyurea-based greases and should not be mixed with lithium based grease. If a lithium based grease is used, the polyurea grease should be purged from the bearing.

NOTE: Bearing regreasing intervals are reduced by 1/2 if shaft is vertical and there are no previsions to keep the grease in the bearings. See manufacturers instructions for regrease intervals on vertical shafts. As previously stated, *do not mix different base greases* as they are not compatible and will not give good lubrication.

Purchase grease in tubes or pint containers. Grease will have less chance of becoming contaminated with dirt as might happen with a 5-gal. (20 liter) pail. Keep grease cans covered when not being used, to keep out contamination. Many manufacturers sell their grease in a

TABLE 17.3 Volume of Grease to be Added

NEMA frame	Largest bearing	Volume (tsp.)
up to 210	6307	2
210–280	6311	3.9
280–360	6313	5.2
360–5000	NU322	13.4

These volumes would apply to pump bearings as well.

cartridge which is inserted into the standard grease gun. This makes reloading grease guns clean and quick.

17.2.3 Flexible couplings (6 month cycle)

The purpose of a flexible coupling is to couple the output shaft of a motor to the input shaft of a pump. The coupling is sized to transmit the torque required. The coupling should be capable of operating with some shaft misalignment, due to temperature changes, without putting undue stress on the shafts and their associated components. The coupling should allow end movement without interference with each other.

17.2.3.1 Types

Spring coupler, Fig. 17.23 (cylindrical helical spring of rectangular cross section with hubs screwed into each end of spring coil). Extension of this coupler is used to load the seal face of the pump water seal as well as to correct for misalignment. This coupler is used on small inline circulators. No maintenance is required except to replace motor resilient rings if they sag, throwing motor out of alignment.

Multispring coupler, Fig. 17.24. This type coupler is made of two pairs of cylindrical tension springs that are positioned 90° from each other. Opposite pairs are linked together and the ends are connected to outreaching arm of an opposing coupler hub. If any of the springs break the whole coupler assembly should be replaced. There are *no* maintenance requirements except to maintain motor alignment. This coupler is used in low horsepower requirements.

Elastomer coupling-lug type, Fig. 17.25. The legs of an elastomer spider are positioned between opposing lugs of metal hubs. No maintenance except maintain alignment. Used on low horsepower pumps.

Shear type coupling, Fig. 17.26. Two opposing hubs connected by an elastomer insert with mating teeth on each end. This coupler design can be sized from fractional through several hundred horsepower. This

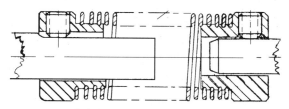

Figure 17.23 Spring coupler.

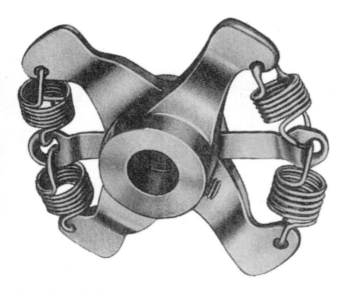

Figure 17.24 Multispring coupler.

coupler should be sized for the actual horsepower. A shear coupler is the most forgiving for misalignment and is flexible in all planes. No maintenance is required except to maintain alignment. Check for rubber dust, which is an indication of insert wear due to misalignment or underloading.

Tire type coupler, Fig. 17.27. This coupler has two hubs. The bead of one side of the tire is clamped to the flange of one hub and the opposite bead is clamped to the opposing flange. This is another shear coupling

Figure 17.25 Lug type elastomer coupler.

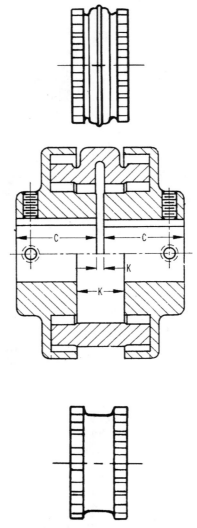

Figure 17.26 Shear-type coupling.

that can take four-way flex action with no maintenance except alignment. Check for flexible member deterioration.

Metallic couplers

Disk type, Fig. 17.28. The flexible elements in this coupler are metal disks laminated together and alternately connected to studs projecting axially from opposing hubs. This coupler is used in higher

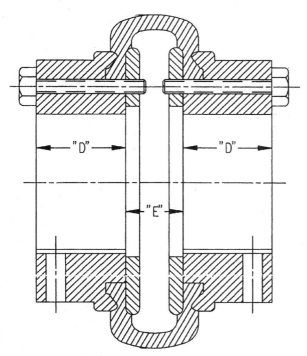

Figure 17.27 Tire-type coupling.

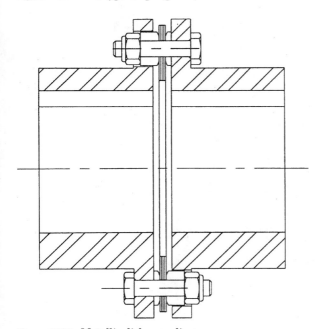

Figure 17.28 Metallic-disks coupling.

horsepower applications. There is no maintenance required except maintain alignment.

Grid type coupler, Fig. 17.29. The flexible member in this coupler is a serpentine metal ribbon that weaves back and forth around radial projections of opposing hubs. The metal ribbon is in shear to transmit torque. The coupling requires lubrication with good quality NLGI #2 grease *once* per year. Maximum misalignment is 0.006″–0.008″ (150–200 μm) parallel misalignment and 0.003″–0.005″ (76–125 μm) angular misalignment.

Gear coupling, Fig. 17.30. The gear coupling consists of two opposing hubs with external gear teeth that engage with internal teeth at each end of a one or two piece sleeve. The flexibility is in the looseness between mating teeth. Gear couplings are used for medium and large pump applications. Maintenance: lubricate every 1–2 years and maintain alignment.

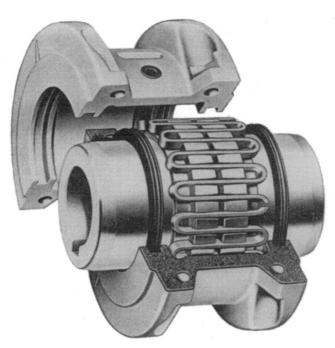

Figure 17.29 Grid-type coupling.

Figure 17.30 Gear-type coupling.

17.2.4 Alignment procedure—
6-month cycle

Warning! Disconnect and lock out power before servicing rotating equipment. Replace all guards before operating equipment.

There are two forms of misalignment between the pump shaft and motor shaft. *Angular misalignment*: when the shafts are not parallel with each other either top and bottom or side to side (Fig. 17.31). *Parallel misalignment*: when the shafts are parallel with each other but not concentric with each other (Fig. 17.32).

A minimum amount of space should be maintained between coupler faces so they do not touch each other when the motor shaft thrusts towards the pump shaft. *Grid* couplers require 1/8-in (3-mm), *Tire couplers* have a factory specified dimension between faces for proper installation of elastomer tire. Consult coupler manufacturer for proper spacing.

Gear couplings gap is 1/8-in (3-mm) for the coupler size used with HVAC size equipment. The *lug-spider* gap should be 1/8-in (3-mm) between the end of the lug and the opposing hub. *Shear* couplings are controlled by the insert. There should be 1/8-in (3-mm) of free space between the coupler flanges and the end of the insert. Note: Before

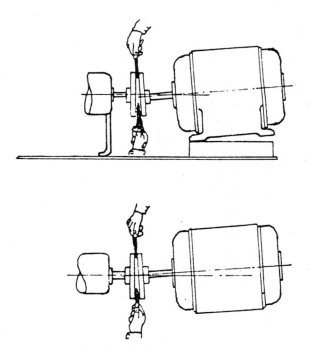

Figure 17.31 Angular misalignment.

assembling any coupler make sure there is at least 1/8-in (3-mm) between the motor shafts and that the shaft keys do not extend out past
the end of the shaft.

Tools: straight edge, taper gauge, or calipers, and feeler gauges.

If the coupler has a cover to contain lubricant or is constructed with
an outer sleeve, it must be removed to get to the hubs. Most hub
outside diameters have been machined concentric and parallel with
the bore so their surfaces can be used for alignment.

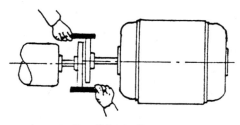

Figure 17.32 Parallel misalignment.

1. Place the straight edge across the top of flange. Hold straight edge flat against the highest flange.

2. Measure distance between opposite OD of flange and straight edge.

3. Move straight edge to bottom side of coupler and measure space between straight edge and OD of coupler. Note: If coupler halves are same diameter, the opposite coupler half will be the coupler you will rest the straight edge against.

4. Measure the gap. It should be equal to the top gap. If the lower coupler is attached to the motor, shims will have to be placed under the motor equal to the measurement taken between the straight edge and the OD of the coupler. If the motor is higher, the pump will have to be raised if there are no shims under the motor to be removed.

5. Repeat steps 1 through 4 on the sides of the coupler. The amount of off set will be the amount the motor will have to be moved one way or the other.

6. Using the taper gauge, insert it into the top gap between the flanges. Note engagement. Remove taper gauge and insert in gap on the opposite side. If gauge goes in deeper, gap is wider. The rear of motor will have to be lowered and/or front raised. Once you get the face of the couplers parallel, you will have to go back and repeat steps 1–4.

7. Using the taper gauge, check side gaps as in step 6. Once you have gotten the side faces parallel you will have to repeat step 5.

8. Once all is aligned, secure all bolts and recheck.

9. Reassemble coupler and regrease, if required, using manufacturer's grease.

10. *Replace coupler guard.*

Note: to achieve alignment in the 0.001-in–0.002-in (25–50 mm) range a dial indicator will be required after using the straight edge and feeler gauges.

17.2.4.1 Dial Indicator Alignment

1. Clamp dial indicator to one of the coupler flanges and rest indicator button on the opposite flange OD. Mark the location where the button rests.

2. Starting at the top, set dial to zero.

3. Rotate both shafts so dial indicator is on the bottom and button is on the chalk mark. Shim motor or pump to achieve a 0.000-in reading. Repeat steps 2 and 3 on the sides.

Note: If the alignment changes from one check point to the next, check the following:

1. Ambient temperature and compare with ambient temperature of last reading.
2. Pump and motor temperature and compare with last pump and motor temperatures.
3. System water temperatures and compare with temperatures taken at previous reading.
4. Check all pump and motor mounting bolts.
5. Check welds on a fabricated steel base;.
6. Check for cracks in metal of base and supports.
7. With a dial indicator attached to a coupler flange, push of suction and discharge piping and note affect on dial indicator reading.

17.2.5 Water seals

17.2.5.1 Mechanical seals. Fig. 17.33. Mechanical water seals do not require any maintenance except making sure that the water in the system remains clean and the chemical balance is maintained to minimize corrosion. *A water seal should not be taken apart until it fails as you will not get the seal back in its exact location again.* If a seal

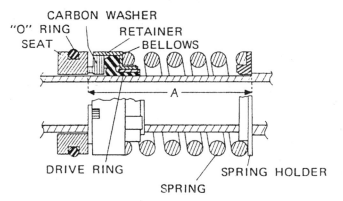

Figure 17.33 Mechanical water seals.

begins to leak in a short period of time, check the following before replacing the seal.

1. Dissolved solids in the water in high concentrations.
2. Suspended solids in water such as iron oxide.
3. System pressure on the suction side of the pump.
4. Check flush lines, fittings, and ports for blockage.
5. Materials of construction of seal with type of system.
   A. open
   B. closed
   C. temperature
   D. chemicals
   E. Impurities

Packed Seals — weekly and 6-month cycles, Fig. 17.34.

Packed type pump seals are different from mechanical seals in that packed seals must leak for proper lubrication and cooling of the packing. *Each week,* adjust packing gland hex nuts to achieve 60–80 drops per minute. If the packing gland has bottomed and there is no adjustment the pump should be repacked. Instructions are detailed repair section.

At the repacking stage, check seal flush lines, making sure line and ports are not plugged and lantern ring is properly located in line with the flushing port. If the life of the packing is short and proper maintenance procedures were followed, contact your seal supplier for a recommendation for a more durable material. It may be advantageous to convert to a mechanical seal. The additional expense will more than

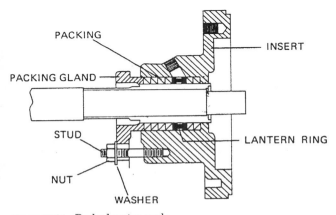

Figure 17.34 Packed water seals.

offset by the amount of man-hours saved in maintenance, fluid savings, elimination of oxygen laden water, and reduced water treatment.

17.2.6   Strainers

Strainers, Fig. 17.35, are inline devices which filter the flowing fluids to protect sensitive downstream equipment. The strainer body is a pressure vessel that supports the screens or filter and is connected to the piping. The strainer can be fitted with a blowdown valve to allow purging the trapped contents. The strainer typically has a cover, which once removed, allows complete removal of the screen or filter for thorough cleaning.

When a system is new, blowdown strainers once per quarter. Note the amount of collected debris. If debris is minimal, extend cycle to 6

Figure 17.35   Strainers.

months and check again. If debris is still minimal, extend to once per year.

17.2.7 Gauges

17.2.7.1 Pressure gauges—yearly cycle A *Bourdon-tube* gauge, Fig. 17.36, is the most widely used pressure indicating device. It is constructed of a flattened tube bent into a circle. One end of the tube is sealed to the pressure connection and the other end is closed with a linkage attached to it. The linkage is attached to a sector gear which meshes with a gear attached to the pointer stem. When pressure increases inside the flattened tube, the tube tends to straighten and its motion can be calibrated to the scale to indicate the pressure. Some gauges are set up to read only (+) positive pressure. Others are set up to read (+) and (−) pressure. These are called compound gauges. Other gauges set up to read only (−) pressure are known as vacuum gauges.

A *diaphragm* gauge, Fig. 17.37, is used to measure low pressures. The gauge is constructed of a diaphragm, which the pressure acts against and an opposing spring. The displacement of the diaphragm is proportional to the pressure and is linked to a pointer to indicate the pressure.

The diaphragm gauge can be used as a differential gauge by having the low pressure enter the spring side of the diaphragm and the high pressure on the opposite side. The difference between the two pressures will be the pressure that displaces the diaphragm. This type of

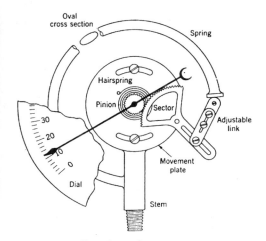

Figure 17.36 Bourdon-tube gauge.

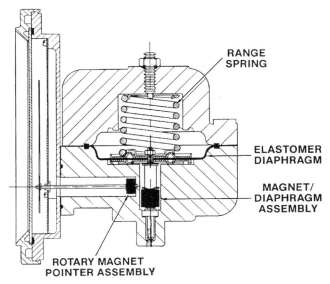

RANGE
SPRING

ELASTOMER
DIAPHRAGM

MAGNET/
DIAPHRAGM
ASSEMBLY

ROTARY MAGNET
POINTER ASSEMBLY

Figure 17.37 Diaphragm gauge.

gauge is used to measure the drop in pressure across a valve, orifice, nozzle, pump, or any other type of differential creating device.

With computer controlled systems, electric pressure transducers are used to transmit the pressure reading. These can be of *single* or *differential* pressure design.

Calibration: Pressure gauges are relied on to indicate toe performance of a system and its components. The pressure gauges are depended on to indicate the true pressure. Due to pulsations and vibrations in a system, the gauges either wear out or go out of calibration. All gauges in a system should be calibrated once per year.

There are three methods of calibration:

1. Using a *dead weight tester*, Fig. 17.38, which is a hydraulic cell with calibrated weights traceable to the Bureau of Standards.
2. Using a *calibrated test gauge*, Fig. 17.39, of the range being tested, with an accuracy of 1 percent of full scale.
3. Using a *"U" tube manometer*, Fig. 17.40, filled with water or mercury.

A range of calibrated test gauges would be the easiest and least time consuming of the methods. The "U" tube manometer is typically used to calibrate low-pressure gauges.

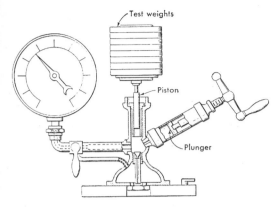

Figure 17.38 Dead-weight gauge tester.

Typically a Bourdon tube gauge is most accurate midrange and slightly lower in accuracy before and after midrange. A pressure gauge should be selected to operate in the midrange + or −1/4 scale.

Dead-Weight Tester Calibration Procedure:

1. Screw gauge to be calibrated into gauge port in tester.

2. Open oil reservoir.

3. Screw in displacement piston until piston bottoms.

4. With oil reservoir open, screw displacement piston out. *Do not allow reservoir to run dry.*

5. Close reservoir valve. Spin weighted piston and screw displacement piston in until weighted piston raises 1/2″ (12 mm). Note gage reading. Note: weighted piston and individual weights have numbers on them. The total of the weights is the pressure being developed.

6. Check if gauge agrees and note reading.

7. Add weight to piston equaling midrange pressure. Screw in displacement piston in until weighted piston rises 1/2-in (12-mm) and spin weighted piston. Note gauge reading and total weight.

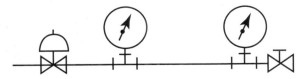

Figure 17.39 Comparison-gauge test setup.

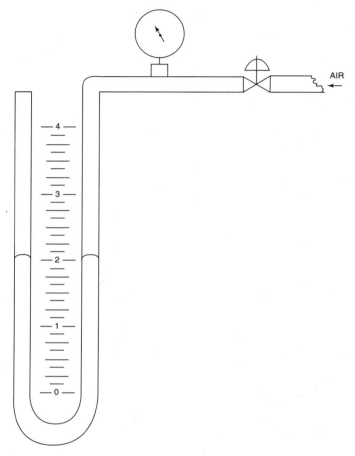

Figure 17.40 Low-pressure "U" tube gauge tester.

8. Add weight to piston to equal full range and proceed as in step 7.

9. Evaluate readings taken.

   A. If all readings are out of calibration the same amount, set weight midrange, remove pointer and reposition to equal weights spinning.

   B. If gauge inaccuracy increases as the pressure is raised, the gage will have to be taken out of its case to gain access to the linkage. The linkage connected to the sector gear will have to be moved closer or farther from the sector gear to make the gauge follow the weights applied. Once the inaccuracy is uniform, the weights can be set for mid scale and the pointer can be reset.

Comparison Calibration: Set up a manifold with a pressure regulator connected to two 1/4-in (6-mm) NPT pipe tees. The tees should be separated by a 1/4″ × 6″ (6 × 150 mm) pipe nipple. Male × female, 1/4-in (6-mm) needle valves, should be screwed into branch connections of the tees and in the end of the last tee (see Fig. 17.39).

1. Install *Calibrated Master Gauge,* with range equal to gauge to be tested, in a branch connection.

2. Install the gauge to be tested in the other branch.

3. Connect an air supply to the inlet to the pressure regulator and close bleed valve at end of manifold.

4. Start adjusting pressure regulator to achieve a mid-range setting and note gauge reading.

5. Increase air pressure to full range and note gauge reading.

6. If gauge inaccuracy is stable between 0 and full scale; readjust to midrange while bleeding off excess pressure.

7. Remove pointer and reposition to agree with master gauge.

8. If gauge is not stable throughout the range, proceed as in 9B of Dead-Weight Tester test.

17.2.7.2 Low-pressure calibrator

1. With a piece of clear plastic 3/8-in (10-mm) diameter tubing, form into a "U" tube long enough to equal gauge requirements (see Fig. 17.40). Fasten to a board, leaving about 3/4-in (19-mm) between tubes.

2. Fasten a steel tape measure between the tubes.

3. Connect an air pressure regulator to a TEE connecting the hose from the manometer and the gauge to be tested.

4. Fill "U" tube half full of water.

5. Slowly adjust air regulator to raise column of water. Note: the water.column connected to the air will drop and the opposite column will rise. The pressure reading is the total distance between the top of the lower column and the top of the upper column. (inches of water), 12 in = 1 ft, 2.31 ft = 1 psi (1 cm = 2.5 kPa).

6. Compare water column reading with gauge being tested. Calibrate gauge as 11A and 11B of Dead Weight Tester.

Note: To calibrate a differential pressure gage, apply pressure to the *high* pressure port and leave low pressure port open to atmosphere. Calibrate as you would a Bourdon tube gauge.

17.2.7.3   Thermometer calibration (yearly cycle).

Thermometers of the liquid filled type do not go out of calibration unless the unit has been exposed to temperatures beyond the range of the thermometer. The red liquid or mercury tends to separate when the column retracts on cooling. Try shaking down the column. If columns do not go together, install a new thermometer.

Calibration Procedure, High Range:

There are two standards available to calibrate thermometers:

A. Boiling water at sea level is 212°F (100°C).

B. Ice water is 32°F (°C).

　　1. Place a can filled with water on a stove. Wrap a layer of fiberglass insulation around can.
　　2. Immerse thermometer to be calibrated in water.
　　3. Bring water to a boil. Keep end of thermometer off bottom of can. From Fig. 17.41 determine boiling point at local elevation.
　　4. Loosen scale screw and calibrate scale. If scale does not move record error on thermometer.

Calibration Procedure, Low Range:

　　1. Insulate a can with fiberglass insulation.
　　2. Fill a can with ice and water.
　　3. Immerse thermometer in ice bath.
　　4. Make a cover of fiberglass insulation and cover top of can.
　　5. Note temperature registering on thermometer when column stabilizes.
　　6. Loosen scale locking screw and adjust scale to read 32°F (0°C) and tighten screw. If scale is not adjustable, record error on thermometer.

Altitude Ft.	Barometer Hg.	Atmos. PSIA	Atmos. Ft.	Boiling Pt. °F	Boiling Pt. °C	Atmos. kPa
0	29.9	14.7	33.9	212	100	101.4
500	29.4	14.4	33.3	211.1	99.5	99.3
1000	28.9	14.2	32.8	210.2	99	97.9
1500	28.3	13.9	32.1	209.3	98.5	95.8
2000	27.8	13.7	31.5	208.4	98	94.5
2500	27.3	13.4	31	207.4	97.4	92.4
3000	26.8	13.2	30.4	206.5	97	91
3500	26.3	12.9	29.8	205.6	96.4	88.9
4000	25.8	12.7	29.2	204.4	95.8	87.6
4500	24.9	12.4	28.8	203.8	95.4	85.5
5000	24.4	12.2	28.2	202.9	95	84.1

Figure 17.41   Water-pressure and boiling points at various altitudes.

17.2.7.4 Bimetallic thermometer. A bimetallic strip is made from two different metals which expand at different rates. The strips are bonded together and when the strips expand, due to there different rates of expansion, the strip bends. In a bimetallic thermometer, the bimetallic strip is wound in a coil. The scale is developed to agree with the motion/degree. The thermometer is calibrated using the same process as the liquid-filled thermometers. Generally there is an adjusting screw on the back of the case to calibrate the dial to the pointer.

17.2.8 Manual valves

Many valves stay in one position for long periods of time before they are to be used. The operating engineers expect to be able to operate the valves when required and do not expect them to leak internally or externally. With some periodic maintenance, the valves can be depended upon.

The maintenance areas of valves are the: (1) stem threads, (2) stem seal, and (3) seat and disk seal.

17.2.8.1 Globe valves, Fig. 17.11. A globe valve has a rising stem when opened. The exposed stem, in the area of the packing, should be smooth and kept free of paint. The stem seal may be packing which can be made to seal tighter by turning down the packing nut. The packing should not be so tight that the stem will not turn. If tightening the packing nut will stop the leak, the valve will have to be repacked. See repair section.

Some large globe valves have their stem threads exposed. With smaller valves, the threads are internal to the valve. The external threads should be kept clean of paint, and lubricated.

The seat and disk can only be checked for leakage by closing the valve and emptying the downstream line. If fluid continues to run from the line the valve is leaking through. Many times it is impractical to drain the downstream line and the only time you will prove the integrity of the seal is when you need to work on the system. If the valve in question doesn't hold, look for a valve upstream of the valve in question. The valve could be repaired at the same time as the other piece of equipment. If the valve is used for throttling, note the leakage problem and schedule the repair during the system overhaul period.

17.2.8.2 Gate Valves, Fig. 17.15. A gate valve has either a rising stem or nonrising stem. Bronze gate valves, whether rising or non rising, the threads are internal. The stem should be treated like a globe valve; clean, smooth, and free of paint. Leakage is controlled by a packing gland as with globe valve. Typically the largest bronze valve is 3-in (80-mm) pipe size.

Iron body gate valves have rising or nonrising stems, Fig. 17.42. The rising stem version has it's threads exposed and passing through a rotating nut mounted in the hand wheel. The hand-wheel and nut assembly are bearing in an outside yoke. The stem and threads should be kept clean, not painted, and lubricated. The hand-wheel/nut bearing should be lubricated. Rising stem gates that are open all the time can have the stem threads protected by installing a loose plastic sleeve over the threads.

The packing gland, on the cast iron valves, is now a flanged device with studs and nuts 180° from each other. The threads of the studs should be lubricated and tightened to prevent gland leakage. Do not overtighten, as this will make operation of the valve difficult.

Seat leakage is difficult to determine until you need to use the valve as a service valve. If a gate valve has not been used for throttling, the fluid is clean, and nonabrasive, the sealing surfaces should not deteriorate. Note: Generally, *do not use a gate valve for throttling.* Nonris-

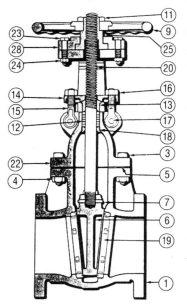

NO.	DESCRIPTION
1	BODY
3	BONNET BOLT
4	BONNET BOLT NUT
5	BONNET GASKET
6	DISC
7	DISC PIN
9	HANDWHEEL
11	HANDWHEEL LOCKNUT
12	PACKING
13	P.KG. GLAND
14	PKG. GLAND FLG.
15	PKG. GLAND FLG. EYEBOLT
16	PKG. GLAND FLG. EYEBOLT NUT
17	PKG. GLAND FLG. EYEBOLT SP.PIN
18	REPACKING SEAT BUSHING
19	SEAT RING
20	STEM
†21	LUBRICATING FITTING
22	YOKE BONNET
23	YOKE CAP BOLT
24	YOKE CAP BOLT NUT
25	YOKE BUSHING
28	YOKE CAP
†29	BONNET
†30	YOKE
†31	YOKE CAP SCREW

† Not Shown

Figure 17.42 Iron-body gate valve.

ing stem valves that are to be opened fully should be closed at least 1/4 turn. This allows the operator to determine if the valve is open or closed at a future date.

17.2.8.3 Ball valves. Ball valves, Fig. 17.12, are quarter-turn valves. The stem seals are either o-rings or adjustable packing. Ball valves are maintenance-free devices, and it is usually less expensive to install a new valve than to repair the old one.

17.2.8.4 Check valves. Check valves, Fig. 17.43, normally have no external maintainable parts. Internal parts which may need service at an overhaul would be a spring, hinge pin, hinge assembly or disk. Note: A check valve that is oversize for the amount of flow, will make a check valve noisy and wear its internal parts due to the disk assembly being in constant motion. If a check valve malfunctions in a short period of time, it would be wise to reduce the size of the check valve one pipe size.

Flow-checks, Fig. 17.17, used in residential heating systems have a manually operated stem which allows lifting the weighted check. The stem seal is an o-ring or packing with an adjustable packing gland. If the gland isn't leaking no maintenance is needed.

Butterfly valves, Fig. 17.13, use o-ring stem seals and require no maintenance.

Lubricated plug valves, Fig. 17.14, should be lubricated and the plug turned periodically. This lubrication and turning assures a good seal and operation of the plug.

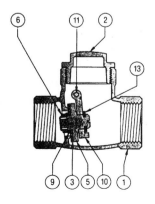

NO.	DESCRIPTION
1	BODY
2	CAP
*3	DISC
5	DISC HOLDER
6	DISC NUT
9	DISC WASHER
10	HINGE
11	HINGE PIN
13	RETAINING RING
†15	SIDE PLUG

Figure 17.43 Check valve.

Pressure reducing valves (PRV), Fig. 17.16, come in all sizes and shapes. They all use a flexible diaphragm made of metal or a reinforced elastomer to contain the controlling fluid. If the control fluid is applied internally an adjusting screw and spring are on the outside of the diaphragm. There is no valve stem packing and the only maintenance is to replace the diaphragm and/or disk if they start to leak. An active PRV may need to be taken apart every couple of years, depending on water conditions, to inspect, clean the interior and clean strainers.

With external diaphragms, Fig. 17.44.*a*, the control fluid is supplied by external tubing. Typically the adjusting spring is external. The valve stem has a packing gland with renewable packing. Note: A PRV usually is an active valve, so weekly checking of the packing gland will be necessary, adjusted frequently and re-packed more often than other valves. If the regulated pressure rises when there is no flow, the seat and disk will have to be resurfaced.

Temperature regulating valves, Fig. 17.44.*b*, are set up with the same configuration as the pressure regulating valves with external diaphragm. The difference is the activating head is filled with a temperature sensitive fluid contained within the head and a capillary tube. The capillary tube end is placed where the temperature

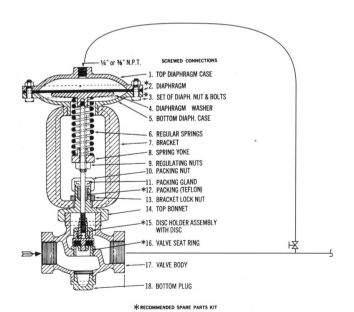

¼″ or ⅜″ N.P.T.   SCREWED CONNECTIONS

1. TOP DIAPHRAGM CASE
*2. DIAPHRAGM
*3. SET OF DIAPH. NUT & BOLTS
4. DIAPHRAGM WASHER
5. BOTTOM DIAPH. CASE

6. REGULAR SPRINGS
7. BRACKET
8. SPRING YOKE
9. REGULATING NUTS
10. PACKING NUT

11. PACKING GLAND
*12. PACKING (TEFLON)
13. BRACKET LOCK NUT
14. TOP BONNET

*15. DISC HOLDER ASSEMBLY WITH DISC
*16. VALVE SEAT RING

17. VALVE BODY

18. BOTTOM PLUG

✳ RECOMMENDED SPARE PARTS KIT

Figure 17.44a Pressure-reducing valve (external diaphragm).

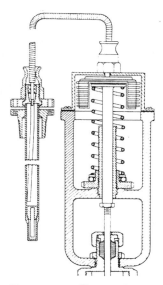

Figure 17.44b Temperature-regulating valve.

is to be monitored. The fluid in the bulb in the end of the capillary expands or contracts with change in temperature. If the fluid expands, it passes that expansion back to the actuator head. The increased pressure pushes the diaphragm and the diaphragm pushes the stem, closing the valve. Maintenance is required to the stem packing and possible seat and disk rework, refacing.

Zone valves, Fig. 17.45, are used to control the heating or cooling fluid going to a heat transfer device supplying a particular zone. The valve can be electrically or nonelectrically, (ambient temperature) activated. The operator is a renewable part, and easily removable from the valve body. Normally these are maintenance free valves. Some manufacturers use a convoluted bellows or a diaphragm as the dynamic seal. Others use o-ring sealing or renewable packing. Depending on the design, some valves use elastomer disks and some use metal to metal.

Relief valves come in two types: (1) relief valve for liquids, Fig. 17.10.*a*, and (2) relief valve-pop type, Fig. 17.10.*b*.

Relief valves for liquids can have metal to metal seats or soft seats. The *liquid relief* is used on systems whose fluid is a liquid and will remain a liquid once the pressure is removed. Because a relief valve protects the components of a system, it should be checked once a year. Relief valves come with manual levers. Lift the lever and the

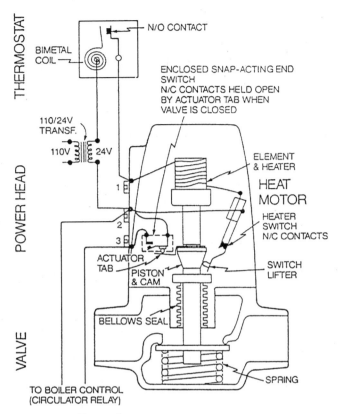

THERMOSTAT

POWER HEAD

VALVE

N/O CONTACT

BIMETAL COIL

ENCLOSED SNAP-ACTING END SWITCH
N/C CONTACTS HELD OPEN BY ACTUATOR TAB WHEN VALVE IS CLOSED

110/24V TRANSF.

110V   24V

ELEMENT & HEATER

HEAT MOTOR

HEATER SWITCH N/C CONTACTS

ACTUATOR TAB

PISTON & CAM

SWITCH LIFTER

BELLOWS SEAL

SPRING

TO BOILER CONTROL (CIRCULATOR RELAY)

Figure 17.45 Zone valve.

system fluid will flow out the discharge port. If the lever binds, will not open the valve, or the valve does not open smoothly, replace the valve. If the valve works satisfactorily but leaks, open the valve several times to purge the contamination from the seat. A few flushes should solve the leakage problem.

Pop type relief valves are used for steam or hot water that would flash to steam. These valves "pop" open fully and stay open until the pressure decreases a predetermined amount and then closes. The outlet port is usually larger than the inlet. This type pop relief should be tested once a year by increasing the boiler or system pressure until the relief "pops." Note the pressure at this point and the pressure at which the relief closes. These figures should agree with the original settings. The valve should be readjusted if the settings do not agree. If the valve leaks after the test, manually relieve the valve several times. If this does not reseat the valve, the valve should be removed for repair.

17.3 Specific Predictive Maintenance Procedures

Dynamic predictive maintenance is different than most preventive maintenance programs in use today. Predictive maintenance monitors the vital life signs of operating equipment to detect early stages of unbalance, misalignment and bearing defects, long before preventive maintenance procedures can do so. This early prediction can allow a pump or motor to be serviced at a preset time when parts will be ready, systems can be secured and proper personnel are available.

Predictive maintenance will achieve higher levels of operating efficiency, higher comfort levels, reduced costs, reduced replacement parts inventories, and lower maintenance man hours.

The biggest plus for predictive maintenance over conventional maintenance programs is in *finding a fault before there is a catastrophic breakdown which could shut down critical services to a building.*

Predictive maintenance is the periodic monitoring, recording and analysis of pump or motor vibration as an indication of its overall condition. If the maintenance personnel consistently record and analyze the vibration readings, trends can be detected and repairs scheduled on a timely basis. Being able to schedule reduces down time and extends the maintenance cycle many times.

Typical vibration levels are taken with a vibration meter at the bearing points of the pump and motor, in the x, y, and z direction. Each recording point has a numbered location so the data can be compiled and analyzed for a particular location. If a check point indicates a change, the frequency can be analyzed and may indicate a known type of pending failure. Corrective actions can be taken to ward off the failure.

When a pump and motor is first installed, the vibration levels at the predetermined points should be recorded. These readings will be the bench mark to judge all others. There is a faint possibility that a manufacturing defect will be detected and can be corrected.

Besides vibration analysis, thermograph, oil analysis, and ultrasonics can identify fault operation, predict the time frame for failure and pinpoint the component degradation. A bearing getting warm or hot will indicate a need for corrective action. The bearing could have been over greased or on the verge of failure due the a lack of grease. When the oil is changed in a bearing housing, and metal chips or water are found, additional maintenance steps can be taken while the pump is down to correct the problem. Bearings and valves tend to emit a certain sound. Using an ultrasonic pick up, the human ear can detect a change from the norm. A lack of lubrication can be noted and regreasing can eliminate a failure.

When vibration levels are analyzed, the associated frequency and the direction of the vibration can indicate the source of the problem. A vibration level present at 1x rotational frequency in a radial direction means, generally, an imbalance. At 2x rotational frequency one can typically observe evidence of looseness. At 4, 6, 7, or 8x rotational frequency, this would indicate vane pass frequency, which is equal to the number of vanes on the impeller.

In analyzing bearing frequencies, there are typically four areas to look at:

1. FTF: fundamental train frequency—the speed at which a ball spins around the shaft.

2. BSF: ball spin frequency—the speed at which the ball spins within the race. Higher than normal amplitude might indicate a bearing flaw.

3. BPFO: ball pass frequency, outer—the speed at which the bearing passes the outside of the race. Higher than normal readings might indicate a nick in the outer wall of the race.

4. BPFI: ball pass frequency, inner—the speed at which the bearing passes the inside of the race. A higher than normal amplitude might indicate a nick in the inner wall of the race.

Before we take the base line reading, the record card should be set up with the pump and motor characteristics: (1) flow & head, (2) no. of stages, (3) power, (4) speed, (5) no. of impeller vanes, and (6) bearing nos.

The characteristics of the pump and motor will help identify the origin of the frequencies being generated. An increase in amplitude at one of the identified frequencies will indicate which part to repair.

A flexible coupling unbalance or misalignment should show up in a radial direction at the bearing closest to the coupler at the 1x rotational frequency.

A pump or motor will not be analyzed with just one reading. The data has to be recorded for several different intervals so that a trend can be established. Recording this data manually or with the new easy-to-use equipment with the user friendly software, makes the data easy to interpret.

17.3.1 Measurements and techniques

1. Low-frequency vibration measurement (i.e., acceleration, velocity, displacement). Identifies unbalance, misalignment, critical resonances, mechanical looseness, and structural problems.

2. Time-based enveloping (high-frequency vibration). Identifies problems undetected in low-frequency vibrations because of masking by other frequencies and mechanical noise. Enveloping isolates and filters very small signals that can indicate problems with bearings. Envelope readings are expressed in RMS (root-mean squared) average.

3. High-frequency acoustics emission measurements (ultrasonic) provide additional information on bearings and valves. One can detect metal-to-metal contact, contamination, lack of lubrication, excessive load, and bearing defects. A valve leak can be detected by operation observation and internal damage.

17.3.1.1 Starting a program. Map out all the pumps and motors in your system and evaluate each pumps significance to the system. Categorize by:

1. critical

2. important

3. general purpose

Based of their maintenance history categorize:

a. no problem

b. normal

c. problematic

Problematic and critical pumps should be monitored first and daily until you get a history. Noncritical and no problem pumps can be monitored monthly or bimonthly. When you see a change you start to monitor more frequently.

17.3.1.2 Human sensors. Even if you don't have fancy equipment, you can still utilize your own natural sense of feel, temperature sensitivity, and hearing. A stethoscope or screw driver to the ear can be a very effective predictive analyzer.

The ultrasonic probes can be used to check for valve leakage. Take note of the sound upstream of a valve and then downstream. When the valve is closed and working properly, there should be no change in the sound.

For temperature measurement a hand-held pyrometer can be used to monitor bearing surface temperatures and motor surface temperatures. An increase in motor surface temperature without an ambient temperature change would be an indication of: increased motor load, electrical problems, low voltage, unbalanced feeder line voltages and currents. The increased motor load when connected to a pump can be attributed to system flow changes or increased mechanical resistance due to pump or motor parts rubbing.

A hand-held voltmeter and clamp-on ammeter is another predictive maintenance tool. The balance of the current draw between electrical feeder lines can be monitored. The current draw has a direct relationship to the load on the motor. The full-load and service factor amps are indicated on the motor nameplate. The readings are compared to the nameplate amps to determine the load on the motor. *Note: If a variable frequency drive is supplying power to a motor a TRUE R.M.S. ammeter should be used to get an accurate reading of the current draw.*

Most motors have an ambient temperature rating of 104°F (40°C). This is the baseline for all the winding insulation classes. This temperature rating of class "B" insulation is 266°F (130°C) and class "F" is 312°F (155°C). These are the common motor insulation classes. These temperatures are a sum of the 104°F (40°C) + 144 F (80°C) winding temperature rise + a 18°F (10°C) hot spot allowance to equal 266°F (130°C). Class "F" insulation would be 104°F (40°C) + 190°F (105°C) + 18°F (10°C) = 312°F (155°C)

As a rule of thumb, insulation life will double for each 18°F (10°C) of unused insulation temperature. If a motor operates above its rated temperature, for every 18°F (10°C), the insulation life is cut in half.

If our ambient temperature is 121°F (50°C) and the motor is operating at full load amps., the life would be halved. If the ambient is lower than 104°F (40°C), operating at full load, the insulation life would be significantly increased.

The motor surface temperatures of modern "T" frame motors do run hot 107°–203°F (75°–95°C) and do not indicate motor failure.

17.3.1.3 Waterside corrosion. The most common water problems in heating and cooling systems can be one or more of the following: (1) corrosion, (2) scale formation, (3) biological growths, (4) suspended solid matter.

Knowledge of these problems is important because each of them can reduce the cooling or heating efficiency of a system and can lead to premature equipment failure and possible damage to persons in the vicinity of a failure.

Controlling waterside problems is complex and involves water chemistry. A company specializing in water treatment should be engaged.

If system water is not treated, corrosion of the steel pipes, fittings and equipment will take place, shortening their life. If a water test indicated a pH level less than 7, the water is acidic and corrosive. If the water is 7 or higher the water is neutral to alkaline. Less corrosion takes place if a system is alkaline.

If a sample of water is taken and the color is brown to black, there is much corrosion taking place. Iron oxide is turning the water black. This material can be very abrasive to pump seals and wear rings. Besides the need for cleaning and treatment, the iron oxide could indicate oxygen laden make up water entering the system due to a leak. The system should be checked to find the leak. If no leak is found, there probably is an air leak in the system. If there are automatic air vents at the top of the system and the system pressure is too low, pressure at the top of the building could be below 0, indicating air is entering the vents. Another source of air could be from a need to recharge a plain steel expansion tank every couple of weeks due to an air leak in a sight glass or automatic vents in the system.

Air is 30 percent oxygen and 70 percent nitrogen. Normally in a closed loop system, the 30 percent Oxygen is used up and then no further corrosion takes place as nitrogen is inert. Every time the plain steel expansion tank is recharged, a fresh supply of oxygen is added. Using a soap solution, check the portions of the expansion tank above the water line. Make sure there are no automatic air vents when a plain steel expansion tank is used. There is no rubber diaphragm to separate the air from the water and so the air is absorbed into the water, goes out into the system, then comes out of solution and back to the tank. If the air is removed while it is out in the system, the tank fills with water or waterlogs.

A sample of system water may not *look* loaded with iron oxide but if allowed to settle, a magnet will pick up the settled iron oxide.

Water analysis done by a water chemist should give a detailed report of the system chemistry. If the dissolved solids and suspended solids are high, pump seal life is due to be shortened

17.3.1.4 Flexible couplings. A flexible coupling is used to connect the motor shaft with the pump shaft. If the coupler uses an elastomer insert that is engaged with a drive hub at each end, any premature wear may show up as rubber dust, collecting on the pump base. This is a good indicator that there is a misalignment problem. If the coupler is all metal, metal filings indicate a lack of lubrication.

17.3.2 Yearly pump performance test

The purpose of the yearly test is to determine if the internal clearances are getting wider. As the internal clearances increase, more and more of the flow will be short circuited back to the suction, the head will deteriorate and energy is being wasted.

The yearly test results can be plotted and compared against the original pump test. With this data the pump overhaul schedule can be predicted.

Equipment needed:

1. pressure gauge readable in 0.5 psig (3.5 kPa) increments and to operate midrange, one-strobe tack or revolution counter, one 1/4-in (6 mm) NPT tee, two refrigeration-charging hoses.
2. 1/4-in (6-mm) NPT × 1/4″ (6 mm) SAE flare shut-off valves, two-1/4-in (6-mm) NPT × 1/4-in (6-mm) SAE flare connectors, one-ammeter, one-voltmeter.

This list of material assumes that there are shut-off cocks at the suction and discharge flanges.

If the original installation was not set up with a single gauge test set up as shown in Fig. 17.46, the parts list above can be used as a portable test rig or the single gauge arrangement can be hard piped.

The single gage arrangement is the most accurate, as you eliminate the inaccuracy of two gauges. To take the readings:

(1) open the discharge tap, read and record the pressure

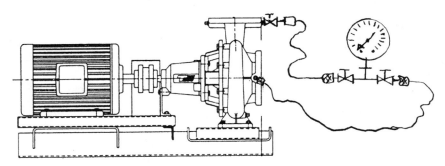

Figure 17.46 Single-gauge setup for pump test.

(2) Close the discharge tap and open the suction tap, read the pressure and record. *Note: When you take the pressure readings, keep the gauge at the same elevation as the centerline of the impeller.*

When each set of discharge and suction pressure readings are taken, the RPM, amperage, voltage, and differential pressure across a flow meter, valve, heat exchange or elbow should be taken. A differential pressure gauge reading in ft (m) of water, 0–25 ft (0–8 m), or 0–100 in (0–250 cm) of water would be recommended.

17.3.2.1 Flow readings. To develop a performance curve for a pump, flow readings or a flow reference reading has to be taken and the head developed by the pump at that flow. If a flow meter is installed in the discharge or suction line, the flow in gal/min (GPM) (liters/min) can be read and the head plotted with this flow. If there is no flow meter to measure the pressure drop across a valve, a clean strainer, heat exchanger or an elbow set up with pressure taps in the center of the inner and outer radius can be used. These devices have to monitor the total flow entering or leaving the pump. If no calibration data is available, GPM will not be determined from these differential pressures. The differential pressures will be reference points. The device should be used exactly the same way each year. A procedure should be written of the exact steps taken so anybody can run the test. Six points should be taken from 0- full flow. 0 and full flow will be two points and the remaining points evenly distributed. Reference points will be 0-differential pressure through maximum differential pressure at full flow.

17.3.2.2 Motor data. The current draw (amps.), and voltage supplying the motor of the pump being tested should be recorded. The current draw will be another good reference point. *Note: Record the motor nameplate data and compare to the original data to make sure the motor has not been changed.*

17.3.2.3 Calibrated balancing valve. If a calibrated balance valve is installed in the discharge line it can be used as flow meter and reference as long as every year the valve is at the same setting. The calibrated balance valve has pressure taps installed at the inlet and outlet of the valve. The manufacturer of the valve supplies a flow chart showing the flow rate at different valve settings and differential pressures, Fig. 17.47.

17.3.2.4 Valve C_v Fig. 17.48. If the C_v rating of a valve is known, a pressure tap can be applied to the upstream and down stream piping and the Differential Pressure (DP) measured at these points. The flow can be calculated by the formula GPM = $C_v \sqrt{DP}$. DP = pressure drop across valve in PSI.

Figure 17.47 Calibrated balancing valve.

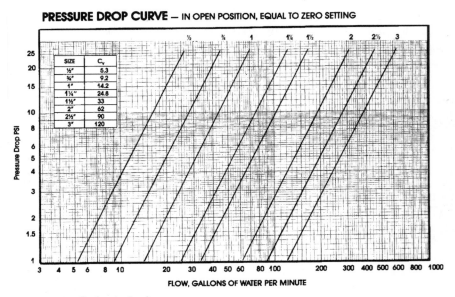

Figure 17.48 C_v chart of valves.

17.3.2.5 Heat exchange flowmeter: (Fig. 17.49). If a heat exchanger is used as a reference, a pressure tap should be set up on the inlet and outlet piping, similar to the valve C_v set up. The heat exchanger manufacturer should be able to give you the pressure drop vs flow rate data. The flow rate at any other point can be determined by the equation:

$$GPM_2 = GPM_1 \sqrt{DP_2/DP_1} \tag{1}$$

DP_2 = measured pressure drop, DP_1 = rated pressure drop at a particular flow (GPM_1)

A 0–100 in (0–250 cm) of water differential pressure gauge would be good to use. Typically the differential pressures are low. The inch reading can be converted to ft. by dividing by 12 to equal ft or by 27.7 to equal PSI.

17.3.3 Field pump test

The performance of a pump can be conducted in the field and compared to the catalog performance curve. The same procedure used in collecting data for the yearly test will by needed except the flow has to be measured with an accurate flow meter.

An averaging pivot tube flowmeter, Fig. 17.50, is a very good flow meter that can be installed easily, is inexpensive, and introduces relatively no pressure drop to a system. If the pipe can be drained, a standard unit can be installed. If the pipe cannot be drained there are wet tap models, Fig. 17.51, that allow you to install the flow meter

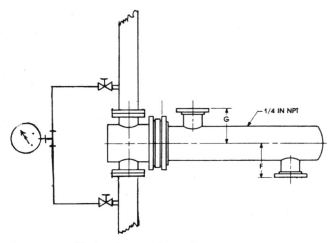

Figure 17.49 Heat exchanger flowmeter.

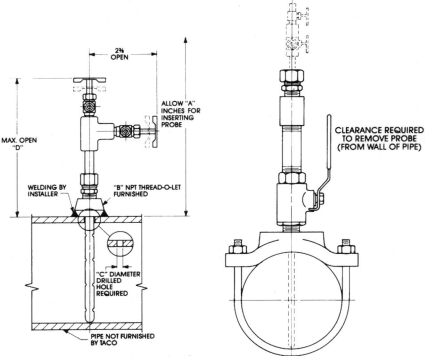

Figure 17.50 Averaging pitot tube flow-meter.

Figure 17.51 Wet tap pitot tube flowmeter.

with the line full of water and pressurized. This type of flow meter has an accuracy of 2 percent or better. The differential pressure recorded (inches [cm] of water) can be plugged into a formula or applied to a chart to read the GPM.

If the pump being tested has the same size suction and discharge connections, there is no velocity head correction. If the suction and discharge are different diameters, a velocity head correction will be added to the pump head. The velocity at the suction and discharge are to be calculated by the following formula.

$$V = .408 \times \text{GPM}/d^2 \tag{2}$$

d = inside diameter (in.) of pump connections. The velocity head is calculated by:

$$H_v = V_o^2 - V_i^2/2g \tag{3}$$

H_v = velocity head (ft of water)
V_0 = outlet velocity (ft/s)
V_i = inlet velocity (ft/s)
g = acceleration due to gravity, 32.2 ft/s^2

If the fluid being pump is something other than water at 60°F (16°C), the gauge readings will have to be corrected for specific gravity (SG) by the following formula: Head (foot of water) = psi × 2.31/SG. The

$$\text{Total head} = H_d - H_s + H_v \qquad (4)$$

H_d = discharge head, H_s = suction head, H_v = velocity head.

This procedure will give you the actual field performance. If the performance is to be compared to the catalog curve, two additional steps will have to be taken. Typically catalog curves are plotted at a constant speed of 1760 or 3450 RPM. Motors do not operate at an exact constant speed, and the full load rated speed varies with size and design of the motor. By measuring the RPM of the motor when each set of data is taken, the data can be corrected to a constant speed by the following pump laws:

$$(\text{RPM}_2/\text{RPM}_1) \times \text{GPM}_1 = \text{GPM}_2 \qquad (5)$$

$$(\text{RPM}_2/\text{RPM}_1)^2 \times \text{H}_1 = \text{H}_2 \qquad (6)$$

$$(\text{RPM}_2/\text{RPM}_1)^3 \times \text{HP}_1 = \text{HP}_2 \qquad (7)$$

RPM_2 = curve RPM, RPM_1 = observed RPM

GPM_1 & Head_1 = observed GPM & Head

GPM_2 & Head_2 = GPM & head at constant speed.

Example: (1760/1740) × 100 GPM = 101.1 GPM, (1760/1740)^2 × 50 ft = 51.2 ft

If a good predictive and preventive program is put in motion, all repairs should be able to be scheduled at a pre-arranged time. Man hours for repair will be low, equipment will last longer with fewer repairs, and no catastrophic failures.

17.4 Repair Procedures

Most pumps and valves of the same type used in HVAC systems are repaired using the same techniques. Standard mechanics tools are all that are needed with the exception of a gear puller, bearing separator, blow touch, packing pullers, and bearing heater.

Maintenance personnel should not be intimidated by pumps and valves. The areas that need close attention are noted in the following procedures. All inline and end suction pumps can be broken down into two categories:

1. A *three piece pump* consisting of a pump casing (VOLUTE), impeller and bearing assembly, and flexible coupled motor assembly.

2. A *two piece pump* consisting of a casing, and motor coupled directly to the motor. The motor shaft is the pump shaft.

Instructions: The following general instructions should guide you through the different types of pumps. If the manufacture's instructions are available, review them for any special details.

17.4.1 Motor replacement or coupler change

WARNING!!! Disconnect power, lock switch closed, and tag before touching electrical connections.

1. Remove junction box cover on motor.

2. Check terminals for power, using a voltmeter or test light.

3. Disconnect wires from terminals and remove from motor. Note: On three phase power note which power line goes to which motor leads to maintain proper rotation. If rotation does reverse, reverse any two feeder wires. Single-phase motors can be wired either way as motor wiring has to be change to make any rotation change.

4. Using an extended shaft allen wrench (typically 1/8-in [3-mm]), loosen the set screw in pumpside coupler hub.

5. Remove motor cradle bolts, attaching cradle to bearing bracket assembly. Pull motor back, and coupler should slide off pump shaft. Note: Some pump manufacturers drill a shallow hole in the side of the shaft where the set screw is to sit to properly position the coupler and assure the set screw does not slip on shaft. Back off a couple of turns on the set screw to clear the shaft.

6. If coupler is worn or has broken parts, replace entire coupler. Note the position of the coupler on the motor end.

7. Before replacing coupler, note condition of resilient motor mounts. If they are sagging due to over oiling, replace. Contact a motor repair shop or the pump manufacturer for new mounts.

17.4.2 Motor mounts

1. Remove motor cradle to mount clamps.

2. Remove motor from cradle and front mounting plate.

3. Pry off deteriorated mount. Note: inner ring of mount is pressed on to motor end bell. Use a screwdriver and fulcrum to pry ring from end bell.

4. Position new resilient ring over stub of end bell. With a ball peen hammer drive new ring home, tapping evenly in a star pattern. Tap only on the inner ring.

5. Reinstall motor in cradel, shaft end to engage in front mounting plate first.

6. Reinstall mount clamps.

7. Install new coupler or reuse old coupler if it is in good condition. Position and tighten set screws to engage with hole in shaft or reposition to original dimension and tighten set screw to flat on shaft.

8. Lift motor assembly into position behind bearing bracket. Engage coupler with pump shaft.

9. Install motor cradle to bearing bracket, engage rabbit and install bolts.

10a. Line up setscrew with hole in shaft and secure. (B&G, Armstrong type).

10b. (Taco and Thrush/Amtrol) Extend pump end of coupler towards pump 3/16 in (5-mm) and tighten set screw. Note: Do not allow pump shaft to move forward towards the pump casing or the seal faces will open and proper seal tension will be lost.

11. Rewire motor and ground connection.

17.4.3 Bearing bracket or seal replace

Warning!!! It is very important to work safely!

1. Secure the power leading to the pump motor, lock it out, and tag.

2. Secure the valves on the suction and discharge of the pump and the fill line to the system. Let pump cool. If there are no pump isolation valves, it may be necessary to drain the system or at least some of the boiler water. *Warning!! Before loosing the pump bolts, allow pump and water to cool to at least 100°F (38°C).* Open drain valve. If no water flows, there is either a plugged drain or a vacuum holding the water. Work carefully while removing bearing bracket

bolts. Break casing-to-bearing bracket seal and allow water to drain before completely removing casing bolts. This is a flat gasket and it may stick. Pry loose with screw driver.

17.4.4   Seal replacement

Note: The motor does not have to be removed.

1. If only the seal is to be replaced: B&G & Armstrong — Hold impeller with a rag and remove impeller nut and lock washer.

2. Hold impeller and tap on pump shaft with screw driver or hammer handle. Impeller should slip off shaft.

3. Remove seal spring.

4. Pry off seal with a screwdriver. A little silicon grease on the shaft sleeve will ease disassembly.

5. On a cast iron pump, check the seal face on the bearing bracket. If the seal face is pitted or grooved, replace the entire bearing assembly. (Some new bearing assemblies have a replaceable seat.)

6. Remove ceramic seat and gasket from units having seat.

7. Clean shaft sleeve.

8. Install gasket in recess of bearing assembly. Install ceramic in recess on top of gasket with side flats engaging with seat ring. Note: Grooved side of seat to go down with lapped surface facing up.

9. Install carbon over shaft with reduced diameter side facing ceramic insert or bearing bracket.

10. Put a few a drops of oil on the I.D. of the rubber inside the seal retainer. Slide seal retainer down shaft sleeve and engage the lugs of the seal retainer with the notches in the carbon.

11. Lift the shaft while pushing seal retainer area closest to the shaft sleeve.

12. Install the seal spring, impeller key, impeller, lockwasher, and impeller nut. Tighten impeller nut using a socket wrench while holding the impeller with a rag.

13. Clean pump body gasket surface, bearing bracket to casing gasket surface, and install new gasket.

14. The bearing assembly, with impeller, is ready to be installed into the pump body. Line up bolt holes and position oil holes facing up. Install bolts and tighten alternately.

17.4.5 Bearing bracket replacement

1. If a new bearing assembly is to be installed, follow steps for motor and coupler removal.

2. Remove bearing bracket as done in seal change.

3. Transfer impeller and key to new bearing bracket if none were supplied.

4. Reinstall bearing bracket assembly, as detailed in steps 12, 13, and 14 of "Seal Change Procedure."

5. Important!! Lubricate bearing assembly per instructions with tube of oil supplied. If oil is not supplied use #20 SAE grade nondetergent oil.

Warning!!! Never operate a pump without water. Make sure pump isolation valves are open, system fill valve is open and system is pressurized to proper pressure.

System pressure is equal to elevation of building in ft (ft) $\times$ 0.43 = PSI + 5PSI = system pressure (in basement).

17.4.6 Seal or bearing bracket change (TACO or Thrush/Amtrol)

The small 3/8-in (9-mm) dia. shaft pumps by Taco differ in that the pump shaft is part of the impeller assembly. The spring coupler not only serves as a flexable coupling but a tension spring for the mechanical seal. To service follow these steps:

1. Follow steps outlined earlier for removing a bearing assembly.

2. For a seal change you don't have to remove the motor. Loosen coupler/pump shaft set screw.

3. Pull impeller and shaft from bearing bracket.

4. Pry plastic seal retainer away from the back of the impeller. Once seal starts to move, the oil on the shaft should allow the seal to slide easily.

5. Clean shaft and re-oil.

6. Assemble seal per Fig. 17.52 and slide down shaft engaging teats of plastic retainer with holes in back of impeller.

7. Pry old seat and rubber cup out of bracket. Clean bore and back stop.

8. Lubricate new cup and o.d. of new seat. Push cup into bore and then push seat into cup. Oil can be used with the Taco small

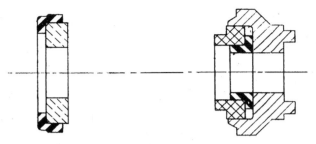

Figure 17.52 Mechanical seal, small pumps.

pumps as the rubber is a buna synthetic. Some pump manufacturers use ethyelyne propelyne, but it swells in the presence of oil. Keep a tube of silicone grease handy when working on seals as silicone grease will not damage the elastomers.

9. Re-install shaft and impeller assembly in bearing bracket.

10. Clean gasket surfaces inbody and bracket, and install new gasket.

11. While holding the impeller, with allen wrench engaged in set screw, slide coupler hub along shaft 3/16-in (5-mm) and lock set-screw in place.

12. Follow previous steps to reinstall.

The assembly of the seal and impeller to the B&G bearing assay is similar to all other pump assemblies whether they be close coupled or large frame mount. The only difference is that the components get larger. The close-coupled pump uses the motor shaft as the pump shaft, and an adaptor cover is attached to the motor, which fits over the shaft, bolts to the motor, and holds the stationary seal. The close-coupled pumps with some manufacturers can be fitted with either mechanical seals, Fig. 17.52, or with packing. The motor frame is known as a *JP shaft*. A motor made only to be fitted with mechanical seals is called a JM shaft motor. The JP shaft is longer than the JM to allow room for servicing. Due to the lower maintenance, more and more pumps are being sold with mechanical seals.

A mechanical seal is a controlled leak device. Fluid is leaking from the seal but as a vapor, or up to 1–2 drips per min.

A packed pump has to leak to lubricate the packing and keep it cool. If a packed pump has no further adjustment and is leaking more the 60 drips per min., it will have to be repacked (see Fig. 17.34). *Warning:* Lock out power to pump to be worked on and tag.

1. Secure the suction and discharge valves.

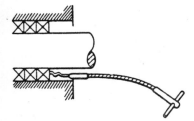

Figure 17.53 Packing removal.

2. Remove gland nuts and slide gland back as far as it will go.

3. Using a corkscrew puller as shown in Fig. 17.53 start removing the old packing. Note the number of rings that come out before the lantern ring is reached. The lantern ring is a water-distributing device to help keep the packing cool and lubricated. Typically its position lines up with the flushing line port. Remove all rings.

4. Inspect the shaft sleeve for scoring. If the scoring is heavy the shaft sleeve will have to be replaced. To remove the shaft sleeve, the pump will have to be taken apart like a mechanical seal change. Some pump manufacturers use Loctite® sealant between the pump shaft and the inside of the shaft sleeve to seal and secure the sleeve. To remove sleeves that have been held with such a sealant, the sealant will have to be destroyed with heat. Once the sealant is heated, the sleeve should slide off.

5. Install the new shaft sleeve with new Loctite® if required and reassemble pump.

6. If new packing rings are to be cut from a coil (see Fig. 17.54) wrap the packing around a wooden mandrel of the same size as the shaft

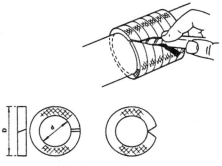

Figure 17.54 Cutting packing.

or a piece of brass bar. The packing can be cut butt-jointed or diagonally. When the packing is installed, the ring joint should be staggered 90° from the previous ring.

7. When all the rings have been installed, tighten packing gland to seat packing. Loosen gland nuts and turn shaft. Tighten gland nuts finger tight and check if shaft rotates freely.

8. Start pump, allowing pump to leak freely. Take up on gland nuts one flat at a time until desired leakage is obtained. Gland should run cool.

17.4.7 Horizontal split case (HSC) pumps

As the name implies, the pump is horizontally split. All of the rotating assembly nests in the bottom half of the casing. A cover fits over the upper half of the rotating assembly, completes the water passages, and completes the watertight compartment. Typically the impeller has two suctions. Because of the double suction, there are two sets of seals, two shaft sleeves, and the impeller is suspended between bearings at each end of the shaft (Fig. 17.55). Servicing the seals on these pumps requires more labor and tools. Some pump manufacturers design their pack design pump so that the stuffing boxes can be converted to a mechanical seal. There is still one problem: the bearings are pressed on the shaft and they have to come off before the seals can be removed. Seal manufacturers have designed seals that you can split in half, so they can be removed from the shaft without removing the bearings. There is one drawback: The seals are expensive.

17.4.8 HSC seal change

Tools required:

■ Bearing separator

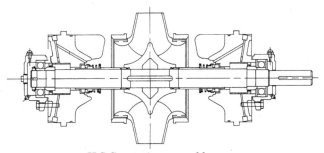

Figure 17.55 H.S.C. rotating assembly.

- Long-armed gear puller
- Spanner wrench
- Bearing heater or a hot plate
- Can of oil, hammer, and wrenches.

Procedure:

1. Remove outside seal flush lines, cover nuts, and bearing caps.
2. Install eye bolt in top of cover, fasten a lifting device, lift off cover, and put aside.
3. Lift rotating assembly from casing and place on a workbench.
4. Fasten bearing separator behind bearing and engage gear puller with bearing separator and end of shaft. Tighten gear puller until bearing slides free. Typically the nondrive end bearing is secured to the shaft with a nut and a star-locking washer. Bend back tab of lock washer, and using spanner wrench remove bearing retaining nut. Fasten bearing separator and gear puller and remove bearing.
5. Apply silicone grease to the outboard end of the shaft sleeve to aid in the removal of the seal (Fig. 17.56).

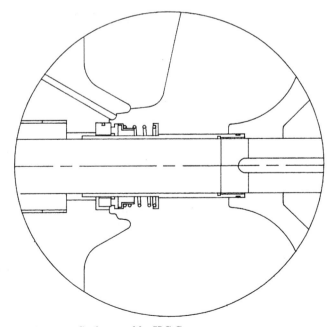

Figure 17.56 Seal assembly, H.S.C.

6. Once the seals are removed, the sleeves can be cleaned up to accept the new seals.

7. Remove the stationary seat from the insert and clean the bore and bottom.

8. If the shaft sleeves were to be replaced, determine the direction of rotation of the pump shaft (Fig. 17.57) as this determines which shaft sleeve is locked with the impeller key. Clockwise, the nondrive sleeve is free. Counterclockwise, the drive end sleeve is free.

Note: At this time the wear rings can be inspected and the clearance measured.

17.4.9   Bearing changing

The bearings in a frame-mounted pump or a motor can be changed in a similar manner as the HSC bearings were removed and installed. The same tools would be used. If the manufacturer cannot supply the bearings, look on the edge of the bearing face and you will find a code number, such as 6310. This code number is utilized by all the bearing manufacturers: 6 = deep groove (conrad) bearing, 3 = medium 300 series, and 10 = shaft size = 1.968 in. Take your bearings to your local machine parts house and/or bearing company and they will match your bearing.

17.4.10   Valve rebuilding

The stem seal on most valves are either packing or o-rings. The packing is removed with a corkscrew, just like repacking a pump. Wrap the packing around a wooden dowel or brass rod the same diameter

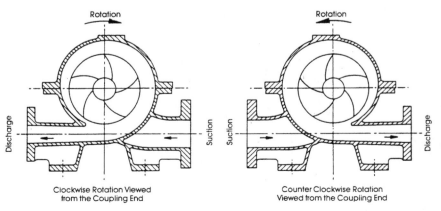

Clockwise Rotation Viewed
from the Coupling End

Counter Clockwise Rotation
Viewed from the Coupling End

Figure 17.57 Pump rotation, H.S.C.

as the valve stem. Cut the packing rings as was done in pump packing, but it is preferable to cut the ends at 45°. Install the rings 90° from each other until the stuffing box is full. Tighten the gland until there is resistance in turning the stem. O-rings can be replaced with O-ring from your local gasket house. O-rings come in standard sizes. Select the material that is compatible with the system fluid and temperature; this is true with any gasket material.

Valve seats and disks can be either metal-against-metal or a soft disk against a metal seat. If the valve has a metal seat and disk, lapping compound can be applied to the disk and the two sealing surfaces lapped together. If there are deep cuts in either seat or disk it may or may not be economical to have a machine shop reface the sealing surfaces. With a soft disk, it is the disk that wears out but it is easily renewable. In relief valves and pressure reducing valves, system debris may be contaminating the sealing surfaces. Normally it is less expensive to *replace* small valves and *rebuild* large valves.

17.5   Repair Versus Replacement Criteria

With the technology today and the materials used in pump construction, pumps and valves in an HVAC system should last a long time. The previous sections of this chapter talked about predictive and preventive maintenance which ideally should allow pumps and valves to last forever. However, there is a point in time, even with the best maintenance, some parts are going to wear out and a decision has to be made as to whether to repair or replace.

Part of the maintenance program was to set up data cards for each piece of equipment. There should have been included with these cards a record sheet indicating work performed, time spent, parts replaced, and costs. The new computer software for maintenance management have the machinery cards and allows you to record all the data once maintained on many pieces of paper. This data can now be stored on a floppy disk. The beauty of these systems is you can compile costs without having to find many pieces of paper. Another advantage is the ease of updating information. Whether with manual cards or with a computer, the information compiled is important in making a repair or replacement decision.

When making a repair or replace decision, the following points should be evaluated.

17.5.1   Existing equipment

- Part cost

- Part availability—if part is no longer made, can the existing part be repaired or duplicated economically?

- Is there an off the shelf bearing, seal, or gasket material that will replace discontinued parts?

- Can this part be repaired or duplicated in house or by an outside machine shop. Note: If the skill level and equipment are available in house, the repair will not be a true expense where outside labor is a true expense.

- Should the entire pump or valve be sent out for repair?

- What time will be spent just to remove the pump or valve and reinstall?

- Is there another parts supplier that has duplicated the parts at a better price or provides better delivery than the original equipment manufacturer? Warning. Is the part as good as the original?

- What is the reputation of this after market part?

- Is there a particular part that fails repeatedly that could be corrected by using a different material?

- Would changing to a mechanical seal solve high maintenance costs?

17.5.2 Replacement equipment

- New product costs vs. repair of old?

- If original product gave good service, is a duplicate still available?

- Is there a new product that is more efficient?

- What will the modifications be to the piping, foundation, and base?

- Can the modifications be done by inside or outside skills (true expense)?

- What additional parts will be needed?

- What material handling equipment will be needed to get pump or valve to its site?

- Are there any accessibility problems?

Most pumps except the horizontal split case pumps (HSC) can be repaired in 1–2 hr. If a pump is an inline in the ceiling, rigging may add to the time.

An HSC pump will take about 6 hr to repair if the rotating assembly has to be disassembled. A complete rotating assembly change would only be 2 hr.

Energy consumption would be a consideration if a pump is more efficient. The following formula can be applied to the old and new pump to determine the yearly energy cost saving.

```
                      Pump Data Card

Equip. #_____Location: Bld_____Floor_____Room_____

Description_____Size_____

Serial #_____Type_____Mfg._____

Original P.O. #_____ Address_____

Phone #_____              _____

Representative:_____

Phone#_____Date Installed_____Date Started_____

____Rated Conditions   Materials of Construction  Mechanical Seal Data

Flow (GPM)_____   Casing_____    Mfg._____

Head (Ft.)_____   Impeller_____ Type_____

R.P.M._____    Wear Ring, csg._____ Shaft Size_____

Fluid_____    Wear Ring, imp._____ Elastomer_____

% Concentration_____  Wear Ring Clearance_____ Seat Mat._____

Operating Temp._____   Shaft Sleeve_____  Seal Mat._____

Viscosity_____    Brg. # Inboard_____ Spring_____

Sp. Gr._____   Brg. # Outbrd._____ Retained_____

Impeller Dia._____   Gasket Matl._____

Rotation_____    Gsk. Thickness_____      Packed Seal Data

Spare Parts In-stock                            Sleeve Dia._____

Part #   Description                   Qty.    Packing width_____

                                                Height_____

                                                # of  rings_____

                                                Lantern Matl._____

                                                Packing Mtl._____
```

Figure 17.58 Pump data card.

BHP = gpm × head × sp.gr. ÷ 3960 ÷ pump efficiency (at operating point)

BHP = brake gorsepower, gpm = gallons per minute, sp.gr. = specific gravity

kilowatt hour (KWH) = BHP × 0.746 × run hours/yr/motor efficiency KWH × cost/KWH = annual energy cost

```
                   Motor Data

Driver_____Mfg._____Ser.#_____

Hp_____RPM_____Frame_____Encl._____Mounting_____

Volts_____PH____Hz____F.L.A._____Service Factor_____

Lubrication: Oil type_____Qty._____S.F.Amps._____

Mfg._____Grease Type_____ Ambient °C_____

Coupling Mfg._____Part #_____Date Strted_____

Pump shaft Dia._____Motor Shaft Dia._____
```

Figure 17.59 Motor data card.

History of Repairs

Equipment # _____ Description _____

Date	Work Order #	Description	Labor Hrs.	Material $

Maintenance Tickler File Card

Equip. #	Name	Location
Interval	Operation	Floor_____ Rm_____

Date Operation Performed						

Figure 17.60 History of repair and tickle file card.

Master Equipment List

Equipment Discription	Equipment #	Location- Room	Floor	Building	

Daily Log Sheet

Month_____ Year_____ Loc._____ Bld._____ Rm._____ Pump #_____

Date	Flow	Disc. Press.	Suct. Press.	Mtr. Amps	Volts	P-Brg. Temp.-I	P-Brg. Temp.-O	Mtr. Brg Temp -I	Mtr Brg Temp.-O	Amb. Temp.	Seal Leak

Figure 17.61 Master equipment list and daily log sheet.

Preventive Maintenance Card

P.M.Cycle_____   Equipment_____

Location_____Man Minutes_____

Operation:_____

_____

_____

_____

Tools/ Spares/ Materials_____

_____

_____

_____

Safty Precautions, _____

_____

_____

_____

Detailed Method,_____

_____

_____

_____

RecordsRequired_____

_____

_____

_____

Figure 17.62 Preventive maintenance card.

The decision to replace a pump or a valve gets more complicated as the pump or valve increases in size. If good maintenance and expense records have been kept, the decisions will be a lot easier.

18.1

Boilers*

Cleaver-Brooks
Division of Aqua-Chem, Inc., Milwaukee, Wisconsin

18.1.1 Inspection

18.1.1.1 New installations

Pressure vessel. Before a new boiler is initially fired, it should be given a complete and thorough inspection on the fireside and the waterside. Examination of all the pressure parts, boiler, drums, tubes, water column, blowoff valves, safety and safety relief valves, baffles, nozzles, refractory setting, seals, insulation, casing, etc., is necessary. Any dirt, welding rods, tools, or other foreign material should be removed from the boiler. The waterside of the boiler should be thoroughly rinsed with water of not less than 70°F (21°C) and with sufficient pressure to effect a good cleaning.

Inspectors. During this period of initial inspection, it is a good time for the customer to call in an insurance inspector to check out the boiler. As a general rule, these inspectors like to inspect the boiler prior to the hydrostatic test. Any state or local inspectors (when required) should also be contacted.

Before any inspecting personnel are allowed to enter a boiler that has just recently been opened up, either the waterside or fireside, the boiler should be ventilated thoroughly. This eliminates any explosive gases or toxic fumes and ensures an adequate supply of fresh air.

---

*Updated from chapters that appeared originally in the Handbook of HVAC Design (First Edition).

Only approved types of waterproof low-voltage portable lamps with explosionproof guards, or battery-operated lamps, should be used within a boiler. Portable lights should be plugged only into outlets located outside the boiler.

Testing. When a hydrostatic test is required to comply with state, local, or insurance company requirements, all of the preparatory steps required and/or recommended by the manufacturer and authorities must be adhered to.

18.1.1.2 Existing installations

An external examination of an existing installation can be made while the boiler is in operation. Close inspection of the external parts of the boiler, its appurtenances, and its connections should be made before the boiler is taken off the line and secured. Notes should be made of all defects requiring attention so that proper repairs can be made once the boiler has been shut down. When conditions permit, the shutdown of an existing installation can be timed to coincide with the annual or semiannual inspection, when the insurance and/or state inspectors can be present. Before an internal inspection can be made, the boiler must be secured in strict accordance with all the recommendations and requirements of the manufacturer and the authorities having jurisdiction.

18.1.1.3 Burner controls

Before making any attempt to start a burner-boiler unit, read the manufacturer's instruction manual thoroughly, to get a good understanding of how the burner operates. The operator should familiarize herself or himself with the various parts of the burner and understand their functions and operation.

Parts. Describing all the different types of fuel-burning equipment is beyond the scope or intent of this chapter. However, the points covered here can, with slight variations be applied to all types of equipment.

The burner should be checked over thoroughly to ascertain that all parts are in proper operating condition. These are some of the points to check:

1. All the fuel connections, valves, etc., should be gone over thoroughly. Make sure that all the joints are tight, pressure gauges and thermometers in place and tight, and valves operative. Fuel lines should be checked for leakage.

2. If the burner has a jackshaft, check that all the linkage is tight. If possible, check the linkage for travel and proper movement.

3. The electric system should be checked very carefully. Make sure that all connections are tight. Check all the terminal strip connections. Sometimes if these are not tight, the vibration from shipping can loosen them. Test plug-in controls and relays. Check the rotation of the fan, pump, and air compressor motors.

Safety limits and interlocks. All the various interlocks and limits should be checked to see whether they are operational. An actual operating test of these controls cannot be made until the burner is running. All or most of the following limits are found on the average packaged burner, depending on the insurance rules being complied with:

1. Operating and high-limit pressure control
2. Operating and high-limit temperature control (hot water)
3. Low-water cutoff
4. Flame failure (part of failure safeguard function)
5. Combustion air proving switch
6. Atomizing media (steam and/or air, proving switch for oil burner only)
7. Low-oil-temperature switch
8. Low-oil-pressure switch
9. High-oil pressure switch
10. Low-gas-pressure switch
11. High-gas-pressure switch
12. Low-fire switch

Other operating limit interlocks can be added to the control system depending on the degree of sophistication; however, this list covers most of them.

18.1.1.4 Auxiliary equipment

Deaerator. All the boiler room auxiliary equipment should be checked thoroughly to make sure everything is in good working order and ready to go.

Boiler feed pumps. Condensate pumps, transfer pumps, all the pumps in the system should be checked out. The coupling alignment should be checked to ensure that no misalignment exists. If the coupling alignment is not up to the pump or coupling manufacturer's recommendations, it must be corrected.

Piping connections to the pump suction and discharge ports should be checked. Adequate provisions should be made for proper expansion so that no undue strains are imposed on the pump casing. All drain lines, gland seal lines, recirculation lines, etc., should be checked to verify that they are in compliance with manufacturer's recommendations. After all these precautions have been taken, check the pump for proper rotation.

Heat-recovery equipment. Generally the heat-recovery equipment for a boiler consists of an air heater and/or an economizer. Simultaneous use of an air heater and an economizer on a boiler is rare; however, on some occasions use of both pieces of equipment can be economically justified.

Manufacturer's drawings and instructions should be followed when the inspection is made of the economizer and/or air heater. The inspection procedure is very similar for a tubular-type air heater and an economizer. Check for proper installation as well as breeching ductwork, arrangement, alignment, and expansion provisions. Make sure all internal baffling, tube arrangements, headers, tubesheets, etc., are installed in accordance with the manufacturer's recommendations. There should be no plugging or obstruction in the tubes and headers. In an economizer, check the feed piping to the inlet header and the piping between the economizer and the boiler. Shut off and check the valves, isolating valves, and relief valves. Test connections, thermometers, and pressure gauges should all be checked and inspected.

If the air heater is a regenerative type, only an operator who is completely familiar with such equipment can inspect and evaluate the appropriateness and condition of the assembly.

The air heater seals should be checked for proper clearance, expansion provisions, and clearances. The drive motor and gear-reducing unit must be lubricated and the rotation checked. All ducts, breeching, dampers, and expansion joints should be checked against installation drawings. Any debris, welding rods, rags, etc., should be removed.

18.1.2 Startup

After the initial inspection has been made, the new boiler is ready for startup. Startup time is a serious period. It can be the most critical period in the life of the boiler. However, it need not be, if proper planning and training have preceded this period. The manufacturer's instruction manual should be consulted and the contents thoroughly understood by all parties involved in the initial startup and eventual operation of the boiler(s). Owners and operating and maintenance personnel should be present during the initial startup period.

18.1.2.1 Checks

No attempt should be made to initially fire the boiler until the following points have been checked:

1. Feedwater supply must be ensured. The boiler should be filled to operating level with 70°F (21°C) or warmer water.

2. Steam lines must be hooked up and ready for operation. If the boiler is equipped with a superheater, then a proper vent should be provided to vent the superheater during boilout.

3. Blowoff and blowdown lines must be checked and ready for operation. Check all piping for adequate support and expansion provisions.

4. All fuel supply lines should be checked out. Make sure that the piping is correct. Check lines for tightness and leaks. Strainers are not normally supplied as part of the package (Cleaver-Brooks supplies strainers as standard on the oil supply line), but these are most important to the safe operation of gas- and oil-fired units. If the boiler is fired with coal, then the coal handling and burning equipment must be checked to make sure an adequate supply of fuel will be available.

5. The electric power supply should be reviewed. The power lines to the boiler-burner unit should be connected and ready for operation. Checking of the burner and controls is covered in Sec. 18.1.1 and 18.1.3.3.

6. Any walkways, platforms, stairs, and/or ladders needed to permit proper access to the boiler should be installed and ready for use.

18.1.2.2 Boilout

Once all the foregoing points have been checked, the next step is to prepare for the boilout. The newly installed boiler must be boiled out because its internal surfaces are invariably fouled with oil, grease, and/or other protective coatings. Boilout will also remove remaining mill scale and rust, welding flux, and other foreign matter normally incident to fabrication and erection.

Existing boilers which have had any tube replacement, rerolling, or other extensive repairs to the pressure parts should be boiled out. The lubricant used for rolling tubes plus the protective coating on new tubes must be removed by boiling out before the repaired boiler can be put back on line.

The boilout chemicals added to the water create a highly caustic solution which, upon heating, dissolves the oils and greases and takes them into solution. Through a series of blowoffs, the concentration of the solution is reduced. After a period of boiling and blowing down, the concentration is diluted enough that practically all the oils and greases and other matter have been eliminated. The boiler manufacturer's recommendations and requirements for boilout should be strictly adhered to.

18.1.3 Routine Operation

Routine operation of a boiler room and all its associated equipment can become boring. Day by day, week in and week out, the same monotonous chores are performed over and over, but diligence must be maintained. It is dangerous to become lackadaisical and let these duties slide.

Safe and reliable operation depends to a large extent on the skill and attentiveness of the operators. Operating skill implies the following:

- Knowledge of fundamentals
- Familiarity with equipment
- Suitable background or training
- Operator diligence in performing these routine functions

18.1.3.1 Boiler room log

During operation certain procedures can and should be performed by the operating personnel. Actually the best way to keep track of these procedures and to help remind the operators to perform these functions is to keep a boiler room log.

In the interest of accident prevention and safe operating and maintenance practices, the Hartford Steam Boiler Inspection and Insurance Company and Cleaver-Brooks offer, without cost or obligation, copies of the boiler room log forms. Consult a Cleaver-Brooks agent or write to Hartford Steam Boiler Inspection and Insurance Company, 56 Prospect Street, Hartford, Connecticut 06102.

The Hartford Steam Boiler Inspection and Insurance Company, in their Engineering Bulletin No. 70, *Boiler Log Program,* explains why a log is desirable and cites the various tests to be performed on the equipment or apparatus. This bulletin recommends both the various tests to be performed and the frequency of log entries as follows:

Low-pressure steam and hot-water heating boiler logs	Weekly readings
High-pressure power boiler log for relatively small plants that do not have operators in constant attendance	Twice daily readings with more frequent checking during operation
High-pressure power boilers, larger units in large plants	Hourly readings, the preferred frequency for all high-pressure boilers

18.1.3.2 Safety and relief valves

Another extremely important safety check to be performed is the testing of all safety and relief valves. How frequently this should be done is an extremely controversial subject. Everyone seems to agree that safety valves and relief valves should be checked or tested periodically, but very few agree on how often.

The American Society of Mechanical Engineers' (ASME) book *Recommended Rules for Care and Operation of Heating Boilers,* in section VI, states that the safety or relief valves on steam or hot-water heating boilers should be tested every 30 days. A try-lever test should be performed every 30 days that the boiler is in operation or after any period of inactivity. With the boiler under a minimum of 5 lb/in^2 (34.5-kPa) pressure, lift the try lever on the safety valve to the wide-open position and allow steam to be discharged for 5 to 10 s (on hot-water boilers hold open for at least 5 s or until clear water is discharged). Release the try lever, and allow the spring to snap the disk to the closed position.

18.1.3.3 Burner and controls

During normal operation, maintenance of the burner and controls is extremely important. If the boilers are firing continuously, most of the maintenance and repairs that can be done under these conditions is of a minor nature.

Spare oil guns should be kept cleaned and ready to be changed as necessary. Check all the valves in the fuel lines, feedwater, and steam lines for leaks and packing gland conditions. Repair, repack, and tighten where needed.

Check out the burner linkage, jackshafts, drive units, cams, etc. Make sure that all linkage, linkage arms, and connections are tight. Lubricate or grease where required. Note any wear or sloppiness in any of this equipment. If something cannot be repaired while the boiler is in operation, then this fact should be noted in the log so that proper repairs can be made when the boiler is taken out of service.

Operation of the burner should be observed. Is the burner firing properly? Check the flame shape and combustion. If the burner is not operating properly, have a qualified serviceperson work on the burner and put everything back into proper condition.

18.1.3.4 Feedwater treatment

Improper water treatment is probably the largest direct cause of boiler failure. When the boiler feedwater and the boiler water do not receive proper preparation and treatment, scale and sludge deposits, corrosion, and pitting of the boiler surface result.

The boiler owner and operator must know that proper boiler feedwater treatment is an absolute necessity. If the boiler does not receive water of proper quality, the boiler's life will be needlessly shortened.

18.1.3.5 Boiler blowdown

When a boiler is generating steam, the feedwater continuously carries dissolved minerals into the unit. This material remains in the drum, causing an increase in the total solids until some limit is reached beyond which operation is unsatisfactory. Many difficulties in boiler operation will occur because of excessive concentrations of sludge, silica, alkalinity, or total dissolved solids.

When these constituents have reached a maximum limit (see Table 18.1), some of the water must be removed from the boiler and replaced with feedwater with a lower solids content.

TABLE 18.1 Specifications for Water Conditions

Pressure at outlet of steam-generating unit, lb/in² (bar)		Total solids, ppm (or mg/L)	Total alkalinity, ppm (or mg/L)	Suspended solids, ppm (or mg/L)
0–50	(0–3.4)	2500	500	150
51–300	(3.5–20.7)	3500	700	300
301–450	(20.8–31)	3000	600	250
451–600	(30.1–41.4)	2500	500	150
601–750	(41.5–51.7)	2000	400	100
751–900	(51.8–62)	1500	300	60
901–1000	(62.1–68.9)	1200	250	40
1001–1500	(69–103.4)	1250	250	20
1501–2000	(103.5–137.9)	750	150	10
2001+	138+	500	100	5

The amount of blowdown or number of times the boiler is blown each day depends on the concentration of solids in the boiler water. Recommendations of the feedwater consultant should be followed regarding blowdown procedures.

18.1.4 Shutting Down

18.1.4.1 Short shutdowns

When a routine shutdown is scheduled, it should be planned so that there is time to perform certain operations in the shutdown procedure.

1. If the boiler is equipped with sootblowers and if any fuel other than natural gas has been fired, all the sootblowers should be operated before the boiler is taken off line.

   All the recommended rules for operating the sootblowers should be followed. One of the most important is that the steam load on the boiler should be 50 percent of boiler rating or greater.

2. After the soot-blowing operation has been completed, the steam flow should be gradually reduced and the burner set to the low-fire position.

3. With the burner in the low-fire position, blow down the boiler along with the water column, gauge glass, and feedwater regulator. Turn off the burner in accordance with the burner manufacturer's instructions. If the boiler is equipped with a flue gas outlet damper, it should be fully closed, to allow the unit to cool slowly.

4. Remove and clean the burner oil gun. Place fuel supply equipment in standby condition (for gas, shut the main supply cock). Open the main electric switch, and take the feedwater regulator out of service. Hand-operate feedwater valves to keep the water level above one-half gauge glass.

5. When the boiler pressure falls below the line pressure, the boiler stop valve should be closed if the setting has cooled enough to prevent any pressure buildup. If the boiler is equipped with a nonreturn valve, this valve should close automatically when the boiler pressure drops below the line pressure. This, of course, isolates the boiler from other units remaining in service. As the drum pressure falls below 15 lb/in^2 (1 bar), the manual closing device (handwheel) of the nonreturn valve (if equipped) should be closed and the top drum vent opened. This prevents a vacuum on the boiler water side which will loosen well-set gaskets and cause future problems. While there is still a small amount of steam pressure available, the

boiler should be blown down and filled back to a safe level with freshly treated hot water in preparation for the next startup.

If the boiler is being shut down just overnight or for the weekend, the previous procedure is generally all that is required. The primary concern is to ensure sufficient water in the boiler. If the boiler is going to be shut down only overnight, probably the boiler can be secured and will have pressure still showing the next morning.

18.1.4.2 Prolonged shutdowns

Up to this point we have been discussing shutting down the boiler for just a few hours or a few days. If the boiler is going to be taken out of service and is expected to be out of service for several weeks or several months, then a different procedure must be followed.

There are two basic methods of laying up a boiler for protracted outages: wet storage and dry storage.

Wet storage method.   If the unit is to be stored for not longer than a month and emergency service is required, wet storage is satisfactory. This method is not generally employed for reheaters or for boilers which may be subjected to freezing temperatures. Several alternative methods have been used.

1. The clean, empty boiler should be closed and filled to the top with water that has been conditioned chemically to minimize corrosion during standby. Water pressure greater than atmospheric should be maintained within the boiler during the storage period. A head tank may be connected to the highest vent of the boiler to keep the pressure above atmospheric.

2. Alternatively, the boiler may be stored with water at the normal operating level in the drum and nitrogen maintained at greater than atmospheric pressure in all vapor spaces. To prevent in-leakage of air, it is necessary to supply nitrogen at the vents before the boiler pressures falls to zero, as the boiler is coming off the line. If the boiler pressure falls to zero, the boiler should be fired to reestablish pressure and drums and superheaters should be thoroughly vented to remove the air before nitrogen is admitted. All partly filled steam drums and superheater and reheater headers should be connected in parallel to the nitrogen supply. If nitrogen is supplied to only the steam drum, the nitrogen pressure should be greater than the hydrostatic head of the longest vertical column of condensate that could be produced in the superheater.

3. Rather than maintain the water in the boiler at the normal operating level with a nitrogen cap, it is sometimes preferred to drain the boiler completely, applying nitrogen continuously during the draining operation and maintaining a nitrogen pressure greater than atmospheric throughout the draining and subsequent storage.

Dry storage method.   This procedure is preferable for boilers out of service for extended periods or in locations where freezing temperatures may be expected during standby. It is generally preferable for reheaters.

1. The cleaned boiler should be thoroughly dried, since any moisture left on the metal surface will cause corrosion after long standing. After drying, precautions should be taken to preclude entry of moisture in any form from steam lines, feed lines, or air.

2. For this purpose, moisture-absorbing material may be placed on trays inside the drums or inside the shell. The manholes should then be closed, and all connections on the boiler should be tightly blanked. The effectiveness of the materials for such purposes and the need for their renewal may be determined through regular internal boiler inspections.

When boilers have been kept in dry storage with any of the aforementioned materials in the fireside and waterside sections, serious damage can be done to the boiler if these materials are not removed before the boiler is filled with water and fired.

We strongly recommend that large signs be placed in conspicuous places around the boiler to indicate the presence of moisture-absorbing materials. These signs could read similarly to the following:

IMPORTANT — Moisture-absorption material has been placed in the waterside and furnace areas of this boiler. This material must be removed before any water is placed in unit and before boiler is fired.

For long periods of storage, inspect every 2 or 3 months and replace with fresh and/or regenerated materials.

3. Alternatively, air dried external to the boiler may be circulated through it. The distribution should be carefully checked to be sure that the air flows over all areas.

4. If the boilers are to be stored in any place other than a dry, warm, protected atmosphere, steps should be taken to protect the exterior components also. Burner components that are subject to

rust—jackshaft, linkage, valve stems, moving parts, etc.—should be coated with a rust inhibitor and covered to protect them from moisture and condensation. Electric equipment, electronic controls, relays, switches, etc., should be protected.

5. Pneumatic controls, regulators, and diaphragm- or piston-operated equipment should be drained or unloaded and protected so that moisture, condensation, rust, etc., will not damage it during a long period of storage. Feedwater lines, blowdown, the soot blower, drain lines, etc., should all be drained and dried out. Valve stems, solenoid valves, and the diaphragm should be protected by lubricant, rust inhibitors, plastic coverings, or sealants.

18.1.5 Auxiliary Equipment

The term "auxiliaries" can be defined as all the attached or connected equipment, used in conjunction with a boiler, which should be operated and maintained in accordance with the regulations established by the equipment manufacturers. Adherence to these rules will help protect the operators from injury and prevent damage to the boiler being serviced and to the auxiliary equipment.

18.1.5.1 Deaerators

When a tray-type deaerator is involved, it can be cleaned with a light muriatic acid solution. The trays can be removed and brushed with the muriatic acid solution until all the scale deposit has dissolved.

Note: The operator should wear protective clothing, goggles, rubber gloves, and a rubber apron when using the acid solution.

After the trays have been cleaned, they should be thoroughly rinsed with water several times to remove all traces of the acid.

The packed column type of deaerator may or may not be cleaned with an acid solution depending on the type of packing in the column.

The spray-type deaerator will generally have a spray valve or spray pipe. With the spray valve it may be necessary to check the valve seat for wear or pitting. The spring compression dimension should be checked and adjusted to the manufacturer's recommendations. If the deaerator has a spray pipe, then the pipe should be inspected to see that all the spray holes are open. If any holes are plugged, they should be opened.

18.1.5.2 Strainers

Strainers in all lines should be cleaned at regular intervals determined by conditions and use; water gauge glasses should be kept clean and replaced if necessary. Tighten all packing glands and fittings to eliminate steam and water leaks. The overflow valve and makeup valve, with their associated level control mechanisms, should be checked. Refer to the appropriate manufacturer's literature for service recommendations of particular components.

18.1.5.3 Air heaters

Tubular air heaters have no moving parts, and therefore one might assume that this type of air heater would need very little maintenance. This assumption would be correct if natural gas were the only fuel being fired. With natural gas or No. 2 fuel firing and a proper air-fuel ratio, the products of combustion are fairly clean. The heat-transfer surfaces will remain fairly clean and will normally not require any cleaning.

The air side of the air heater should be inspected to see that the baffles are all secure and in place. Sometimes air flow through these areas may cause a baffle to vibrate and break loose. Proper repairs should be made and the baffles welded or clamped in place.

Check for signs of corrosion in the low-temperature end. If the temperatures in the low-temperature end are below the dew point, the flue gases will condense and corrode the metal. Installations that are firing coal or heavy oil with high sulfur content are particularly subject to corrosion in this area. The higher the sulfur content of the fuel, the higher the dew point of the flue gases.

Where coal or residual oil is the main fuel, the air heater tubes become plugged with fly ash and soot. When this happens, brush or steam-clean the tubes.

Maintenance on a regenerative air preheater involves checking the radial and circumferential seals. The seals should be inspected for wear and proper clearance. If any of the seals are damaged or so worn that they cannot be adjusted for proper clearance, they should be replaced. Manufacturer's recommendations should be consulted regarding making the proper clearance adjustments.

18.1.5.4 Economizer

In general, the economizer is subject to the same types of troubles as the air heater. Corrosion in the low-temperature ends due to condensation of gases and erosion due to action of the flash or soot deposits are the more common problems.

18.1.5.5 Fuel systems

For any boiler to be able to operate at peak efficiency and as free of trouble as possible, the fuel-handling system must be designed properly and must be kept in top condition at all times.

Full and effective use should be made of manufacturer's instruction books on operation and maintenance. Of special importance are written procedures prepared expressly for each installation. Fuel-burning equipment should never be operated below the safe minimum level at which a stable furnace condition can be maintained. In like manner, operation should not be tolerated at rates of fuel input which are excessive in relation to the available air supply.

This chapter cannot cover the size of industrial and utility boilers that normally use pulverized fuel. Those interested in obtaining further information on this type of equipment should consult the *ASME Boiler and Pressure Vessel Code,* section VII, "Recommended Rules for Care of Power Boilers."

See Bibliography 18.1.10 for additional information sources.

18.1.6 Safety Controls

18.1.6.1 Introduction

The purpose of safety controls is to minimize the possibility of an overpressure condition of the heating device and of the unintended ignition of a combustible mixture. The type and the number of devices used to protect and control heating equipment are, to a large extent, mandated in the United States by national standards such as the American Society of Mechanical Engineers' (ASME) *Boiler and Pressure Vessel Code,* American Gas Association (AGA), Underwriters' Laboratories (UL), National Fire Protection Association (NFPA), insurance carriers, and state and local authorities.

Operating controls determine the cycling and/or firing rate of the equipment based on the imposed load. In addition to the required devices, controls to meet a specific need may be necessary. Items such as efficiency, automation, response time, downtime, and economics affect the design of a control system.

18.1.6.2 Safety devices

Flame safeguard. Probably the most important safety device on a boiler is the flame safeguard control, whose primary function is to monitor the flame, to allow the fuel valves to remain open or closed, depending on the presence or absence of flame.

Fuel interlocks. The temperature and pressure of the fuel(s) must be maintained within a narrow range to ensure proper combustion.

When gas is burned, two pressure switches are typically provided, one to detect a drop and the other to sense a rise in pressure. If either switch detects an improper pressure, the boiler is shut down. If the pressure were allowed to go above the normal operating range, the fuel-air ratio would go fuel-rich and/or overfire the boiler. If the pressure is too low, efficiency will be lowered and flame properties will be adversely affected, possibly causing flame outage.

The pressure of oil fuels must be maintained for the same reasons as for gas. But, in addition, low oil pressure may cause improper atomization of the oil, which in turn causes poor combustion.

Air interlocks. The combustion air system should contain a device to shut down the burner when combustion air is not present. This is done with an air pressure proving switch. If combustion air is not present, the main fuel valves are prevented from opening. If combustion air is lost during operation (e.g., due to belt breakage), the main fuel valves are closed before a dangerous air-fuel ratio can develop. Dampers may be added to control the air-flow rate.

Low-water cutoffs. The water-level (water-column) control (optional on hot-water boilers) is a device which shuts down the burner when the water level approaches a dangerously low condition. On low-pressure firetube boilers, the low-water cutoff prevents firing when the waterline is at or below the lowest visible point of the water glass (*ASME Code* states that the lowest visible part of the water glass shall not be below the lowest safe waterline).

High-limit controls. On steam boilers, a pressure-sensitive switch monitors boiler pressure and will immediately shut the fuel valves(s) if an abnormally high pressure is detected. The switch is set at some valve below that of the steam safety valve so that the burner shuts down before the safety valve "pops." However, it is set higher than any other device which controls steam pressure. Once the high-limit control locks out, it must be manually reset before the burner will fire. A similar function control is provided on low-pressure hot-water boilers. The control is responsive to boiler water temperature.

Safety and relief valves. The single device used to protect the pressure vessel is a *safety valve* for steam boilers or a *relief valve* for hot-water boilers. Details of the application of steam safety valves and water relief valves are given in the latest editions of the *ASME Boiler and Pressure Vessel Code*.

18.1.7 Operating Controls

18.1.7.1 Introduction

Operating controls determine the cycling and/or firing rate of the equipment based on the imposed load. In addition to the required devices, controls to meet a specific need may be necessary. Items such as efficiency, automation, response time, downtime, and economics affect the design of a control system.

18.1.7.2 Fuel system controls

Gas-fired boilers must have safety shutoff valves that respond to the action of the various limit controls and combustion safeguards. Safety shutoff valves are tested by recognized testing agencies for reliability and safety. Strainers may be added to ensure clean fuel gas. Hand-operated valves (such as lubricated plug cocks) are provided so that the gas supply to the burner can be shut off manually. Pressure regulators are added to provide regulated gas pressure at the burner. Metering valves and/or orifices may be added to control the fuel rate.

Snap-acting solenoid valves are used in gas pilot lines or as shutoff valves in the main line to the boiler; however, these are generally used on small boilers only. On larger boilers, motorized valves are used. These are slow-opening valves which produce an easy lightoff of the main burner and minimize pressure surges in the gas supply line. Some testing agencies (e.g., Underwriters Laboratories', Inc., Canadian Standards Association) or insurance carriers may require multiple safety shutoff valves and/or other valving arrangements.

Each gas-fired appliance should have a pressure-reducing valve (PRV) or regulator to maintain a constant gas pressure to the burner. Oil-fired boilers must have fuel oil safety shutoff valves that respond to the action of the various limit controls and combustion safeguard. Strainers to ensure clean fuel oil at the burner are recommended.

One or more solenoid valves control the flow of oil to the burner. However, motorized valves may be used to achieve certain lightoff characteristics or if valve-closure-proving switches are required. In either case the valve(s) chosen must have temperature and pressure ratings suitable for the type of oil being burned.

18.1.7.3 Limit devices

A limit device is a temperature- or pressure-actuated switch which controls the cycling of the boiler. If the sensed boiler temperature or pressure rises above the setpoint, the switch opens and the boiler is kept off until the measured variable again falls below the setpoint. Next, the switch is closed, allowing the boiler to begin firing. The switch mechanism is automatically reset, unlike that described earlier under high-limit controls.

Each boiler must have a limit control in addition to the high-limit device. The high-limit setpoint is always higher than the limit control setpoint. If the boiler is designed for on/off operation, the limit control cycles the boiler on and off. If the burner is capable of off/low/high operation, a second limit control is used to control the firing rate between low fire and high fire. The setpoint of this control will be lower than that of the primary limit control.

If the boiler is equipped with a modulating burner, a device is required to detect changes in boiler temperature or pressure over a continuous range. This device transforms a changing process variable to a mechanical or electric output. Depending on the type of device, a potentiometer or slidewire output may be produced, or the output may be a voltage or current. These output signals are used to drive an actuator which positions air dampers and fuel-flow control valves.

18.1.7.4 Firing rate control

There are three major firing rate control methods: on/off, off/low/high, and modulating. In the on/off method, the burner is either off or firing at its rated capacity. This type is used primarily on small boilers when control of output pressure or temperature is not critical. The on/off/ high mode allows the boiler to more closely match the load. Given a demand for heat, the boiler begins firing at approximately 30 to 50 percent of capacity (low fire). Upon a further increase in demand, the burner fires at maximum capacity (high fire). When the demand is satisfied, the firing rate is reduced and held at low fire. If the demand increases, the burner again goes to high fire; on a demand decrease, the burner will turn off.

The modulating firing rate control method provides the greatest flexibility in matching the boiler output to the load. On a call for heat, the boiler is brought on at low fire, which typically is 15 to 30 percent of capacity.

As a greater load is imposed on the boiler, an increasing amount of fuel and air is introduced to the burner. The output of the boiler varies continuously between low and high fire. In this way the boiler output matches any load between 25 and 100 percent of boiler capacity.

Gas-fired burners use a butterfly valve to vary the amount of gas to the burner. When oil is the fuel, a variable-orifice or characterizeable valve is used to meter the flow of oil. Typically, the butterfly valve or variable-orifice valve is mechanically linked to the combustion air damper to provide the proper fuel-air ratio at all firing rates.

18.1.7.5 Optional features

Many controls can be added to the standard system to accomplish a specific task. Boiler temperature or pressure may be reduced at night or on weekends by adding a second limit control at a reduced setpoint which is selectable with a switch.

Each installation must be considered individually. When a control system is being designed, it is important that the load requirements be met, but at the same time the integrity of the boiler must be maintained by avoiding undue stresses or thermal shock.

18.1.7.6 Feedwater control

Feedwater controls are used to maintain a safe and acceptable water level in the boiler. Too high a water level in the boiler can lead to water carryover with the steam. Too low a water level in the boiler can lead to a safety shutdown of the burner or a ruptured pressure vessel, should the low-water safety control fail to operate.

In the following paragraphs we discuss a few of the water level sensing controls and operating schemes used for controlling water level. Selection of the correct ones depends on load and operating conditions, type of boiler, and economics.

There are several different methods of sensing the water level in steam boilers. The most common uses a float located in an external chamber whose position changes in response to water-level variations. This float is either mechanically or magnetically connected to an electric or pneumatic device which actuates a feedwater valve and/or motor.

A second method of sensing the water level uses electric probes. When the probes make contact with the water, an electrical path is established which actuates a relay to power a feedwater valve or pump. This method is used only for on/off or open/close operation of a feedwater motor or valve.

A third method of sensing the water level involves inserting a special probe into the boiler shell or external chamber to measure the variance in electric capacitance as the water level changes. The variance in capacitance is converted to an electric or pneumatic output signal for control.

A fourth method of sensing the water level uses a thermohydraulic system that has a "generator" mounted at normal water level which sends a pressure signal to a feedwater valve in response to the amount of steam the generator is exposed to. This system is not generally used on low-steam-pressure applications.

In conjunction with the water-level sensing, several different control schemes are used to maintain the proper water level in the boiler. One way is to turn a feedwater pump on and off in response to a water-level signal. This system can be modified to include opening and closing a feedwater valve.

A second control scheme is to have the water-sensing element send a proportionate signal to a feedwater valve to throttle the valve in response to the water level. This system may or may not turn a feedwater pump on and off as the feedwater valve opens and closes.

On boilers where water levels are hard to maintain due to rapid and large load swings, firing rates, and low water volume, a more elaborate proportioning control scheme may be in order. Such a system would include additional sensors besides water-level sensors, such as steam flow or steam flow and feedwater flow sensors. The proportioning feedwater valve would be positioned in response to the two or three inputs it receives.

18.1.8   Summary

A wide variety of control devices ensure the safe and reliable operation of heating equipment. *The use of these devices, however, does not preclude the proper use and maintenance of the equipment.* The wide selection also allows control systems to be designed which can achieve many operational objectives.

18.1.9   Abbreviations of Organizations

ABMA   American Boiler Manufacturers Association
AGA   American Gas Association
AMCA   Air Moving and Conditioning Association (fans)
ANSI   American National Standards Institute
API   American Petroleum Institute
ASHRAE   American Society of Heating, Refrigeration, and Air-Conditioning Engineers
ASME   American Society of Mechanical Engineers
ASTM   American Society for Testing and Materials
AWS   American Welding Society
CSA   Canadian Standards Association
HI   Hydronics Institute (a division of GAMA, Inc.)
HSB   Hartford Steam Boiler Inspection and Insurance Company
IEEE   Institute of Electrical and Electronic Engines
NBS   National Bureau of Standards
NEMA   National Electrical Manufacturers Association
NFPA   National Fire Protection Association
SBI   Steel Boiler Industry
SMA   Stoker Manufacturers Association
UL   Underwriters' Laboratories, Inc.

18.1.10   Acknowledgments and Bibliography

The material contained in this chapter represents countless hours of preparation, including reading several books and sources of reference material. We wish to acknowledge those who contributed some of the information contained in this chapter. In addition, we wish to acknowledge the many Cleaver-Brooks employees for their knowledge, expertise, and dedication to this project.

American Boilers Manufacturer's Association (ABMA)
950 N. Glebe Road, Suite 160
Arlington, VA 22203

American Society of Mechanical Engineers (ASME)
United Engineering Center
345 E. 47th Street
New York, NY 10017

Babcock and Wilcox Company
20 S. Van Buren Ave.
Barberton, OH 44203-0351

ABB, Inc. (formerly Combustion Engineering, Inc.)
Windsor, CT 06095

Factory Insurance Association
85 Woodland Street
Hartford, CT 05102

Factory Mutual Insurance Association
1151 Boston-Providence Turnpike
P.O. Box 688
Norwood, MA 02002

Hydronics Institute (a division of GAMA)
35 Russo Place
Berkeley Heights, NJ 07922

Bibliography

ABMA: *Handbook of Power Utility Terms & Phrases,* 6th ed., ABMA, Arlington, VA, 1995.

ASME: *ASME Boiler and Pressure Vessel Code* (Section I, "Power Boilers," Section IV, "Heating Boiler"; Section VIII, "Pressure Vessels"), ASME, New York, New York 1995.

Babcock and Wilcox: *Steam: Its Generation and Use,* The Babcock and Wilcox Company, Barberton, OH, 40th ed., 1992.

Cleaver-Brooks: *Cleaver-Brooks Packaged Firetube Boiler Engineering Manual (Q38),* Cleaver-Brooks, Milwaukee, WI, 1986.

Singer, Joseph (ed.): *Combustion Fossil Power: A Reference Book on Fuel Burning and Steam Generation,* ABB, Inc., Windsor, Connecticut 1991.

18.2

Burners and Burner Systems

Roland W. Brown

V.P., Engineering Services, Power-Flame, Inc.,
Parsons, Kansas

18.2.1 Introduction

Most of the burners described in this chapter use a motor driven blower to provide air for combustion which may also provide sufficient static pressure to overcome positive pressure within the heat exchanger precluding the need for an induced draft fan or high chimney. The motor/fan speed may be 1725 RPM or 3450 RPM.

18.2.2 Burner Functions and Types

All burners must perform five functions:

1. Deliver fuel to the combustion chamber.
2. Deliver air to the combustion chamber.
3. Mix the fuel and air.
4. Ignite and burn the mixture.
5. Remove the products of combustion.

Regardless of the type of fuel used, the burner must perform all five functions. In the case of liquid and solid fuels, the first function also includes preparation of the fuel so it will burn. The most common methods of accomplishing these functions are discussed in the following sections for gas, oil, coal, and combination burners. Classification of burners is generally based on the means of accomplishing one or more of the burner functions. Classifications used in this chapter are typical; many others may be encountered. At times, combinations of types may be required to describe a burner completely:

A. GAS BURNERS

The gases considered in this discussion are natural gas (nominally 1000 BTU/ft^3) [8900 Kcal/m^3] and propane gas (2500 BTU/ft^3) [22250 Kcal/m^3].

B. OIL BURNERS

Pressure atomizing burners for #2 fuel oil and atomizing fuel pressures from 100 psig (690 kPa) to 300 psig (2070 kPa) are given consideration in this chapter. Air atomizing burners require compressed air from a designated burner air compressor (or plant air system). Although this discussion is limited to #2 fuel oil, many burner manufacturers provide burners capable of firing #4, #5, or #6 fuel oils. Due to the complexity and variation of the systems of fuel oil delivery and temperature control [these fuels (#4, #5, #6 oil) require pre-heating for proper burning] limited discussion of #4, #5, & #6 fuel oils is provided.

C. DUAL-FUEL BURNERS

These burners may be used with natural gas and propane, or natural gas and #2, #4, or #6 oils. The changeover from one fuel to the other may be made with the use of a manual switch or a temperature control to automatically change from one fuel to the other. Most applications with well designed dual fuel burners will permit fuel selection without changes in fuel pressure or air settings.

BURNER SAFETY SYSTEMS

An essential element of the burner safety system is the flame detector which proves the presence or lack of a burner pilot flame and/or main flame. Gas pilot flames are usually ignited by an electric spark produced by an ignition transformer of 6000 to 10,000 volts output capacity and may be sensed (proven) by a flame rod (gas burners only) or optical sensor such as an ULTRAVIOLET or INFRARED Scanner for gas or oil burners. The intermittent pilot flame continues to burn with the spark shut off during the main flame operation. Intermittent pilots generally do not meet codes for inputs over 2,500,000 BTU (2638 KJ) per hr. Interrupted pilots provide for shut off of pilot and spark when main flame is proven. Inputs over 2,500,000 BTU (2638 KJ) per hr require interrupted pilots. Gas pilots for oil burners should be of the interrupted type.

18.2.3 Flame Safeguards

The modern Primary Flame Safeguard Device (PFSD) control system on a commercial or industrial boiler performs the following functions:

- Provides a safe means of starting and stopping the burner, either manually or automatically.
- Provides the proper sequencing and operation of burner components and supervises the burner flame during operation.
- Guards the system against excessive pressure or temperature conditions.

In addition, some systems also perform these functions:

- Regulate the burner firing rate.
- Maintain burner readiness during the burner-off cycle.

18.2.3.1 Types of controls

The following types of controls are used to perform basic flame safeguard functions:

A. CONTROLLERS

The controller may be an automatic temperature or pressure sensing device designed to operate the burner to maintain (depending on the medium being heated) set limits of pressure, water temperature, or air temperature. The controller may also be a start/stop station which allows the operator to manually start and, if desired, manually stop the burner.

B. LIMIT AND SAFETY CONTROLS

As a check upon the controller and to provide a maximum limit beyond which the burner should not be allowed to operate, a limit control must be provided, responding either to pressure or temperature. On steam boilers it is also necessary to provide a low water cutoff (and, if code requires, an additional auxiliary low water cut off) to prevent burner operation should the water level drop below safe limits. In addition, if the boiler has a feed water or condensate return pump, this is started prior to low water cutoff action in order to return the boiler water to the desired level.

C. BURNER INTERLOCK CONTROLS

Burner interlock controls perform two functions:

- Prove that the conditions for combustion are established and that the burner is ready to be started.
- Prove that conditions are satisfactory for burner operation to continue.

Start interlocks include valve-closed interlocks, damper positioning controls, fuel pressure switches, electrical, start interlocks, and oil preheater controls. Running interlocks include fuel pressure switches, combustion air controls (airflow switch), and draft controls (low and high fire proving switches).

Lockout interlocks include airflow and fuel pressure switches. These would be used with flame safeguard controls incorporating a lockout interlock circuit. They would replace similar running interlocks.

Smaller burners do not require all of these interlocks in their control systems, but larger burners would incorporate most of these interlocks.

D. FIRING RATE CONTROLS

The majority of commercial and industrial burners have a means of varying the firing rate according to the load demand, or firing rate modulation. This is usually accomplished electrically (but may be done pneumatically) and involves changing the firing rate (both air and fuel quantities simultaneously) from high fire to low fire.

When firing rate controls of the high/low or modulating type are provided, a guaranteed low fire start may be required to provide a smooth startup, (i.e., the fuel/air components are proven to be in the reduced flow light off position).

E. FUEL VALVES

Fuel valves open, close, and sometimes modulate the fuel supply. On oil burning systems, some models have delayed opening features. Gas valves may be of the slow or fast opening type, and may incorporate double seats with valve seal overtravel interlocks (also called "proof of closure"). Oil valves for large burners may also have a valve seal overtravel interlock (proof of closure).

F. PRIMARY FLAME SAFEGUARD DEVICE (PFSD) OR PROGRAMMING CONTROLS

The PFSD or programming control, with its associated flame detector, is the heart of the control system and provides a means of starting the burner in its proper sequence, sensing that the flame has been established, and supervising the flame during burner operation.

An electronic flame safeguard control performs two functions which are interlocked such that, without completion of the first, the second cannot continue:

1. Sensing the presence of a satisfactory flame.
2. Sequencing the operation of the burner system.

In addition, the flame safeguard system must provide a third function to provide safe operation-

3. Checking its own components and circuitry.

Sensing the presence of a satisfactory flame. A flame safeguard must distinguish between a safe and stable flame, an unsafe flame, and something (for instance, hot refractory) that is not a flame at all but merely has radiative characteristics of flame. And it must do this faster and more accurately than you could do it or the result may be a different kind of fireworks.

Sequencing the operation of the burner system. Burner system operation, requires its motors, blowers, ignition, and fuel valves to be energized in the proper sequence. On a burner shutdown, either normal or because of flame failure, components must also be deenergized in the proper sequence to prevent unburned fuel from accumulating in the combustion chamber where it could cause a hazardous condition.

When a start button is pressed or an automatic controller calls for heat, a system sequencing device must make sure that the burner gets fuel and ignition at the proper time to assure a good light off, and power must be supplied to start motors, open valves, and to provide ignition. If some component or interlock shows that the system is dangerous, this system sequencing device must stop the sequence and shut down the burner system safely. The system sequencing device that must accomplish all this is the primary flame safeguard device or PFSD. It may perform other duties too, such as purging or sounding alarms, but its prime duty is to assure safe operation. To do this, it must not only sense the presence or absence of a satisfactory flame but also control the entire operating sequence of all components of the burner system in proper order.

Checking its own components. By means of a built-in safe-start check, the flame safeguard checks itself for unsafe failures on every start and every time power is reapplied. If a flame simulating failure is present, the safe-start check will prevent the burner from starting.

One may be inclined to consider all shutdowns as nuisance shutdowns. But a shutdown is a nuisance shutdown *only if continued operation would have been safe beyond all question.* A shutdown on a marginal flame condition is not a nuisance shutdown; the flame safeguard is functioning as it should. A flame safeguard system considers safety first and convenience second.

The *primary* control contains the necessary relays and contacts for starting and stopping the burner under orders from the operating controller, limit controller, interlocks, and flame detector. A generalized Flame Safeguard control system is shown in Fig. 18.2.1.

This generalized system is representative of all flame safeguard control systems. The figure makes it apparent that the primary control starts the burner system on signal from the controller and permits operation to continue under control of two feedback loops. The flame detector monitors flame conditions and signals the burner to stop if flame is lost, or if a failure occurs in the detection system. The limits and interlocks monitor conditions other than the presence of flame shutting down the system if conditions exceed limit or interlock settings.

A *programming* control is basically the same as a primary control except that, as the word "programming" implies, it provides timed sequencing of the burner functions. It may also include control of additional functions, such as prepurge, postpurge, timed trial for ignition, and firing rate selection.

18.2.4 Why We Need Special Control Systems for Large Burners

A burner is simply a device for converting fuel into usable heat energy. Burners are rated by the amount of heat they produce in Btu's per hour [BTUH] (Kcal). This is determined by the amount of fuel being burned, and the Btu content of the fuel. A space heater in a residence might be rated for 30,000 Btuh (7560 Kcal); a domestic furnace for 80,000 Btuh (20,160 Kcal/hr) or 100,000 Btuh (25,200 Kcal/hr).

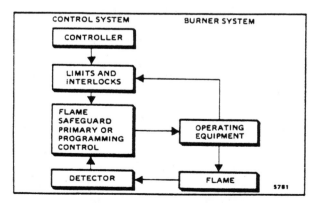

Figure 18.2.1 Flame safeguard control/burner system — Courtesy Honeywell, Inc.

Equipment of this size is usually controlled by heat sensing flame detection systems such as thermocouples for gas and cadmium sulfide cells for oil, which stop the operation when they sense an unsafe condition. In a thermocouple system fuel is cut off if no pilot is present. In the case of a cadmium sulfide cell, oil fuel is cut off if ignition does not occur within 15 to 30 sec depending on the size of the burner. These systems, generally requiring from 1 to 3 min to actually cut off gas fuel supply after flame loss or failure to ignite, are perfectly satisfactory for the types of burners on which they are used. Assume that a gas burner using 120 ft³ (3.3984 m³) of 1000 Btu/ft³ (8900 Kcal/m³) gas per hr [120,000 Btuh (30,240 Kcal/hr) burner] has a flameout. Its heat sensing protection system shuts off fuel to the burner in about a minute. What happens? In the minute following flameout, 2 ft³ (0.057 m³) of gas are released into the combustion chamber. Mixed with enough air for combustion—about 10 times the amount of gas in this case—we could have 22 ft³ (0.62 m³) of explosive mixture. But what would happen if this same control system were applied to a burner rated for 1,200,000 Btuh (302,400 Kcal/hr) burning 1200 ft³ (33.984 m³) of gas every hour? Within a minute following flame out, this system would introduce 20 ft³ (0.57 m³) of gas or, when mixed with combustion air, 220 ft³ (6.23 m³) of explosive mixture. If delayed ignition takes place, the results will be disastrous.

Obviously

1. We need faster methods of flame failure detection for large burners. For safety, the system must actually shut off the fuel within 2 to 4 sec of a flame failure.

2. We need reliability—both for safety and for economic reasons, since loss of burner use may be expensive.

3. We need automatic control capable of controlling all the functions of a complex burner system safely and reliably, without constant attention from an operator.

4. We need to protect the system from other dangers beside loss of flame—such as excessive temperature or boiler pressure.

In order to meet all these requirements, we need sophisticated electronic flame safeguard control systems.

18.2.5   Gas Burners

18.2.5.1   Gas burner functions

This section includes a general discussion of the methods of performing the five basic functions. The section "Gas Burner Types" includes details of the methods used in common types of gas burners.

1. DELIVERING FUEL TO THE COMBUSTION CHAMBER

Gas may be blown into the combustion chamber, pulled in (aspirated) by high velocity air, or permitted to escape into the chamber under its own distribution pressure. Since gas is always distributed under pressure, the last method is widely used. The quantity of gas escaping into the combustion chamber is usually controlled by a pressure regulating valve (PRV) in the gas supply line. It can also be controlled by an orifice or by a manual valve in the gas line.

Gas distribution gauge pressures at the burner vary from a few oz/in^2, gauge (osig) (a few millibar gauge) to as many as 50 lbs/in^2 gauge (psig) (344.74 kPa). The pressure is generally classified as low, intermediate, or high as follows:

- Low—2 to 8 osig (.8618 to 3.4472 kPa).
- Intermediate—8 osig to 2 psig (3.4472 to 13.79 kPa).
- High—2 to 50 psig (13.79 to 344.75 kPa).

When the gas distribution pressure is high, many burners can be satisfactorily adapted to a wide range of capacities by installing different sizes of gas orifices. Another advantage of high pressure gas is that smaller orifices may be used. Smaller orifices and higher gas velocities develop higher burner-head pressures, which give greater working range or turndown. A turndown ratio of about 5 to 1 is usually required. For higher ratios, several small burners rather than a single large one may be installed so that one or more burners may be turned off completely.

If the gas is delivered from the side(s) of the combustion chamber, resulting in a horizontal flame, the burner is called an *inshot burner* (Fig. 18.2.2). If the gas is delivered vertically from below the combustion chamber, resulting in a vertical flame, the burner is called an *upshot burner.*

2. DELIVERING AIR TO THE COMBUSTION CHAMBER

A gas burner may simply rely on atmospheric pressure to bring in combustion air, in which case it is called an atmospheric burner, or it may use machinery to bring in combustion air, in which case it is called a mechanical-draft burner.

Mechanical draft burner. A mechanical-draft burner uses machinery to deliver air to the combustion chamber. Two general methods are used, forced-draft or induced-draft.

A forced-draft burner (Fig. 18.2.3), commonly called a *power burner,* uses a motor driven fan or blower at the inlet of the combustion chamber to blow air into the combustion chamber. If required, additional air may be provided by natural draft or by venturi action. Many of the larger industrial and commercial burners are power burners.

An induced-draft burner (Fig. 18.2.4) uses a motor-driver fan or

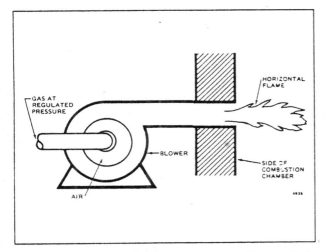

Figure 18.2.2 Typical inshot burner—Courtesy Honeywell, Inc.

blower at the outlet of the combustion chamber to create a slight partial vacuum within the chamber. This causes a suction condition which draws in air.

Closed and open burners. The amount of combustion air supplied through the burner gives rise to additional terms regarding the delivery of air.

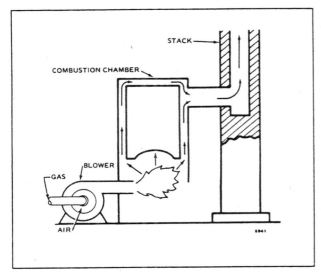

Figure 18.2.3 Forced draft power burner—Courtesy Honeywell, Inc.

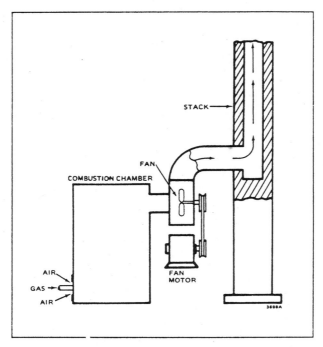

Figure 18.2.4 Induced draft burner—Courtesy Honeywell, Inc.

A *closed* or *sealed-in* burner is mounted on a furnace in an airtight manner so all of the air required for combustion is supplied through the burner. This permits accurate control of the air-fuel ratio, a wide range of furnace operating pressure, and a wide range of firing rates (turndown).

An *open* burner has an opening around it through which secondary air can enter the combustion chamber. Open burners permit greater capacities by virtue of this additional air. However, a correspondingly larger fuel capacity must also be available.

Many burners have an air register around them with an adjustable shutter to permit operation as either an open or a closed burner. The shutter can also be adjusted to control the volume of secondary air.

3. MIXING THE FUEL AND AIR

One method of classifying gas burners is by the method where the gas and air are mixed. In premixing gas burners, the gas and primary air are mixed before they reach the burner port(s). If secondary air is also required, they are sometimes called partial-premix burners.

As the name implies, in nozzle mixing gas burners the gas and air are kept separate within the burner itself, but the nozzle orifices are designed to provide rapid mixing of the gases as they leave. In delayed mixing gas burners, the rate of mixing the gas and air is very low so that they travel a considerable distance from the burner before mixing and burning.

The thoroughness of mixing determines to a great extent the characteristics of the flame. Turbulence and high velocities result in good mixing, and produce a short, bushy flame. Delayed mixing and low velocities produce a long slender flame.

The manner in which the gas and air are mixed results in several types of premixing, nozzle mixing, and delayed mixing gas burners. These are discussed in the section entitled *Gas Burner Types*.

4. IGNITING AND BURNING THE FUEL-AIR MIXTURE

A small gas burner (called a *pilot*) is usually used to ignite large industrial and commercial gas burners. A continuous pilot (sometimes called *constant, standby,* or *standing* pilot) burns without turndown throughout the entire time the burner is in service, whether or not the main burner is firing. An interrupted pilot (sometimes called *ignition* pilot) is automatically turned on each time there is a call for heat and automatically cut off at the end of the main burner flame-establishing period. An intermittent pilot is automatically turned on each time there is a call for heat and maintained during the entire run period. It is shut off with the main burner at the end of heat demand.

In a manually ignited burner, fuel to the main burner is turned on only by hand and ignited under the supervision of an operator. In an automatically ignited burner, fuel is automatically turned on and ignited without the presence of an operator. An automatic burner starts, runs and stops automatically, while a semiautomatic burner requires some manual operations — starting, igniting, and stopping. (Approval agencies' definitions vary; refer to the specification/standard for the particular agency.) Both types are purged, sequenced, and modulated automatically, with certain steps and conditions supervised by safety interlocks. Industrial burners are usually semiautomatic or manually ignited, while commercial burners are usually automatic.

Considerable control of the flame characteristics can be obtained by regulating the relative amounts of primary and secondary air.

- *Rich* mixtures (too little air) do not burn completely until secondary air mixes with the fuel. Therefore a rich mixture results in a long, cool, yellow flame.
- *Lean* mixtures (too much air) burn more rapidly than perfect mixtures because they produce a higher flame propagation rate.

However, the excess air decreases the efficiency of the burner because the rapid burning allows too much heat to escape up the stack. A lean mixture results in a short, cool, blue flame.

5. REMOVING THE PRODUCTS OF COMBUSTION

The products of combustion are usually removed from the combustion chamber as a result of supplying combustion air. In review, these methods include:

- Natural Draft—products of combustion escape up the chimney or stack and pull in secondary air.
- Inspirating—venturi action brings in primary air which mixes with the gas and displaces the products of combustion.
- Forced-draft—a fan or blower blows air into the combustion chamber which forces out the products of combustion.
- Induced draft—a fan or blower creates a suction which draws in air and removes the products of combustion.

18.2.5.2   Other factors affecting gas burner design

In addition to the five basic functions just discussed, several other requirements may affect the design of gas burning equipment.

1. MODULATION OF BURNER INPUT TO MATCH LOAD

In many cases it is desirable to modulate the fuel input to the burner in response to changes in load. To maintain efficiency under modulating control it is necessary to modulate the fuel and air supplies simultaneously. The range of firing rates over which satisfactory combustion may be obtained is referred to as the turndown range of the burner. Modulation may be continuous throughout this range, or it may involve only high and low fire positions. Several means are used to obtain modulation.

- Simultaneous adjustment of the air register or draft damper and the gas valve. It usually is difficult to obtain a wide range of turndown by this means, chiefly because of the nonlinear characteristics of dampers and valves.
- Varying the gas pressure. This means is particularly effective where all air for combustion is inspirated in a venturi section by the energy of a high pressure gas stream. It is also effective in other atmospheric burners.
- Varying the air pressure. This means provides a large turndown ratio in aspirating type burners where the gas is entrained by the high velocity air stream.
- Proportional mixing in a blower. A large turndown ratio can be obtained by this means with units having specially designed air

and gas valves controlling fuel and air input to a mechanical blower.

2. PREVENTION OF FLASHBACK

Valves and premixing burners are subject to flashback under some conditions. Flashback occurs when the velocity of the air–gas mixture through the burner ports is less than the flame propagation rate. It often occurs at minimum firing rates. Flashback may result from:

- Low gas pressure
- Leaking gas valve
- Lean gas air mixture
- Burner ports or pipes too large
- Excessive temperature of the gas-air mixture
- Excessive temperature of the burner ports or pipes
- Insufficient draft
- Insufficient combustion space

3. PROVISION OF SMOOTH IGNITION

Among factors influencing gas burner ignition are the following:

- Location and number of pilots
- Opening characteristics of the gas valve
- Manifold charging time
- Burner draft conditions
- Combustion chamber temperature
- Flame propagation rate

The following are a few of the many means used to provide smooth, quiet ignition:

- Slow-opening gas valves
- Multiple pilots for simultaneous ignition at several points
- Pilot location at point of first gas delivery
- Low-fire operation until stack and combustion chamber are still warm.

18.2.5.3 Gas burner types

There are many types of gas burners and there is not always a sharp distinction between types. In fact, most burners are a combination of two or more types depending upon the methods of classification. For a given application, several satisfactory designs can be worked out.

Gas burners range in capacity from about 20 (0.5664 m^3) to more than 100,000 ft^3 (2832 m^3) of gas per hr. Depending on the heating value of the gas used, this range is from about 10,000 btuh (Btu's per hr) (2520 Kcal/hr) for small domestic burners to over 300 million Btuh

TABLE 18.2.1 Typical Oil Burner Characteristics*

Type of burner	Approximate size (input)					Atomizing pressure						Turn down ratio[b]	Usual applications
	Gal per hr	Liters per hr	BTU per hr	Kcal per hr		Oil		Air or steam					
						psig	kPa	psig	kPa				
Air-atomizing low pressure	10 to 500	37.85 to 1892.7	1.4 million to 75 million	0.35308 million to 18.915 million		—	—	25 to 150	172.4 to 1034.2		3 to 1 up to 8 to 1	Major sized industrial plants utilizing compressed air in their process applications. Particularly adaptable for converting to combination gas-oil burners.
Steam-atomizing	10 to 500	37.85 to 1892.7	1.4 million to 75 million	0.353080 million to 18.915 million		—	—	25 to 150	172.4 to 1034.2		3 to 1 up to 8 to 1	Major-sized industrial plants using steam generators, particularly water-tube boilers. Particularly adaptable for converting to combination gas-oil burners.
Mechanical-atomizing, nonrecirculating simplex nozzle	0.5 to 80	1.8927 to 302.83	70 thousand to 12 million	17.654 thousand to 3.024 million		75 to 300	517.1 to 2068.5	—	—		2 to 1 (on-off control only)	Domestic warm air furnaces, boilers, and small industrial furnaces.
Mechanical-atomizing, recirculating (return-flow or bypass)	25 to 1200	94.64 to 4542.5	3.5 million to 180 million	0.8827 million to 45.396 million		100 to 1000	689.5 to 6895.0	—	—		3 to 1 up to 10 to 1	Most economical atomizing burner. Wide range, from domestic oil burners to major-sized boiler plants, including marine boilers.

*Courtesy Honeywell, Inc.
[a]British thermal units per hour, based on a heat content of 140,000 Btu per gal. or 9321 Kcal/Liter for light (distillate) oils, and 150,000 Btu per gal. or 9987 Kcal/Liter for heavy (residual) oils.

424

(75.6 million Kcal/hr) for large industrial and commercial burners. Most gas burners are adaptable to wide ranges of capacity, gas pressure, and gas heating value merely by changing orifice sizes and air register openings. Therefore it is difficult to assign a size range to a particular type of burner.

Some convenient classifications of gas burners are by draft type, port type, and mixing type. Table 18.2.1 shows the mixing types commonly used with the various draft types and port types. Another convenient classification that is sometimes used is the physical shape of the burner. The most common of these types will be described in this section.

18.2.5.4   Draft types

These types have already been described in section 18.2.5.1 on *"Gas Burner Functions"* (Function No. 2 — Delivering Air to the Combustion Chamber). In review, they include:

A. Atmospheric Burners
  - Natural draft burners.
  - Inspirating (venturi) burners.
B. Mechanical Draft Burners
  - Forced draft (power burners).
  - Induced-draft burners.

18.2.6   Oil Burners

18.2.6.1   Oil burner functions

This section includes a general discussion of the methods of performing the 5 basic functions. The next section, 18.5.6.2, *"Oil Burner Types,"* includes details of the methods used in the common types of oil burners.

1. DELIVERING FUEL TO THE COMBUSTION CHAMBER
   Oil is almost always delivered to the burner under pressure provided by a fuel pump. Exceptions are small domestic burners and one type of vertical rotary burner. In these instances, the oil tank is mounted higher than the burner so the oil flows downward to the burner by gravity.
   A heavy oil (No. 4, 5, or 6) may require preheating in a tank to lower its viscosity so it can be pumped to the burner. A No. 5 or 6 oil may also require additional preheating near the burner to further lower its viscosity so it can be atomized easily, and to raise its

temperature closer to the ignition point. No. 6 oil always requires preheating, both for handling and for burning.

Unlike gas, oil must be prepared for burning. It must be vaporized (converted to the gaseous state) before it can be burned. Some small burners accomplish this vaporization in a single step by the application of heat alone. Such burners are called *vaporizing* burners. Large capacity commercial and industrial burners use two steps to get the oil into combustible form-atomization plus vaporization.

Atomization is the reduction of the oil into a multitude of tiny droplets which can then be vaporized at a much higher rate than if heat alone were used. Burners which use high pressure or an atomizing medium to accomplish this are called atomizing burners. Others which use centrifugal force are called rotary burners. The common types of vaporizing, atomizing, and rotary burners will be described in section 18.2.6.2 entitled *"Oil Burner Types."*

2. DELIVERING AIR TO THE COMBUSTION CHAMBER

Oil burners use the same methods as gas burners to bring in combustion air, with the exception of the inspirating burner and the fanmix burner. The methods used are natural-draft (Fig. 18.2.2), forced-draft (Fig. 18.2.3), and induced draft (Fig. 18.2.4). Vanes (turbulators) are often used to give the air a swirling motion; this provides a more intimate mixture of the air and oil and aids in flame shaping.

3. MIXING THE FUEL AND AIR

In vaporizing burners, mixing usually takes place by diffusion. Air pulled in by a natural draft surrounds the flame and intermingles with the vaporized oil. Burners may use induced draft to increase their oil-burning capacity. In atomizing and rotary burners, air is supplied by a blower or fan. The air mixes with the droplets of atomized oil just beyond the burner nozzle(s). Vanes or turbulators give the air a rapid swirling motion for more thorough mixing. In air-atomizing burners, atomizing air mixes with the oil before and during atomization.

4. IGNITING AND BURNING THE FUEL-AIR MIXTURE

Small domestic burners of the vaporizing type are lit by a match. Pot type vaporizing burners used in central heating systems, atomizing burners, and rotary burners usually use direct spark ignition. A high voltage electric spark (10,000 volts minimum), is induced between 2 electrodes located in close proximity to the nozzle(s), which ignites the fuel-air mixture.

Larger industrial and commercial burners using heavy oil require a gas pilot that is ignited by a spark.

5. REMOVING THE PRODUCTS OF COMBUSTION

Small oil burners seldom require more than natural draft to remove
the products of combustion from the combustion chamber. Larger
oil burners usually require induced- or forced-draft fans. Some
newer burners recirculate some of the products of combustion to
reduce smoke.

18.2.6.2 Oil burner types

Oil burners are classified by the method used to prepare the oil for
burning. In a vaporizing burner, vaporization occurs from the surface
of a pool or layer of liquid lying on the bottom of a combustion chamber. In an atomizing burner, vaporization occurs from the surface of
minute droplets or globules of liquid floating in air as a spray or cloud,
within the space enclosed by the combustion chamber. The rotary
burner is similar to the atomizing burner except that the vaporization
is accomplished by centrifugal action.

Table 18.2.1 lists typical characteristics of oil burners which will be
described in this section.

18.2.6.3 Atomizing oil burners

All large capacity commercial and industrial oil burners burning #4,
#5, and #6 oil use two steps to get the oil into combustible form-
atomization plus vaporization. Large commercial industrial burners
for #2 oil do not usually require heating of the oil. By first atomizing
the oil into millions of tiny droplets, the exposed surface area is increased many times so the oil can be vaporized at a much higher rate.
For good atomization and vaporization, a large volume of air must be
intimately mixed with the oil particles. The air must be turbulent to
produce a scrubbing action on the surface of the oil particles. *Atomizing* oil burners use nozzles to accomplish atomization. Those using
centrifugal action are called *rotary* burners.

Low pressure air-atomizing oil burner. This type of burner uses a large
quantity of air at low pressure (½ to 2 psig) (3.4475–13.79 kPa) as the
oil atomizing medium. An oil pump delivers oil to the burner at low
pressure (usually 1–5 psig) (6.895 to 34.475 kPa)—just enough for
positive delivery and flow control. There are about as many special
types of nozzles and variations in mechanical operation as there are
manufacturers, although in principle they are about the same.

In the operation of a typical low pressure burner, the oil and air
meet prior to leaving the nozzle. Primary air is introduced into the
inner tube with the oil through tangential slots, which give the air a

rotary motion. (Some types inject whirling oil into the air stream.) The air picks up the oil, partially atomizing it. When the mixture leaves the inner tube at the nozzle, a second stream of atomizing air strikes it, completing the atomization and mixing with it. Some types even use a third stream of atomizing air. Ignition occurs a few inches (centimeters) beyond the nozzle outlet. Additional secondary air is supplied to complete the combustion. The low pressure air-atomizing oil burner is perhaps the most versatile of all oil burners. It is simple in design and construction, has no moving parts, has relatively large air and oil orifices, and requires no high pressures. This makes it rugged, simple to install and operate, flexible to load and fuel changes, easy to maintain, trouble-free, and economical. This type of oil burner is used in most installations to burn No. 2 or No. 4 oil, but with a properly designed nozzle it can burn No. 5 oil.

High pressure air-atomizing oil burner. This type of burner uses compressed air (25–150 psig) (172.375 to 1034.25 kPa) as a medium to tear droplets from the oil stream and propel them into the combustion space. It is practically identical with the steam-atomizing oil burner; the same nozzles can be used for both. Oftentimes with a boiler, compressed air is used for starting, and then steam is used when the boiler is up to pressure. High pressure air is seldom used as the principal atomizing medium because of its cost, except in plants which have an adequate supply of compressed air available because of its use in other process applications. Refer to the description of the steam-atomizing oil burner for more details.

Steam-atomizing oil burner. This type of burner uses high pressure steam (25–150 psig) (172.375–1034.25 kPa) as a medium to break up the oil into fine particles. As mentioned previously, high pressure air can be substituted for steam in nearly all installations, but it is not common because of the high cost of producing the power required to compress the air. Depending upon the point of mixing the oil and steam, there are 2 general designs — *outside-mixing* and *inside-mixing*.

In the outside-mixing burner, the steam and oil don't make contact until their release into the combustion chamber. The action of the atomizing steam shears the oil from the oil outlet orifice and delivers it in a spray. The nozzle causes both the steam and the oil to swirl before making contact, resulting in better mixing. This type generally requires more steam than the inside-mixing burner, however, it is less susceptible to clogging and requires a lower oil pressure.

In the inside-mixing burner, the steam makes contact with the oil inside the burner, resulting in a suspension of very finely divided oil in the steam. The mixture is called an *emulsion*, so the atomizer is sometimes called an *emulsion atomizer*. When the emulsion is discharged through the nozzle into the combustion chamber, further atomization occurs. The steam nozzle or tip is interchangeable to permit a variety of flame shapes. This type is more commonly used than the outside-mixing burner because is uses less steam and is therefore more economical.

A variation of the inside-mixing burner uses a *tip emulsion atomizer* (Fig. 18.2.5), which has been found to be the most flexible for commercial and industrial boilers. In it, the steam and oil make contact just inside the tip of the atomizer. The resulting emulsion is forced through the nozzle orifice. The total pressure is almost completely converted to velocity pressure so that viscous friction separates the emulsified oil into minute droplets. The tip can be designed for a desired spray angle. This type does a better job of atomizing with lower steam pressure and less steam consumption, so it is the most economical.

In both types of steam-atomizing oil burners, air for combustion is introduced into the combustion chamber concentrically with the burner through a fixed or adjustable air register.

Since steam is required, steam-atomizing oil burners are generally used with boilers that are fired continuously. If the boilers are allowed to lose their pressure, compressed air can be used for starting.

Steam atomizing burners are capable of burning any grade of fuel oil. They are convertible from light to heavy oil with the addition of proper heating and pumping equipment. Their chief advantage is their simplicity of construction and operation. The slim compact nature of these burners makes them readily adaptable for converting gas burners to combination gas–oil burners.

This *mechanical-atomizing* oil burner (Fig. 18.2.6) atomizes the oil by delivering it at a high pressure (a minimum of 75 psig to as high as 1000 psig) (517.125–6895 kPa) to a specially designed nozzle which breaks it into a spray of fine droplets. Its characteristic parts are the oil pump and nozzle which are machined to close tolerances to finely atomize the fuel oil to provide excellent combustion characteristics.

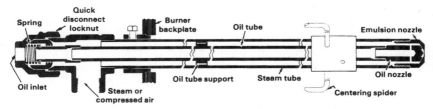

Figure 18.2.5 Tip emulsion atomizer—Courtesy North American Mfg.

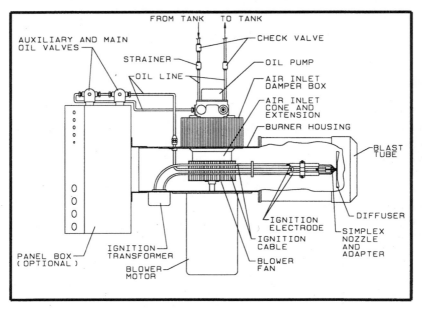

Figure 18.2.6   Mechanical atomizing oil burner — Courtesy Power Flame, Inc.

The term mechanical-atomizing is synonymous with pressure-atomizing. This pressure-atomizing gun type burner is the simplest, and probably the most common, oil burner. Besides the oil pump and nozzle, it consists of a fan to deliver the air necessary for a burnable oil–air mixture, a motor to drive the pump and fan, an ignition transformer and ignition assembly, and a burner housing to which all the components are mounted. Vanes (turbulators) give the air a whirling motion. Action of the nozzle, plus the impact of the rotating air stream from the fan, mixes the oil and air. A high voltage electric spark (10,000 volts minimum), supplied by the ignition transformer to the ignition electrodes, ignites the mixture.

The nozzle used in a mechanical-atomizing oil burner consists of a system of slots tangential to a whirl chamber followed by a small orifice. In passing through the slots, the liquid velocity is increased (pressure head is converted to velocity head). In the whirl chamber in a tangential direction which imparts a centrifugal effect forcing the oil against the confining walls of the nozzle. From here, it passes on through the orifice in the nozzle tip and into the combustion chamber. Once the walls cease to restrain the oil within a constricted space, the liquid fairly tears itself apart, fanning out into a cone-shaped spray of extremely small particles. The design of the nozzle determines the pattern of the oil spray, the size of the droplets, the angle of the cone, and the distribution of the droplets.

There are two types of nozzles—the *nonrecirculating* type (simplex) and the *recirculating* type (by-pass). In the nonrecirculating (simplex) nozzle, the oil flows through the strainer, spinning slots, whirl chamber, and orifice. The firing rate is controlled by the oil pressure, which is usually varied between 100 and 200 psig (689 and 1378 kPa); with a maximum of about 300 psig (2068.5 kPa). By the square root law, this results in a turndown ratio of less than 2 to 1, so this type of nozzle can be used only for on-off operation. Doubling the velocity of the oil also quadruples the work of pumping and requires smaller orifices, which may clog more easily. Lack of flexibility and maintenance problems limit the use of this nozzle, and it is seldom used in new installations today.

The recirculating nozzle (Fig. 18.2.7), also known as a wide-range, return flow, or *by-pass* nozzle, partly or wholly overcomes the difficulties of the nonrecirculating nozzle. Oil flows at a constant, high rate (at pressures as high as 1000 psig) (6895 kPa) into the whirl chamber. Here, some of the oil is recirculated back through the central pipe. The firing rate is controlled by regulating the back pressure on this return line, which controls the quantity of oil recirculated. With this type of nozzle, the turndown ratio can be as high as 10 to 1. Atomizing

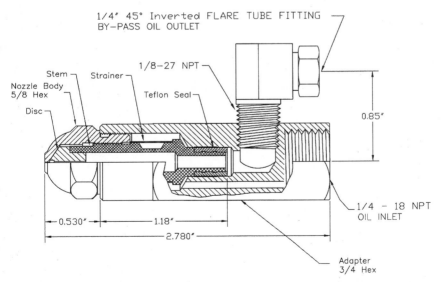

Figure 18.2.7 By-pass nozzle assembly also installation schematic—Courtesy HAGO Mfg.

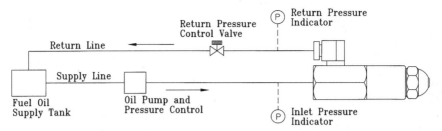

Figure 18.2.7 *Continued.*

pressure does not have to be reduced at turndown, so the effectiveness of atomization is maintained throughout the entire range of firing.

The mechanical-atomizing oil burner is the most economical type of atomizing burner, as far as initial and operating costs are concerned. It is used in small, residential heating plants as well as in some of the largest boiler plants requiring high firing rates. The smallest models burn light oil (No. 2) at pressures of about 100 psig (689.5 kPa) while the largest can burn the heaviest oil (No. 6), if properly pre-heated at pressures from about 200 psig (1378 kPa) up to as high as 1000 psig (6895 kPa).

18.2.7 Installation

A pressure regulating valve is used to control the by-pass pressure. With the by-pass valve fully closed the nozzle discharges at nominal flow rate. The nozzle flow rate is adjusted by controlling the by-pass pressure.

18.2.7.1 Gas–oil pressure atomizing burners

This type of burner is widely used and is commonly available as a packaged automatic burner. (Some companies have had U.L. Listed units since the late 1940s.) A single nozzle or multiple nozzle system may be used in the center of the burner with gas ports evenly spaced around the burner combustion head.

The gas mixing may be nozzle mix type or partial pre-mix type. An integrally mounted blower is used to provide the combustion air. A typical gas–oil unit is shown in Fig. 18.2.8.

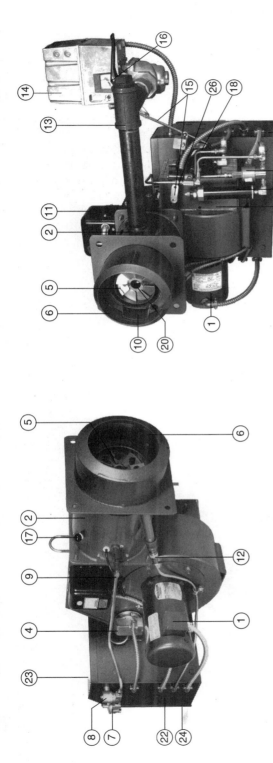

Burner Component Identification
Typical for Model C-GO with Low-High-Off or Low-High-Low Fuel/Air Control Modes of Operation.*

01. Blower Motor
02. Blast Tube
03. Air Inlet Housing
04. Air Flow Switch
05. Air Diffuser
06. Flame Retention Firing Head
07. Gas Pilot Regulator
08. Gas Pilot Solenoid Valve
09. Gas Pilot Test Tee
10. Gas Pilot Assembly

11. Gas Pilot Ignition Transformer
12. Flame Scanner (Detector)
13. Orifice Tee with Gauge Test Port
14. Motorized Gas Valve
    (Low-High-Off or Low-High-Low)
15. Air Damper Drive Linkage Assembly
16. Leakage Test Clock
17. Gas Premix Adjustment (Optional Feature)
18. Oil Pressure Pump
19. Hydraulic Damper Actuator

20. Oil Nozzle
21. Low-High-Off or Low-High-Low Oil Control Train
22. Control Panel
23. Hinged (Total Access) Top Section
24. Removable Total Access Door

* The components and arrangements shown are typical for a Model C combination gas/oil burner. Gas only or oil only units will have similar components relating to their specific fuel. In some cases, the type of components and/or their arrangements may vary from this depiction. For specifics on your system, refer to the technical information supplied with the burner.

Figure 18.2.8 Burner component identification — typical for Model CGO with Low-High-Low Fuel/Air control modes — Courtesy Power Flame, Inc.

18.2.7.2 Gas burner with low-high-off or low-high-low system (Figs. 18.2.9 and 18.2.10)

Mechanical operation: The low-high-off system uses a motorized gas valve* (1) to control the low-high-off operation of gas to the firing head (2), as well as movable air dampers (3) by means of the mechanical linkage (4). Gas flow control rate is accomplished by adjustment of the Main Gas Pressure Regulator (5) and by a limiting orifice (when installed) located in the side orifice tee fitting (7) at the inlet piping to the gas manifold. A proven spark ignited gas pilot** provides ignition for the main flame. When the gas pilot** has been proven by the flame detector (scanner),*** the motorized gas valve begins to open, allowing a controlled fuel/air mixture to the Firing Head for low fire light off—and continues to open, increasing the fuel/air flow until the high fire position has been reached. Firing head gas pressure are measured at the ¼" (6 mm) plugged gas test port (8) in the Side orifice tee. The burner operates at high fire until the system load demand is satisfied,

---

*Valve shown has valve seal overtravel. Auxiliary valve required if valve seal overtravel not furnished.

**Not shown in this depiction. (See Fig. 18.2.8.)

***Not shown in this depiction. (See Fig. 18.2.8.)

Figure 18.2.9 Typical gas burner with low-high-off or low-high-low fuel/air control mode (Model C-G) (See section 18.2.7.2 for component identification). Courtesy Power Flame, Inc.

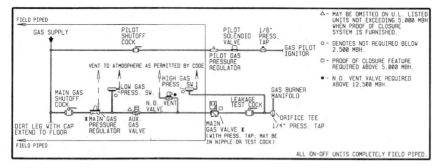

Figure 18.2.10 Typical schematic gas piping for C burner, on-off, low-high-off and low-high-low fuel air control.

at which time the motorized gas valve closes and the air dampers are returned to the light off position in preparation for the next operating cycle. This depiction shows the linkage in the low fire start position.

The low-high-low system is identical to the low-high-off system except that the motorized gas valve (1) has a low fire operating position adjustment in addition to the light off and high fire operating positions. (See manufacturer's bulletin included with the burner.) An additional temperature or pressure controller is added to the system, which at a selected preset point will electrically switch the motorized gas valve and air dampers (3) to either the low fire or the high fire position, as the system load demand required. Depending on system load conditions, the burner can alternate indefinitely between the low and the high fire positions without shutting down. When the system demand is satisfied, the motorized gas valve closes (normally the burner will be in the low fire position at this time) and the Air Dampers are returned to the light off position, in preparation for the next operating cycle. The driver arm (10) connected to the motorized gas valve will increase the travel of the air damper arm (13) as the linkage rod ball joint (11) is moved away from the gas valve crank shaft (12). The travel of the air damper driven arm will be increased as the linkage rod ball joint (14) is moved toward the air damper axle shaft (15). When adjusting linkage travel, make certain that the driven arm linkage return iron weight (16) does not interfere with the linkage operation — and that all linkage components are free from binding.

Note 1. Component operational sequencing will vary with the specific flame safeguard control being used. Refer to the specific flame safeguard control bulletin supplied with the burner for complete information.

Figure 18.2.11 Typical gas burner with full modulation fuel/air control mode (Model C-G)—Courtesy Power Flame, Inc. (See section 18.2.7.3 for component identification)

18.2.7.3 Gas burner—full modulation (Figs. 18.2.11 and 18.2.12)

Mechanical Operation: This full modulation system uses a motorized gas valve to ensure opening and positive closure of the gas source to the Firing Head (2). A modulating motor (3) controls the positioning of a butterfly gas proportioning valve (4) and movable air dampers (5) through mechanical linkage (6). The gas flow control rate is accomplished through adjustment of the main gas pressure regulator (7) and the butterfly valve. A proven spark ignited gas pilot* provides ignition for the main flame. When the gas pilot has been proven by the flame

---

*Valve shown has valve seal overtravel. Auxiliary valve required if valve seal overtravel not furnished.

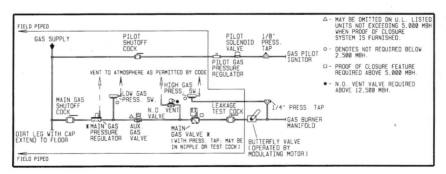

Figure 18.2.12 Typical schematic gas piping for type C burner, modulating system—Courtesy Power Flame, Inc.

detector,** the diaphragm or motorized gas valve opens and allows gas at a rate controlled by the Butterfly Valve to go to the burner head for main flame low fire light off. After a short period of time at the low fire position, the modulating motor will drive the butterfly valve and the air dampers to the high fire position. The burner will stay at high fire until the system pressure or temperature increases to a selected present point, at which time a modulating type controller will drive the modulating motor to low fire, or whatever firing position between low and high fire is required to match the system load demand. The modulating motor will continually reposition the firing rate in an effort to exactly match system load demand. Firing head gas pressures can be taken as the ¼-in. (6 mm) plugged test port (8) located between the Butterfly Valve and the gas Firing Head. Refer to the burner specification computer printout supplied with the burner, for specific high fire gas pressure values. When the system pressure or temperature cutoff point is reached, the diaphragm or motorized gas valve closes (normally the burner will be at the full low fire position at this time) and the air dampers will go to the low fire light off position in preparation for the next firing cycle. This depiction shows the linkage in the low fire light off position. Refer to Fig. 18.2.18 for information on linkage adjustment.

Note 1. Component operational sequencing will vary with specific flame safeguard control being used. Refer to the specific flame safeguard control bulletin supplied with the burner for complete information.

18.2.7.4 Oil burner with on-off fuel/air control (Fig. 18.2.13)

Mechanical Operation: The on-off system uses a single stage, high suction lift Oil Pump (2) with a Simplex oil nozzle. A direct spark oil ignition system will normally be supplied, but certain insurance company codes could require a spark ignited gas pilot* to provide ignition for the main oil flame. The nozzle oil flow rate is set by adjusting the oil pump pressure regulating valve (3) (⁵⁄₃₂-in. [4 mm] Allen wrench fitting). Turn clockwise to increase the pressure and counter-clockwise to decrease the pressure to the nozzle. Normal nozzle pressure will be 100 to 300 psig (689–2068 kPa). Nozzle pressures are taken at the plugged nozzle pressure gauge port (6). The oil flow to the nozzle is controlled by the oil solenoid valves (1). The air dampers (4) are ad-

---

**not shown in this diagram. (See Fig. 18.2.8.)

*not shown in this depiction.

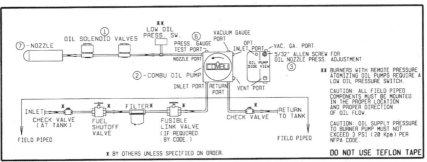

Figure 18.2.13   Typical oil burner with On-Off Fuel–Air Control Mode—Courtesy Power Flame, Inc. (See section 18.2.7.4 for component identification)

justed and locked in place with the air damper arms (5). The burner operates at one fixed firing rate. See pump manufacturer's bulletin packed with the burner for more information.

Note 1. Component operational sequencing will vary with specific flame safeguard control being used. Refer to the specific flame safeguard control bulletin supplied with the burner for complete information.

Note 2. The system depicted above is based on the use of an oil pump manufactured by COMBU, Inc. If your system uses other than a COMBU pump, refer to the oil piping diagram and oil pump manufacturer's bulletin supplied with the burner for specifics pertaining to your system.

18.2.7.5 Oil burner with Webster 22R oil pump (Fig. 18.2.14)

Mechanical Operation: This low-high-off system uses a two-stage oil pump (2) with an internal bypass oil nozzle (14) in conjunction with movable air dampers (4) to provide a low fire start and a high fire run sequence. A direct spark oil ignition system will normally be supplied at firing rates up to 45 GPH (170.34 liters/hr), with a spark ignited gas pilot* to ignite the main oil flame above that point. Certain insurance company codes will require the gas pilot system on lower input sizes. Nozzle supply pressure is set by adjusting the oil pump pressure regulator ⅛-in. Allen wrench fitting (3). Turn clockwise to

---

*not shown in this depiction. (See Fig. 18.2.8.)

Figure 18.2.14a Typical oil burner with low-high-off or low-high-low fuel/air control mode using Webster 22R Oil Pump.—Courtesy Power Flame, Inc. (See section 18.2.7.5 for component identification)

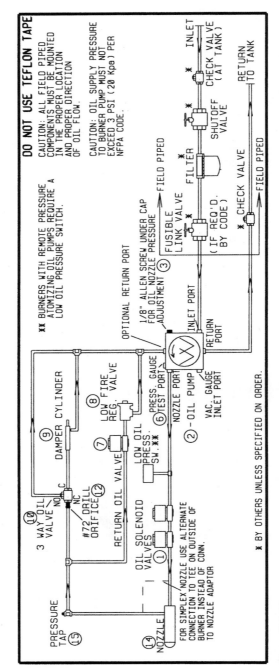

Figure 18.2.14b

440

increase the pressure and counter-clockwise to decrease the pressure to the nozzle. Nozzle supply pressure is taken at the plugged pump nozzle pressure gauge port (6). Nozzle supply pressure will normally be approximately 300 psig (2065 kPa) at both high and low firing rates. Flow rate pressure for both high and low fire is taken at bypass pressure gauge tee (15). Low fire pressures are set by adjusting the low fire regulating valve (8). Turning the low fire regulating valve adjustment nut clockwise will increase the pressure at the bypass pressure test tee gauge (increasing the low fire input) and counter clockwise will reduce the pressure at the gauge (decreasing the low fire input). Low fire pressure will normally be in the 60 to 100 psig (413.7–689.5 kPa) range and at high fire in the 180–225 psig (1241.1–1551.4 kPa) range, but both pressures will vary according to the specific nozzle being used, as well as job conditions. At light off, the main oil solenoid valve (1) is energized, allowing fuel to flow to the nozzle. At the same instant a position of the oil bypasses the nozzle through the adjustable low fire regulating valve, reducing the pressure at the nozzle as required for low fire rates. When the low fire flame is proven by the flame detector,* the return oil solenoid valve (7) is de-energized, putting full high fire pump pressure on the nozzle. Simultaneously, the three-way solenoid valve (10) is energized, allowing oil into the hydraulic cylinder (9) which mechanically drives the air damper arm (13) to the high fire position. The burner operates at full high fire until the system demand is satisfied. This depiction shows the air dampers and hydraulic cylinder at the low fire light off position. The low-high-low system is identical to the low-high-off system, except that an additional pressure or temperature controller is added to the system, which at a selected present point will electrically switch the burner to either the high or low fire position. When the burner is running at high fire and the controller calls for low fire, the normally closed oil solenoid return valve (7) (closed at high fire) is energized, reducing nozzle pressure to the low fire rate. Simultaneously, the three-way solenoid valve (10) is de-energized, allowing oil to flow out to the hydraulic cylinder (9) back to the pump and driving the air dampers (4) to the low fire position. Responding to load conditions, the burner can alternate indefinitely between the low and high fire positions without shutting down. When system load demand is satisfied, all fuel valves are de-energized and the air dampers are placed in the light off position in preparation for the next firing cycle. The opening distance of the air dampers is controlled by positioning the air damper drive arm (13) relative to the acorn nut (16) mounted on the end of

---

*not shown in this depiction.

the hydraulic cylinder piston rod. The maximum travel is with the Damper Drive Arm positioned to be in contact with the hydraulic oil cylinder acorn nut at all times. If less travel is desired, set the air damper drive arm to allow a gap between it and the acorn nut. (Depending on air damper positioning, it may be necessary to loosen its set screws to attain proper air damper opening distance.) The wider the gap (when the burner is off), the less the overall travel when going to the high fire position. When setting the drive arm position, relative to the acorn nut, make certain that the air damper's travel is correct for proper combustion at all firing positions and that there is no binding of the linkage or dampers. Make certain the cast iron linkage return weight (5) is secure on its air damper arm (17).

Note 1. The system depicted in Fig. 18.2.12 uses a Webster model 22R oil pump. If your system uses a Suntec H model pump, the sequence of operation and the oil components would be identical to the Webster 22R system. For additional information on your specific system refer to the oil piping diagram and the oil pump manufacturer's bulletin supplied with the burner.

Note 2. Component operational sequencing will vary with the specific flame safeguard control being used. Refer to the specific flame safeguard control bulletin supplied with the burner for complete information.

18.2.7.6   Oil burner with 2-step oil pump (Fig. 18.2.15)

Mechanical Operation: This low-high-off system uses a two-step oil pump with simplex oil nozzle (14) in conjunction with movable air dampers (4) to provide a low fire start and a high fire run sequence. A direct spark oil ignition system will normally be supplied, but certain insurance company codes could require a spark ignited gas pilot to provide ignition for the main oil flame. Nozzle flow rate pressure is taken at the $\frac{1}{8}$-in (3.2-mm) plugged pump pressure gauge port (6). The low fire oil rate is set by adjusting the oil pump low pressure regulator (8). The high fire oil flow rate is set by adjusting the oil pump high pressure regulator (3). For both high and low fires turn the adjustment screws clockwise to increase the pressure and counterclockwise to decrease the pressure to the Nozzle. Approximate low fire oil pressures are 100–125 psig (689.5–855 kPa) and high fire, 200–300 psig (1379–2065 kPa). Both settings will vary depending upon the specific nozzle size selected and job conditions. See manufacturer's data for specific nozzle pressures and flow rates. At light off the main oil Solenoid valve (1) is energized, allowing fuel to the nozzle. A normally open pump mounted oil solenoid valve (7) allows a controlled flow of oil to the nozzle in accordance with the pressure setting of the pump low fire adjustment. When the low fire flame is proven by the flame

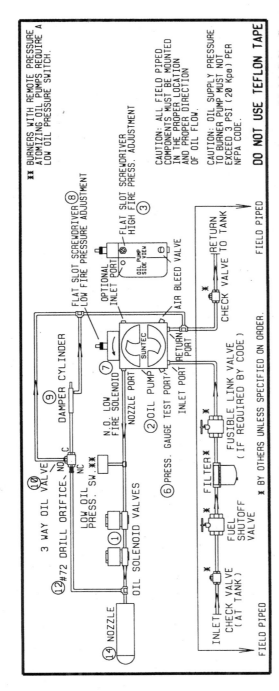

Figure 18.2.15 Typical oil burner with low-high-off or low-high-low fuel/air control mode using a two-step oil pump (Model C-O)—Courtesy Power Flame, Inc.

detector,* the pump mounted, normally open solenoid valve is energized (closes), putting full high fire pump pressure on the nozzle. Simultaneously, the three-way solenoid valve (10) is energized, allowing oil into the hydraulic oil cylinder (9) which mechanically drives the air damper arm (13) to the high fire open position. The burner operates at full high fire until the system demand is satisfied. This depiction shows the air dampers and the hydraulic cylinder at the low fire light off position.

The low-high-low systems are identical to the low-high-off system, except that an additional temperature or pressure controller is added to the system. At a selected preset point, it will electrically switch the oil valves and air damper components to place the firing rate either in the low or the high fire run position. When the burner is running at high fire and the controller calls for low fire, the normally open pump mounted solenoid valve (7) (which is closed at high fire) is de-energized (opens), reducing nozzle pressure to the low fire rate. Simultaneously, the three-way solenoid valve (10) is de-energized, allowing oil to flow out of the hydraulic cylinder back to the pump (2) and driving the air dampers (4) to the low fire position. Depending on load conditions, the burner can alternate indefinitely between the low and the high fire positions, without shutting down. When system demand is satisfied all fuel valves are de-energized and the air dampers are placed in the light off position for the next start up. The air damper position for low fire run and light off position are on and the same in this system. The opening distance of the Air Dampers is controlled by positioning the air damper drive arm (13) relative to the acorn nut (16) mounted on the end of the hydraulic cylinder (9) piston rod. The maximum travel is with the damper drive arm positioned to be in contact with the hydraulic oil cylinder Acorn Nut at all times. If less travel is desired, set the air damper drive arm to allow a gap between it and the acorn nut. (Depending on air damper position, it may be necessary to loosen its set screws to attain proper air damper opening distance.) The wider the gap (when the burner is off), the less the overall travel when going to high fire position. When setting the drive arm position relative to the Acorn Nut, make certain that the air damper travel is correct for proper combustion at all firing positions and that there is no binding of the linkage or dampers. Make certain the cast iron linkage return weight (15) is secure on its linkage arm (17).

Note 1. Component operations sequencing will vary with the specific flame safeguard control being used. Refer to the specific flame safe-

---

*Not shown in this depiction. (See Fig. 18.2.8.)

guard control bulletin supplied with the burners complete information.

8.2.7.7 Oil burner with full modulation (Fig. 18.2.16)

Mechanical Operation: The full modulation system uses a two-stage Oil Pump (2) with an internal bypass type Oil Nozzle (see Fig. 18.2.7). A modulating motor (4) controls the positioning of the air dampers (6) and the modulating oil valve (5) in the nozzle return line through mechanical linkage. A direct spark oil ignition system will normally be supplied at firing rates up to 45 GPH (170.3 liters/hr). Above that rate burners will be supplied with a spark ignited gas pilot* to light the main oil flame. Certain insurance company codes will require the gas pilot system on all input sizes. At main flame light off the normally closed Oil valve (1) is energized, allowing oil to flow to the nozzle. The modulating oil valve is adjusted to allow a controlled amount of oil to bypass the nozzle, which keeps the pressure reduced to the nozzle for low fire light off. Nozzle oil supply pressure is set by adjusting the oil pump pressure regulating ⅛-in. Allen wrench fitting (7). Turn clockwise to increase the pressure and counter-clockwise to decrease the

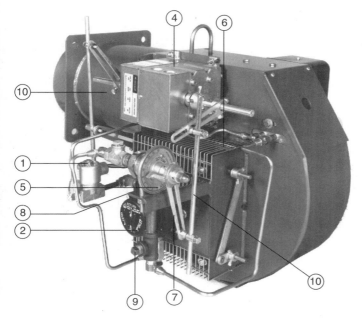

Figure 18.2.16a Typical oil burner will full modulation fuel/air control (Model C-O)—Courtesy Power Flame, Inc. (See Section 8.2.7.1 for component identification)

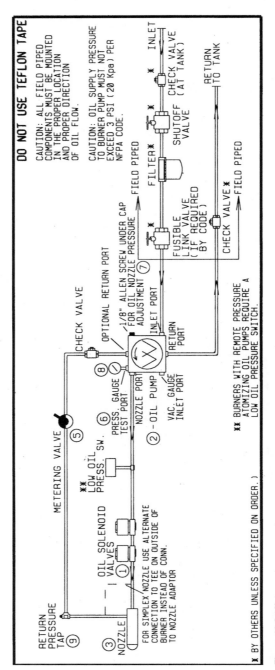

Figure 18.2.16b

446

pressure to the nozzle. The low fire nozzle pressures should be taken at the plugged oil pump gauge port (8) and should be approximately 300 psig (2068 kPa) (but could be as low as 240 psig (1654 kPa) on certain inputs of the C4 and C5 model(s) with pressure at the nozzle bypass gauge port (9) from 60 to 100 psig (418–690 kPa), these pressures varying with nozzle size and job conditions. A typical low fire oil flow setting on the modulating oil valve would be number 7, but will vary with job conditions. After a brief period of time, for the low fire flame to stabilize, the modulating motor will drive the fuel/air linkage (10) to the high fire position. At this point the Air Dampers will be full open (or as required for good combustion) and the modulating oil valve will be at the "closed" position and the nozzle bypass line will be fully closed, putting full oil pressure to the Nozzle. The oil pump pressure gauge port pressure reading will show approximately 300 psig (2069 kPa) and pressures at the bypass pressure gauge port will be 180–225 psig (1241–1551 kPa), although this will vary with the specific nozzle size being used. A modulating temperature or pressure controller will now modulate the firing rate to match the load demand of the system, while maintaining proper fuel/air ratios. Prior to reaching the system pressure or temperature operating control cut off point,the burner should be at or near the low fire operating position. At the end of the firing cycle, the normally closed oil valve will be de-energized and the modulating motor will position the air dampers and modulating valve to the low fire position, ready for the next start up sequence. This depiction shows the linkage in the low fire light off position. See Fig. 18.2.7 for linkage adjustment information.

Note 1. Some modulating low-high-off and low-high-low burners will be supplied with simplex, rather than internal bypass type, oil nozzles. The mechanical operation of the simplex nozzle system is essentially the same as the internal bypass system except that low fire oil pressures should be set at 100–125 psig (1931–2069 kPa) (adjust to suit job conditions) and high fire oil pressures at 280–300 (1930 to 2069 kPa) psig at the oil pump nozzle pressure gauge test port. Refer to the burner specification sheet shipped with the burner for high fire oil pressures and flow rates.

The oil pump depicted in the oil flow schematic above is as manufactured by Webster Electric Company Inc. If the pump on your burner is not Webster, refer to the oil pump bulletin shipped with the burner for specific adjustment information.

Note 2. Component operational sequencing will vary with the specific flame safeguard control being used. Refer to the specific flame safeguard control bulletin supplied with the burner for complete information.

18.2.7.8   Gas or gas/oil fuel/air premix adjustment (Fig. 18.2.17)

FUEL–AIR PREMIX ADJUSTMENT (OPTIONAL)

The adjustable premix blast tube (optional) incorporates an adjustable gas/air premix within the burner firing head. The premix configuration is primarily used for cylindrical combustion chambers or high heat release pressurized fireboxes. Moving the adjustment knob back increases the premix air; moving it forward decreases the premix air. Generally, the best (quietest/smoothest) operation is in the full forward position with minimum pre-mix air. The premix adjustment is set at the factory in the forward position. To attain the best combustion results for specific job conditions, change position in small increments.

Diffuser adjustment. Moving the blast tube diffuser assembly fore or aft on gas or oil firing will move the flame front (point of retention) in order to attain the best (quietest/smoothest) combustion for specific job conditions. If the initial midway point factory setting does not provide satisfactory results, move fore or aft in small increments to achieve the best combustion results. If unit is oil or combination gas/oil, the attached, flexible copper oil nozzle line will move fore or aft with the assembly. When firing on oil, moving the assembly forward will tend to broaden the flame pattern and moving it back will narrow the flame pattern. Similar results are obtained on gas, but observation of sound and combustion tests are the best determinants of results on either gas or oil.

18.2.8   Information of Fuel–Air Modes of Operation for Combination Gas–Oil Units (Fig. 18.2.18)

18.2.8.1   General information

Specific adjustments and mechanical operation of the various modes of fuel/air control for straight gas and straight oil burners are included in this chapter. This information can be used to properly adjust each fuel for combination gas/oil units as well. The following information is offered as additional guidance.

18.2.8.2   Gas on/off system combined with oil on-off system

The air dampers are adjusted and locked in place for the most efficient operation for both fuels. Refer to the mechanical operation of the gas on/off and oil on/off systems for specific adjustment details.

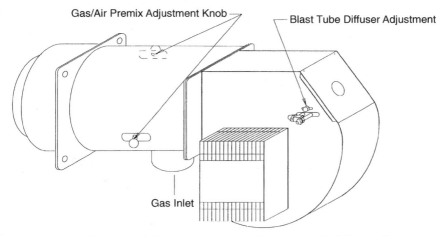

Gas/Air Premix Adjustment Knob

Blast Tube Diffuser Adjustment

Gas Inlet

Figure 18.2.17 Gas or gas/oil fuel/air premix adjustment and diffuser adjustment—
Courtesy Power Flame, Inc.

18.2.8.3 Gas low/high/off system with oil low/high/off system

Gas-movable air dampers are adjusted to provide a smooth light off position and then moved to the high fire position through mechanical linkage from a motorized gas valve. The air dampers are adjusted to open to provide maximum combustion efficiency at the gas high fire input rate. For oil, the same air dampers are operated by a hydraulic oil cylinder which, through mechanical linkage, is adjusted to provide a smooth light off and then open to a point where the highest combustion efficiencies will be achieved at the high fire input rate. Smooth oil light off is further achieved by the use of a solenoid oil valve bypass system, which allows a reduced amount of oil to flow at light off and then switches to the high fire rate (simultaneously energizing the hydraulic oil cylinder) once low fire has been established.

The mechanical linkage from the gas valve is physically arranged so that the hydraulic oil cylinder (which is nonoperational when burning gas) has no effect on the gas linkage adjustments. Similarly, the gas valve (which is nonoperational when burning oil) has no effect on oil linkage adjustments. It should be noted that when the hydraulic oil cylinder moves the air dampers, the movement of the air dampers will cause the motorized gas valve linkage to move up and down with the opening and closing of the hydraulic oil cylinder. The motorized gas valve linkage is "free floating," and even though it moves with the oil cylinder operation, it cannot cause any gas flow to pass through the motorized gas valve. Refer to the mechanical operation of the gas

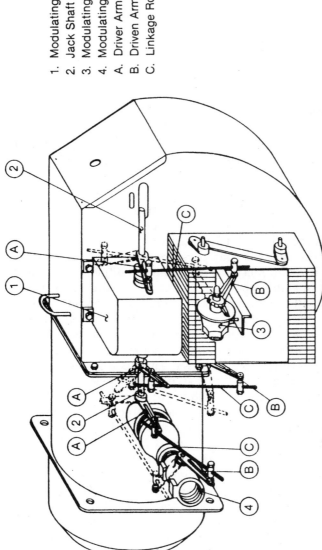

1. Modulating Motor
2. Jack Shaft
3. Modulating V Port Oil Valve
4. Modulating Butterfly Valve
A. Driver Arms
B. Driven Arms
C. Linkage Rods

Typical general linkage arrangement for combination gas/oil full modulation burner, shown in low fire light off position. Dotted lines indicate approximately high fire position. When making adjustments, make certain the motor can make its full 90° stroke without any linkage binding.

Driver Arms (A) connected to the Modulating Motor (1) Jack Shaft (2) will increase the travel of the Driven Arms (B) as the Linkage Rod (C) ball joint is moved away from the Jack Shaft. The travel of the Driven Arms will be increased as the Linkage Rod ball joint is moved toward the shaft of the driven device.

Figure 18.2.18 Gas/oil linkage adjustment for full modulation standard system—Courtesy Power Flame, Inc.

TABLE 18.2.2 Periodic Check List

Item	Frequency	Checked by	Remarks
Gages, monitors, and indicators	Daily	Operator	Make visual inspection and record readings in log
Instrument and equipment settings	Daily	Operator	Make visual check against heat exchanger manufacturer's recommended specifications
Atomizing air	Weekly	Operator	Air atomizing burners; check filter, drain moisture from traps
Firing rate control	Weekly	Operator	Verify heat exchanger manufacturer's setting
		Service technician	Verify heat exchanger manufacturer's setting
		Service technician	Check with combustion test
Flue, vent, stack, or outlet damper	Monthly	Operator	Make visual inspection of linkage, check for proper operation
Combustion air	Monthly	Operator	All sources remain clean and open
Ignition system	Weekly	Operator	Make visual inspection, check flame signal strength if meter-fitted (see "Combustion safety controls")
Fuel valves			
Pilot and main	Weekly	Operator	Open limit switch, make aural and visual check, check valve position indicators, and check fuel meters if so fitted
Pilot and main gas or main oil	Annually	Service technician	Perform leakage tests — refer to valve manufacturer's instructions
Combustion and safety controls			
Flame failure	Weekly	Operator	Close manual fuel supply for (1) pilot, (2) main fuel cock, and/or valve(s): check safety shutdown timing; log
Flame signal strength	Weekly	Operator	If flame signal meter installed, read and log; for both pilot and main flames, notify service organization if readings are very high, very low, or fluctuating; refer to flame safeguard manufacturer's instructions

TABLE 18.2.2 (Continued).

Item	Frequency	Checked by	Remarks
Pilot turndown tests	As required/annually	Service technician	Required after any adjustments to flame scanner mount or pilot burner; verify annually—refer to flame safeguard manufacturer's instructions
Refractory hold in	As required/annually	Service technician	See "Pilot turndown tests"
High limit safety control	Annually	Service technician	Refer to heat exchanger manufacturer's instructions
Operating control	Annually	Service technician	Refer to heat exchanger manufacturer's instructions
Low draft, fan, air pressure, and damper	Monthly	Operator	Refer to this manual and control manufacturer's instructions
High and low gas pressure interlocks	Monthly	Operator	Refer to instructions in this manual
Low oil pressure interlocks	Monthly	Operator	Refer to instructions in this manual
Fuel valve interlock switch	Annually	Service technician	Refer to valve manufacturer's instructions
Purge switch	Annually	Service technician	Refer to fuel–air control motor manufacturer's instructions
Low fire start interlock	Annually	Service technician	Refer to fuel/air control motor manufacturer's instructions
Automatic changeover control (dual fuel)	At least annually	Service technician	Under supervision of gas utility
Inspect burner components		Service technician	Refer to this manual and control component manufacturer's instructions
Oil filter	Monthly	Operator	Remove and clean or replace
Remove oil drawer assembly	Annually	Service technician	Remove and clean
Check blower motor and blower wheel for cleanliness. Remove and clean as necessary	Annually	Service technician	Remove and clean
Remove, inspect, and clean gas pilot assembly	Annually	Service technician	Remove and clean

Refer to heat exchanger manufacturer's instructions for general inspection procedures and for specific testing and inspection of all liquid level controls, pressure/temperature relief and other applicable items.

low-high-off system and the oil low-high-off system for specific adjustment details.

Note. The oil side operation can be supplied with either a Suntec or a Webster oil pump. Refer to the mechanical operation for the specific system for adjustment details.

18.2.8.4 Gas low/high/low system with oil low/high/low system

Refer to the above explanation of gas low/high/off system and oil low/high/off system. The gas low/high/low Systems are identical in operation, except that an additional temperature or pressure control is added to the system, which at a selected preset point will electrically switch the motorized gas valve (which is a different model number than the low/high/off motorized gas valve) to an adjustable "low fire" position. This low fire position is an adjustment that is designed internal to the gas valve and, depending upon the specific manufacturer of the valve, will be found either in the valve wiring compartment or under a removable cap on the top of the valve. Refer to the valve manufacturer's product bulletin supplied with the burner for specific details.

For the low/high/low System, oil side operation the additional temperature or pressure control will also cause the solenoid oil valves, hydraulic oil cylinder, and air dampers to go to a "low fire position" at the same present temperature or pressure as the gas side operation. The oil side low fire position is one and the same position as the light off position, i.e., the air dampers and oil pressures have identical settings (as compared to the gas side-which can be adjusted, if desired, to have different settings for "light off" and "low fire"). Refer to the mechanical operation of the gas low-high-low system and the straight oil low-high-low system for specific adjustment details.

Note. The oil side operation can be supplied with either a Suntec or a Webster oil pump. Refer to the mechanical operation for the specific system for adjustment details.

18.2.8.5 Gas full modulation system with oil full modulation system (Fig. 18.2.18)

The Gas system uses an automatic diaphragm or motorized gas shutoff valve to control the on-off flow of the gas. The oil system employs an oil solenoid valve to control the on-off flow of oil to the oil nozzle. A modulating motor controls the modulated positioning of a butterfly type Gas Proportioning Valve, while a V ported metering oil valve provides the modulating function in the oil nozzle return line. The modulating motor also controls the positioning of the combustion air dampers, through appropriate sequencing, providing low fuel-air input for a smooth low fire start and an infinite number of fuel-air positions between full low and high fire. Additional finite fuel/air adjustments are provided when the optional Power Flame Varicam™ characterized fuel metering system is used (optional). When firing gas the oil metering valve will open and close because it is linked to the modulating motor; however, the oil solenoid shutoff valve remains closed, and so no oil is allowed to flow to the nozzle. Similarly, when firing oil, the butterfly gas valve will open and close, because it is linked to the modulating motor; however, the main; automatic gas supply shutoff valve (motorized or diaphragm type) remains closed, and so no gas is allowed to flow to the burner head. Refer to the mechanical operation of the gas full modulation system (Figs. 18.2.11 and 18.2.12) and the oil full modulation system (Fig. 18.2.16 as well as Fig. 18.2.18 entitled "gas/oil—linkage arrangement full modulation system") for specific adjustment detail.

18.2.8.6   Burner start up and service test equipment required

The following test equipment is required to ensure proper start-up and adjustment of burner equipment to obtain maximum efficiency and reliability of operation.

For Any Fuel
CO_2 indicator or O_2 analyzer
Stack thermometer
Draft gauge or inclined manometer

Combination volt–ammeter
D.C. microammeter or D.C. voltmeter, as required by Flame Safeguard programmer selection

For Gas
CO indicator
U-tube manometer or calibrated 0'10" and (25.40 cm) 0–35" (88.89 cm) W.C. pressure gauges.

For Oil
Compound vacuum/ pressure gauge. 0–30" (76.2 cm) vacuum 0–30 psig (206.85 kPa) 0–400# (2758 kPa) oil pressure gauge (two required for internal bypass type oil nozzles)
Smoke tester

Note. When firing gas fuels, it is possible to attain CO_2 readings that appear to be acceptable (i.e., 8 percent, 9 percent, 10 percent, etc.) while actually producing an unsafe condition. At such CO_2 readings, a deficiency of air will create the formation of CO (carbon monoxide) in the flue gases. Therefore, when firing gas, test for CO to make certain that the burner is adjusted so that it has an "excess" rather than a "deficiency" of air. Carbon monoxide is a dangerous product of incomplete combustion, and is associated with combustion inefficiency and increased fuel cost.

18.2.9 Inspection and Servicing

Table 18.2.2 is a periodic chart listing all items to be periodically inspected.

18.2.10 Bibliography

HAGO Mfg. Co.—*Installation Manual*, 1993, Mountainside, NJ 07092.
Honeywell, Inc.—*Flame Safeguard Controls* (A Honeywell textbook), 1989, Golden Valley, MN 55422-3992.
North American Mfg.—*Combustion Handbook*, Vol. I, 3rd ed., 1988, Cleveland, OH 44105-5610.
Power Flame, Inc.—*Installation & Operation C Manual*, #C888 Rev 895, Parsons, KS 67357.

18.3

Heat Exchangers

David O. Seaward, P.E.

Vice President
After Sales & Service Division
Alfa Laval Thermal Inc.
Richmond, Virginia

18.3.1 Introduction

Heat exchangers have long been recognized as an effective means of transferring heat from one fluid to another while protecting delicate and costly HVAC system components from damaging contact with dirty and sometimes corrosive transfer fluids. Maintaining and operating the heat exchanger within its original design parameters and implementing an effective preventative O&M program will result in lower heat exchanger maintenance cost, better heat exchanger performance, reduced system down time and lower overall HVAC system operational cost.

The guidelines that follow are intended to help maximize those results.

18.3.2 Heat Exchanger Overview

Since its introduction in 1930, the plate heat exchanger (PHE) has gained wide acceptance as a viable means of transferring heat between liquids such as water to water and steam to water. In recent years its acceptance has spread to numerous fluid applications such as water to coolants in applications such as refrigerant and ammonium chiller systems. Typical applications include:

- Cooling Tower and Natural Cooling Water Isolation Applications where the PHE is used to transfer heat and isolate the HVAC system from the dirty, potentially fouling cooling liquids

- Free Cycle Cooling - Chiller By Pass Applications where the PHE is used to by pass the chiller, transferring heat from the condenser cooling media to a natural, free cooling source such as a local lake or river

- High Rise Pressure Interceptor Applications where the PHE is used in high rise building designs to create multiple pressure loops which allows the entire system to be designed for normal design pressures

- Thermal Storage Applications where the PHE is used in both the thermal charging and discharge modes

- Water Heating Applications where the PHE is used to transfer heat from an impure heat source to the pure water source

- District Heating and Cooling Applications where the PHE is used to reliably divide up the heating and cooling network with minimal energy loss

- Coolant Evaporation and Condensing Applications where the PHE is used as a compact means of enhancing refrigerant coolant evaporation and condensation

In all applications, the primary function of the plate heat exchanger (Fig. 18.3.1) is to create the optimum conditions for the transfer of

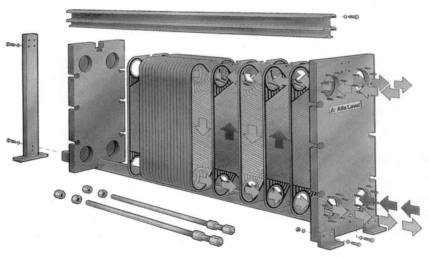

Figure 18.3.1 Typical Plate Heat Exchanger. (Courtesy of Alfa Laval Thermal, Inc.)

heat between two fluids. This is accomplished by passing the fluids through a series of thin, corrugated plates which are pressed together and sealed with gaskets in the traditional gasketed PHE, by welds and gaskets in a semi-welded PHE, or by brazing in a brazed heat exchanger (BHE). The fluids are pumped counter currently between the thin channels that are formed between the plates. Hot fluid flows on one side and the cold fluid on the other. The results are the effective transfer of heat between the fluids through the plates without inter-mixing of the fluids.

The advantages of plate heat exchangers over other types of heat exchangers, such as shell-and-tube exchangers, include:

- *Compact heat exchange:* One-sixth the weight and 1/3rd the foot print for the same heat load capacity

- *Greater transfer efficiency:* High heat transfer coefficients can provide transfer efficiencies in excess of 95 percent and can achieve 1°F(0.6°C) degree approaches between the process fluids

- *Flexible design:* Plates can be easily added to accommodate future expansions and changing load demands

- *Low maintenance cost:* Fast and easy cleaning and change out of parts, if needed.

- *Lower initial cost:* Low capital equipment cost and two-thirds less valuable building space required.

These advantages account for the plate heat exchangers ever growing popularity in HVAC and refrigeration type applications.

If properly operated and maintained, plate heat exchangers can provide many years of trouble free operation. The keys to this success lie in properly selecting and designing the exchanger for the application, operating it within its design limits and properly maintaining it. *A deficiency in any one of these areas will result in reduced thermal performance, reduced operating life and needless repair, maintenance and operational cost.* The impact can range from an increase in operating and maintenance cost to full system failure, loss of building comfort and sometimes loss of valuable, refrigerated products.

18.3.3   PHE and System Operation

All equipment should be operated and maintained in accordance with the instructions provided by the original equipment supplier. A preventative maintenance program which incorporates those recommendations is the most cost effective, long term means of realizing low maintenance and operating cost. It also minimizes overall system operational cost by maintaining the exchanger and associated system components operating at peak performance levels.

18.3.3.1 Start

Before initial start-up of the HVAC system, it (including the interconnecting piping) must be flushed of all foreign matter prior to final tie in of the exchanger and operation of the system. The same holds true if the system has been worked on or modified after initial start-up.

The high turbulence design of the plate heat exchanger resists fouling between the plates. However, a plate heat exchanger will filter out any solids that are larger then the spacing between the plates. Thus, any large materials left in the system will likely end up being filtered out by the heat exchanger.

Depending on the type and quantity of material, the consequences of using the exchanger as a filter can vary from no operational impact to total system performance loss. The loss of performance is typically due to the loss of heat transfer area resulting from the plugging of the inlet ports and/or the heat exchanger. Tell tale signs of pluggage include a lack of or reduction in thermal performance, high or increased pressure loss across the exchanger, no or reduced flow through the exchanger or a combination of all three. The system should be designed with an in-line basket filter or inlet port filter to prevent any large debris from entering the exchanger.

All systems should be equipped with pressure surge suppression devices and slow operating valves to protect the heat exchanger and other system components from the potentially devastating effects of pressure shocks. During start-up, fluid systems need to be charged slowly to prevent pressure shocks from occurring. Water hammer must be completely avoided. External leaks, reduction of thermal performance, internal leaks and intermixing of fluids are indicators of pressure surge damage.

Prior to start-up, the isolation valve on the inlets to the heat exchanger must be closed completely and those on the outlets and the exchanger vents, completely opened. Also, the inside distance between the exchanger pressure plate and the frame plate, the "A" dimension, should also be checked and verified against the name plate data or with the manufacture (Fig. 18.3.2).

Unless specified otherwise by the manufacture or the system operating procedures, the low pressure side of the exchanger should be charged first followed by charging of the high pressure side. Pumps should be started slowly and the exchanger inlet valves opened slowly after pump start-up to prevent pressure shocks.

The outlet vents should be left opened until all the air in the side of the exchanger which is being charged is vented. After which, the

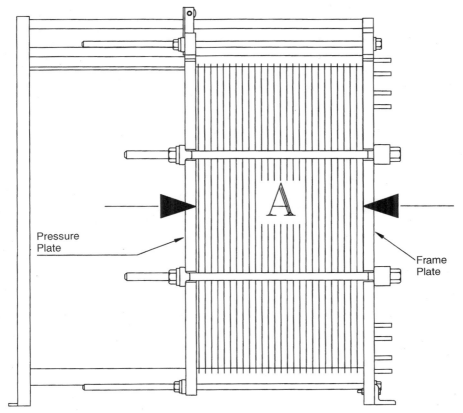

Figure 18.3.2 The "A" Dimension, the distance between the pressure and frame plate. (Courtesy of Alfa Laval Thermal, Inc.)

vent is closed and charging of the other side of the exchanger is completed in the same manner.

18.3.3.2 Shut down

Prior to shut down, review the system shut down procedures and those provided by the exchanger manufacture. Be sure to establish whether one side should be shut down before the other. Slowly close the exchanger inlet valve or the pump discharge valve for the side of the exchanger which will be shut down first. After the valve is completely closed, shut the pump down. Do the same for the other side.

If the exchanger is to be shut down for more then a few days, it should be drained. Depending upon the process fluids, it may be advisable to rinse and dry the exchanger as well.

Before draining, close the exchanger outlet valves to prevent the loss of fluid downstream of the valve. Once the outlet valve is closed, open the exchanger vent followed by slowly opening the PHE drain valve. If the unit is not equipped with drain valves, draining can be accomplished by disconnecting the lower PHE flange connection. The PHE vent connection can be used to introduce sequentially, rinse water and dry hot air when rinsing and drying is recommended.

18.3.3.3 Normal operation

During normal operation, the heat exchanger must also be protected against large solids and pressure surges. Solids build-up results in a loss of transfer surface and an associated loss of thermal performance. Pressure surges will result in damage plates, gaskets, weld seams and/or brazed seams which in turn results in loss of thermal performance, leaks and potential inter-mixing of fluids.

In open systems, such as cooling tower and cooling water isolation applications, the cooling liquid is typically dirty and may contain high, unpredictable solids levels and debris. To prevent loss of exchanger performance, the heat exchanger should be protected against large solids through the use of an in-line strainer or similar device. The strainer should be installed upstream of the heat exchanger to capture any large solids prior to introduction to the exchanger.

Any strainers or filters need to be inspected and cleaned on a regular basis (Fig. 18.3.3). Some strainers are supplied with automatic

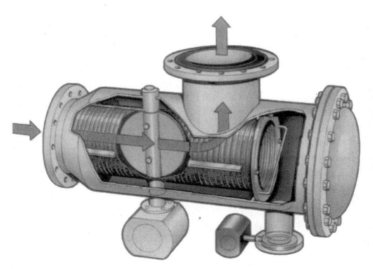

Figure 18.3.3 In-line strainer with automatic flushing and backflushing system. (a) Normal operation, (Alfa Strainer, courtesy of Alfa Laval Thermal, Inc.)

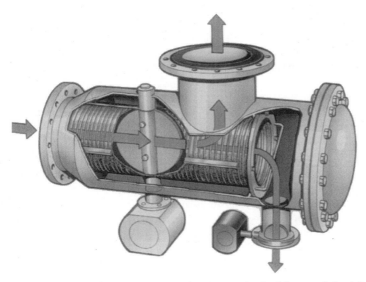

Figure 18.3.3 In line strainer with automatic flushing and backflushing system. (b) Flushing, (Alfa Strainer, courtesy of Alfa Laval Thermal, Inc.)

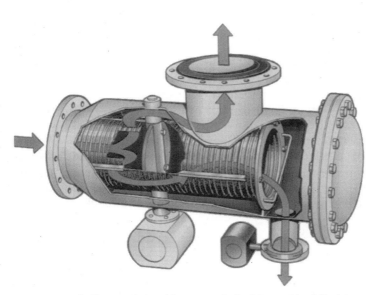

Figure 18.3.3 In line strainer with automatic flushing and backflushing system. (c) Backflushing. (Alfa Strainer, courtesy of Alfa Laval Thermal, Inc.)

flushing mechanisms which are either electronically or pneumatically operated. The mechanism periodically activates to flush the strainer with a small amount of process water to remove accumulated solids, limit the pressure drop build-up in the strainer and to maintain its overall effectiveness.

In high-solids applications, heat exchanger back flushing valves are also used to back flush the exchanger. The back flushing valve is hard-piped into the exchanger's inlet and outlet piping. It is used to reverse the flow of liquid through the exchanger, on an "as needed" basis, to flush out any materials which may have accumulated in the unit.

Pressure surges are one of most devastating operation upsets. As with any piece of equipment, the heat exchanger must be protected from these high energy surges. *Unless properly protected, the plates and gaskets in an exchanger can be irreparably damaged.* Internal and external leakage of fluids may occur, as well as cross contamination of fluids due to inter-leakage. Both conditions may require that gaskets, welded cassettes or whole brazed units be replaced. Plates can also be deformed and require replacement. In extremely severe situations, the complete plate pack may need replacement.

Operational changes must be made slowly and smoothly. Sudden starting and stopping of pumps and the quick opening or closing of valves can send instantaneous high energy shocks through the system. Quickly closing valves downstream of the exchanger can also send devastating shock waves back through the system. These shocks waves can damage the heat exchanger and other system components.

Pressure surges can also occur when using steam as in water heating applications. If the steam outlet of the heat exchanger is located on the top of the heat exchanger or if a steam trap has been installed above the elevation of the exchanger steam outlet, condensate will accumulate in the exchanger. When it does, severe and repeated hammering of the system and the exchanger occurs. The results are the same as with any other form of high pressure surge; plate deformation, gasket blowout, and leakage.

For steam applications, the steam outlet of the heat exchanger should always be located on the bottom of the exchanger and the outlet piping sloped away from the outlet to facilitate condensate drainage. A steam trap should also be located below the elevation of the exchanger outlet, to trap the condensate and avoid its back up into the exchanger.

18.3.4   PHE Maintenance

A heat exchanger is a static device which *can not* create pressure, pressure surges, energy, temperature, plugging solids or scale. If properly designed for the service and operated within its original design parameters, it will provide many years of trouble free, high efficiency service. When it is abused and operated outside of the conditions it was design for, problems may occur. When problems occur, it is important that the source of the problem be identified and eliminated as quickly as possible. Doing so will minimize the impact of the problem and prevent it from growing into a bigger, costly problem.

In addition to the operating procedures outlined above, simple preventative techniques such as periodic visual inspections, performance monitoring and an appropriate cleaning regimen can further help extend the operating life of the exchanger.

18.3.4.1 Inspection and monitoring

The exchanger and the area immediately adjacent to it should be inspected periodically for external signs of problems. Leaks, drips and/ or dry solids accumulated on the outside of the exchanger or the surrounding area are indications that there may be problems with the sealing system, the gaskets, the welded seams or the brazed seams.

The performance of the unit should also be monitored. Flow, pressure and temperature gauges or monitors should be installed in the inlet and outlet piping of the exchanger. These gauges should be checked on a regular basis and any changes noted. Changes in exchanger flow rates, temperature differences and changes in pressure are primary indicators that a problem may be developing. An operational log should be maintained so trends and operational changes in pressure, temperature and flow rate can be readily spotted.

Any problems or suspected problems should be reviewed with the exchanger manufacture or a qualified PHE service organization. This review should include the operating history of the exchanger, the overall system and the specifics of the performance variance. All are necessary to help identify the cause of the change and to develop recommendations to prevent the cause from happening again. Once the cause is isolated, an action plan needs to be developed and implemented to eliminate the problem and to repair any associated damage.

18.3.4.2 Cleaning

Depending upon the nature of the transfer fluids and the application, performance of the heat exchanger may degrade over time. This decline in performance is typically due to the build-up of scale, sediment and/or biological mass on the plates. Fouling of the exchanger manifests itself as a decrease in thermal performance, an increase in pressure drop across the exchanger and/or a reduction in the flow through the exchanger.

Two methods are currently available for cleaning the exchangers: the plates are either removed from the exchanger and mechanically cleaned or the plates are chemically cleaned while still installed in the exchanger. The first method can be done either at the site by or under the guidance of qualified field service personnel or off-site in a qualified PHE service center. Either way, the plates are removed from the exchanger and are cleaned external to the system.

An effective alternative to external cleaning is chemically cleaning the plates while installed in the exchanger. Cleaning In Place (CIP) is an economical method of maintaining the exchanger at peak performance and extending its operating life. It is recommended that regular CIP cleaning be included in a preventative maintenance program to maximize exchanger performance and minimize system operational cost and overall maintenance cost.

Under no circumstances should hydrochloric acid be used to clean stainless steel plates, nor should hydrofluoric acid be used to clean titanium plates. If these acids are used on these plates, the plates will be corroded and need to be replaced. Cleaning agents containing ammonia and organic acids such as nitric acid must not be used to clean copper brazed plate exchangers. Doing so will result in corrosion and pitting of the brazed joints and ultimate failure of the exchanger.

Quality water of known make-up should be used in preparing all cleaning agents. Water with a chlorine content of 300 PPM or higher must not be used.

In all cases, care must be taken to properly dispose of all materials used in the cleaning process which can sometimes complicate on-site cleaning.

18.3.4.3 On-site cleaning

On-site cleaning can also be cumbersome, messy and not necessarily the safest or most reliable alternative. The cleaning as well as the disassembly and reassemble of the exchanger should be done under the direction of a qualified field service engineer. If not, it should be done by a service organization which is specifically qualified in plate exchanger cleaning. The leading exchanger manufactures provide such services and should be consulted. They can tailor a service program to best suit your needs.

Nonmetallic brushes, high pressure washing and various cleaning agents can be used to clean the plates on-site. The combination used will depend on the nature and degree of fouling. Common cleaning agents for encrusted scales and sedimentation include:

- Hot water
- Nitric, sulfuric, citric or phosphoric acid
- Complexing agents such as EDTA or NTA
- Sodium Polyphosphates

It is recommenced that the concentration of these agents not exceed 4 percent and that a maximum temperature of 140°F (60°C) be used.

For biological growth and slime, alkaline cleaning agents such as sodium hydroxide and sodium carbonate are usually effective. Recommended maximum concentrations and temperatures for these agents is respectively, 4 percent and 176°F (80°C). Cleaning can sometimes be enhanced by the addition of small quantities of complex forming agents or surfactants.

Care must be taken during the cleaning process *not to damage the gaskets*. All gaskets should be thoroughly inspected after cleaning and any damaged gaskets replaced. If more then a couple of gaskets need to be replaced, *all* gaskets should be replaced to assure uniform gasket hardness, sealing force and extended operation. If the exchanger has been in operation for a number of years and/or the unit is opened frequently, all the gaskets should be replaced at the same time.

For glue-free, clip-on or snap-on gaskets (Fig. 18.3.4) make sure the gasket grove under the gasket is free of any foreign matter before installing the plates back in the unit. Regardless of gasket type, after hanging the plates in the exchanger, the gasket (and the groove it will

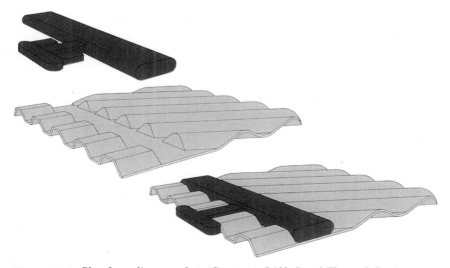

Figure 18.3.4 Glue free, clip-on gasket. (Courtesy of Alfa Laval Thermal, Inc.)

seat in) should be wiped down prior to tightening of the exchanger to remove any foreign matter that may have gotten on the gaskets and/ or grooves during hanging.

18.3.4.4   Off-site cleaning

A recommended cleaning alternative to on-site cleaning is to have the plates cleaned off-site in a qualified PHE service center. The well qualified service centers provided by leading PHE manufacturers are suitably equipped and staffed to cost-effectively clean, inspect and replace, as necessary, plates and gaskets. In those situations where the plates and gaskets need to be replaced, the manufacture's service centers can also provide plate and gasket failure analysis and subsequent recommendations on how to improve long term operation of the exchanger.

Qualified service centers should be used whenever gasketed plates are regasketed. Nondestructive gasket removal techniques such as the use of low temperature nitrogen are used to remove the gasket without damaging the valuable plates. Once the gaskets are removed, a combination of high pressure washing and chemical baths are used to clean the plates.

Using dye penetration techniques, the centers then inspect the plates for pin holes and cracks that would not otherwise be detected. Plates passing the dye penetrant inspection are fitted with gaskets using oven cured glues and two-part epoxy to assure a high quality, high strength fit. The reconditioned plates are shipped from the centers, in plate pack order to facilitate reinstallation in the frame and minimize the time required to do so at the site.

Where access has been provided for removal of the heat exchanger, the need to remove and reinstall the plates in the frame can be eliminated by sending the entire exchanger to the service centers for reconditioning and repair. The exchanger is inspected, disassembled, plates reconditioned and the exchanger reassembled. The exchanger can also be hydrotested at the center prior to it being shipped back to the site.

18.3.4.5   Cleaning in place (CIP)

An alternative to external plate cleaning is on-site cleaning of the plates while still installed in the exchanger. Cleaning In Place (CIP) is a procedure used to clean the heat exchanger while it is installed in the system without having to open the exchanger. CIP procedures are particularity effective in maintaining the performance of brazed exchangers and needs to be performed before the exchanger is highly fouled. During CIP cleaning, specific chemicals at prescribed temperatures, typically 140 to 170°F (60 to 77°C), are circulated at low velocities through the fouled side of the exchanger to chemically clean the unit of fouling deposits. Sometimes CIP methods are used to clean entire systems.

CIP cleaning is achieved through a number of mechanisms, namely:

- Chemical or partial dissolving of the deposit thru chelation or chemical dissolution

- Decomposition of the deposit containing sand, fiber or other foreign particles

- Reducing the adhesive force between the deposit and the plate.

- Killing the biological mass present in the mass

CIP cleaning is accomplished at low velocities, typically 10 to 20 percent of the exchanger's normal operating velocity. Thus it is important to place the exchanger back in service shortly after CIP cleaning to completely flush the unit of loosed deposits. If the exchanger is placed in stand-by service, the deposits loosed during CIP, may harden and reattach to the plates. In that case, the full benefit of the cleaning will not be achieved.

When there is only an occasional need to clean the exchanger, a portable CIP unit can be rented from some of the leading PHE Manufacturers, (Fig. 18.3.5). For installations where there are a large number of installed exchangers or where regular cleaning is required, purchased of a suitably sized, potable unit should be considered.

The CIP unit is connected to the inlet and outlet piping of the exchanger after shut down, isolation and drainage of the exchanger. The cleaning solution is circulated through the exchanger in reverse flow

Figure 18.3.5 Mobile CIP unit. (Courtesy of Alfa Laval Thermal, Inc.)

to the operating flow direction, at a prescribe temperature, until the scale has been loosened and removed. The type, concentration and temperature of chemicals to be used are dependent upon the type of scale to be removed.

To achieve the maximum benefit from CIP, the fouling deposit and circulating fluid should be analyzed to determine the best chemistry to be used. This analysis and CIP technical support is provide by some of the leading exchanger manufactures. These manufactures also provide, CIP chemicals, CIP units and field service specialists to assure that the methods and chemistry to be used are right for the application and exchanger.

18.3.4.6 Exchange plates and strat packs

A time saving external cleaning alternative that is offered by some of the leading exchanger manufactures is an exchange plate program. In the exchange plate program, reconditioned plates of the same model and materials as those in the exchanger are sent to the site and swapped out for the existing plates. The existing plates are then sent to the local center for cleaning and replenishing of the exchange plate inventory. In this approach, downtime is limited to the time it takes to change out the plate pack.

In critical applications where the time to receive the exchange plates can not be tolerated, it may be advisable to have a "Strat Pack," plate pack on-site. A Strat Pack is made up of enough plates to change out the complete plate pack for one of the critical heat exchangers at the site. By having the Strat Pack on site, the pack can be change out as soon as the need arises and the existing plates cleaned later at a more convenient time.

18.3.5 PHE Gasket and Plate Protection

If operated within design conditions and not abused by improper operation and maintenance, an exchanger should provide many years of trouble-free, high-efficiency heat transfer. A heat exchanger should *never be opened except for repair or cleaning*. The fewer times it is opened the longer the gaskets will last before needing to be replaced and the lower the potential the plates will be damaged.

18.3.5.1 Gaskets

In most cases, external leaks are a sure sign of gasket failure. Intermixing of transfer fluids is a possible sign of gasket failure but because of the design of the gasket system, it more commonly results from a plate problem. The most common causes for gasket failure include:

- Gasket blow-out
- Crushing
- Permanent set in old gaskets

In all three failure modes, the affected gasket(s) need to be replaced. In the case of gasket blow-out and crushing, the cause of the failure needs to be established and eliminated. Gasket blow-out is commonly caused by the exchanger not being tightened enough or by pressure surges. Both causes must be eliminated to protect the exchanger and other system components.

Gasket crushing can be caused by swelling of the gasket or plate misalignment. If swelling is suspected, compatibility of the gasket material with the transfer fluid should be checked to make sure the right gasket material is being used. The composition of the transfer fluid should be checked against the design specs to establish whether there have been any changes. Small changes in some components, even trace amounts, can result in swelling and failure of the gasket.

Crushing of gaskets due to plate misalignment is typically caused by failure to install the plates correctly after cleaning, damaged hangers positioning the plate incorrectly during tightening and/or installation of deformed plates. The first can be avoided by taking care during the installation process, using qualified personnel to reassemble the exchanger.

During regasketing and prior to reinstalling any plates, the hangers should be inspected. Any damaged hangers need to be repaired or replaced. Care should also be taken not to damage the hangers during the hanging of the plates and exchanger tightening process.

Plate deformation, particularly in or around the gasket grove, can cause crushing of adjacent gaskets, plate misalignment and leakage. In most cases plate deformation can not be repaired and the plate and effected gaskets have to be replaced.

With time, all gaskets take on a permanent set, which causes a reduction in sealing force. The decline in sealing force usually takes place slowly, over many years. This decline in sealing force is a result of natural oxidation and/or chemical attack. It is also dependent upon the system operating conditions, pressure, temperature and thermal cycling. The greater the extreme the quicker the decline.

Gasket service life varies with each application. Replacement of the gaskets should be considered every time an exchanger is opened, particularly in high temperature, high pressure applications; if the exchanger has been in operation for a number of years or the exchanger is frequently opened.

Because of the difference in gasket sealing force between an old and a new gasket, the mixing of old and new gaskets should be avoided.

Replacing all the gaskets at the same time assures a good, long lasting seal. Planned regasketing of the PHE prior to gasket failure should be included in all maintenance programs for gasketed exchangers.

18.3.5.2 Plates

Intermixing of transfer fluids is often a sign of plate failure. The most common forms of plate problems include:

- Corrosion
- Erosion
- Fatigue
- Deformation

Corrosion is caused by the presence or concentration of corrosive components in the transfer fluids. Examples are: chlorine attack of stainless steel, fluoride attack of titanium and ammonia attack of copper in a copper brazed heat exchanger. The most common forms of corrosion are pitting, crevice and stress.

The rate of corrosion is primarily dependent upon the concentration of the corrosive component, the fluid pH and fluid temperature. Plate corrosion is generally more predominant in the high temperature areas of the plate such as in the inlet and flow distribution area.

Chloride corrosion is directly proportional to the chloride concentration, the fluid temperature and inversely proportional to the fluid pH. Chloride corrosion of stainless steel becomes more aggressive with the accumulation of materials on the plate such as calcium carbonate and silica scale. As the oxygen levels under fouling deposits decreases, the corrosion rate accelerates.

Erosion of plates can take place if high concentrations of an abrasive solid such as sand is present in the transfer fluids. Tell tale signs of erosion are a smooth, shiny finish on the plates, typically in the inlet neck and distribution area. In severe cases the inlet port of the plate can be eroded away, which can lead to leakage.

The best way to prevent erosion is to eliminate the abrasive solids from the transfer fluid using a strainer or similar device. If erosion is encountered, the fluid velocity through the inlet port and neck area of the plate should also be reviewed by a qualified service engineer. Adjustments in the fluid flow rate may be necessary to reduce the velocity in these high velocity areas by either reducing the fluid flow and/or by increasing the number of plates in the exchanger.

Repeated flexing of the plate will result in fatigue cracks and interleakage of the transfer fluids. The most common cause of fatigue cracking is due to subjecting the plates to frequent and regular pulsations. Fatigue-causing pulsations are typically associated with piston and chemical dosage pumps or reoccurring water hammer. The best means of preventing this form of fatigue is by protecting the exchanger and other system components from the pulsations.

Another potential cause of fatigue in plates is tightening of the exchanger to an improper "A" dimension, the inside distance between the frame plate and pressure plate of the exchanger (refer back to Fig. 18.3.2). Plates are designed to have metal-to-metal contact once tightened to the proper "A" dimension. This metal-to-metal contact strengthens and helps support the plate to minimize flexing and vibration during operation and assure proper sealing by partially encapsulating the gaskets on gasketed exchangers. Whenever the exchanger is opened, it is important that the plate pack be retightened to the proper "A" dimension designated by the manufacturer. If the plates are too loose, premature failure of the plates and blow-out of gaskets resulting in leaks are likely results.

It is just as important not to over tighten the exchanger. Tightening the plate pack down to a dimension which is shorter than 1 percent below the specified "A" dimension may result in gasket crushing and plate deformation. Leaks may result and plates and gaskets will likely need to be replaced.

Care must also be taken when reinstalling plates to make sure the plate hanger has not been damaged and the plates hang properly in the frame. The hanger serves to position the plate so that it seats properly against the next plate's sealing surface. In doing so, the plates and gasket are suitably positioned and compressed preventing plate deformation and leaks.

Another cause of plate deformation is high pressure surges such as water hammer. A sudden high pressure surge can render a plate useless. The best prevention is to protect the exchanger and other system components from pressure surges and pulsations, as noted earlier.

Once deformed, a plate is rarely salvageable and needs to be replaced. Exceptions are plates manufactured from high cost, exotic materials such as titanium. Depending upon the extent of the damage, it may be cost effective to have the plates re-pressed. Due to the high cost of re-pressing, it is typically less expensive to replace standard type 304 and 316 stainless steels with new plates then to have them re-pressed. This option and the cause of the deformation should be discussed with the exchanger manufacture and the cause of the deformation eliminated.

There is no reliable means of repairing corroded, eroded or fatigued plates. The only way to handle this kind of plate problem is to eliminate the cause of the problem and replace the damaged plate(s). If plates are replaced without identifying the cause and correcting it, money should be budgeted for the future replacement of the plates.

18.3.6   Shell and Tube Heat Exchanger Operation and Maintenance

In many ways the operation and maintenance procedures of a shell and tube (S&T) heat exchanger are similar to those of a PHE. Both are started-up, operated and shut-down in the same manner and the same operating precautions such as protecting the exchanger from pressure shocks apply. As with a PHE, a S&T exchanger should not be operated outside the operating conditions originally specified without prior consultation with the exchanger manufacturer. In all cases, the S&T exchanger should be operated and maintained in accordance with the instructions provided by the original equipment supplier.

18.3.6.1   Operation

During start-up, the flow of the coldest fluid should be established first. After which the hot fluid should be gradually introduced to prevent thermal shock. All vents should remain open until all air has been purged from both sides.

In shutting the exchanger down, the flow of the hot fluid should first be reduced gradually, followed by the systematic shut-down of the cold side. The speed with which each side is started-up or shutdown is depend upon the nature of the fluids involved, operating conditions and the differential temperatures and pressures.

If the fluids are corrosive or susceptible to freezing the exchanger should be drained completely after shut-down. All fluids drained from the exchanger should be recovered, stored or disposed of properly.

Prior to start-up and during operation, the external bolts should be checked to assure they are tightened to the proper specification. Bolted joints should be tightened uniformly in a diametrically staggered pattern to the torque values specified by the original equipment manufacturer. Overtightening may damage the gaskets resulting in leaks.

18.3.6.2   Maintenance

The interior and exterior of the exchanger should be inspected on a regular basis. The exterior shell, flanges, bonnets and hub should be inspected for signs of damage such as, dents, bulges, stress marks and corrosion. Gaskets should be checked for leaks and displacement. Instrumentation should be checked and operating conditions noted for comparison with previous reading. All variations and problems should be noted and corrected.

Interior inspections should be attempted only after the unit has been fully drained and allowed to cool. A visual inspection of the tube ends for thinning, nozzle threads for damage, and erosion of the tubes should be done during the interior inspection. One or all of these conditions could be a indication of high fluid velocities, excessive particles entrained in the fluid and/or corrosion. Depending upon the damage found, the unit may need to be replaced.

Where removable bundles are used, removal and replacement of the damaged bundle should be done by a qualified service organization. Similarly, the detection and repair of tube joint leaks and tube splits, as well as the internal cleaning of the exchanger should be done by a qualified service organization.

Mechanical cleaning of the tubes may be required to restore the operating efficiency of the exchanger. Rotary, nonmetallic, electric or pneumatically driven brushes should be used to mechanically loosen tube deposits and scale. Use of metallic brushes may damage the interior surface of the tubes, which could accelerate tube corrosion and failure. Once brushed, a tube should be flushed with clean water to remove lose deposits and scale.

During the re-assembly of the exchanger, all gaskets should be replaced and all bolts tightened per the manufacturer's recommendations. Care should be taken to insure that all gaskets and seals are properly positioned and that the bolts are not overtightened to prevent gasket crushing and leaks.

In some applications, the exchanger can be chemically cleaned without having to disassemble the exchanger. The cleaning in place (CIP) techniques described above for a PHE can also be used on S&T exchangers. CIP cleaning should be performed on a regular basis prior to the tubes becoming heavily fouled and in accordance with the recommendations provided by the CIP specialist.

The use of in-line strainers and back-flushing valves as described above for PHEs may also have a beneficial effect on S&T operation and maintenance. Each should be evaluated on a case by case basis and incorporated into the overall system layout and designed.

19.1

Refrigerant Management and Maintenance

Don Batz
Mike Batz
Comm Air Mechanical Services Co., Oakland, CA

19.1.1 Introduction

One of the most pressing issues facing the refrigerant user today is the maintenance and management of the refrigerant itself. This issue reaches not only the contractors, but also the end users, whether as the actual equipment owner or a property management provider. The economics of today's refrigerant issues have forced the user to consider its maintenance and management long before the equipment is turned on. The choices of manufacturers, equipment, refrigerant, and installation practices all must be considered in the development of a refrigerant maintenance program. The development of a refrigerant maintenance program will involve several steps that all become critical in implementing an effective program. The major steps in this development process are: 1) evaluating the goals of the refrigerant maintenance program; 2) selecting the equipment, installation, and maintenance procedures consistent with the goals of the program; 3) writing out and communicating the plan to all involved; 4) enforcing the maintenance program, and finally; 5) reevaluating the program regularly.

19.1.2 Evaluating Goals

Evaluating the goals of the refrigerant maintenance program is the first and most critical step in establishing a successful program. The demands of not only government and environmental regulation, but also determining the long-term goals of the program help set the agenda for success. A number of factors that will influence the process from a regulation standpoint are the effects of future regulation, refrigerant supplies and cost of replacement. It is well advised that each end-user utilizes all sources available for information on government trends and industry lobby efforts. The uncertain nature of future additional regulation and/or taxes requires diligence in monitoring this arena. The factoring of long term goals in relation to cost and capitalization of replacement equipment, expected life of equipment and cost of refrigerant maintenance procedures versus replacement of all must be evaluated according to each individual user requirements. The choice of service contractors, and consultants in the case of a property manager/building owner, or the levels of training and equipping of a contractor must be considered to successfully develop a workable program.

19.1.3   Selection Equipment and Procedures

Selecting the equipment, installation and maintenance procedures consistent with the goals is the next step in a successful program. Each individual owner, manager or contractor should develop a written procedure for the equipment and refrigerants being dealt with. This should include the procedures an owner/manager will expect from not only their own staff but also from outside contractors and consultants. This must include a method of notification of work to be done, report of any refrigerant system problems and a detailed record of refrigerant additions or losses, with causes. The days of sending a contractor to the roof and signing the tag on the way out are long gone. A higher level of expectation and communication must be demanded and received to protect the refrigerant resources. This communication also will build a greater understanding of the factors that influence the refrigeration equipment and maintenance. Each contractor should develop a training program of procedures to be followed with work on the refrigerant system. The potential liability and losses when an accidental release occurs has risen to the point that a contractor can not ignore this critical area. Together with the proper record keeping of maintenance practices and constant training, a contractor can become an ally to the building owner/manager in the refrigerant maintenance program. In considering the installation of new equipment, consideration should be made of the level of refrigerant containment desired. The cost is typically much less if the additional maintenance tools are built in during new construction, rather than retrofitted later. These tools include isolation valving, transfer vessels, relief containment measures and high efficiency purges on larger equipment. If these are ordered and built in when the equipment is ordered new, vs. retrofitted in the field, a substantial savings can be realized. This planning of procedures when weighed and balanced by the long term goals of the program sets the stage for success.

19.1.4   Communicating the Plan

Writing out and communicating the plan to all involved is the third step in developing process. Establishing the processes and reporting procedures in good, but useless if not communicated to all that come into contact with the equipment. Whether these be the owner/managers staff or the contractors, letting all know that refrigerant conservation is the heart of the maintenance program is critical. Writing these steps and goals expected, and communicating them to all levels is critical. The plan should be simple but comprehensive. Feedback on program improvement should be encouraged. It is a wise idea to post these refrigerant maintenance procedures and goals in the equipment areas or roof access and it is a good way of communicating the seriousness of the program to the service personnel as they work. Be specific in the procedures, simple but direct.

19.1.5   Implementing the Maintenance Program

Enforcing the maintenance program is where the rubber meets the road. A user must be proactive in monitoring compliance with the goals of the program. If the procedures require acid testing, purge maintenance and use of clean recovery equipment/drums, for example, then follow up during a service visit and spot check the methods of work followed. It is, after all, the asset of the end user that is at potential risk for loss or contamination. If you consider the cost of replacement of 100 percent of R-12 refrigerant due to mixing of refrigerants from the use of dirty recovery drums, then 15–20 minutes of time to visit with and observe the contractor or maintenance staff becomes a good investment. That time spent observing and evaluating the programs effectiveness allows constant monitoring of the success of the methods chosen.

19.1.6   Program Reevaluation

Reevaluating the program is the step that separates programs with one time success from a program that continues to improve over time and remain consistent in the face of changing situations. The gathered input of service personnel, owners, consultants and others involved in the refrigerant maintenance program allow constant tuning to ever-changing situations. For example, an owner may decide that 5 years of use of the existing equipment will be required before replacement can be considered. This may be due to the lack of capital funds, future planning of remodeling the building, or expectations that the loads in the building may change significantly in the future. The existing equipment is mechanically sound, so evaluation will center on refrigerant maintenance. The program will include regasketing, adding purge containment, increasing acid testing, and requiring all service personnel to follow the guidelines chosen. Reporting of any venting or releases and work requiring opening of the refrigerant system will be required if two years into the program the internal motor burns out and the refrigerant is contaminated. It may make economic sense to opt for change-out than to rebuild, due to the loss of the refrigerant charge and the mechanical damage.

19.1.7   Setting Up the Program

As you start to set up a refrigerant management and maintenance program, there are many items that must be taken into consideration. Start with the main information needed to give you the basis for a program that can work for you:

1. *Equipment Type:*
   - New high efficient packaged equipment with environmentally friendly refrigerant.
   - New built-up system with environmentally friendly refrigerant.
   - Existing equipment with currently manufactured refrigerant (HCFC — hydrochlorofluorocarbon).
   - Existing equipment with CFC (chlorofluorocarbon) refrigerant.
   - Existing equipment converted to one of the new alternate refrigerants.

2. *Type of System*
   - Package units
   - Built up system
   - Chilled water
     - Reciprocating
     - Centrifugal
     - Screw
     - Absorption
   - Thermal storage

3. *Type and Amount of Refrigerant*

	high pressure	low pressure
a. CFC-lbs (kg)	_____	_____
b. HCFC-lbs (kg)	_____	_____
c. HFC-lbs (kg)	_____	_____
d. Alternates-lbs (kg)	_____	_____

4. *Type and Amount of Oil*

- Mineral oil - gals (ltrs)   _____
- Alkylbenzene oil- gals (ltrs)   _____
- Polyester oil - gals (ltrs)   _____

5. *Piping and Accessories*

- Driers   #_____ Type   _____
- Sight glass
  - Standard
  - Moisture indicating
- Relief valve
  - standard relief
  - Manifolded double relief
  - rupture disk
  - Non-fragmenting double-resetting valve
- Purge unit
  - Standard purge
  - High efficiency purge
- Hot water pressure pack
- Electric pressure blanket
- Quick disconnect service valves
- External oil filters with isolation valves
- Oil reclaim educator
- Oil quality monitor
- Receivers
- Pump out systems
- Recovery and recycle units
- Refrigerant monitoring system
- Record keeping system
- Certified technicians
  - In-house
  - Outside contractor

After completing this inventory of your equipment, systems and other resources, put together your plan to manage and maintain your current supply of refrigerant and to budget ahead for replacement of equipment or conversion of units using CFC's or HCFC's as the need arises. (Most people haven't realized that with a large unit containing

800 (360 kg) to 1000 pounds (450 kg) of C.F.C. 11 or 12 that within a few years it could cost between $50,000 to $100,000 (1995 dollars) just to recharge a unit if the charge was lost or badly contaminated). Set up a plan and then work the plan to save a lot of headaches.

Basic refrigerant management and maintenance just requires using common sense. After you have completed your inventory as mentioned before, it is time to start your program.

1. Start a history card for each piece of equipment and system. This should include the following:
    - Make and model of unit.
    - Capacity of unit.
    - Start-up date of unit.
    - Warranty status/date.
    - Type of refrigerant.
    - Amount of refrigerant in system.
    - Location of equipment.
    - Complete service records as performed including date and technician.

2. All systems should be clearly marked as to type of refrigerant and amount of charge. These markings should be visible at any place that someone can access the system.

3. Refrigerant in larger systems should be checked for moisture or acid at least annually.

4. Oil should be checked for contaminants and wear metals at least annually.

5. Refrigerant pressures and temperatures should be checked on a regular basis.

6. Systems should be checked for visible signs of leaks on a regular basis. Repairs should be made immediately when a leak is discovered.

7. Complete leak check of system with a compatible leak detector should be accomplished annually.

8. Driers should be replaced any time a system is opened for service or repair.

9. Consider installing refrigerant monitoring equipment in equipment rooms. This will give you 24-hour leak checking.

10. Make certain that any technician working on your equipment is trained and certified to work on your units.

11. Take every effort possible to be certain that you don't get cross contamination of refrigerants in your units. Be careful especially when using recovery equipment.

12. Plan any conversions to alternate refrigerants to coincide with annual teardown or when a major failure occurs. Do your homework ahead of time.

13. If you do convert a system to one of the new refrigerants, plan to have the old refrigerant reclaimed and then save for use on other equipment that may use that type of refrigerant.

14. Make sure that you keep records of refrigerant purchases and usages in case you are ever audited by the EPA.

15. Whenever you add or replace equipment, make sure that you look at system efficiency, type of refrigerant and serviceability of the unit before you purchase a new piece of equipment.

16. Install refrigerant containment accessories on existing equipment whenever possible.

You could probably add to the list suggested above for starting of your refrigerant management and maintenance program. *Now* is the time to start your program.

19.1.8 Information Sources

Refrigerant technology is continually changing. Contact the following organizations and publications to obtain important up-to-date information concerning refrigerants:

1. Refrigerant manufacturer's catalogs and application manuals

2. Equipment manufacturer's catalogs and application manuals

3. American Society of Heating Ventilating and Refrigerating Engineers (ASHRA) 1791 Tullie Circle, Atlanta, GA 30329

4. Mechanical Service Contractors of America (M.S.C.A.) 1385 Piccard Drive, Rockville, MD 20850

5. Building Owners & Managers Association (B.O.M.A.) 1201 New York Av., N.W. Ste 300, Washington, DC 20005

Compressors and Condensing Units

Billy Langley, Ed.D., CMS
Consulting Engineer, Azle, Texas

Compressors and condensing units are usually one unit. Understanding how each fits into the overall design and operation of the air conditioning system is important to understanding how to service and maintain them.

Compressors: The purpose of a compressor is to move the refrigerant through the system by causing a difference in pressure between the high side and low side. This pressure difference is also aided by the flow-control device mounted on the indoor coil, or evaporator.

Several types of refrigeration compressors are used; the most popular are the reciprocating, scroll, and rotary.

Currently the reciprocating is the most popular in most all but very large units. The scroll compressor is gaining popularity in the smaller units (up to 5 tons capacity). Rotary compressors are popular in domestic refrigerators and freezers.

The reciprocating and rotary types each have a large segment of the domestic part of the industry.

Heat: The enemy

Heat is the biggest enemy of compressors, especially hermetic and semi-hermetic types.

---

*Reprinted by permission from the Air Conditioning, Heating and Refrigeration News. Copyright © 1996.

Heat can cause the motor windings to overheat and the compressor lubricating oil to break down, causing a mechanical failure. Should there be any moisture inside the refrigerant circuit, heat will increase its effects by each degree of temperature increase.

The heat will cause the moisture to change into acids that will attack every part in the system. This is indicated by motor windings breaking down and copper plating on all the steel components.

Copper plating will cause the close tolerances to change, affecting system operation.

An overheated compressor can be caused by several circumstances, such as:

- A dirty or scaled condenser;
- Shortage of refrigerant;
- Overcharge of refrigerant;
- Too high superheat setting on the flow-control device;
- Low or high voltage;
- Dirty evaporator;
- Low load; and
- Dirty filters (in an air conditioning system).

Clean condenser

A dirty condenser must be cleaned before the compressor and system will function properly again.

When an air-cooled condenser is used, the coil may be washed out with a high-pressure nozzle on a garden hose.

Occasionally a non-acid cleaning solvent must be used to remove all the dirt and oil that have accumulated on the surface. Be sure to protect all the electrical components from the water by placing a plastic sheet over the condenser fan motor and the electrical compartment.

Water-cooled condensers have a tendency to collect scale from the water on the heat-transfer surfaces.

This type of condenser acts much like a tea kettle. When water is boiled in it, some of the impurities in the water are left on the surface of the tea kettle as scale. This scale must be periodically removed from the heat-transfer surface so that the system will maintain its efficiency and keep the compressor running as cool as possible.

Care must be exercised when removing this scale because an acidic chemical is used. When too much of the chemical is used, the condenser refrigerant tubes could possibly be damaged.

When the scale is extremely thick, it must be removed with wire brushes. The brushing procedure requires that the condenser shell be opened. Therefore, it is desirable to remove the scale more frequently, to prevent a heavy buildup.

Hermetic and semi-hermetic compressors are cooled by the returning refrigerant. The motor windings are cooled as the refrigerant flows over them.

Should the system lose part, or all, of its refrigerant charge, the compressor will not receive the much-needed cooling. The refrigerant vapor will already have more superheat than the system is designed to have.

Therefore, the leak must be found, repaired, and the system recharged to the proper amount.

An overcharge of refrigerant will cause the compressor to pump against a higher-than-normal head pressure. This higher pressure causes the compressor to draw excessive current, producing more heat inside the motor housing.

An overcharge is usually indicated by a higher-than-normal head pressure and a higher-than-normal current draw. Also, the system will not be cooling as designed.

The refrigerant must be properly removed from the system until the correct charge remains.

When the flow-control device is set for too high a superheat setting, the refrigerant returning to the compressor will be too warm to cool the compressor properly.

Too high a superheat setting will generally be indicated by a lower-than-normal suction pressure, the suction line will be warmer than normal, and the discharge line will be warmer than normal. The superheat setting must be changed to the correct temperature before the compressor will operate at its design temperature.

When a non-adjustable orifice such as a capillary tube is used for flow control, check to make certain it is not partially plugged. Then check the system charge and bring it to the proper amount.

Voltage problems

Either low or high voltage will cause the compressor to draw excessive current. This in turn causes the compressor motor to run warmer than normal.

After a time, the motor will probably burn out. The voltage condition must be corrected to prevent motor and other component damage.

An under-loaded evaporator will also cause the compressor motor to run warmer than normal. This condition may be caused by several things:

- A dirty filter (in an air-conditioning system);
- Dirt collected on the surface of the evaporator coil; and

■ A lack of products (in a commercial refrigeration cabinet).

These conditions will cause the evaporator to ice and frost to the point that no, or little, refrigerant is being vaporized and is being held in the evaporator. The returning refrigerant is warmer than required to cool the compressor.

The condition causing the problem must be found and corrected.

Compressors are like any other system component. They must be sized for the load they are to handle.

A compressor that's too small will not remove the refrigerant from the evaporator to produce the temperature required for proper operation.

A compressor that is too large will cause a lower-than-normal suction pressure. The evaporator will probably ice over, reducing heat transfers to the refrigerant.

Compressors require lubrication for proper operation. System operation usually determines the type to be used.

Some newer refrigerants require special lubricants to operate properly. Follow the compressor manufacturer's recommendations when adding to or changing the lubricant.

Condensing units.   The purpose of the condensing unit is to house high-side system components.

Generally, they are installed outside at a place where they will not interfere with the normal use of adjacent areas.

Two types of condensing units are the air-cooled and water-cooled. The particular type used will depend on the size and use of the equipment.

Most of the larger air conditioning units use water-cooled condensing units because they use less electricity than air-cooled units.

Condensing units usually contain the compressor; the condenser coil; an accumulator when used; a liquid receiver when used; condenser fan and motor; and controls for the condensing unit and its components, such as the contactor, pressure switches, fan speed control, reversing valve on heat pump units, and defrost controls on heat pump systems.

Keeping it clean

To keep the system operating at peak efficiency, certain maintenance procedures must be occasionally performed.

The condenser coil must be cleaned periodically to remove the accumulation of dirt and oil from the surface. This can be done with a high-pressure nozzle on a garden hose. Sometimes a non-acid cleaner will be needed. Be sure to keep the water out of any motors and electrical panels.

The condenser fins will sometimes get bent and should be straightened. This is usually best done with a fin comb (available at most supply houses).

Bent fins will restrict the airflow through the coil and reduce heat transfer from the refrigerant to the air. Any reduced heat transfer at this point will cause higher-than-normal head pressure and higher-than-normal current draw.

Contactor contacts also become dirty and need to be cleaned. This should be done carefully to prevent damage to the contact surface.

When the contacts need to be filed, replacement is usually more desirable; the filed contacts will only become pitted in a short time, requiring replacement.

All debris must be kept from inside the condensing unit. When debris is left inside, it will generally find its way to the coil and cause high head pressures.

Occasionally it may get into the condenser fan, causing damage to the motor and fan blade.

Installation advice

Flowers, shrubs, and other plants that shed their leaves should be grown a reasonable distance from the condensing unit. Otherwise the leaves could collect on the condenser coil surface, preventing it from properly cooling the refrigerant.

Condensing units should be installed so that the prevailing wind will help the condenser fan motor in cooling the condenser. When the fan and prevailing wind are opposed to each other,the condenser will not be cooled as it should be.

When possible, place the condensing unit away from patios and bedroom windows. The noise may interfere with normal usage of these places.

Always install the condensing unit with the access panel where the technician can easily remove it for service.

Also, the technician must be able to easily reach the components inside to check and replace them. Otherwise, any repairs or service procedures will take longer than normal.

Compressor, condensing unit maintenance do's and don't's

Do:

- Keep coils clean of dirt and scale buildup.
- Make sure the system has the correct refrigerant charge.

- Make sure the flow-control device has the correct superheat setting.
- For a non-adjustable flow-control orifice, such as a capillary tube, make sure it is not partially plugged.
- Make sure the unit is receiving the proper voltage.
- Make sure filters are changed regularly.
- Make sure the unit is the correct size for the application.
- Make sure the unit has proper lubrication.
- Straighten bent fins.
- Clean the contactor contacts.
- Install the condensing unit with the access panel where the technician can easily remove it for service.

Don't:

- File the contactor contacts; it's a waste of your time and the customer's money. Just replace them.
- Let debris get inside the condensing unit; it generally finds its way to the coil and can cause high head pressures.
- Grow flowers, shrubs, and other plants that shed their leaves too close to the condensing unit.
- Install the condensing unit so that the fan and prevailing wind are opposed to each other; the condenser will not be cooled as it should be.
- Place the condensing unit near patios and bedroom windows; the noise may be unwelcome.

19.3

Centrifugal Chillers

George White, P.E.
Aggreko, Inc.
Benicia, California

19.3.1 Introduction

In today's industrialized world, air conditioning is no longer a luxury; it is a necessity. Businesses are dependent on both comfort cooling and process cooling. Most of the large air conditioning applications (high-rise office buildings, hospitals, universities, industrial facilities, manufacturing plants, chemical plants) involve centrifugal chillers. Centrifugal chillers have been used for heat loads of over 100 tons because of their reliability and performance. With the use of computerized design and manufacturing techniques among other tools, centrifugal chillers have become even more reliable and efficient. This trend of increased reliability should continue.

Most of these facilities are constructed by a General Contractor, with the mechanical equipment being purchased through the Mechanical Contractor. A great deal of attention is focused on the price of the equipment. The more enlightened client will take this process one step further and consider the *total life cycle cost*. The life cycle analysis is important because the operating expenses (maintenance, service, repair, utilities) over the life of the equipment can be as much as 10 to 100 times the initial installed cost of the equipment. If risk and/or down time is considered, the figure could even be orders of magnitude larger.

In this chapter we will address philosophies, strategies, and tasks to manage operating costs specifically related to centrifugal chillers and large chilled water plants. The proven results of these strategies are reduced downtime, increased reliability, full design capacity, and maximum efficiency.

In section 19.3.2 we will discuss the various types of centrifugal chillers. We will also compare and contrast the components and materials that comprise the different styles and models of these chillers. At the end of the section we will describe a process that we have been using over the past few years in office buildings, hotels, sports arenas, manufacturing plants, hospitals, and universities to reduce operating and maintenance costs.

In section 19.3.3 we will describe the various factors that affect reliability in centrifugal chillers. Reliability involves keeping the machine operating to meet the desired load. The reliability of a chiller is improved through maintenance and a good maintenance program can pay for itself through cost avoidance and downtime prevention. Energy savings provide additional cost reduction as a collateral benefit of a good preventive maintenance program.

In section 19.3.4, we will discuss machine performance. Once a machine is running reliably, we can then start to look at its performance. The two measures of performance are capacity and efficiency. Capacity is the total refrigeration capacity of the machine. It is not unreasonable to expect a chiller that is 10, 20, or even 30 years old to operate at or near its full load design capacity, *if it has been properly maintained.* Once the machine is operating at its maximum capacity, an equipment owner can check for the second measure of performance: efficiency. Efficiency is often expressed as the electrical cost (kilowatts) per ton of air conditioning delivered. Even though a chiller could be meeting the load requirements, it could also be consuming 20 percent or 30 percent more energy to deliver the design tonnage—therefore wasting money. We will document specific methods and processes to identify and improve machine inefficiencies.

In section 19.3.5 we will document specific upgrades that allow the equipment owner to squeeze additional reliability, capacity, or efficiency from a chiller plant. Equipment manufacturers are continually creating new products for use on existing systems. We would recommend these upgrades and improvements for chillers and central plants that are running reliably while efficiently meeting the capacity requirements.

In section 19.3.6, we will list criteria useful in repair versus replace decisions. These decisions usually surface before or after expensive repairs. There are *external* factors to consider: government regulations, industry codes, utility costs, and interest rates. There are also *internal* factors to consider: building use, future needs, the availability of funds, internal rate of return, and other financial criteria.

In this era of "Do more with less," it is a challenge to balance cutting costs while maintaining performance. While the cost and risk associated with operating centrifugal chillers have increased, the quantity and quality of resources available to equipment operators has also increased. While we may not be able change environmental regulations or macroeconomic conditions, there are a multitude of tools, tasks, and resources readily available that we can use to keep mechanical equipment functioning reliably while reducing overall maintenance expenses. Implementing the procedures in this chapter is a good start toward improving your central plant performance.

19.3.2 Components and Materials

In this section, we will describe the common components and materials that comprise centrifugal chillers. We will describe what each part or component is, what it does, and how it can fail or cause other parts and components to fail. Prediction and prevention of these failures will be addressed in a later section on reliability.

19.3.2.1 Refrigerant

First, a word concerning the refrigerant issue. Since December 31, 1995, CFC-based refrigerants are no longer manufactured in the United States, with some exceptions. Even with the phaseout, some equipment owners have chosen the strategy of containment to delay making a choice between converting or replacing their chillers. Regardless of the refrigerant class, for maintenance purposes there are two types of refrigerants: *high pressure* and *low pressure.*

For low pressure refrigerants (HCFC-123, CFC-11, CFC-113), moisture infiltration is a key concern. In most equipment rooms there will be a vacuum inside the chiller while the chiller is not operating because these refrigerants have boiling points above room temperature. When the chiller is operating, the evaporator is in a vacuum. When the chiller is under a vacuum, air can infiltrate the chiller if there is a leak.

To remove air, moisture, and other non-condensibles, equipment manufacturers add *purge units.* Problems associated with purge units can vary from incomplete moisture or non-condensible removal, purge compressor failure, or in a few extreme cases, older style purge units (pre-1990) could even pump out an entire refrigerant charge. The new style purge units, called high efficiency purge units, have safety features to prevent losing an entire refrigerant charge. If a purge unit underperforms or fails, then at the very least moisture removal will be reduced and at the very worst a chiller will trip off on high head pressure.

For the high pressure refrigerants (HFC-134a, HCFC-500, HCFC-502, HCFC-22, CFC-12,), air infiltration is not as serious an issue. Since these refrigerants are under positive pressure, a leak will cause refrigerant to be expelled and the chiller will slowly lose its charge. To remove moisture in chillers that use high pressure refrigerants, some equipment manufacturers apply a *dehydrator*, which is similar to a purge unit on low-pressure machines. If a dehydrator fails, moisture removal will be reduced, and contaminants will enter the system. Since there are fittings and screens on the dehydrator, it is another leak source.

High pressure chillers are often equiped with a *pump out unit* (condensing unit) and a *utility vessel* for storing refrigerant. The pump out unit is often comprised of a reciprocating compressor and a condensing unit. The utility vessel is essentially a large pressure vessel with a relief valve. The utility vessel is usually sized large enough to hold an entire refrigerant charge and is used to store the refrigerant charge while the chiller is being serviced.

There is also an issue of safety with both low- and high-pressure refrigerants. Although refrigerants have been thoroughly tested and used safely for years, *they must be handled properly to avoid injury.* The most immediate *safety* concern with any refrigerant is the displacement of oxygen. The risk is that if refrigerant leaks into an equipment room that isn't properly ventilated, it can fill up a room and displace enough oxygen to suffocate someone who enters the room. This can be solved by complying with the ASHRAE standard for equipment room ventilation and sensors. There are other safety related issues pertaining to short-term exposure and long-term exposure such as cardiac sensitivity. Table 19.3.1 compares the short-term exposure risk for various refrigerants:

To appreciate this table, it must be understood that the higher the number, the safer the product is, i.e., it takes a higher level of refrigerant to produce the effect. With proper training and equipment, *refrigerants can be used safely.* For more information, consult the refrigerant manufacturers or your local refrigerant distributors.

Regardless of what refrigerant is used, *leak prevention is key to reliable performance.* The main source of leaks in chillers are on *fittings,* (external refrigerant and oil lines, filter housings, valves) gaskets, and o-rings. The chillers should be checked frequently for refrigerant leaks. When minor leaks are found, they can be temporarily repaired by either reflaring a copper line or applying some type of high vacuum sealant, such as glyptol. For significant leaks, the machine must be regasketed. When the machine is reassembled, an additional sealant

TABLE 19.3.1 Comparison of various refrigerants.

	Low Pressure			High Pressure	
	CFC-11	HCFC-123		HCFC-22	HFC-134a
Acute					
LC_{50} (40 hr)	26,000 ppm	32,000 ppm		>300,000 ppm	>500,000 ppm
Cardiac Sensitization	5,000 ppm	20,000 ppm		50,000 ppm	75,000 ppm
Anesthetic Effect	35,000 ppm (10 min)	40,000 ppm (10 min)		140,000 ppm (10 min)	205,000 ppm (4 hr)
Pressure at 72 F	1.5" Hg vacuum (Liquid)	5.6" Hg vacuum (Liquid)		126 psig pressure (Gas)	74 psig pressure (Gas)

(like Loctite® R #515) can be applied on the gaskets and o-rings to provide additional leak protection. Refrigerant leaks can cause several problems if left untreated. Leaks can cause low-pressure chillers to trip out on high condenser pressure and high-pressure chillers to trip out on low evaporator pressure.

With respect to leak management, your equipment must comply with the Clean Air Act of 1990 which documents acceptable leak rates. It specifies that chillers used for comfort cooling can lose 15 percent of their refrigerant charge per year, and chillers used for process cooling can lose 35 percent of their refrigerant charge per year. Any leak rate over these thresholds must be repaired within 30 days, or the machine must be shut down.

Leaks in centrifugal chillers are one of the most common problems and the root cause of secondary and tertiary damage. When air infiltrates a refrigerant system, it contains a certain amount of moisture. Some of the water (from the moisture in the air) mixes with refrigerant to form acids. Some of the water oxidizes the iron in the shells and castings to form rust. This rust and acid can attack surfaces inside the chiller and cause premature wear of bearings, seals, gears, and any other moving or sensitive parts. This can cause premature failure of other components or in an extreme case shorten equipment life. Moisture and contaminant removal are equally important in the lubrication circuit for the same reasons as in the refrigerant circuit.

Regarding the refrigeration cycle, all refrigeration machines, from refrigerators to car air-conditioners and 1,000-ton industrial chillers, have four components in common. These four components are as follows:

- Compressor
- Condenser
- Metering device
- Evaporator

Figures 19.3.1 and 19.3.2 are diagrams of a typical refrigeration cycle for a two-stage and three-stage centrifugal chiller along with its corresponding P-H (pressure vs. enthalpy) diagram. In the next few pages, we will describe the various styles of the aforementioned four components with respect to centrifugal chillers, including their function and any potential problems that an operator may experience. The solution of these problems will be addressed in the reliability section.

19.3.2.2 Compressor

The heart of a centrifugal chiller is the *compressor*. There are several different styles of compressors and each type is comprised of several

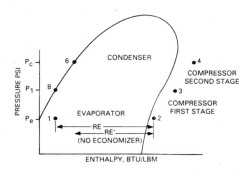

Figure 19.3.1 Two-stage chiller, P/E diagram. (Courtesy of The Trane Company.)

components. In the following paragraphs, we describe the most common types of compressors and their major components.

The main function of the compressor is to take the low temperature, low pressure gas from the evaporator, raise its temperature and pressure, and send a high temperature high pressure gas to the condenser. The *centrifugal* compressor creates this pressure differential between the evaporator and condenser. This pressure differential is referred to as *lift*. A centrifugal compressor is designed with a specific lift that at a specific suction pressure, it will deliver a specific discharge pressure. Unlike a screw compressor, which is a positive displacement device, a centrifugal compressor cannot deliver a certain discharge pressure at a given suction pressure. This means that at too high condensing pressure or too low evaporator pressure a centrifugal compressor will surge. Both types of surging (low side and high side) can damage a centrifugal compressor.

One of the main components in the compressor is the the *motor*. On chillers with *hermetic motors,* the refrigerant is sprayed directly on the motor windings to cool the motor. Moisture and acid levels are important to prevent motor burnout, because contaminants can erode the insulation on the motor windings. On chillers with an *open-drive*

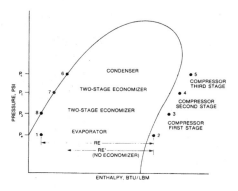

Figure 19.3.2 Three-stage chiller, P/E diagram. (Courtesy of The Trane Company.)

motor, the motor is located outside of the refrigeration circuit and is cooled by ambient air. Although the risk of motor failure is reduced, the motor must be kept clean. Dust and dirt in the equipment room can accumulate on the windings and cause overheating and potentially even motor failure, although this is very rare, according to my contacts at a chiller manufacturer. Also, the hot air coming off of the open-drive motor must be directed away from sensitive electronic components in the starter and control panel to prevent component failure.

In the United States, the most common voltage for these motors is 460/480 V. There are also some 208/230 V motors, and some 4160 V motors. These motors normally operate at 60 Hz power, although there are some chillers that utilize a variable frequency drive which varies the frequency of the power to adjust motor speed according to the load. In Europe, 50 Hz is standard with several different voltages. Canada, with 60 Hz power, often has non-standard voltages.

Almost all of the motors used in centrifugal chillers are three-phase induction motors. The two main components of these motors are the *stator* and the *rotor.* The stator is group of windings which are coiled, insulated wire, wrapped around a core of iron laminates. These coils are either form wound or random wound. The insulation on these motors is critical, especially on hermetic motors. In the case of hermetic motors, which are sealed inside the refrigeration circuit, the windings are cooled by liquid refrigerant. In the case of open motors, which are exposed to the atmosphere, the windings are cooled by ambient air. The insulation integrity is tested through a resistance test, or *megging* the motor. The term "megging" refers to measuring the resistance of the stator, and the units are megohms. (A megohm is 1,000,000 ohms). When a motor "burns out," typically one set of windings (single phasing) or all three sets of windings (complete burnout) are shorted and go to ground. There can also be a phase to phase short. If the motor failure is severe enough, part of the iron core could be melted and damaged beyond repair, requiring the purchase of a new motor. Motor failure can also be caused by a starter malfunction, such as a contactor welding shut, or (in a star-delta starter) the motor staying in transition too long. Excessive cycling—turning on and off too many times in a certain period—can also cause motor failure.

The rotor fits inside of the stator with a small air gap in between. The rotor shaft is constructed of hardened, alloy steel. On the outer diameter of the rotor are aluminum or copper bars. The most common type of rotor failure is an open rotor. This condition exists when one of the bars becomes loose and the current is discontinuous (or open) at that bar as it rotates. If left unrepaired, an open rotor can cause premature bearing wear.

Another cause of rotor failure could occur after a power outage. After power failure, while the shaft is coasting down, the motor acts as a generator. If power is returned during this coast-down period, and the power from the utility is out of phase with the current generated by the motor, the rotor is stopped instantly. This can break the shaft and/or destroy the impeller—not an easy feat considering the shaft is hardened steel 1–3-in. (25–75 mm) in diameter.

One advantage that open-drive motors have over hermetic motors is the reduced risk of motor failure. In the event of motor failure, it is generally less expensive and faster to repair an open-drive motor than a hermetic motor. Another advantage of open-drive centrifugal chillers is that the conversion cost from CFC-11 to HCFC-123 or CFC-12 to HFC-134a is significantly less expensive. One disadvantage of open-drive motors is the *shaft seal*. Shaft seals are typically a mechanical seal around the low-speed compressor shaft. An additional sealant is provided by a thin film of pressurized oil between the shaft and the mechanical shaft seal. The oil continually drips out of the shaft seal. The rate of dripping must be checked and if it increases beyond an acceptable, predetermined rate, the shaft seal may need to be replaced. During the annual inspection, the pressure of this oil can be checked and adjusted.

The motor shaft to impeller connection further segments the different styles of centrifugal compressors. Some chiller manufacturers attach the impeller (or impellers in a multi-stage compressor) directly on the motor or compressor shaft. These compressors are referred to as *direct drive compressors*. Figure 19.3.3 is a schematic diagram of a hermetic, direct-drive compressor.

Since most of the motors rotate at 3,600 rpm, the impeller also rotates at 3,600 rpm. Some chiller manufacturers attach a bull gear at

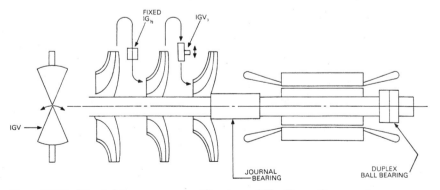

Figure 19.3.3 Direct-drive compressor. (Courtesy of The Trane Company.)

the end of the motor or compressor shaft which turns one or more smaller pinion gears that are attached to the impeller shaft. This increases the speed of the impeller to 5000, 7000, or 10,000 rpm based on the transmission ratio. These compressors are referred to as **gear-driven or high-speed compressors.** Figure 19.3.4 is a cross-section of an open-drive, gear-driven, centrifugal compressor.

On *high-speed compressors,* the gears (like the rotor) are made of hardened steel and the teeth are precision machined. Gears rarely fail, but when they do, it is generally due to one of several causes:

- *Open rotor*—The damage from an open rotor was discussed in the rotor section.

- *Halogen fire*—a non-aerobic flame (occurs in the absence of oxygen) caused by the refrigerant burning—can cause aluminum to melt and or burn.

- *Wear from rust*—again the age-old culprit—rust—the enemy of mechanical equipment

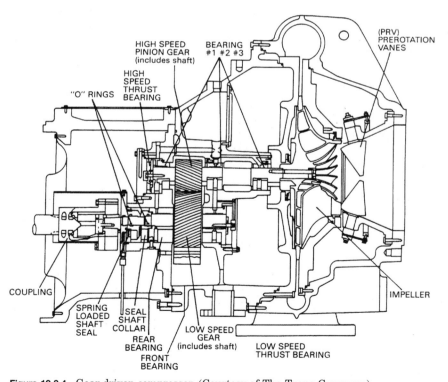

Figure 19.3.4 Gear-driven compressor. (Courtesy of The Trane Company.)

- *Viscosity breakdown* —lubrication fails, usually due to either old oil, acid, moisture, contamination of the lube oil, or less viscosity due to higher temperature

- *Bearing failure*—this damages the wear surface on the gears

The **impeller** imparts kinetic energy into the refrigerant gas stream and thereby compresses the refrigerant. Impellers on direct drive machines are typically larger (2′–4′ [60–120 cm] diameter) than impellers on high speed machines (6″–2′ [15–60 cm]). Also, impellers on direct drive machines are usually shrouded while the impellers on high speed machines are often unshrouded, where the blades are exposed. The impeller to volute clearance on unshrouded impellers is smaller and more critical than on direct drive machines with shrouded impellers. Direct drives machines are often multi-stage (2–3) while high speed machines are often a single stage. Figures 19.3.5 and 19.3.6 are P-H diagrams comparing single-stage vs. three-stage compressors. Impellers rarely fail. The few instances of impeller failure are listed below:

- Bearing failure (journal or thrust)
- Metal fragment/liquid refrigerant drop hitting impeller face
- Halogen fire
- Impeller eye seal or rear hub seal wear

Again, since compressor oil lubricates bearings and gears, rust and contaminants can damage these parts. Proper lubrication is another

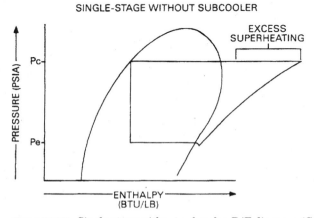

Figure 19.3.5 Single-stage without subcooler, P/E diagram. (Courtesy of The Trane Company.)

THREE-STAGE WITH TWO-STAGE ECONOMIZER

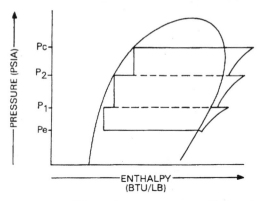

Figure 19.3.6   Three-stage compressor with two-stage economizer, P/E diagram. (Courtesy of The Trane Company.)

key to reliability and performance. The lubrication system is comprised of an oil pump, oil filters, oil sump(s), oil heater, oil cooler, temperature regulator, and oil lines and channels throughout the compressor. The oil must be clean and maintained at the proper temperature, pressure, and viscosity. It must be kept hot enough to boil off refrigerant, but it must be kept cool enough to maintain its viscosity. Consult the manufacturer's recommendations on the type of oil to use and its optimal temperature range. The manufacturer should also provide information how to monitor the oil and when to change it. Typically the oil in centrifugal chillers should be changed annually, but more often if contaminants are found in an oil sample.

Internal to the compressor are seals and bearings. The seals keep refrigerant out of the lubrication circuit and keep oil inside the compressor. Often the seals inside the compressor are referred to as *labrynth seals*. Labrynth seals are usually made out of aluminum and rarely wear. When they do wear, it usually is due to improper alignment with respect to the shaft (eccentricity) or severe bearing wear. Wear can be caused by the shaft or the impeller rubbing against the seals.

There are several types of bearings used in centrifugal chillers. The most common type of bearings are *journal bearings*. These bearings support the shaft or gears vertically and allow proper rotation. Another type of bearing is a *thrust bearing*. Thrust bearings allow for horizontal movement of the impeller as the compressor loads and unloads. Thrust bearing adjustments are critical and leave little margin for error. If the thrust is too tight, it can reduce machine capacity since the clearance between the impeller and the volute is too great. If the thrust is too loose, it can cause impeller damage because the clearance between the impeller and the volute is too small. Bearings are *designed to wear* and are made of a soft metal, like babbit (tin and lead) or aluminum. The frequency of bearing replacement may be from every 3 years to 10–15 years depending on machine style and manufacturer.

One cause of bearing failure is from the loss of lubrication which could be caused by blockage of flow from debris. Oil pump failure can also cause loss of lubrication, although the low oil pressure cutout should shut the chiller down before wear occurs. Another cause of bearing wear (related to the loss of lubrication) is oil laden with refrigerant. Refrigerant acts as a perfect degreaser and strips the oil from the wear surfaces which allows contact with the shaft. Also, any particulate matter or corrosion can prematurely wear the bearings.

Steam driven centrifugal chillers are rare in the western United States but are more common in the eastern United States. There are even a few engine-driven (either natural gas or diesel) chillers.

The last component in the compressor to be discussed is the vane assembly. There can be multiple vane assemblies in a centrifugal compressor, with typically one set of vanes for each stage of compression. Essentially the vanes perform two functions: one is to provide a means for capacity control and the other is to pre-rotate the gaseous refrigerant entering the impeller. The vane assembly controls the load on the compressor by regulating how much refrigerant is allowed to enter the compressor. There are several styles of vanes assemblies and several designs that have evolved. Common in all designs is that that the individual vanes are linked, either through a cable or through a series of mechanical levers or joints.

Vane assemblies rarely fail, but one cause can be moisture (or rust) in the refrigerant circuit. Symptoms of vane failure are improper loading/unloading, cycling, and/or current fluctuation.

19.3.2.3 Condenser

After the low-temperature, low-pressure gaseous refrigerant leaves the compressor as a high-temperature, high pressure gas, it is pumped into the condenser. While there a few centrifugal chillers with air-cooled condensers, most centrifugal chillers are water-cooled. On most water-cooled chillers, to diffuse the refrigerant and prevent tube failure, there is a *baffle plate* inside the shell, directly in line with the refrigerant path. The condenser receives the high temperature, high pressure gas from the compressor and condenses it into a liquid. When the refrigerant comes in contact with condenser tubes, it condenses. This phase change releases the heat of vaporization picked up in the evaporator along with the heat of compression picked up in the compressor. This heat is transferred from the refrigerant across the tube surface, and into the condenser water, where the heat is then transferred from the condenser water to the atmosphere in the cooling tower.

One useful rule of thumb concerns flow rates through condensers and evaporators. On typical comfort cooling conditions (45F [8°C] supply and 55F [13°C] return chilled water temperature, 85F [25°C] supply and 95F [35°C] return condenser water), the condenser water flow rate is typically 3 gpm (11 L/m) per ton, and the chilled water flow rate is typically 2.4 gpm (9 L/m)per ton. For example, a 200-ton water cooled centrifugal chiller should have a flow rate through the condenser of about 600 gpm (2200 L/m), and a flow rate through the evaporator of about 480 gpm (1650 L/m). Remember, this is only a *rule of thumb*. Lower condenser water temperatures and system load conditions greatly change this specific rule of thumb.

There are several problems that can occur in the condenser. One relates to leaks. Since the condenser is the high pressure side of the machine, any air or other non-condensibles will end up in the condenser. One of the main problems associated with condensers is *high head pressure* or high condenser pressure, which can be caused by the following:

- Leaks
- Fouled tubes
- Refrigerant stacking
- Low condenser water flow
- High condenser water temperature

Since the condenser and evaporator are so similar, (they are both shell and tube heat exchangers with similar construction) we will discuss their individual components in the heat exchanger section.

19.3.2.4 Metering device

The function of the **metering device** is to create a pressure differential between the high pressure side of the system (condenser) and the low pressure side of the system (evaporator). The metering device allows the liquid refrigerant from the condenser to flow into the evaporator. It is usually located on the liquid line between the condenser and the evaporator. On older centrifugal chillers, the metering device is usually a float valve. One common problem associated with the **float valve** is the tendency for refrigerant to stack in the condenser if the float valves sticks and doesn't allow enough refrigerant into the evaporator. On newer centrifugal chillers the metering device is usually an *orifice plate*.

19.3.2.5 Evaporator

The evaporator in a centrifugal chiller receives the liquid refrigerant from the condenser through the metering device. As the liquid refrigerant boils in the evaporator, it absorbs heat from the water in the chilled water loop. Inside the evaporator is a set of screens called *eliminators*. The eliminators are designed to prevent any drops or droplets of liquid refrigerant from entering the compressor. These eliminators are subjected to corrosion from acid in the refrigerant circuit. If the eliminators are corroded, it is a symptom of a serious corrosion problem and you've got more to worry about than just eliminators. At the bottom of some evaporators is a liquid distribution system. Its function is to evenly distribute the liquid refrigerant throughout the entire evaporator.

Heat exchangers.   Heat exchangers are just like they sound — surfaces where heat transfer takes place. Since the heat transfer surfaces play such an important role in capacity and efficiency, we will examine those aspects of their function in the performance section. In this section, we are more concerned with reliability.

Both the condenser and evaporator are comprised of the following components:

- Vessel
- Tubesheets
- Support sheets
- Water Box Covers
- Tubes

Vessels.   Each vessel is cylindrical in shape and constructed of cast iron, usually ½ in. to 2 in. (1.3–5 cm) thick. Inside the shell are **tubes** and their **support sheets**. At the end of each heat exchanger are the **tube sheets**. The tube sheets are welded to the ends of the condenser and evaporator. Connecting the water side of the tubes to the chilled water and condenser water pipe is the water box cover. Inside the water box may be none, one, or several divider plates for a one, two, or multiple pass heat exchanger.

In addition to the condenser and evaporator there is one other vessel we haven't mentioned. This other vessel is called an **economizer** and it is only present on multi-stage centrifugal compressors. Although economizers have no tubes, some have internal features, such as eliminators, orifice plates, and weir devices.

The function of the economizer is to provide additional cooling and subcooling of the liquid refrigerant between the stages. It also may provide motor cooling. It is essentially a liquid receiver that gets liquid refrigerant from the condenser and is vented to the compressor and possibly the motor.

While we're dealing with vessels, now is a good time to discuss refrigerant safety again. Since most of the refrigerant is present in the vessels, as a safety measure, both low pressure chillers and high pressure chillers have a pressure relief system. Low pressure (HCFC-123, CFC-11, CFC-113) machines were designed with a *rupture disk* while high pressure (HFC-134a, HCFC-22, CFC-12) machines have *relief valves*. Both devices will release refrigerant to relieve pressure in an emergency. The relief valves reseat, while the rupture disks do not. With the increased cost of the low pressure CFCs (CFC-11, and CFC-113) and the health and safety concerns with HCFC-123, equipment owners have been retrofitting their machines with relief valves. Instead of losing an entire charge with a rupture disk bursting, they have a chance of keeping at least part of the refrigerant charge if the relief valve reseats. Which ever system is used (rupture disk or relief valve), the refrigerant exiting the chiller must be vented outside, per ASHRAE guidelines. This prevents the refrigerant from burning and creating poisonous gases.

Why a pressure relief device? If a boiler explodes, or a fire is generated in the equipment room, or any emergency causes the temperature in the equipment room to rise quickly, the chiller can become a locomotive-sized hand grenade. If the pressure was great enough, it could explode, sending metal fragments through walls. Besides the projectile danger, the burning refrigerant could turn into phosgene gas, or another of several types of deadly gases that could cause harm to building personnel or the general public. To prevent these catastrophes, equipment designers have designed pressure relief systems.

Several other things can cause a pressure relief device to release refrigerant: a heat source in the equipment room — like a fire, or a boiler explosion, or a large steam leak. Also, too high of an entering chilled water temperature can cause excessive pressure in the evaporator and cause the rupture disk to rupture. Water flowing through an idle, low pressure chiller can also cause a refrigerant release.

Tubesheets.   The tube sheets secure the tubes at each end and form a seal between refrigerant on the outer diameter of the tube and water on the inner diameter of the tube. On boilers, tubes are welded into tubesheets, but with chillers, the soft copper tubes are mechanically expanded into holes in the tubesheet. The tubesheets are usually 1 in. to 2 in. (2.5–5 cm) thick.

Tubesheets rarely fail. When they do, it's usually do to poor water treatment. If the tube sheet is eroded or corroded, water can get in between the tube and tube sheet mechanical seal. To prevent this from happening, you can coat your tube sheets with some type of epoxy or ceramic polymer metal.

Support sheets.   The support sheets do exactly what they sound like they do — they support the tubes. Since evaporator and condenser tubes can be from 8–20 feet (2.4–6 m) long, without any support along the way, they would sag in the middle if not supported. They are usually the size of the diameter of the vessel, and ¼ in. to ½ in. (6–12 mm) thick, with several holes drilled for the tubes. The longer the heat exchanger, the more support sheets. Typically, they are spaced every 2–3 feet (0.6–0.9 m). In most chillers, each heat exchanger would have 3–5 support sheets. The only time that support sheets fail is when the refrigerant circuit is badly contaminated from acid and rust.

Support sheets are especially critical in evaporators, since the turbulent action of the refrigerant boiling can rattle the evaporator tubes. Their function is not as critical in the condenser.

Water box covers.   The water box covers cover the entire face of the tube sheet. The water box covers on at least one should be removed annually to allow for inspection of the tubes and tubesheets. There is also a gasket that prevents water from the leaking out. If the tubesheet is corroded, the water box is probably also corroded and should be coated with a protective coating if corrosion is present. Some water boxes (with multiple passes) may have a divider plate that separates one pass from another. If the divider plate leaks, the flow could be short-circuited and cause high condenser or low evaporator pressure since not enough heat is being transfered between the refrigerant and the water.

Tubes.   The tubes in the condenser and the evaporator are the heat transfer surface in the refrigeration system of almost all centrifugal chillers. Since they are the heat transfer surface, they have a big impact on machine performance. If the tubes are fouled, capacity is reduced, as is efficiency. Maintenance items related to tubes will be addressed in the reliability section, and their effect on capacity and efficiency will be addressed in the performance section.

The tubes in both the condenser and evaporator are almost always copper. Occasionally you find cupronickel, brass, carbon steel, or stainless steel tubes in corrosive environments, process cooling loops, and marine/saltwater condensers. The tubes may be prime surface copper or they may have internal or external enhancements.

Tubes can fail in several ways—almost all of them catastrophic. In considering the cost of a retube, include secondary and tertiary costs. Listed below are the most common tube failures:

Freeze-ups.   Freeze-ups are usually caused by either operator error or control failure. Occasionally it is due to a design problem. To freeze up a chiller while its running, three components have to fail: flow switch, low water temperature cutout and the low refrigerant cutout (either pressure or temperature). Another way to freeze up a chiller is when a refrigerant charge is lost quickly, and the sudden drop is pressure causes a sudden drop in temperature. Yet one more creative way to freeze up a chiller is to remove the refrigerant charge without operating the chilled water pump to add the heat of vaporization.

One unusual case of a tube freeze-up occured on a low temperature screw chiller application at an ice skating rink. While one chiller was idle, it had a 12F (−12°C) glycol brine flowing through its evaporator and there was no positive shutoff between the evaporator and condenser. The refrigerant in the evaporator was chilled by the brine and then thermal migration ocurred between the condenser and evaporator through the liquid line. Since the machine was idle, there was no condenser water flow and the low temperature refrigerant (12F) [−12°C] took heat out of the water in the condenser tubes until they ruptured.

Inner diameter pitting, erosion.   This is caused by incorrect or incomplete water treament, sand, grit or another abrasive substance circulating through the condenser loop, or too great of a flow rate. This happens more frequently on condensers and open-loop chilled water circuits.

Outer diameter pitting.   Outer diameter pitting is usually caused by water in the refrigerant that forms acids and damages the tube surface from the outside. This can occur on either the condenser or evaporator, but would only happen in a machine with a lot of water infiltration over a long period of time. This could only occur under a state of negligence.

Saddle damage.   This usually occurs only in the evaporator. In the evaporator, the refrigerant is boiling. In most chillers, the tube is rolled into a support sheet. If the tube is overrolled, the soft copper will be deformed and worn just from overrolling. If it is underrolled, the tube is too loose. In both cases, the boiling action of the refrigerant will cause the tube to rub against the support sheet and if left unrepaired, can cause tube failure and consequential damage.

Manufacturing defects. Tube can also be caused by the any of the following manufacturing defects: zipper cracks from seams, welding slag on tube outer diameters, over rolling, or pin holes. One large real estate developer I have worked with has every chiller eddy current tested by an independent testing contractor before he accepts it.

Another trend in chiller manufacturing (to squeeze additional efficiency out of the chillers) is to decrease the tube wall thickness and to add enhancements on both the inside diameter and the outside diameter. Internal rifling, similar to rifling in a gun barrel, causes more turbulence and increases the heat transfer. Also, the number of fins per inch have steadily increased from 19 or 26 fins per inch (8–10 cm) to 40 or 50 fins per inch (6–20 cm). Then, the fins are cross-hatched, or folded over to get additional surface area and hence better heat transfer.

Other chiller components and systems.

Controls. There are two types of controls—*operating controls* and *safety controls.* Operating controls start and stop the chiller and allow for capacity control. Safety controls prevent the chiller from destroying itself. Controls can get out of calibration and not perform properly, or they can even outright fail. Below is a partial list of common operating and safety controls:

Operating Controls

Start/stop/status

Capacity—vanes/VFD (variable frequency drive)
        load limit
        current limit

Safety Controls

Flow switch

High pressure cutout

Low pressure cutout

Low refrigerant cutout

High/low oil temperature

Low oil pressure

High motor temperature

Low water temperature

This list of operating and safety controls is not all inclusive. With additional automation, new controls and sensors are continually being developed, tested, and manufactured.

Starter. In older chillers, starters came separate and were usually mounted on a wall near the chiller. Now you can order the starter chiller mounted in the factory, saving installation time. Also, with solid state features, starters have a multitude of additional features offering the following:

- Soft start to reduce start-up current

- Solid state overloads and voltage, current, and frequency sensors to aid in motor protection

If your chiller will sit outdoors, care must be taken to have the proper NEMA-approved enclosure. If the chiller will operate in a classified area like a refinery or chemical plant, further steps may be need to be taken to meet the requirements for the classified area. Typical classified areas are spaces that may contain explosive hydrocarbon gases like propane or butane. Any spark producing component (e.g., starters and switchgear), must be sealed to prevent gas from entering the device where it can come into contact with a spark and cause an explosion.

Other system components and systems external to a chiller that affect chiller performance are as follows:

- *Chilled water and condenser water piping and pumps* — delivering consistent and unvarying designed water flow at the proper pressure.

- *Water treatment* — keeping tubes clean and clear of inorganic and organic fouling.

- *Cooling tower* — Rejecting heat from the condenser water loop and delivering design temperature condenser water supply

- *Controls* — Proper sequencing, staging, loading, coordination, and operation of chillers and ancillary equipment.

- *Seismic bracing* — requirement for seismically sensitive areas, for example, California and Japan

- *Power* — correct and unvarying voltage, current, and frequency. Quality power is especially important for chillers utilizing microprocessors and SCRs (silicon-controlled rectifiers) in control panels and starters.

Chillers are designed and installed to do one thing — remove heat from something. It's about energy transfer — taking heat from a conditioned space or a process and transferring it to the chilled water loop, which then transfers it to the refrigerant which is then compressed. In the condenser, the refrigerant condenses and releases the heat picked up in the evaporator and the heat of compression. If the chiller either fails to operate or fails to remove enough heat, either the people, equipment or substance being cooled will overheat. The cost could be minimal at best, or expensive at worst, and could possibly include the loss of life.

The most common denominator in good chiller performance has been organizations that not only maintain their equipment well, but also took ownership of the entire chilled water plant. These organizations also paid attention to their equipment to prevent things from becoming worse and acted fast when a problem was encountered. Here is a procedure:

- Perform equipment survey and needs analysis
- Make repairs to return equipment to design conditions
- Choose a strategy of either containment, conversion, or replacement
- Chiller optimization — either incrementally or all at once
- Chilled water plant optimization
- Feedback

The first step is to find out where you're at. We perform an equipment survey and a needs analysis, so you know where you stand with your equipment condition. Then you get the chiller plant repaired and performing up to design conditions. Then you choose a strategy of either *containment, conversion,* or *replacement.* Containment involves reducing refrigerant emissions. Conversions involves converting your chiller to operate on one of the new refrigerants. The next step involves optimization. Then you choose the schedule of implementation, either all at once or incrementally. As you install upgrades that generate energy savings, you use those savings to fund your next upgrade. Then you need a monitoring tool for feedback. This ensures that what you did worked and it also justifies the expenditures.

Now that we've looked at the different parts and components in centrifugal chillers and you know most of the ways they can fail, you might ask yourself how to prevent some of these catastrophes from happening? Sometimes you can't — you get a current or voltage surge, or you get an imbalance among the phases, and your motor fails — even though you've done everything humanly possible to prevent it from happening. Some things are out of your control. These are only about $\frac{1}{2}$ of 1 percent of all failures. The other 99.5 percent can be prevented.

In the next section, we'll look at specific methods and tasks to prevent chiller failure.

In the next section we will list specific tasks to keep machines operating on line continually and reliably. *With proper planning, tools, and equipment,* most of the equipment damage and failures can be prevented.

19.3.3   Reliability

In this section we will discuss both the general and the specific along with the philosophical and the pragmatic aspects of maintenance, not only for centrifugal chillers, but as it applies to your entire physical plant. We will also address the solution and the prevention of problems that an operator may encounter.

Preventive maintenance on mechanical equipment is sometimes neglected. In a lot of organizations it is one of the first things to be reduced during budget cuts. When additional capital or operating funds are available it is one of the last things added back into a budget.

In some instances, a temporary "deferral" of maintenance is acceptable. In other instances, such as when downtime is critical, it can be catastrophic. For certain industries, like semiconductor manufacturing, chemical plants, refineries, and data centers, a single day or sometimes even a single hour of downtime may cost an organization more than an entire year of preventive maintenance.

Here are four other factors that affect machine reliability:

- Design
- Application
- Installation
- Operation

19.3.3.1   Design

Regardless of your specific skills, background, and experience, we recommend hiring a consulting engineer to design or at least help design the mechanical systems. Listed below are some criteria for evaluating different designs:

- Comfort vs. process cooling
- Future needs and growth
- Cost: installed and life cycle
- Proper clearances/access

- Redundancy
- Ease of use
- Automation (BAS/EMS)

19.3.3.2 Application—machine selection and sizing

Even with an excellent design, if equipment is not sized or selected properly with adequate features, problems can arise with the best of maintenance programs.

Whether for new construction or retrofitting an existing building, *machine selection* is the beginning. A large factor related to reliability is the application of the correct equipment for the specific purpose. Depending on the use of the chiller—whether for comfort or process cooling, low or medium temperature, high or low run time, high or low cycling (starting and stopping), several makes and model may work. Or possibly, only one make or only one model will deliver the desired performance. It may require the use of another style of compressor—e.g., screw, reciprocating, or scroll. Due to the availability and price of steam or natural gas, an absorption chiller may be the best choice.

If the chiller is sized improperly, it may not deliver the desired tonnage. If undersized, the chiller may not meet the load or provide redundancy. If oversized, the chiller could fail prematurely due to excessive cycling. Other factors in machine selection:

- Usage
- Budget
- Future Growth
- Building Purpose

19.3.3.3 Installation

Another factor that impacts reliability is the installation of the equipment. Once the mechanical room is designed, the equipment is selected and ordered. The next step is installation. Besides the pertinent laws, codes, and regulations, each equipment manufacturer will have specific installation instructions for their whole line of equipment, and possibly various installation requirements for various machine models.

In most comfort cooling applications, the chillers are piped in parallel. Two benefits of this arrangment are redundancy and serviceability. In process cooling applications, there are several possibilities. The chillers may be piped either in parallel or series, depending on

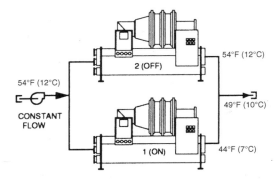

Figure 19.3.7 Parallel piping configuration. (Courtesy of The Trane Company.)

several conditions. For example, if the load has a high temperature difference or if it is a batch or erratic process where the heat load can vary, it may be better to pipe machines in series, staging the machines as the load changes. Or if there is a high flow rate, it may be better to pipe machines in parallel, splitting the flow between two chillers. Figures 19.3.7 and 19.3.8 are schematic diagrams of both parallel and series piping configurations:

A chiller in a manufacturing facility in California often tripped out on high head pressure because the water returning from the cooling tower was too hot. The reason was that the piping was not plumb and did not distribute the hot water coming from the condenser evenly in the cooling tower. The chiller and cooling tower operated perfectly, but the installation caused reliability problems. In addition to obeying codes and piping the equipment properly, there is a myriad of other considerations to a proper installation.

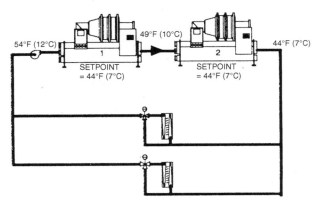

Figure 19.3.8 Series piping configuration. (Courtesy of The Trane Company.)

19.3.3.4   Operation

Listed below are four approaches to operating your chiller plant:

- In-house
- In-house with service contracts
- Outasking
- Outsourcing

In-house is just that: operating and maintaining your chillers with in-house labor. Another variation is to do all of the operation in-house and contract for all or part of the maintenance. The third approach is contracting out for an entire discipline—all HVAC, or all fire and life safety maintenance. Outsourcing is contracting for *all* disciplines—both for operations and maintenance.

After the chilled water plant has been designed and installed, nothing affects machine reliability more than proper operation. Whether the chiller plant is operated manually or through some type of automation or control system, the operation of the equipment is the responsibility of the operators.

Some buildings do not have dedicated operators on site for their centrifugal chillers, but someone must operate the equipment. It can be started and stopped remotely using a building automation system (also known as an energy management system). With full integration, log readings can even be taken automatically. No control system can duplicate what an experienced and trained operator can do—notice a strange sound, notice a drop of oil on the floor or on the machine that wasn't there yesterday. Not only can highly skilled operators save the property owner several times their salary in energy expenses, they can prevent a lot of expensive failures or catch the failures *before* they cause secondary or tertiary damage.

Set a high priority for getting a skilled crew of operators. Operators must be trained. While you may not need a journeyman chiller mechanic on site (you may if you have a critical operation or a large plant—10,000 tons or more), you need operators who understand the refrigeration cycle, basic electricity, some HVAC and/or controls knowledge, plumbing skills, and especially troubleshooting skills. They must know how to take readings and decipher which readings are abnormal or out of range. They also should know about what the temperatures and pressures should be in the condenser, evaporator, and compressor. They should be able to spot trends and take corrective action. *If your operators don't know this,* they need training. If it is beyond their skills or if they are limited by time, they should be instructed to call in a specialist for further diagnosis or for repair.

Another skill in operators often is overlooked is communication skills. If you depend on outside contractors for chiller repairs, you can save yourself a lot of money if you can communicate effectively with your contractors' personnel. If you have a good relationship with a chiller repair company (either the manufacturer's local service department or a good independent mechanical service contractor), you can often describe the chiller's symptoms to a qualified mechanic and get suggestions on remedial action, saving the expense of a service call. Even if a chiller mechanic must make a service call, the more information you can provide to him or her the more you can save in troubleshooting expenses.

A strong operator crew can do most of your maintenance on your chillers. In the lists of tasks discussed in this chapter, a qualified stationary engineer can do all of them up to the annual inspections. All daily, weekly, monthly, quarterly, and semi-annual tasks can be performed with in-house personnel. For annual inspections, it is worth the expense of bringing in an outside specialist like a journeyman centrifugal chiller mechanic. For major repairs or special testing, we recommend contracting out.

As a minimum, the following is a list of what an operator should do:

Operator Duties and Tasks

- Start-up — daily, weekly, seasonally
- Shutdown — daily, weekly, seasonally
- Extended shutdown (seasonal, temporary)
- Start-up after extended shutdown
- Documentation — log readings, federal regulations and codes, EPA Refrigerant compliance, OSHA, State and local documentation
- Inspecting, monitoring, and maintaining chillers — lubrication, minor adjustment, painting and corrosion prevention, insulation repair
- Monitoring and inspecting ancillary equipment
- Ensuring proper flows and temperatures in and out of the chiller on both the condenser and chilled water loop

Although the operation of a centrifugal chiller can sometimes appear to be mundane since it is such a reliable piece of equipment, proper operation is crucial for machine reliability. For example, one important aspect that could be taken for granted is proper flow through the condenser and evaporator tube bundles. Lack of flow or an elevated entering temperature in the condenser water loop can cause high head pressure and even surging. This causes damage to the chiller from possible bearing or impeller failure. Surging could also cause the chiller to use 10 percent to 20 percent more energy.

If you have a limited operator staff, you may have to contract out for your monthly and quarterly inspections. As a manager or director, there is a trade-off you may have to make. You can trim your staff to save expenses, but the maintenance tasks on your chillers and other equipment and systems doesn't go away. Although you may save money on overhead and benefits, you could lose some control of your maintenance. If done properly, you can gain additional control through contracting out. Another choice is to outtask or outsource, turning over even the operation of your facility to an outside contractor. Whether you contract out for major repairs, regular maintenance, or all or part of your operations, there are benefits, costs, and risks associated with each choice or combination of choices. Choose carefully.

19.3.3.5 Maintenance

There are a lot of philosophies and approaches toward maintenance. (See Section D of this book). Some organizations put a lot of effort and energy toward maintaining their equipment while others do little. *Maintenance is the key to reliability.* Overall, there are three main approaches to maintenance listed below:

- Breakdown maintenance (also called "run to failure")
- Preventive maintenance
- Predictive maintenance

Some organizations still have a breakdown maintenance plan. Their reasoning is that even after spending a large amount of money on preventive and predictive maintence, chillers still fail and need major repairs or replacement. To them, a large expenditure is just around the corner, so why waste money on all of this maintenance when mechanical failure is inevitable? Breakdown maintenance is the least expensive method to implement since there is no cost, but is the most expensive, once the life-cycle costs are considered. These life-cycle costs include (but are not limited to) the following:

- Major equipment failure—repair or replace
- Energy
- Downtime
- Risk and liability

There *is* a grain of truth to the philosophy of the breakdown maintenance supporters. Even with performing every feasible preventive and predictive task, chillers do require a major repair sometime in their life, and eventually they are replaced. But breakdown maintenance people are short-sighted in that a good preventive maintenance program can pay for itself in energy costs alone, while avoiding downtime and providing better control of when repair or replacement dollars are spent. Since I do not recommend a breakdown maintenance philosophy, I will not elaborate on that approach. We will describe preventive and predictive maintenance approaches in more detail.

Preventive maintenance is just that; it is maintaining equipment to prevent secondary and tertiary damage. The result of preventive maintenance is increased reliability, reduced downtime, and reduced loss of energy efficiency.

Predictive maintenance is taking preventive maintenance one step further. The key word is predictive — predicting when chillers need major service. Once up and running, over a period of time it can help you stretch your preventive maintenance dollars so that you don't overspend on maintenance. A predictive maintenance program is most expensive up front, but saves money in long-term secondary and tertiary damage and energy. A thorough predictive maintenance plan also has some beneficial side effects: it reduces downtime further, gives the owner additional control and time for planning, and can reduce preventive teardown inspections.

Whichever philosophy you subscribe to, there are other tools to help plan and implement your maintenance program. One of these new tools is a computerized maintenance management system (CMMS). There are several CMMS programs commercially available, the best of which offer a large database of tasks, labor and material planning, cost tracking, inventory control, and especially refrigerant tracking. Besides a CMMS, there are other types of software for a specific task, such as refrigerant tracking and EPA documentation. See Chapter 12, *Computerized Maintenance*.

Whether you contract out all of your maintenance or do it all with in-house labor, and track it with either a card file or use a sophisticated software program, there are three ways it should be implemented to get the most benefit out of it. It should be a *proactive* tool and used for planning. It should also address and meet your *documentation* needs. Finally, someone should see that it is *standardized* and integrated into your department. If you manage a large facility with several shops or several locations, then standardization is crucial for you.

Being *proactive* is the key to success.

Documentation is getting more important every year — both from a code compliance standpoint and a legal standpoint. Besides protecting yourself and your company, it's also helpful to document your successes and prove that the money invested in your maintenance program is justified. This documentation should reflect uptime, capacity, maintenance, and energy savings. The following information should be included in your documentation:

Start-up Information
- Submittals — specifications and drawings
- Original model and serial numbers
- Design conditions
- Logs

Maintenance History
- All logs
- Preventive and predictive maintenance tasks performed
- Refrigerant tracking
- Monthly/quarterly/annual inspection checklists
- Major service
- Failure history — machine, component, part
- Financial information
- Capacity and efficiency reports

Standardization is crucial for a large facility or an organization with multiple locations; it goes hand-in-glove with organizational quality initiatives such as Total Quality Management (TQM) or any other statistical-based quality process. One of the basic tenets of a quality process is that quality is not the most expensive path — it is the most *cost-effective* path. It is not overspending or underspending. That this is true for preventive maintenance. You can also create a "Best Practices" process for your maintenance or facilties department.

Intimately related to standardization and documentation is the subject of logging your machines. It is important to take log readings on your centrifugal chillers. At a bare minimum it must be done once per day, and the more readings the better. Hourly readings would be ideal. With new building automation systems, this logging can be done automatically. You also can design a program to signal an alarm if a reading is too high or too low. The reason for this is when there is a machine problem, it make take a while to manifest itself. Big problems usually start as little problems. Of course, it matters who analyzes the readings. Operators need to know what to look for, and it saves time and money to provide a set of log readings when a contractor shows up.

The following preventive maintenance tasks are recommended for your centrifugal chillers. Since there are many makes and models of centrifugal chillers, these tasks are generic. When they differ from the manufacturer's recommendations of your specific machine, use the manufacturer's recommendations.

Hourly
- Logs, if possible—manually or automatically through building automation system

Daily
- Start/stop
- Load/unload
- Look, listen, and feel for any unusual visual cues, sounds, temperatures, or vibrations

Weekly
- Inspect all sight glasses—record refrigerant and oil levels
- Inspect system piping and pumps on both condenser and chilled water systems

Monthly
- Record hours of run time and the number of starts of machine
- Record minutes of run time on purge unit
- Inspect all refrigerant and oil piping, fittings, flanges, gaskets, o-rings for leaks
- Check operating and safety controls
- Inspect cooling tower
- Inspect lubrication system

Quarterly
- Inspect all components and subsystems
- Adjust and lubricate where necessary
- Inspect all electrical connections
- Leak check (Some people feel that monthly or quarterly leak checks are necessary. On a well running machine with a history of few leaks, it's probably not necessary. If you have had a lot of leaks, you probably should check for leaks quarterly. Some local codes require semi-annual leak checks.)

Annual Tasks

Nothing is more crucial than a thorough annual inspection by a qualified technician. The following tasks are the *minimum* that should be performed *annually* on your centrifugal chillers:
- Pull oil sample and analyze for wear metals and contaminants
- Leak check
- Meg main motor and oil pump motor
- Check and calibrate operating and safety controls

- Service purge unit — or — Service dehydrator and/or pump out vessel
- Pull heads and inspect tubes and tubesheets
- Clean tubes (if necessary) This may need to be performed semi-annually or quarterly depending on water conditions
- Change oil and oil filter
- Cut oil filter in half and inspect for bearing shavings or fragments
- Thoroughly inspect and test lubrication circuit
- Change refrigerant filters and strainers
- Inspect vane linkage
- Inspect and clean starter
- For open motors: Inspect coupling alignment
- On gear driven machines: Inspect gears for wear
- Inspect shaft seal assembly
- Start-up, test, and run chiller
- Log machine and document all tests and readings

Consult the equipment manufacturer for additional or more specific tasks on your chiller.

Water treatment.   Quality water is important not only for reliability, but also for energy efficiency. Improperly treated water will foul condenser tubes and increase head pressure. If the head pressure gets too high, the chiller could trip off on the high head pressure control. This increased pressure can also damage the thrust bearing on high-speed machines.

The two main groups of culprits in tube fouling are dissolved solids and microorganisms. Chlorine and ozone (and other biocides) work well on microorganisms, but do nothing for dissolved solids and scale. Depending on the hardness of your water, you may need one or more chemicals to keep the dissolved minerals suspended. Regardless of the quality of your water treatment you need to make sure the blow down mechanism is functioning on your cooling tower. Since water quality varies, contact one or two water treatment companies, have them test your water, inspect your machines, water boxes, pipe, and cooling towers. They can then recommend a water treatment program. You also can buy chemicals yourself and perform this service in-house. We would recommend an automatic type feeder to add the proper chemicals based upon water hardness and conductivity.

Water treatment is so important for reliability and performance. If your concentration of solids builds up too much, it doesn't take long to scale up condenser tubes. If you *don't* maintain the proper chemical levels, the condenser can scale up in a matter of hours or even minutes. Since the chilled water loop is usually a closed loop, it needs to be treated only as often as you add make-up water.

Multiyear service. Even with the best preventive maintenance program, most machines require a major internal inspection on some interval. Some manufacturers have 3- and 6-year major inspections. Others recommend 5-year teardowns. With new manufacturing technologies and closer tolerances, some chillers can go for 8, 10 or maybe 15 years before a major service is required. Follow the manufacturer's recommendations and as a double check compare the machine's performance along with the analyses from predictive maintenance tools to decide when a machine needs a major inspection or service.

In addition to internal inspections, you may need some or all of the following major repairs:

- Overhaul
- Retube
- Motor rewind or replacement
- Compressor or Driveline replacement

Predictive maintenance. We recommend using predictive maintenance tasks. If you don't want to launch into a full blown predictive maintenance plan, one strategy may be to start with the simple, low cost items first, like oil analysis and vibration analysis. Those tasks can point you in the right direction to allocate money to either repair the worst chillers or prevent the worst failures.

Over a short time frame (less than 1 year), predictive diagnostics can be used to monitor symptoms. Each analysis can give you a snapshot of the machine's condition at one particular point in time. If your machine is experiencing a particular problem, a predictive diagnostic tool or a combination of diagnostics may be used to uncover an acute problem or confirm a suspicion. Predictive tools are not always 100 percent exact or accurate.They should be used in conjunction with each other and with the operators' and chiller mechanics' input. For example, oil analysis and vibration analysis are a good check for each other.

Over a longer time frame (2–10 years), predictive diagnostics can be used for *trending* to plan for major inspections or repairs 1, 2, or maybe even 3 years in advance. An entire booklet of snapshots can give you an idea of where a machine's condition is headed by looking where it came from. It may take 2–4 readings per year in the first year or two to establish a trend, but after a trend has been established, it may require only one or two analyses per year.

There are two types of predictive maintenance tasks — *non-intrusive* and *intrusive*. Nonintrusive tasks can be performed while the machine is running. Intrusive tasks require the machine to be shut down or in some instances, partially disassembled. The following is a list of predictive maintenance tasks, its purpose, and the recommended frequency:

Nonintrusive Tasks	*Frequency*
Oil analysis	**At least once per year**
Wear metals — Tin, Lead, Aluminum	
Contaminants — water, refrigerant,	
acid	
Refrigerant analysis	**On suspicion of difficulty**
Water/Moisture	(contamination due to a tube
Acid	leak, air leak, urgent
Oil	malfunction)
Thermography	**Every three years**
Electrical wiring and connections	
Switchgear, starters, control panels	
Motors	
Vibration analysis	**Once or twice per year**
Motors, bearings, shaft	
Impeller	
Gears	
Motor current analysis	**Once per year**
Open rotor bar	
Intrusive tasks	**At least once per year**
Megger test	
Insulation integrity	
Eddy current analysis　Tube	**Every three years**
condition	
Special occasions	
Boroscope/fiber-optic inspection	**On suspicion of difficulty**
Metallurgical analysis	**On suspicion of difficulty**

Even with the best operators and the most complete preventive and predictive maintenance program, your chiller could fail. You can always rent a piece of equipment — not only your chiller, but pumps, cooling towers, air handlers, electrical cable and distribution panels, ancillary eqiupment — even electrical generators to power your entire facility. You can rent until your equipment is repaired or new equipment is installed.

We suggest forming a relationship (maybe even a partnership) with an equipment rental company to provide this equipment on an emergency basis. Listed are criteria important in evaluating a rental equipment partnership:

■ Contingency plans

- Experience
- Equipment availability and diversity
- Disaster recovery expertise
- Local, 24-hour service
- Age and condition of fleet
- Ability to deliver a complete, turnkey solution for power, cooling, dehumidification, heating
- Equipment with modem capability for remote starting and monitoring
- Support — engineering and managerial

Maintenance Review: 19 Questions to ask yourself

1. What per cent of the time is my equipment available for operation?
2. How many minutes of downtime was there this past year?
3. Is our chiller plant meeting the demands of the load?
4. What is the capacity of each chiller? 4a. chilled water plant?
5. What is the efficiency of each chiller? 5a. chilled water plant?
6. What is our annual utility expenses?
7. Has it been increasing or decreasing with time?
8. What is our annual maintenance expense? (Including internal costs like overhead, allocated labor, overtime)
9. Is it increasing or decreasing?
10. What is the condition of our chillers?
11. Is it improving?
12. How many unscheduled breakdowns did we have?
13. How fast did we get the chillers back on line?
14. How much money was spent on corrective actions and repairs vs. preventive maintenance?
15. Do we have enough people?
16. Are we getting the kind of value-added services we need from our vendors?
17. If we're not where we need to be, how are we going to get there?
18. What does 1 day (or 1 hour or even 1 minute) of downtime cost me or my organization?
19. Why do those Dilbert cartoons remind me so much of my workplace?

Every now and then it helps to take a step back and question some of the assumptions you have made. You must analyze some of the things you've been doing and how you've been working. We also strongly recommend performing a risk analysis. In the refinery and petrochemical businesses, engineers refer to this as *Hazard Operations,* or HazOps for short. In this exercise, everyone plays devil's advocate with each other to come up with every possible scenario of equipment failure, and then a plan to handle the consequences of that failure. It usually starts off with a bunch of what if questions — what if this chiller fails?, what if this pump fails?, what if this valve fails to close or fails to open? Then, when all possible scenarios are examined and contingency plans are created to handle every possible outcome.

This risk analysis can also help justify a preventive maintenance program once the following costs are considered and quantified:

- Downtime
- Rentals
- Overhead

Regardless of what you decide to do, we suggest you have a strategy and a plan to implement that strategy.

Sometimes preventive maintenance is not performed often enough or thoroughly enough. On the other hand, you can perform *too much* preventive maintenance and waste money, or worse yet, degrade the performance of the machine. You want to do enough maintenance, but not love your chiller to death. A good yardstick is to do things that add value your chiller. If a task doesn't extend the useful life of the chiller or if it does not increase the capacity or efficiency of the chiller, don't do it. Spend your money elsewhere.

Energy savings provide additional cost reduction as a collateral benefit of a good maintenance program. The first step in optimizing performance is a good reliable machine. Once a machine is operating properly, we can start to improve its capacity and efficiency. We will investigate this in the next section.

19.3.4 Performance

The performance of a chiller is measured in two ways: capacity and efficiency. Capacity is the total refrigeration capability of the machine. When the chiller is started up, it should be operated at its design conditions, to assure that it will perform as expected. Most chiller manufacturers offer a witness test when you purchase a chiller, whereby you can have the chiller tested at the factory under controlled conditions.

After your chiller is installed at your plant, and after the flows have been adjusted, you can do a rough check of the chiller capacity. If you don't have flowmeters installed, you can at least install pressure gages and check the pressure drop across the evaporator and the condensers. You can then interpolate the flow rate from a pressure drop vs. flow rate chart. Once the flow rate is determined, the temperature change, or ΔT must be checked. Measure the difference between entering and leaving chilled water temperatures. Do the same for the condenser water temperatures.

English Units

$$\frac{\Delta T \times \text{flow rate}}{24} = \frac{\text{tons of refrigeration}}{\text{(at 45°F leaving chilled water temp)}}$$

where: $\Delta T = \text{temperature}_{in} - \text{temperature}_{out}$ (°F) flow rate is measured in gal/min

SI Units

$$\Delta T \times \text{flow rate} \times 5.13 = \text{watts of refrigeration}$$

where: $\Delta T = \text{temperature}_{in} - \text{temperature}_{out}$ (°C) flow rate is measured in L/sec

1 ton of refrigeration = 12,000 BTU/hr = 3,516.2 watts

The calculation from there is simple if you're at 45°F (9°C) leaving chilled water temperature and 85°F (30°C) entering condenser water temperature. If your temperatures vary from that, you may have to derate your calculation. You take the flow rate (in gpm) and multiply it by the temperature difference. You then divide that number by 24 and that gives you your tons of refrigeration. If you multiply your tonnage by 12,000, you get the total BTUs/hr. The tonnage through the condenser should be larger than through the evaporator because the condenser not only has the heat from the load that was transferred in the evaporator, but also the heat of compression. The total heat rejected through the condenser will always be larger than the heat picked up through the evaporator.

As far as capacity is concerned, a well running, reliable machine is probably operating at or near capacity. If the tubes are clean, the clearances in the compressor are all within specifications, there is proper flow and pressure drop across the condenser and evaporator, there is no leakage through the divider plate, there is no flow blockage in the piping systems, there is the proper refrigerant charge in the chiller, and maybe a few other factors, the chiller should be operating at or near capacity.

The most common factor affecting capacity occurs at the heat transfer surfaces: the condenser and evaporator tubes. That is why proper water treatment is so important. With fouling factors of less than $1/1000$ (25 μm) of an inch, it doesn't take much fouling to affect capacity and efficiency. Fouled tubes are often misdiagnosed when something else is causing high head pressure, much to the dismay of the water treatment companies. It could be caused by fouled tubes, or any of the following conditions:

- Air or other noncondensibles in the machine
- A leaking or broken division plate
- Low condenser flow
- Elevated condenser water temperature
- Condenser water by-passing tubes
- Blocked tubes
- Improper refrigerant charge
- Refrigerant stacking

If you are operating at less than design capacity or efficiency, it could also be caused by any of the following low-side problems:

- Low refrigerant charge
- Dirty evaporator tubes
- Excessive oil in the refrigerant
- Refrigerant "stacking"
- Chilled water bypassing tubes
- Low chilled water flow

If the chiller checks out O.K., it could be any of the following system problems:

- Low flow
- Excessive flow
- Erratic flow

- Cooling tower
- Pumps
- Flow blockage, like a plugged strainer
- Controls/sequencing

Efficiency is measured as the energy output per energy input. Most chiller manufacturers express efficiency as kW per ton of refrigeration. At full load, efficiencies below 0.6 kw/ton are not uncommon, and some chiller manufacturers claim efficiencies below 0.6 kw/ton and some below 0.5 kw/ton. Typically, this is referenced at full load at ARI conditions. What is not typical is chillers running at or near full load. Chillers often run at partial load, so full load efficiency is only part of the story. Some chillers that may be efficient at full load, are not nearly as efficient at partial load, while other chillers maintain their efficiency at partial loads and some others actually increase in efficiency at partial loads. Since chillers operate so often at partial load, the part-load efficiencies should also be considered.

Another fact to consider is that chillers may not operate at or close to design conditions. The machine selection process may have selected a machine that was efficient only across a small band of operating parameters. Once those conditions changed, efficiency dropped.

Once capacity has been determined and efficiency has been calculated, there may be several things to improve efficiency. If you are already operating very efficiently, there may not be much you can do. Here are some ideas to wring out additional efficiencies.

19.3.4.1 Optimization

Controls. There are several control strategies and sequences of operations for your chilled water plant that you can implement to increase efficiency. With the state-of-the-art controls systems, you can program into your system almost anything you can imagine, often only with software changes. Sometimes you may need to add sensors and possibly valves, but often it's only a change in your sequence of operations.

Chilled water reset. At lower loads, you can let the chilled water supply temperature climb 1 or 2 or more degrees. This saves energy because it takes less energy to make warmer chilled water. Of course, you must check your load to make sure it won't be adversely affected. If you are in a humid area, this may not work because you need a lower chilled water temperature for dehumidification. Also a process, may not tolerate a 1 or 2 degree change in temperature.

Condenser water reset. Similar to chilled water reset, you can adjust the condenser water temperature to reduce energy. This is also referred to as *floating head pressure,* or *floating condenser pressure.*

Compressor head reduction. Similar to floating head pressure — if you can run cooler condenser water temperatures the compressor is pushing against less pressure and this translates into less horsepower which is less current which will save you money.

Sweet spot. Chillers have a point at maximum efficiency, and it's not necessarily at full load. Every chiller has a set of curves that show how efficiency varies with the load percentage. Once the sweet spot for each chiller is found, you can program your controls to keep each chiller at its sweet spot. Some chiller manufacturers iteratively find this sweet spot at all systems parameters and keep the chiller operating as close as possible to these sweet spots as often as possible for maximum energy savings. If you have multiple machines, you can locate the sweet spot for your entire chilled water plant and set your controls to sequence your chillers at that sweet spot.

Free cooling. There are several ways to achieve "free cooling." One way is to run the cooling tower water through the chilled water system. Another way is to cool the chilled water with cooling tower through a heat exchanger. The first way is more effective, but if you're not careful you can contaminate your entire chilled water piping system. With plate and frame heat exchangers, you get approaches close enough to make this a very attractive payback.

VFDs (variable frequency drive.) VFD's have been put on supply fans, exhaust fans, chillers, pumps, and cooling towers — essentially anything with a motor. As a load decreases, you can slow down a motor by reducing the frequency. As it slows down and operates at slower speeds, it draws less amps and consumes less energy. Since there is a cubic relationship between horsepower and energy, at $\frac{1}{2}$ speed the motor uses $\frac{1}{8}$ the energy of full speed. Listed below are some examples of VFD applications:

Pumping — You can convert your piping system to primary loop through your chillers at aconstant flow with a secondary loop through the load at variable flow. Reduced flow meansreduced pump horsepower. You vary the flow on the secondary loop only because youdon't want to vary the flow through the chiller because you can cause reliability problemsand machine damage. Figure 19.3.9 is a typical primary/secondary pumping schematic:

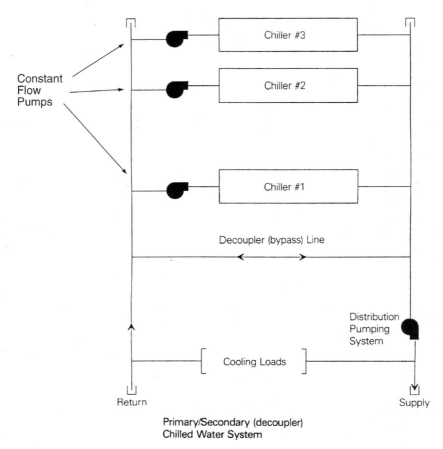

Figure 19.3.9 Primary/secondary chilled water system. (Courtesy of The Trane Company.)

Chiller Driveline—Although more involved than a putting a drive on a pump or a fan motor, substantial energy savings can be achieved through applying a VFD on a chiller motor.

Cooling Tower—Putting a VFD on a cooling tower has a very good payback—often under 1 year. It also gives an additional collateral benefit in that it also increases chiller reliability because you can gain much better control of your condenser water temperature. Some chillers are susceptible to oil migration at low condenser water temperatures.

Now that we've looked at capacity and efficiency, in the next section we will detail additional upgrades for increased reliability.

19.3.5   Improvements and Upgrades

Over the years, the chiller manufacturers, equipment owners, consultants, and independent service companies have dealt with a variety of problems in centrifugal chillers related to reliability and performance. Luckily, you can benefit from this vast knowledge base. With a few, rare exceptions most of the problems you will encounter as an operator have been experienced, dealt with, agonized over, and solved by someone else. It is just an issue of information access. We will document a few of the upgrades available for centrifugal chillers. For additional upgrades or more detailed information, consult your chiller manufacturer or your local independent mechanical service company.

19.3.5.1   Motor protection system

Reliability. Motor failure, although rare, is an expensive event. Not considering downtime and overhead expenses, a motor failure will cost between $15,000 and $100,000 (1996 dollars) in primary and secondary damage. In addition to the electrical distribution system (switchgear, circuit breakers, starter) you can install additional sensors and overloads to protect the motor on your centrifugal chillers.

19.3.5.2   Epoxy coating tubesheets

Reliability. If you have eroded or corroded tubesheets, we recommend sandblasting them to bare dry metal and applying either an epoxy or a ceramic metal coating to prevent tubesheet failure. There are several qualified companies in every major metropolitan area that specialize in this type of work.

19.3.5.3   Additional refrigerant filter

Reliability. If you have either a wet machine or a machine that is full of rust, I recommend installing an additional refrigerant filter consisting of four 1 gal (4 lit) filter cores to clean up the moisture and/or rust. This is also commonly referred to as a clean-up kit.

19.3.5.4   Solid state starters

Reliability and energy. Besides offering additional motor protection, solid state starters offer energy savings and reliability improvement with the ability of soft starting. Solid start starters also offer addi-

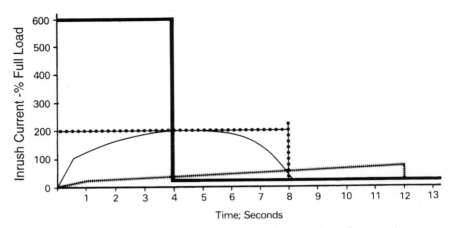

Figure 19.3.10 Inrush current versus time. (Courtesy of The Trane Company.)

tional documentation, as most will have the capability to save and print out information. Figures 19.3.10 and 19.3.11 depict the energy savings due to the reduced current and torque.

Solid-state starters achieve these increases in efficiency through solid-state switching. This is accomplished through 3 pairs of SCR's (Silicon Controlled Rectifiers), one pair for each phase or motor leg. Figure 19.3.12 is an electrical schematic of a six-SCR configuration. Figure 19.3.13 shows the current waveforms of reduced power consumption at various stages.

By using a solid-state starter, reliability is increased due to the following features:

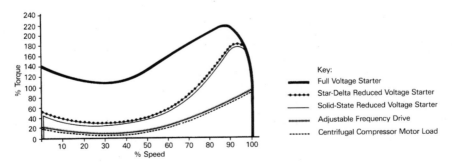

Figure 19.3.11 Torque versus speed. (Courtesy of The Trane Company.)

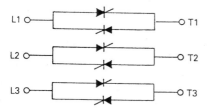

Figure 19.3.12 Six-SCR configuration. (Courtesy of The Trane Company.)

- Arc-less switching, longer contact life
- Reduced mechanical shock to equipment due to less torque
- Reduction or even removal of current and voltage transients and harmonics

Applying a VFD (Variable Frequency Drive) yields similar reliability and efficiency gains, but saves even more energy since chillers often run at partial load.

19.3.5.5 Electronic chiller control panel

Reliability and energy. A lot of older centrifugal chillers were built using pneumatic controls for part of the operating and safety controls. The early electric and electronic controls are not as sophisticated as modern electronic control panels. Upgrading your chiller with an electronic control panel will offer you more diagnostics, better control, more reliability, and possibly energy savings.

19.3.5.6 Integration with building automation system

Reliability and energy. To benefit from all the advances in hardware and software, you're not getting the most out of either your chiller

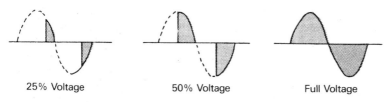

Figure 19.3.13 Wave forms. (Courtesy of The Trane Company.)

plant or your control system if the chillers are not integrated with your Building Automation System. By integration, we don't mean just on/off, start/stop, and status, but *true integration*. True integration involves getting all of the information from the chiller control panel to the control systems's user interface or workstation, along with the ability to monitor and affect changes.

If you still have a chiller that operates on either CFC-11 or CFC-113, with no plans to convert or replace it, we strongly recommend the following items. (If you don't have an CFC-11 or CFC-113 chiller, you can skip to the next section).

The low pressure chiller you have was designed when CFC-11 cost less than $1.00 per pound. By the late 1980s and early 1990s, CFC-11 approached $10 per pound. As you read this, it could be $15, $20, or more per pound. (Or even less.) Refrigerant prices have been unstable and could vary up or down based on demand and supply. If you have lost an entire refrigerant charge the following refrigerant containment devices could pay for themselves many times over:

- High-efficiency purge unit
- Vacuum prevention
- Rupture disk relief valve retrofit
- Leak detection system
- External oil filter
- Isolation valves on oil and refrigerant filters

If you are interested in additional upgrades or if you have a problem or situation that was not addressed in these sections, consult you chiller's manufacturer or a qualified independent mechanical service contractor.

19.3.6 Repair Versus Replacement Decisions

Deciding whether to repair or replace a centrifugal chiller or any expensive piece of capital equipment can often be a difficult choice. We have assembled the following list of potential criteria used in evaluating repairing or replacing a centrifugal chiller, or any other large expensive piece of capital equipment. This list of criteria is not necessarily all inclusive - based on your specific situation any or all of the following criteria should be considered when faced with these decisions:

Age

Type and amount of refrigerant

Capacity

Present and future load requirements
  comfort vs. process
  redundancy

Internal rate of return

Budget needs, availability of capital

Equipment history
  compressor
  starter
  motor
  bearings
  tubes

Present machine condition:
  compresor
  starter
  motor
  bearings
  tubes

Make and model

Downtime costs and risks

Efficiency

Hours of service/run time

Primary vs. backup

Safety

Reliability

Repair parts costs

Repair parts availability

Noise level

Access/space limitations

Cooling tower capacity

Expected operating life of equipment

Codes, laws, and requirements (EPA, etc.)

You may want to add additional criteria for your particular situation. There are more resources to help you in your repair/replace decisions. You can consult the manufacturer of your equipment, a trusted mechanical consultant, internal technical experts, independent mechanical service contractors, mechanical contractors, controls companies, and even your boiler and machinery insurance carrier.

If you choose the repair option, it is straightforward. You just repair the chiller and while you're at it, it may be a good time to convert your chiller to a new refrigerant, or add performance-enhancing components. If you choose the replace option, the choices are more numerous. You can choose an exact replacement, you can redesign your entire chilled water plant, or something in between.

The basis of this technical evaluation is to quantify and justify the costs of each decision. These costs can be divided into two areas: capital expenditures and operating expenses. In most organizations, replacing a piece of equipment would fall under the capital expenditure while servicing and repairing a piece of equipment would be categorized as an operating expense and come out of a separate budget.

If there is ample capital dollars and operating expenses are tight, an investment in a new chiller could reduce operating expenses because a new chiller comes with a warranty. Repairs will cost very little the first 5 to 10 years. Also, energy expenses will drop because a new chiller is more efficient. Conversely, if capital dollars are tight and operating dollars are easy to come by, you could increase maintenance and add energy saving upgrades to extend equipment life and avoid or delay a capital expenditure. I have several clients that have succesfully chosen one path or the other.

There is a third option in addition to the traditional choice between repairing or replacing a chiller: leasing, and for certain clients this may be a viable alternative. If the cooling need is seasonal, capital spending is being reduced, the heat load increases or decreases faster than permanent equipment can be delivered, or there is a lack of maintenance personnel, leasing or renting a chiller may be the most cost effective path. It may seem ludicrous at first glance, but renting a chiller or even an entire chilled water plant may be a financial tool usable to solve a technical problem.

Funding a chiller replacement is another matter. A creative option that helps with expensive retrofits is a concept called *performance contracting*. In a performance-based contract, a chiller retrofit can be paid for out of the energy savings created by replacing or upgrading the lighting, controls, and other mechanical equipment. Some performance contracts even guarantee the energy savings so there is little or no risk for the owner. Several HVAC controls and chiller manufacturers, energy service companies, and mechanical contractors offer this type of program.

Acknowledgments

The author acknowledges the help of the following individuals: Wesley Wojdon and Ken Mozek of York International; Dave Bishop and John Bauernfeind of Johnson Controls; and Cristi Johnson and Todd Elmgren of The Trane Company.

19.4

Absorption Chillers

Willis Schroader
Robur Corp., Evansville, Indiana*

Absorption chillers are machines that utilize heat energy directly to chill the circulating medium, usually water. The absorption cycle utilizes an absorbent and a refrigerant.

19.4.1 Description of the Cycle

The absorption cycle is not much different than the more familiar mechanical refrigeration cycle.

19.4.1.1 Mechanical refrigeration cycle

In the mechanical refrigeration cycle (Fig. 19.4.1), refrigerant vapor is drawn in by the compressor (1), compressed to high temperature and high pressure, and discharged into the condenser (2). In the condenser, the vapor is cooled and condensed to a high-pressure, high-temperature liquid by the relatively cooler water flowing through the condenser tubes.

The heat removed from the refrigerant is absorbed by the condenser water and is rejected to the atmosphere by the cooling tower.

The hot refrigerant liquid is metered through an expansion valve (3) into the low-pressure evaporator (4). The lower pressure causes

---

*Paragraph 19.4.1, written by Nick S. Cassimatis of Gas Energy Corp, reprinted from Handbook of HVAC Design, 1st ed., Chapter 41.

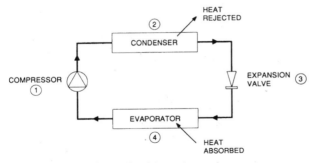

Figure 19.4.1 Mechanical refrigeration cycle.

some of the refrigerant to evaporate (flash), chilling the remaining liquid to a still lower temperature.

Heat is transferred from the warm system water (which is flowing through the evaporator tubes) to the cooler refrigerant. This exchange of heat causes the refrigerant to evaporate and the system water to cool.

19.4.1.2 Absorption cycle

The gas fired absorption chiller operates on an ammonia-water (ammonium hydroxide-NH40H) absorption refrigeration cycle (See Fig. 19.4.2). Ammonia is the refrigerant and distilled water is the absorbant. The solution charge in the unit is approximately two parts water to one part refrigerant. *It is a two pressure system.* Exact pressures are primarily controlled by the temperature of the ambient air drawn across the air cooled condenser-absorber section. Pressure separations are maintained during operation by restrictors in both the refrigerant and solution lines and a check valve on either side of the solution pump. Keep in mind that water has a very high affinity for the *refrigerant vapor. Strong solution* is solution which is strong in its refrigerant content. *Weak solution* is solution which is weak in its refrigerant content.

On cooling demand from an external control, heat from the *gas burners* is applied to the *generator,* causing the solution inside to boil. Since refrigerant vaporizes at a much lower temperature than water, a high percentage of the *refrigerant vapor* and a small amount of water vapor rises to the top of the generator. The remaining solution (*weak solution*) tends to gravitate to the bottom of the generator. As this boiling action takes place, high side pressure is increased. This pressure forces weak solution through a strainer and *restrictor* where it is reduced to *low side pressure,* and the weak solution is then metered into the solution-cooled *absorber.*

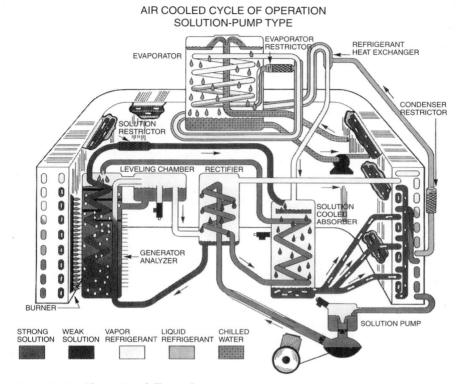

Figure 19.4.2 Absorption chiller cycle.

Now, lets go back and follow the path of the refrigerant vapor as it rises to the top of the generator. This vapor leaves the *generator* and enters the *leveling chamber*. The leveling chamber is designed to slow down the flow of vapor causing it to drop some of the water which is carried from the generator. Leaving the leveling chamber, the vapor now passes into the *rectifier*. The rectifier is simply a heat exchanger. Inside the rectifier is a coil through which strong solution is flowing to the generator. This solution is cooler than the vapor passing across it. When the hot vapor makes contact with the cooler coil it will condense any water vapor which it has carried from the generator and the leveling chamber. This weak solution drains to the bottom of the rectifier and returns through the condensate line to the generator. This is accomplished by thermal-siphoning action. Refrigerant vapor leaves the rectifier at *high pressure* and temperature. This vapor enters the U-shaped *condenser* coil. The condenser fan is moving ambient air across the outside of this coil, removing heat from the refrigerant vapor, causing a change of state from vapor to liquid. The liquid refrigerant is sub-cooled as it passes from the condenser through the first of two refrigerant strainer restrictor assemblies into the refrigerant heat exchanger. The liquid refrigerant then passes through a second refrigerant strainer restrictor assembly where it is reduced to low side pressure as it enters the *evaporator*. Flowing over the outside of the evaporator coil is water containing heat removed from the conditioned space. Heat from this water causes the refrigerant to boil at a low temperature. As the water gives up heat to the refrigerant, the refrigerant vaporizes and the water is chilled as it drains into the bottom of the chiller tank to be recirculated to the conditioned space. The refrigerant vapor leaves the evaporator and enters the inner section of the refrigerant heat exchanger. Here it picks up heat from the liquid refrigerant moving counterflow in the outer section of the *refrigerant heat exchanger*. The vapor then enters the solution cooled absorber where it is reunited with the weak solution. In the solution cooled absorber, the hot weak solution is distributed over a large coil area through which flows cooler strong solution. This temperature difference aids the absorption process, allowing refrigerant vapor to be entrained in the weak solution. This mixture now leaves the solution cooled absorber through multiple feeder tubes into the *air-cooled absorber*. This hot solution now passes through the air cooled absorber, giving heat to the cooler ambient air being drawn across the coils by the condenser fan. To insure complete absorption, the absorber coils cross over into the last passes of the outside coil exposing this circuit to the coolest ambient air. By the time the mixture passes through the absorber, the absorption process is completed, and we again have strong solution.

This strong solution enters the inlet tank of the *solution pump*. The solution pump is a volume displacement pump utilizing an inlet check valve, a discharge check valve and a Teflon® diaphragm operated in conjunction with a hydraulic pump designed to deliver a pulsating pressure from 0 to 400 psig (0 to 2757.9 kPa). At 0 psig, from the hydraulic pump, the inlet check valve is opened by unit pressure allowing solution to fill the upper cavity of the solution pump flexing the diaphragm downward. When the pressure above and below it are equalized, the spring loaded valve will close. The hydraulic pump now delivers a positive pressure to the bottom side of the diaphragm forcing it upward. As it does so, it will force the solution out through the discharge check valve and into the discharge line. As soon as the pressure is equalized above and below it, the spring loaded valve will close.

As the strong solution leaves the solution pump, it enters the inside coil of the rectifier where it removes heat from the hot vapor being passed across the coil. The strong solution enters the inside coil of the solution cooled absorber where additional heat is gained from weak solution being distributed over the outside of this coil. The strong solution then returns to the generator, completing the solution and vapor flow cycle at design conditions.

19.4.2 Maintenance

19.4.2.1 Care of service tools and equipment

For maximum safety, it is imperative that all service tools and equipment be kept in a safe and in good operating condition.

Periodic inspections should be made on all service tools and equipment which handle refrigerant 717 (ammonia). All gauges, gauge manifolds, and hose fittings must be steel. Refrigerant 717 is corrosive to brass and copper.

After each use, all hoses, gauges, manifolds, and buckets should be flushed and cleaned with water. They should be stored in a manner to prevent damage when carried on the service truck.

Hoses should be checked at the beginning of each cooling season and periodically during use. They should be checked for the following:

- Possible weak spots
- Cuts and cracked areas
- Condition of the threads and fittings
- Condition of the gasket

Gauge Manifold: should be checked at the beginning of each cooling season. Check for leaks around the valve stems, gauge connections, and fittings.

Gauges: should be checked prior to each cooling season for leaks and calibration.

Solution Tank: should be tested yearly by a company with the necessary test facilities. It should withstand a test pressure of 500 psig (3447.4 kPa). If there is a D.O.T. number stamped on the tank, it must be tested in accordance with federal regulations.

Refrigerant 717 Tank: *must* have a D.O.T. number stamped on the tank. It must be suitable for this type of refrigerant and tested in accordance with federal regulations.

Purging Bucket: should be no less than a two gallon (7.57 liters) bucket, with a handle, preferably strong plastic, and in a color other than yellow. Some method should be provided to hold the end of the purge line under water while purging, other than by hand.

It is recommended that a gas mask, approved for Refrigerant 717, be available for use if needed, particularly at service shops.

19.4.2.2 Safety information

Proper service and repair procedures are vital to the safety to those performing repairs. This manual outlines procedures for servicing and repairing chillers and chiller-heaters using safe, effective methods. These procedures contain many notes, cautions, and warnings which should be followed along with standard safety procedures to eliminate the possibility of personal injury.

It is important to note that repair procedures and techniques, tools, and parts for servicing chillers and chiller-heaters, as well as the skill and experience of the individual performing the work, vary widely. Standard and accepted safety precautions should be used when handling refrigerant 717 (ammonia) fluids and safety goggles or other protection should be used during cutting, grinding, chiseling, prying, or any other process that can cause material removal or projectiles.

Some procedures require special tools. Before substituting another tool or procedure, you must be completely satisfied that personal safety will not be endangered.

While working on chillers or chiller-heaters, or any other equipment, *safety should be foremost in our minds and actions.* Using good common sense can prevent many unnecessary accidents. Certain safety precautions should be taken when working on the sealed refrigeration unit. To be prepared for any unexpected incidents, the following precautions should be taken:

- Any time the sealed refrigeration unit is entered, safety goggles must be worn to protect the eyes.
- All service tools and equipment should be in good condition.
- Any time the sealed refrigeration unit is entered, consult the chemical safety information supplied in this book.
- Any time the sealed refrigeration unit is entered, a garden hose attached to a nearby faucet or a bucket of clean water should be handy.
- All personnel not involved in the actual work on the equipment should be kept at a safe distance.

Like all refrigerants, refrigerant 717 (ammonia) should be treated with respect. If refrigerant or refrigerant solution gets into eyes, on the skin, is inhaled or ingested, follow emergency and first-aid procedures in this section.

The excess refrigerant solution and/or water used for purging must be disposed of in accordance with local, state, and federal regulations.

If working on the chilled or chilled-hot water system and anti-freeze is spilled, wash the area thoroughly with water as it can be toxic.

19.4.2.3   Data on aqua-ammonia-chromate charging solution

A. **Composition**   **CAS No.**
   Ammonia   1336-21-6
   Sodium Chromate Tetrahydrate   10034-82-9
   Sodium Hydroxide   1310-73-2
   Water   N/A

B. *Physical Characteristics*
   Aqua-ammonia-chromate is yellow in color and has a strong ammonia odor. Solution is 100 percent soluble in water.

C. *Fire and Explosion Information*
   No special fire extinguisher required. Use any fire extinguisher required for surrounding fire. Use pressure demand self-contained breathing apparatus. Flash point, auto-ignition, and flammable (explosive) limits are not applicable to solution.

   Unusual Fire and Explosion Hazards
   Vapors in the range of 15 to 28 percent ammonia in the air can explode on contact with a source of ignition. Use of welding or flame-cutting equipment on a container *is not recommended* unless ammonia has been purged and rinsed with water.

D. *Emergency and First Aid Procedures*
   Eyes: Immediately flush eyes with plenty of water for at least 15 minutes holding eyelids apart to assure complete irrigation. Washing eyes within one minute is essential to achieve maximum effectiveness. Seek medical attention immediately.
   Skin: Flush area with soap and water for at least 15 minutes. Remove contaminated clothing and footwear and wash before re-use. Contact a physician for ammonia burns, if any, or any skin irritation.
   Inhalation: Get to fresh air immediately. If breathing has stopped, give artificial resuscitation. Oxygen may be administered by qualified personnel. If conscious, irrigate nasal and mouth with water. Seek medical attention immediately.
   Ingestion: Do not induce vomiting. If conscious, give the person large quantities of water or milk to drink immediately. If vomiting occurs spontaneously, keep airway clear. Seek medical attention immediately.

E. *Chronic Overexposure Effects*

Sodium Chromate is listed as a carcinogen as per NTP (1981) and IARC (1982) and is on the TSCA inventory.

F. *Reactivity*

Aqua-ammonia-chromate is considered stable.

G. *Acute*

Sodium Hydroxide: Corrosive to all body tissues with which it comes in contact. The effect of local dermal exposure may consist of multiple areas of superficial destruction of the skin or of primary irritant dermatitis. Similarly, inhalation of dust, spray, or mist may result in varying degrees of irritation or damage to the respiratory tract tissues and an increased susceptibility to respiratory illness. These effects only occur when the TLV is exceeded.

Ammonia: Ammonia is a severe irritant of the eyes, respiratory tract and skin. It may cause burning and tearing of the eyes, runny nose, coughing, chest pain, cessation of respiration, and death. It may cause severe breathing difficulties which may be delayed in onset. Exposure of the eyes to high concentrations may produce temporary blindness (cataracts and glaucoma) and severe eye damage. Exposure of the skin to high concentrations of the gas may cause burning and blistering of the skin. Contact with liquid NH_3 may produce blindness and severe eye and skin burns.

Sodium Chromate Tetrahydrate: Causes severe irritation to eyes and may cause blindness. May cause deep, penetrating ulcers on skin. May cause severe irritation of the respiratory tract and nasal septum, and possible perforation.

H. *Chronic Effects:*

Sodium Hydroxide: No known chronic effects.

Ammonia: No known chronic effects.

Sodium Chromate Tetrahydrate: Prolonged or repeated eye contact may cause conjunctivitis. Prolonged or repeated skin contact, especially with broken skin, may cause "chrome sores."

I. *Cancer Hazard* (*Sodium Chromate Tetrahydrate*)

Hexavalent chromium compounds as a group of chemicals have been listed as there being sufficient evidence of carcinogenicity to humans and animals.

Reference: IARC (International Agency for Research on Cancer) Monographs on the Evaluation of Carcinogenic Risks to Humans, Supplement 7, 1987. NTP (National Toxicology Program) Annual Report on Carcinogens, 1983.

There is laboratory evidence that aqueous sodium dichromate administered directly into the lung, at the highest tolerated dose, over the lifetime of rats, causes a significantly increased incidence of lung cancer. It is suspected that if sodium chromate were tested in the same manner as aqueous sodium dichromate, it would give a similar response. Other laboratory animal tests indicate that this product is carcinogenic to laboratory test animals.

Reactivity:

Hazardous polymerization will not occur. Avoid contact with organic materials, oils, greases, and any oxidizing materials. The solution is incompatible with strong acids. Ammonia reacts with acetaldehyde, acreoin, chlorine, fluorine, bromine mercury, gold, silver, silver solder, and hypochlorite to form explosive compounds. Avoid use of nonferrous metals. When ammonia is heated above 850°F (454°C), hydrogen is released. The decomposition temperature may be lowered to 575°F (300°C) by contact with certain metals such as nickel.

J. **Spill or Leak Procedures**

Contain spill, do not allow it to enter sewers or waterways. Ventilate area and allow ammonia to dissipate, if possible. Soak up with inert absorbent (sand) and place in labeled container(s) for proper disposal.

K. **Waste Disposal**

The solution must not be discharged into sewers or navigable waters or allowed to contaminate under ground water sources. Waste solutions should be reclaimed, if possible. If reclamation is not possible, contact local waste disposal contractor for proper disposal.

L. **Safety Equipment and Health Tips**

a. Use chemical goggles, a face shield, and wear protective clothing: solution is very irritating to the eyes. Overexposure to chromate in the solution has the potential to permanently damage the eyes.

b. Wear rubber gloves: aqua-ammonia can damage skin allowing exposure to chromium.

c. Work in a well ventilated area, or if in a confined space, use a NIOSH/MSHA approved respirator. Ammonia is a strong irritant and may damage mucous membranes.

d. Do not swallow solution. This material is alkali and may damage tissue.

e. Good hygiene dictates washing hands after handling this material. Wash protective clothing before reuse.

f. If an emergency shower and eyewash are not available, keep a bucket of fresh water or a garden hose readily available.

19.4.3   Unit Analysis

19.4.3.1   Checking externals

THE FOLLOWING EXTERNAL CHECKS SHOULD BE MADE IN THE EXACT ORDER THAT THEY APPEAR BELOW, PRIOR TO CHECKING THE INTERNAL OPERATION OF THE UNIT. ANY ATTEMPTS AT SHORTCUTS OR FAILING TO MAKE THESE CHECKS IN THE ORDER THAT THEY APPEAR BELOW COULD RESULT IN AN INCORRECT ANALYSIS OF UNIT OPERATION.

General appearance.   Check the general appearance of the equipment for such things as proper support, clearance, etc. *A yellow residue or an ammonia odor, if detected, could indicate a sealed system leak.*

Air flow through the condenser.   Since the unit capacity is reduced by failure to properly remove heat from the unit by air passing over the condenser-absorber coils, it is recommended that these coils be inspected and cleaned as required.

When inspecting for debris on the condenser-absorber coils, *check the inner coil as well as the outer coil.*

Unit level.   The unit should be level both front to back and side to side. The level can be checked by placing a level on the top of the unit. If the unit is not level, metal shims are recommended for use under proper corners to level. If more than ½-in. (12.7-mm) thickness of shims is needed under one corner or end, support shims should be inserted at center of long frame rail.

The unit must be level to provide even water flow over the surface of the evaporator.

Air flow—air handler or surface.   Rated air flow across the coil or coils is 1200 CFM (34 m³/min) for 3-ton (11 kW) units, 1600 CFM (45.3 m³/min) for 4-ton (14 kW) units, and 2000 CFM (56.6 m³/min) for 5-ton (17.5 kW) units. For all sizes of air handlers, the air flow rate is 400 CFM (11.3 m³/min) per ton.

For the correct adjustment, set blower to obtain an 18–20°F (-7.78 to -6.67°C) drop in air temperature across the coil, with an 8–10°F (-13.33 to -12.22°C) drop in water temperature across the chiller.

Water level and flow rate.

a. Turn off power and gas to unit.

b. With electrical supply to the unit turned off, disconnect condenser fan motor wires at control panel and tape exposed ends.

c. Remove top panel. Remove four bolts attaching fan support to side panels and set condenser fan assembly on generator/leveling chamber assembly.

d. Remove the styrofoam chiller tank and the distribution pan cover. **NOTE:** *The chiller tank top is held in place with metal tabs pushed under a plastic band. To remove these tabs, place a small piece of metal, such as a six inch (152 mm) ruler between the plastic band and metal hold-down tab. This releases the band from a notch on the metal tab. The tab can now be raised straight up and away from the top.*

e. Start the chiller or chiller-heater electrically.

f. In order to establish correct chilled water level, with the unit operating electrically, drain excess water from chiller tank. This is done by extending fill test hose to horizontal position at the "0" notch (see Fig. 19.4.3).

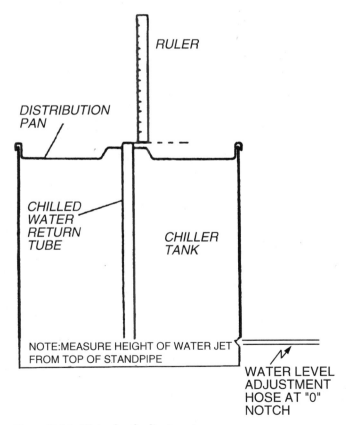

RULER

DISTRIBUTION PAN

CHILLED WATER RETURN TUBE

CHILLER TANK

NOTE:MEASURE HEIGHT OF WATER JET FROM TOP OF STANDPIPE

WATER LEVEL ADJUSTMENT HOSE AT "0" NOTCH

Figure 19.4.3 Water level adjustment.

g. After correct water level is established, flow rate must be adjusted.

h. Place a ruler on the chilled water return tube and adjust height of water in accordance with Table 19.4.1 below.

i. After setting correct water flow, turn power *off.* Check wiring diagram and replace condenser fan wires on correct terminals.

19.4.3.2 Purging unit of noncondensibles

Caution *Always wear proper safety equipment and know safety data on the refrigerant solution charge.*

Note.   The unit should be purged only after all externals have been checked and corrected and the unit is still not cooling properly, or anytime the unit has been off on the generator high temperature limit switch. Purging should be done with the unit in operation (burner on) and with at least a purge valve gauge pressure of 5 psig (34.5 kPa). Proceed as follows:

1. Remove cap from purge valve on top of purge chamber (See Fig. 19.4.4). Check to be sure valve is completely closed. Remove pipe plug from valve.

2. Connect ¼-in. (6.35 mm) steel adapter fitting (¼-in. [6.35 mm] pipe thread to ¼″ [6.35 mm] flare thread) to purge valve on solution pump.

3. Connect purge hose to adapter and immerse loose end at least 6-in. (152.4 mm) below surface of water in bucket. If purge hose is not available, a ¼-in. (6.35 mm) steel flare nut and a length of ¼-in. (6.35 mm) aluminum tubing can be used.

4. Open valve just enough for bubbles to come out the end of the hose into the water. If noncondensible gases are present, bubbles will rise to surface of water. Ammonia vapor will be absorbed by the water and this is indicated by a cracking sound. Close purge valve immediately if yellow solution appears in the purge bucket.

5. Close purge valve when bubbles do not rise to surface of water.

6. Remove purge hose and adapter. Replace cap and plug.

TABLE 19.4.1 Water Weight Adjustment

Unit size	Flow rate	Water column
3 ton (11 kW)	7.2 GPM (27.4 liters/min)	4″ (101.6 mm)
4 ton (14 kW)	9.6 GPM (36.5 liters/min)	7″ (177.8 mm)
5 ton (17.5 kW)	12 GPM (45.6 liters/min)	6¼″ (158.8 mm)

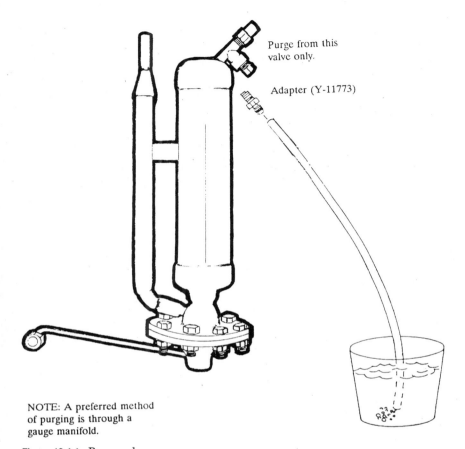

Purge from this valve only.

Adapter (Y-11773)

NOTE: A preferred method of purging is through a gauge manifold.

Figure 19.4.4 Purge valve.

NOTE.   Dispose of removed solution in accordance with local, state, and federal regulations.

19.4.3.3 Solution pump operation

Turn the unit off at the thermostat, or disconnect the "Y" wire at the low voltage terminal strip. Allow the unit to go through a normal Time Delay shut-down and leave out of operation for a minimum of ten minutes.

During the ten minutes "off" period, connect gauges:

- Use gauge manifolds with charging hoses.
- Connect the low pressure gauge to the low side valve located on the side of the SCA.

- Connect the high pressure gauge to the high side valve located on the leveling chamber.

- Connect the low pressure gauge to the purge valve located on the solution pump.

Close valves on gauge manifolds.

Open the high side, low side, and purge valves on the unit ¼ turn.

This test simply determines the ability of the solution pump to move solution. If the purge gauge on the unit has continued deflection for more than 5 minutes, replace the solution pump before continuing to further unit analysis as stated below. If the purge gauge on the unit shows either of the other patterns shown below in Fig. 19.4.5, continue the diagnostic procedures.

1. On pilot burner models, turn the gas valve to the "pilot" position. On DSI models, disconnect the RED wire from the chilled water limit switch.

2. Start the unit electrically.

3. Observe gauge patterns.

19.4.3.4 Solution restrictor flow check and gauge readings

TURN THE GAS VALVE TO "ON" POSITION.

Having checked the operation of the solution pump, turn the gas valve on and reconnect the red (#33) wire to the chilled water limit switch.

Immediately grasp the solution strainer-restrictor line, as illustrated below (see Fig. 19.4.6). If there is weak solution flow through the line, it will get hot as heat travels from the generator toward the solution cooled absorber. This is a preliminary step to be used in fur-

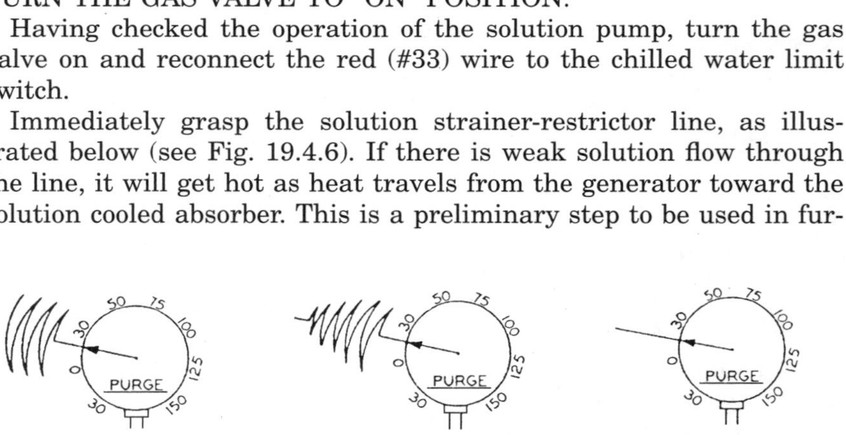

Deflection Stops within 5 minutes
Diagnosis:
Solution pump capable of emptying, continue check out procedure.

Wide Deflection - more than 5 minutes.
Diagnosis:
Solution pump malfunction, replace solution pump.

No Deflection within 5 minutes.
Diagnosis:
Possible store out, continue check out procedure.

Figure 19.4.5 Purge gauge patterns.

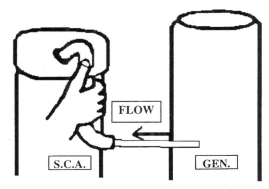

Figure 19.4.6 Solution restrictor check.

ther analysis. The indication of solution flowing through the line does not eliminate the possibility of a partially plugged solution restrictor.

After determining that solution is flowing through the solution restrictor, observe the pressure patterns on the gauges that are connected to the unit. Once the pressure patterns are established, record both those pressures and the ambient temperature at the time the pressures were taken. The next section (Trouble-Shooting with Gauges) includes a temperature/pressure chart (see Tables 19.4.2 and 19.4.3) and gauge illustrations (Fig. 19.4.7) which, with this information, will enable you to determine the problem.

19.4.2.5 Trouble-shooting with gauges

Use Table 19.4.2 to determine what the pressures should be at the temperature you recorded in the Solution Restrictor Flow Check and Gauge Readings section. Also, at the time the ambient temperature was noted, you recorded the pressures shown on the gauges. Comparing those readings with pressures found in the table, determine whether the unit is operating at normal pressures, or above normal pressures.

If the pressures are above or below normal, the next step is to compare them with the gauge readings on the next two pages, eliminating all that do not match as possible problems. Once you narrow it down, observe the solution pump gauge to see how it is operating: normal with 2–4 lbs (.9–1.8 kg) of deflection, normal to above normal, normal to erratic to normal, or erratic. That should pinpoint the problem. After determining the problem, you will be interested in how high or how low the gauges are reading and how they are acting. Make notes on your worksheet as to how the pressures are acting.

TABLE 19.4.2 Pressure/Temperature

Ambient air °F	Low side ±5 PSIG		Pump pressure normal deflection 2 to 4 pounds			High side ±10 PSIG			Ambient air °F
	3 Ton 5 Ton	4 Ton	3 Ton	4 Ton	5 Ton	3 Ton	4 Ton	5 Ton	
60	20	17				195	175	180	60
65	25	22				210	190	195	65
70	29	26				225	205	210	70
75	33	30	8–12	7–10	10–15	240	220	225	75
80	38	35	PSIG Below	PSIG Below	PSIG Below	225	235	240	80
85	43	40	Low Side	Low Side	Low Side	270	250	255	85
90	48	45				285	265	270	90
95	53	50				300	280	285	95
100	57	54				320	300	305	100
105	61	58				335	315	320	105
110	65	62				350	335	340	110
115	68	65				365	350	355	115

TABLE 19.4.3 Pressure/Temperature (metric)

Ambient air °C	Low side ±34.5 KPA		Pump pressure normal deflection .9 to 1.8 kg			High side ±10 PSIG			Ambient air °C
	3 Ton 5 Ton	4 Ton	3 Ton	4 Ton	5 Ton	3 Ton	4 Ton	5 Ton	
15.6	137.9	117.2				1344.5	1206.6	1241.0	15.6
18.3	172.4	151.7				1447.9	1310.0	1344.5	18.3
21.1	200.0	179.3				1551.3	1413.4	1447.9	21.1
23.9	227.5	206.8	55.2	48.3	69.0	1654.7	1516.8	1551.3	23.9
26.7	262.0	241.3	to 82.7	to 69.0	to 103.4	1758.2	1620.3	1654.7	26.7
29.4	296.5	275.8	KPA Below	KPA Below	KPA Below	1861.6	1723.7	1758.2	29.4
32.2	331.0	310.3	Low Side	Low Side	Low Side	1965.0	1827.1	1861.6	32.2
35.0	365.4	344.7				2068.4	1930.5	1965.0	35.0
37.8	393.0	372.3				2206.3	2068.4	2102.9	37.8
40.6	420.6	399.9				2309.7	2171.8	2206.3	40.6
43.3	448.2	427.5				2413.1	2309.7	2344.2	43.3
46.1	468.8	488.2				2516.5	2413.1	2447.6	46.1

NO DEFLECTION   DOES NOT RISE   DOES NOT RISE

DIAGNOSIS:
REFRIGERANT STORE OUT

CORRECTION:
STORE OUT CORRECTION PROCEDURE
SEE TROUBLE-SHOOTING MANUAL:[1] PAGES 15-16

NOTE: Dotted arrow denotes approximate
pressures at start-up.

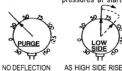

NO DEFLECTION   AS HIGH SIDE RISES   MAY NOT RISE BUT IF
           LOW SIDE FALLS   IT RISES IT WILL
                        THEN FALL

DIAGNOSIS:
SOLUTION STORE OUT

CORRECTION:
STORE OUT CORRECTION PROCEDURE
SEE TROUBLE-SHOOTING MANUAL:[1] PAGES 15-16

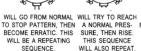

WILL GO FROM NORMAL   WILL TRY TO REACH   WILL FALL BEFORE
TO STOP PATTERN, THEN   A NORMAL PRES-   REACHING A NORMAL
BECOME ERRATIC. THIS   SURE, THEN RISE.   PRESSURE AND
WILL BE A REPEATING   THIS SEQUENCE   ESTABLISH A
SEQUENCE.   WILL ALSO REPEAT.   FLUCTUATING
                              PATTERN. THE HIGH
                              POINT OF THE RISE
                              WILL BE THE SAME IN
                              EACH SEQUENCE.

DIAGNOSIS:
DEFECTIVE FLOW CONTROL VALVE

CORRECTION:
REPLACE THE HYDRAULIC PUMP FLOW CONTROL VALVE
SEE TROUBLE-SHOOTING MANUAL[1] PAGES 20-22

NO DEFLECTION   DOES NOT RISE   DOES NOT RISE

DIAGNOSIS:
PARTIALLY PLUGGED SOLUTION STRAINER RESTRICTOR

CORRECTION:
CHANGE SOLUTION STRAINER RESTRICTOR ASSEMBLY
SEE REPAIR MANUAL:[2] PAGES 100-102 (3 TON),
                       PAGES 111-113 (4 & 5 TON)

NO DEFLECTION   HIGH AND STEADY   LOWER THAN
                              NORMAL AND MAY
                              YO-YO

DIAGNOSIS:
PLUGGED SOLUTION STRAINER RESTRICTOR

CORRECTION:
CHANGE SOLUTION STRAINER RESTRICTOR ASSEMBLY
SEE REPAIR MANUAL:[1] PAGES 100-102 (3 TON),
                       PAGES 111-113 (4 & 5 TON)

AS UNIT CONTINUES TO   EXCEEDS NORMAL   RISES BUT FALLS
OPERATE, PURGE   OPERATING   BEFORE REACHING
DEFLECTION GRADUALLY   PRESSURE   NORMAL PRESSURE
INCREASES (BEYOND   FOR PREVAILING
NORMAL 2-4 LBS.)   AMBIENT

DIAGNOSIS:
IMPROPER SEATING OF HYDRAULIC FLOW CONTROL VALVE OR
SOLUTION PUMP CHECK VALVE

CORRECTION:
CHECK FLOW CONTROL VALVE FIRST. IF PRESSURE PATTERN
CONTINUES, REPLACE SOLUTION PUMP.
SEE TROUBLE-SHOOTING MANUAL[1] PAGES 20-22

Figure 19.4.7a   Trouble shooting with gauges.

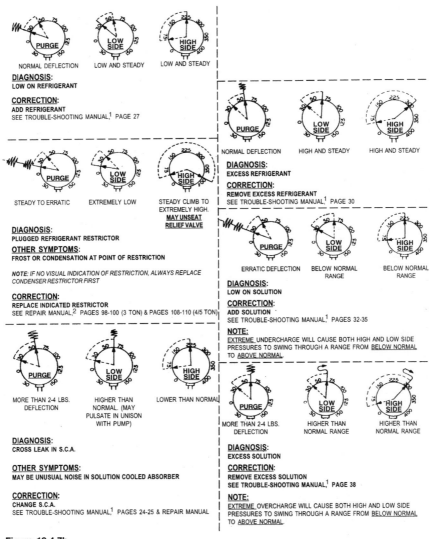

Figure 19.4.7b

CAUTION: Should the high-side gauge ever reach 350 lbs (157.5 kg), shut the gas off immediately.

In Table 19.4.3 the pressures and temperatures of Table 19.4.2 have been converted to metric measures.

19.4.2.6   References

1. Trouble-Shooting Manual, form AS42502, published by Robur Corp., Evansville, IN.
2. Repair Manual, form AS 42451, published by Robur Corp., Evansville, IN.

19.5

Cooling Towers

Joe Gosmano

Western Regional Manager, Marley Cooling Tower,
BREA, California

19.5.1 Role of the Cooling Tower in the HVAC System

In the simplest of terms, a cooling tower is the vehicle that dissipates heat that is generated by equipment to the atmosphere. During the process of working, most air conditioning and industrial systems produce various amounts of heat. To extract this heat from the exchanger or condenser, water is the most commonly used cooling mechanism.

Figure 19.5.1 is a simple look at the cooling tower and its relationship to the air conditioning unit.[1] The cooling water flows from the cold well of the tower to the condenser, thus removing heat from the condenser. As the unit works and generates heat, water flows through the unit. The heated water is then returned to the tower for cooling. The water circulates over the top of the unit, and as it falls through packing inside the tower, it is contacted with air to facilitate the removal of the heat.

In the past, cooling was accomplished by using streams of water, either from a well or nearby water source. Once the water collected the heat to be removed, it returned to the stream or river and the process was repeated. This circular process required an unlimited supply of water and quickly became expensive. The role of environmental agencies emerged when concerns arose about the heated water and its return to rivers and streams. The heated water, it was concluded, may have an effect on the wildlife.

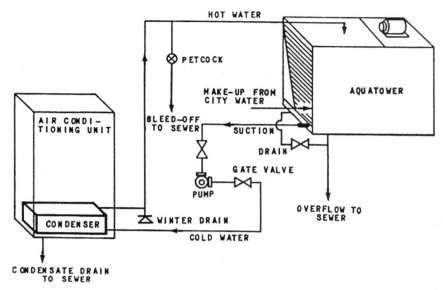

Figure 19.5.1 Typical piping diagram.

Because of these environmental concerns, and most importantly, because the supply of water is limited, the cooling tower was introduced as an efficient means to control a precious resource.

On average, cooling towers reduce the amount of water necessary for cooling a once-through system to less than 5 percent.[2] Towers can also cool to within 5°F (3°C) of the ambient wet bulb temperature. A cooling tower cools by combining water and air. Typically, water is distributed over such things as a heat transfer media, a packing material of wood or pvc (polyvinylchloride). As it falls through the media, air is drawn across or past that fill material. The water droplets are exposed to the passing air and the transfer of heat is accomplished. The air can be introduced either through fans or by natural draft. Either method requires an adequate amount of air to properly contact the water. When using a motorized fan to introduce the air, issues of energy consumption arise.

When water flows and horsepower moves the cooling tower fan, the tower must be clean and properly maintained to be efficient and in the best possible working condition.

19.5.2 Impact of the Cooling Tower on the System

Cooling tower manufacturers design the unit for a given set of performance conditions, such as by the type of chiller used, ambient temperatures, location and specifications. As the system ages it may lose efficiency. *The cleanliness of the tower and its components become crucial to the success of the system.* If unattended, the cold water temperature will rise, sending warmer water to the chiller. When the chiller kicks out on high head pressure, the system may shut down. Certain precautions can be taken to keep this closely shut down from occurring.

Cooling towers are usually remotely located; it becomes necessary to regularly inspect and clean the tower according to the manufacturers' recommendations. The few hours each month spent on inspecting at the cooling tower and maintaining it will pay dividends. The life of a tower varies according to materials of construction, location within the system and the location of the city or country.

Typically, the premium materials of construction are wood, concrete, stainless steel and fiberglass. These units are expected to last from 20 to 30 years if properly cared for. The less expensive units, made of galvanized steel, will operate for 8 to 20 years. Of course, tower life will vary due to the extremes of weather, number of hours used each year and type of water treatment. It is sufficient to say that in order to get the most out of the tower, cooling tower manufacturers want to make that tower last as long as possible.

19.5.3 Types of Cooling Towers

In the air conditioning world there are considered four types of cooling towers that play key roles: induced draft crossflow (single flow), induced draft crossflow (double flow), induced draft counterflow and counterflow forced draft. All serve the same purpose, but meet various specified needs.

The induced draft crossflow (see Figs. 19.5.2 and 19.5.2a) are propeller fan towers. Both are mechanical draft, meaning they use a fan to introduce air to cool the water. With the induced draft design the fan is located in the exiting air stream and air is drawn through the tower. In the crossflow configuration, the water flow in the fill (packing) is perpendicular to the flow of air. Water is delivered to the top of the tower and as it falls, air is drawn across the fill. The *singleflow* has the packing and water flow on one side only, while the *doubleflow* takes advantage of two sides.

The induced draft counterflow (Fig. 19.5.3) also uses the fan in the leaving air stream. In counterflow towers the air moves counter, or vertically upward, through the fill. The air and water make contact as they cross paths. This type of tower is characterized by the high-pressure spray system, taller design and (compared to the crossflow design) may use much more power.

Figure 19.5.2 Induced draft singleflow.

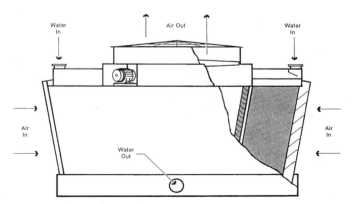

Figure 19.5.2a Induced draft doubleflow.

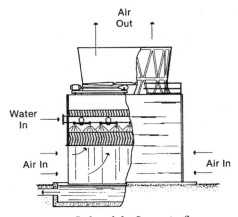

Figure 19.5.3 Induced draft counterflow.

The counterflow forced draft (Fig. 19.5.4) is the fourth type of conventional air conditioning tower. It is characterized by the fan located on the side in the airstream, but the air is blown through the tower. These towers are more susceptible to hot discharge air recirculation and may have less performance stability than induced draft towers. Usually these towers are equipped with centrifugal blower type fans which require considerably more horsepower than the propeller fan. These types of towers are often located in smaller enclosures or indoors and ducted to the atmosphere.

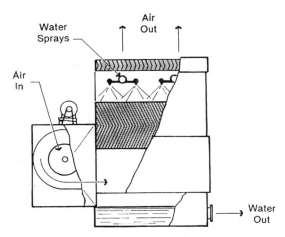

Figure 19.5.4 Forced draft counterflow.

19.5.4   Materials of Construction

Industrywide, there are four types of typical exterior construction materials: *galvanized steel, fiberglass, stainless steel and wood.* All are usually factory assembled and delivered to the jobsite ready for installation. And in some cases, the larger designs are pre-fabricated and assembled at the location.

For packaged tower designs that range from 10 tons (9 metric tons) of air conditioning to 1000 tons (900 metric tons) in a single unit, steel is the material of choice. Steel's strength makes it economical. In most cases, with normal circulating water conditions, the structure, hardware, sides and top and bottom pans are made of mill galvanized material. For severe water service many components must be coated in addition to being galvanized. Coatings offer the advantage of added protection, but with that comes the need for added maintenance. Coatings are susceptible to peeling, thereby exposing bare metal to the severe water. Added steps are necessary to achieve long-term protection.

When the process is severe, or when the tower owner wants extended life, stainless steel should be considered as a construction material. Although it is the premium material for cooling towers, stainless steel represents significant added expense. Another reason to consider stainless is the tower location. If in an area that is not easily accessible for change-out, such as on the roof of a tall building, stainless steel offers the one-time expense for equipment and installation.

Fiberglass is the latest material to enter the market. The capability to be molded into single parts of complex shapes and dimensions offers distinct advantages. Generally used only for the fan blades, louvers and external casing, fiberglass has become a successful material for the tower structure. The lighter weight components allows for installation in areas sensitive to high concentrations of load. Typically, fiberglass is less susceptible to abnormal water conditions because it is inherently resistant to microbiological attack, corrosion and erosion. It is more expensive than the standard galvanized steel version, but the trade-off with extended life and lighter weight offer advantages unavailable when using other construction materials.

For larger installations, the use of wood as the primary building component has existed for years. It is readily available, relatively low cost and extremely durable under a variety of weather and water conditions. Such a structure is delivered ready for field installation. Douglas fir and redwood, because of their external structural capabilities, are the most common types of wood used. Most wood towers are built in the crossflow and counterflow induced draft designs. Regardless of the type of wood, it must be treated for long-term usage and protection against bacteria and fungal attack. Pressure treatment takes the form of chromate copper arsenate or acid copper chromate. This preservative, which offers life to the tower, is injected into the wood under pressure. Although a wood tower is the most expensive, it offers the longest service life and easiest to repair.

19.5.5   Maintenance Procedures

As the seasons change, the maintenance staff is offered a chance to inspect and take necessary steps to ensure the cooling tower will remain trouble free for many years to come. *Late fall is the best time to start inspections.* Beginning inspections around November gives the staff a chance to make repairs during the early winter months. The milder days allow crews to make repairs before the bad days of winter set in. It also guarantees the unit will be ready for the hot summer months and peak cooling period.

Figure 19.5.5 is the anatomy of a cooling tower. Regardless of size, materials of construction or manufacturer, all towers have similar parts. Structure, water distribution, heat transfer media and mechan-

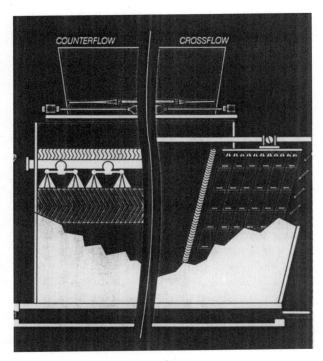

Figure 19.5.5 Anatomy of a cooling tower.

ical components are the mainstay of all units. These are the major areas to inspect, repair and keep in good operating condition.

Before any attempt is made to begin the cooling tower inspection, safety issues must be addressed. No inspection should be conducted with safety gear that fails to meet local safety codes. If the inspection is conducted with the fan running and water flowing, use caution when opening inspection doors or walking around the top of the unit. Always be aware of the hazards of being around operating mechanical components. If any preliminary corrections are going to be made or work performed, always shut off the electrical power to the tower fan motor. All electrical switches should be locked out and tagged to prevent others from turning the power back on. This is especially true if the power can be turned on from a remote source, such as the main control panel which may be away from the cooling tower.

19.5.6 Tower Inspection

Walk around the operating cooling tower and visually inspect the structure. Look for signs of leaks, cracks or deterioration. Notice any separations in the end panels and make certain connections are tight and firmly attached to the structure. In most cases the end panel casing, or skin material, is fiberglass, steel or wood. In older units it may be an asbestos cement board or corrugated material. If the budget allows or if corporate policies require, consider replacing it immediately — regardless of condition. Inspect wood casing for signs of rot or plywood separation, commonly referred to as delamination. When inspecting a steel tower, look for corrosion or signs of pitting. If the skin is fiberglass, look for brittleness, cracking or fiber separation.

Next, conduct a thorough inspection of the structure, making sure the fan and water are turned off. This allows easy, safe entry and ease in inspecting. When inspecting a steel tower, look for corrosion. Most importantly, loss of metal or major areas of discoloration. In addition spot check tightness of bolts, making certain the size of bolt hole has not created an area where water can leak through. Check any welds for cracks. This will mean cleaning an area near welds to get an up-close look. If possible go under the tower. The cold water collection basin in most towers and the associated structure are the least inspected. They are not easy to access, yet a leak or failure in these locations creates major problems. Like an automobile, anything above the frame can be replaced or repaired, but trying to fix the frame itself is a very expensive proposition. If there are leaks or pin holes, use an oversized rubber-backed washer and sheet metal screw to fix. If the area is large, use a sheet metal panel and silicon-based caulk to cover the affected area. Of course, the area must be cleaned and ultimately recoated with a high grade epoxy coating.

The inspection process for a wood tower is very similar. First, look for signs of rot or decay. Inspect with a small hammer and tap the wood member. A dull, low-pitched sound indicates softness; a sharp, high-pitched sound indicates good, solid wood. If areas of rot exist, probe them carefully with a knife or ice-pick, paying attention to the depth of penetration of the probe. Especially examine the areas around the bolts and where there is contact with steel or cast iron.

When checking the tightness of bolts, look at the indentations in the wood. Be sure they are not excessive. The joint connectors, either steel or fiberglass, must be inspected. Look for deterioration, cracking or corrosion. Some areas in the cooling tower are inaccessible. For example, in the heat transfer media, material must be moved before there is access to the unit. Reach as many bolts as possible and along the way check all wood pieces with the tap of the hammer.

When checking fiberglass or coated steel towers, check the assembled joints for tightness. Look for cracking or signs of distress in the coating. Many minor scratches and scrapes which do not require extraordinary corrosion protection can be touched up with a repair kit.[3] If the extent of damage or protection is uncertain contact the manufacturer.

The most often overlooked area of the structure is the top surface or fan deck. It is always a good idea to get on top and take a firsthand look. On steel towers, look for pitting or corrosion. On wood units watch for softness of the lumber, delamination of the plywood or holes caused by rot. A lot of information can be gained by lightly jumping up and down. The spring action under your feet is an indication of softness. Remember, it may not be the deck itself, but the supports underneath. They should be accessed from the bottom and ends. Be aware of uneven walking surfaces. The sheets of plywood on wood towers or connections on steel towers can come apart or shift over time, presenting a tripping hazard. In all cases, the connections should be tight and in line.

19.5.7   Water Distribution System

The heart of the cooling tower is the *water distribution system*. Before the cooling can be accomplished, the water must flow evenly over the heat transfer media (fill). Regardless of the type of tower and internals, a pattern of water fall must be established and maintained throughout the life of the unit. Keeping the components of the distribution system clean and operational is a primary function of maintenance.

Starting at the top of the tower, inspect the hot water distribution basins, also referred to as *water pans*. These metal or wood boxes are located on the top of the tower and hold the water before it cascades over the fill. In steel units check for corrosion and loss of metal. If the basin is wood, usually plywood, look for delamination or wood decay. In both cases check for leaks between joints. Inspect the integrity of all basin support members. Check the tightness of bolted joints in steel or wood basins.

Each tower basin contains nozzles. In crossflow designs, they are a plastic holes with an extended stem that hangs off the bottom. From the basin top, all that is visible is the plastic hole. Carefully clean all debris that may be clogging the nozzle opening. Do not use a probe that enters the nozzle, to clean away sludge, leaves and the like. Too long a probe, like a broom handle, could break off the bottom part of the nozzle, reducing the effectiveness of the nozzle.

Counterflow type towers contain a spray-type nozzle. To properly inspect these nozzles, it is necessary to gain access into the tower and remove them. They generally screw into the piping system and can be easily removed. Clean away debris and reinstall. It is always important to have the proper number of nozzles installed. Do not insert a plug in place of a nozzle—the performance of the system could be affected.

On some towers, flow control valves are located on top of these basins. Their purpose is to even the water flow across the entire area. Each valve has a grease fitting which keeps the valve disc and stem lubricated. Follow the vendor's maintenance instructions and lubricate at least once a year. The flow control valves have a locking bar which must be properly turned to lock the valve disc in place. Under water pressure the disc tends to shift and, if not locked down, will eventually fail.

The water is brought to the tower by means of a distribution pipe, generally located on top of the tower. Inspect the pipe, usually made of iron, for signs of corrosion or loss of coatings. Examine all supports for structural integrity. Spot check for leaks and check the tightness of all bolts. On fiberglass pipe look for signs of cracking or distress.

Many towers are equipped with a single inlet pipe arrangement, whereby water enters into a chamber before being distributed to the upper basin. This arrangement is used when there are no balancing valves involved. Usually it is equipped with an internal strainer and blow-down connection. The strainer can be easily removed and cleaned. The bottom of the chamber has a plugged connection, allowing for dirt and debris to be collected for removal.[4]

The lower water basin, where the water is collected, accumulates sludge and debris. Periodically, when the tower can be shut down, it should be cleaned. This removal of deposits keeps the sediment from getting corrosive and attacking the basin metal or wood. It also removes a potential breeding ground for harmful bacteria. The sumps and screens should be inspected and cleaned of trash to allow for the free flow of return water.

19.5.8 Heat Transfer Media

Inside every cooling tower are heat transfer media, commonly referred to as "fill." Its only function is to ensure that the water droplets are broken up to allow for easy contact with the incoming air. The evaporation of water provides the cooling. There are various types of fill arrangements and materials, the most common being splash and film, either constructed of wood or polyvinylchloride (pvc). See Figs. 19.5.6a and 19.5.6b.

There are several types of splash fill. Generally they consist of wood or plastics bars supported in wire or fiberglass hangers. Ceramic tile is also a type of splash fill. When inspecting splash fill bars, which can be viewed from the outside of the tower, look for sagging, broken or decayed bars. These bars can be seen from the outside of the tower.

Figure 19.5.6a Splash-type fill.

Figure 19.5.6b Film-type fill.

There may also be scale buildup on the bars. Also, pay attention to the pattern of the bars—misplaced bars will not act as efficiently as designed. If water is running over the fill, look at the pattern of water flow, making certain the flow is even.

When inspecting the supporting grids or hangers, see that they are in place and not sagging. Check the coating, if there is any, on steel grids and the condition of the welds on stainless steel grids. If the grids are fiberglass, check the condition and look for any breakage. Any section that shows excessive deterioration should be replaced. Most importantly, always look at the supporting structure where the grids are attached. Pay particular attention to sagging or damaged pieces.

Film fill, which consists of several sheets of pvc, is a second type of packing. Film fill is either hung from tubes or resting on the bottom of the structure. For this type of packing, look for buildup of algae or scale on the surface and within the gaps of the sheets. Also check for sagging, tears, erosion or ice damage. Continue to inspect the supporting members just as you would for the splash type.

Some older towers contain fill types that were manufactured from asbestos material. When possible, *consider replacing them immediately.* Besides removing the environmental concerns, the new fills of today are more efficient than the older type. The benefit of increased performance could pay for these repairs.

The next major component is the drift (mist) eliminators, which are located inside the doors at each end of a crossflow tower and above the spray system of a counterflow tower. Usually wood or pvc, a visual inspection can be done from either inside or outside the tower.

To work properly, all air passages must be clear of mud and debris. Look at the arrangement of the eliminator and note any shifting or dislodging of material. Check for visible gaps between the packs or frames. These seals should be caulked, and in some cases, replaced to prevent air and water from exiting the tower. Most eliminators are wood or plastic bars supported in a framework of wood, or pvc cellular packs.

Lastly, go outside and look at the front of the tower. The louvers, which are attached to the structure, are angled pieces that direct the air into and keep water droplets from splashing out of the tower. Usually of wood, steel or pvc, these pieces should be in place and free of algae or scale buildup. They should be clean and sitting properly in the supports. Check the condition of both the individual louvers and the supporting members. Look for signs of deterioration. In newer units the louvers are pvc and molded into the face of the fill sheets. Inspect and clean as you would the fill.

Older models used asbestos-containing materials. *Consider replacing them immediately.*

19.5.9 Mechanical Equipment

Cooling towers usually have two methods of turning the fan: either through belts or a gearbox (transmission). For belt drive systems, check the pulleys and belts at least monthly. Look for signs of corrosion or loss of metal on the pulleys. On the belts, identify wear patterns. If nicks or gouges exist consider replacing the belt. The procedure for belt tightening depends on the manufacturer. Review the operation and maintenance manuals provided for recommendations and frequency.

The pulleys are driven through a bearing housing or pillow block bearings. Lubricate as instructed by the manuals. Check for loose or damaged bearings. Does the input or output shaft have any play? Look at the support for this housing. Identify signs of wear or corrosion and repair as necessary. The bolted connections should be inspected and tightened.

Because of their design, gearbox driven units have more areas to pay attention to. The primary concerns are the oil level and quality of the oil during operation. Like an automobile, the oil level should be checked when the unit is off and cold. Each manufacturer will provide a minimum time interval between oil changes, but as a minimum, it should be semiannually. Some of the newer gearboxes have a new sealed design that uses, synthetic oil, which requires an oil change every 5 years.[5]

When examining the oil, drain a little off and look for water, sludge or metal shavings. Check for leaks at both the input and output seals. If may be necessary to replace these seals. Rotate the input shaft and be sure there is no slippage. The gear teeth should engage without excessive movement. The endplay on the gears can be determined by pulling up and down on the fan blade. When uncertain about the amount of movement, a service contractor or cooling tower manufacturer can help determine the desirable amount of movement.

In observing the gearbox housing, as with any metal part, check for corrosion or loss of metal. It's easy to see if the seals are holding by spotting any oil buildup on the case. The hardware holding the box to the support should be tight. Some boxes are equipped with an external oil fill/drain line and dipstick, which usually extends outside the fan and are easily accessible from the deck. Look at the connection to the back of the gearbox. Is the connection cracked? Is it leaking? This connector hose can be a hydraulic type and over time will crack or break due to exposure to the elements and vibration. A sudden failure of this hose causes loss of oil and the gearbox will cease working.

In direct drive arrangements, that is a motor attached directly to the gearbox, make sure the fan bushing is tight and in good condition. There are no areas to lubricate in this bushing. It should simply be checked to be sure if fits tight to both the motor and gearbox.

If the unit operates through a driveshaft and coupling, attention should be paid to the connection and alignment. Examine the metal parts, looking for corrosion and tightness of hardware. Some units have flexible metallic elements that allow for minor amounts of vibration or misalignment. The bushings should be checked for cracking or brittleness. The driveshaft itself is either one or two piece. If the shaft has mid-span bearings, they should be lubricated regularly. Newer towers will have a single metal or fiberglass shaft. The alignment of the shaft is critical and should be checked with a dial indicator. All hardware should be properly torqued according to manufacturers' instructions.

The two piece shafts can easily be converted to a single shaft. Consult with a mechanical contractor or cooling tower manufacturer. Having only one integral shaft reduces maintenance costs and possibility of failures.

The fan attaches to the top shaft of the gearbox or pulleys on a propeller fan arrangement. Fans are either fixed or adjustable pitched. Fixed blades are permanently attached to a hub, while the adjustable type are clamped and can be moved to accommodate changes in design and air flow. Regardless of type, look for metal corrosion. The bolts holding the blades in place should be torqued at least twice a year. Blades must not have excessive buildup, chips or cracks. Look at the leading edge of each blade for any nicks or separation of material.

When working with adjustable pitched fans, the blade angle can be measured and changed. At any fan speed, the pitch determines the amount of air flow. By loosening the bolts and changing the pitch, more air can be drawn and, in some cases, improve cooling. However, consult your cooling tower manufacturer, first.

Each fan operates within the confines of a cylinder or shroud, which can be made of wood, fiberglass or steel. In each case check the material for loss or corrosion. The blade tracks inside that shroud and must be sufficiently close to minimize air loss but not too close that it hits the sides. Adjust as necessary to accomplish this. The shroud can be realigned by adjusting the bolts and tightening.

For a fixed, centrifugal fan, similar inspections are necessary. Look at the blades for corrosion, nicks or breaks. Check the welds of the blades to the hub and the tightness of all bolting.

The entire drive train rests on a mechanical equipment support made of steel. As with all the other components, the metal on the drive train should be looked at for corrosion and tightness. If this support is on a wood tower, look at the connection to the wood and see if iron rot of the wood exists. If the tower is fiberglass, check the bolted connections.

The last member of the mechanical package is the fan motor. Most motors are open drip proof, totally enclosed or fan cooled. All three types have grease fittings for the lubrication of the bearings. The housing metal should be checked for corrosion, while the bolts that attach it to the framework should be checked for tightness. Listen for sounds that indicate a bad bearing or fan hitting the motor housing.

19.5.10 Safety

Most areas of the tower must be inspected for safe working and operating conditions. These items should be inspected yearly and repaired immediately to guarantee the safety of maintenance personnel.

Most towers are equipped with handrails around the top. Inspect the members and the hardware to be sure they are properly connected. If the tower has a ladder or stairway to the top, inspect the members for corrosion or breakage. If corrosion or breakage exists, it should be corrected immediately. Stairway landings should be sturdy and tightly bolted.

The end of many cooling towers have access through a door. Check the operation of that door and its attachment to the structure. Keep all hardware tight and the door in good working condition. Towers also have a guard, generally make of steel, that protects the fan. Again, check for corrosion and tightness of the hardware. Be sure that individuals can not stick their fingers into the fan and that flying debris can not enter the fan.

To protect against failure of the fan due to a loss of blade, the tower can be equipped with a vibration limit switch. There are many types, but to be effective it must be capable of shutting the fan down the moment excessive vibration is noticed. A remote reset should be included, requiring the maintenance person enter the tower and inspect for damage. This switch can be wired to the main control panel, lighting up when a problem exists.

19.5.11   Frequency of Inspection

Recommended frequency of inspections depend on tower location and number of available staff. The following are minimums that should be considered:

Daily: Observe, listen, and walk around the unit. Become accustomed to the tower's appearance, sound and level of vibration.

Weekly: Look at the motor, drive shaft, gearbox, belts, and fan. Shut off the fan and, if it is equipped with a gearbox, check the oil for proper operating levels.

Monthly: Check for silt buildup in the basin sump, check the operation of the float valve, inspect the upper distribution chambers for cleanliness of the basin and nozzles, and drain gearbox oil sample.

Semi-Annually: Drain and refill the gearbox. Check the condition of belts and replace if necessary. Re-lubricate the fan motor. Check all bolting for tightness in the mechanical region. Visually inspect tower structure; heat transfer media; drift eliminators; and condition of stair, ladder, and handrail.

Annually: Inspect the tower thoroughly, taking advantage of the operation and maintenance manuals as a guide. If the tower is equipped with a protective finish, inspect it. Take the time to make corrective repairs. Ask for an annual inspection by a service contractor or the local cooling tower manufacturer.

The operation and installation manual provided with the equipment should be used as a guide to efficiently operate the tower. Each tower is supplied with a serial number. That number is the key to maintenance history and will always be the first thing a supplier will ask for when ordering parts. It should be kept handy.

Figure 19.5.7 is a standard troubleshooting guide. Most problems with cooling towers are in the mechanical equipment. This two-page guide offers examples of a problem, possible cause and remedy. Figure 19.5.8 is an easy-to-use cooling tower inspection checklist. It covers all aspects of the cooling tower and can be used by maintenance personnel to keep a record of cooling tower repair activity. Such records

Troubleshooting

Trouble	Cause	Remedy
Motor Will Not Start	Power not available at motor terminals	• Check power at starter. Correct any bad connections between the control apparatus and the motor.
		• Check starter contacts and control circuit. Reset overloads, close contacts, reset tripped switches or replace failed control switches.
		• If power is not on all leads at starter, make sure overload and short circuit devices are in proper condition.
	Wrong connections	Check motor and control connections against wiring diagrams.
	Low voltage	Check nameplate voltage against power supply. Check voltage at motor terminals.
	Open circuit in motor winding	Check stator windings for open circuits.
	Motor or fan drive stuck	Disconnect motor from load and check motor and Gearereducer for cause of problem.
	Rotor defectve	Look for broken bars or rings.
Unusual Motor Noise	Motor running single-phase	Stop motor and attempt to start it. Motor will not start if single-phased. Check wiring, controls, and motor.
	Motor leads connected incorrectly	Check motor connections against wiring diagram on motor.
	Bad bearings	Check lubrication. Replace bad bearings.
	Electrical unbalance	Check voltages and currents of all three lines. Correct if required.
	Air gap not uniform	Check and correct bracket fits or bearing.
	Rotor unbalance	Rebalance.
	Cooling fan hitting end bell guard	Reinstall or replace fan.
Motor Runs Hot	Wrong voltage or unbalanced voltage	Check voltage and current of all three lines against nameplate values.
	Overload	Check fan blade pitch. See Fan Service Manual. Check for drag in fan drive train as from damaged bearings.
	Wrong motor RPM	Check nameplate against power supply. Check RPM of motor and gear ratio.
	Bearings overgreased	Remove grease reliefs. Run motor up to speed to purge excessive grease.
	Wrong lubricant in bearings	Change to proper lubricant. See motor manufacturer's instructions.
	One phase open	Stop motor and attempt to start it. Motor will not start if single-phased. Check wiring, controls, and motor.
	Poor ventilation	Clean motor and check ventilation openings. Allow ample ventilation around motor.
	Winding fault	Check with Ohmmeter.
	Bent motor shaft	Straighten or replace shaft.
	Insufficient grease	Remove plugs and regrease bearings.
	Too frequent starting or speed changes	Limit cumulative acceleration time to a total of 30 seconds/hr. Set on/off or speed change set points farther apart. Consider installing a Marley VFD drive for fine temperature control.
	Deterioration of grease, or foreign material in grease	Flush bearings and relubricate.
	Bearings damaged	Replace bearings.
Motor Does Not Come Up To Speed	Voltage too low at motor terminals because of line drop	Check transformer and setting of taps. Use higher voltage on transformer terminals or reduce loads. Increase wire size or reduce inertia.
	Broken Rotor bars	Look for cracks near the rings. A new rotor may be required. Have motor service person check motor.
Wrong Rotation (Motor)	Wrong sequence of phases	Switch any two of the three motor leads.

Figure 19.5.7a Troubleshooting guide. (continued, next page)

Troubleshooting

Trouble	Cause	Remedy
Gearducer Noise	Gearducer bearings	If new, see if noise disappears after one week of operation. Drain, flush, and refill Gearducer. See Gearducer Service Manual. If still noisy, replace.
	Gears	Correct tooth engagement. Replace badly worn gears. Replace gears with imperfect tooth spacing or form.
Unusual Fan Drive Vibration	Loose bolts and cap screws	Tighten all bolts and cap screws on all mechanical equipment and supports.
	Unbalanced drive shaft or worn couplings	Make sure motor and Gearducer shafts are in proper alignment and "match marks" properly matched. Repair or replace worn couplings. Rebalance drive shaft by adding or removing weights from balancing cap screws. See Drive Shaft Service Manual.
	Fan	Make certain all blades are as far from center of fan as safety devices permit. All blades must be pitched the same. See Fan Service Manual. Clean off deposit build-up on blades.
	Worn Gearducer bearings	Check fan and pinion shaft endplay. Replace bearings as necessary.
	Unbalanced motor	Disconnect load and operate motor. If motor still vibrates, rebalance rotor.
	Bent Gearducer shaft	Check fan and pinion shaft with dial indicator. Replace if necessary.
Fan Noise	Blade rubbing inside of fan cylinder	Adjust cylinder to provide blade tip clearance.
	Loose bolts in blade clamps	Check and tighten if necessary.
Scale or foreign substance in circulating water system	Insufficient blowdown	See "Water Treatment" section of this manual
	Water treatment deficiency	Consult competent water treating specialist. See "Water Treatment" section of this manual
Cold Water Temperature Too Warm (See "Tower Operation")	Entering wet bulb temp. is above design	Check to see if local heat sources are affecting tower. See if surrounding structures are causing recirculation of tower discharge air. Discuss remedy with Marley representative.
	Design wet bulb temp. was too low	May have to increase tower size. Discuss remedy with Marley representative.
	Actual process load greater than design	May have to increase tower size. Discuss remedy with Marley representative.
	Overpumping	Reduce water flow rate over tower to design conditions.
	Tower starved for air	Check motor current and voltage to be sure of correct contract horsepower. Re-pitch fan blades if necessary. Clean louvers, fill and eliminators. Check to see if nearby structures or enclosing walls are obstructing normal airflow to tower. Discuss remedy with Marley representative.
Excessive Drift Exiting Tower	Distribution basins overflowing	Reduce water flow rate over tower to design conditions. Be sure hot water basin nozzles are in place and not plugged.
	Faulty drift elimination	Check to see that integral fill, louvers, and eliminators are clean, free of debris, and installed correctly. If drift eliminators are separate from fill, make sure they are correctly installed in place. Clean if necessary. Replace damaged or worn out components.

Figure 19.5.7b Troubleshooting guide. (continued)

are important to itemize work to be completed and justify those repairs or replacements. It also offers a track record so multiple occurrences of a problem can be identified and corrective action taken.

19.5.12 Repair Criteria

Given the expected life as discussed above, it is the responsibility of the chief engineer and cooling tower contractor to make the most out of the equipment. The decision to repair or replace becomes a matter of dollars. In most cases the life can be extended by following the operation and maintenance procedures and doing repair work as

Cooling Tower Inspection Checklist

Date Inspected:_____   Inspected By:_____
Owner:_____   Location:_____
Owner's Tower Designation:_____
Tower Manufacturer:_____   Model No. _____ Serial No._____
Process served by tower_____   Operation: Continuous ☐ Intermittent☐ Seasonal ☐
Design Conditions: GPM_____ , HW _____°F, CW _____°F, WB_____°F
Number of Fan Cells:_____   Tower Type: Crossflow ☐   Counterflow ☐

Condition: 1 — Good, 2 — Keep an eye on it, 3 — Needs immediate attention

	1	2	3	Comments
Structure				
Casing Material _____				
Structural Material _____				
Fan Deck Material _____				
Stairway? _____ Material _____				
Ladder? _____ Material _____				
Handrails? _____ Material _____				
Interior Walkway? _____ Material _____				
Cold Water Basin Material _____				
Water Distribution System				
Open Basin System				
Distribution Basin Material _____				
Inlet Pipe Material _____				
Inlet Manifold Material _____				
Flow Control Valves? _____ (Size _____ inches)				
Nozzles (Orifice diameter _____ inches)				
Spray Type System				
Header Pipe Material _____				
Branch Pipes Material _____				
Spray Nozzles (Orifice diameter_____inches)				
Up Spray ☐ Down Spray ☐				
Heat Transfer System				
Fill (Type & Mat'l _____)				
Drift Eliminators (Type & Mat'l _____)				
Louvers_____				

Use this space to list specific items needing attention: _____

Figure 19.5.8a Inspection checklist. (continued, next page)

needed. As with any product, *waiting until the tower is in unusable shape will only hasten the need to replace.* Use the resources available, conduct regular inspections and keep good records. The actual need to repair or replace will vary with the owner and the contractor.

Common sense and loss of cooling efficiency will dictate when repair work needs to be done.

Two additional topics should be discussed to help extend the service life of the tower and protect it from airborne bacteria. The whole issue of water cleanliness and tower life and efficiency are tied together. A cooling tower will accumulate dirt and debris because of the nature of drawing air into the path of falling water. This buildup of sludge can

Cooling Tower Inspection Checklist (Page 2)

	1	2	3	Comments

Mechanical Equipment

Speed Reducer Type: Belt ☐   Gears ☐   Direct Drive ☐

Belt Drive Units

Belt (Designation _____)				
Fan Pulley (Designation _____)				
Motor Pulley (Designation_____)				

Gear Drive Units

Manufacturer _____ Model _____ Ratio_____

Oil Level: Full ☐   Add Immediately ☐   Low, check again soon ☐

Oil Condition: Good ☐   Contains Water ☐   Contains Metal ☐   Contains Sludge ☐

Oil Used (Type_____)				
Seals				
Back Lash				
Fan Shaft End Play				
Any Unusual Noises? No ☐ Yes ☐ Action Required:_____				

Drive Shafts (Mfr. & Mat'l _____)

Fans

Propeller ☐   Blower ☐   Wheel Diameter_____

Manufacturer_____   Fixed Pitch ☐   Adjustable Pitch ☐

Dia. (_____ feet _____ inches)	Number of Blades_____

Blade Material _____				
Hub Material _____				
Hub Cover Material _____				
Blade Assembly Hardware _____				
Blade Tip Clearance_____ " min. _____ " max.				
Vibration Level				

Fan Stacks (Dia.at Fan _____ ft.; Height _____ ft.)				
Mech. Eqpt. Support (Mat'l _____)				
Fan Guards				
Oil Fill & Drain Lines				
Oil Level Sight Glass				
Vibration Limit Switches				
Make-up Valves				

Other Components:_____

Motor Mfr.:_____

Name Plate Data: HP_____ RPM_____ Phase_____ Cycle_____Volts_____

F.L. Amps_____ Frame_____ S.F._____ Special Info._____

Last Lubrication (Date_____)

Grease Used (Type_____)

Any Unusual Noise?   No ☐   Yes ☐   Action Required:_____

Any Unusual Vibration?   No ☐   Yes ☐   Action Required:_____

Any Unusual Heat Build-up? No ☐   Yes ☐   Action Required:_____

Figure 19.5.8b Inspection checklist. (continued)

collect on the condenser, heat exchanger and other equipment, forming an insulating layer that decreases the heat transfer capacity.[6] A water filtration system can reduce this buildup. Such a piece of equipment not only maintains component efficiency but can reduce the amount of chemicals necessary in the circulating water system. A single important by-product is the reduced maintenance time to clean the heat exchangers, cooling tower basins and condenser.

A final issue related to water cleanliness is the *possible breeding ground for bacteria*. The single most difficult issue is stagnant water. System piping should be free of "dead legs" and tower flow should be maintained. When dirt accumulates in the collection basin of a tower, the combination of moisture, oxygen, warm water and food supply[7] *creates the possibility of Legionella bacteria*. This bacteria can be found in water supplies and around rivers or streams. They are contained in water droplets and become airborne making humans susceptible to breathing the contaminated air. There are no chemicals which can positively eliminate all bacteria from the water supply in a cooling tower. However, evidence suggests that good maintenance along with a comprehensive treatment program will dramatically minimize the risk.[8]

One final comment on cooling tower maintenance: Take frequent walks around the tower and listen to its sounds. Become aware of the way it operates and looks on a regular basis. There is no substitute for common sense and using the knowledge gained on working with equipment of all types. Rely on the cooling tower manufacturer and service contractor as a source of skilled labor and parts. Being in charge of keeping the cooling tower running efficiently does not have to be a difficult task.

19.5.12   References

1. *Cooling Tower Fundamentals*, Published by Marley Cooling Tower Company, second edition, 1985.
2. *ASHRAE Handbook*, 1996 Systems and Equipment Edition, Atlanta, GA. p. 36.1.
3. Baltimore Aircoil Company, Operation and Maintenance instructions, bulletin M24/1-0AB, p. 8, Baltimore, MD.
4. Ibid, p. 5.
5. Marley Cooling Tower Company, Owners Manual OM-NC2 A, Number 92-1327B, p. 20, Overland Park, KS.
6. Process Efficiency Products, Inc., Cooling Tower Filtration Systems, p. 3, Mooresville, NC.
7. L. S. Staples Company, article written July 1996, "Legionnaires' Disease-Still With Us," p. 1.
8. Ibid, p. 3.

20.1

Indoor Air Quality

Travis West
President, Building Air Quality
The Woodlands, Texas

20.1.1 The Issue of Indoor Air Quality

Concerns over Indoor Air Quality (IAQ) have propelled it into the forefront of the workplace environment, and it appears it will become a dominant issue confronting facility management and maintenance personnel well into the next century. As the public has become more aware of the health and comfort implications of IAQ, pressure has increased on commercial facilities to maintain an acceptable level of air quality.

Like many other building issues, IAQ problems can result in tensions between the occupants affected and those with the responsibility for the building's management. We are now learning that successfully resolving IAQ problems is based as much on a person's understanding that something is being done to resolve the problem, as it is on the actual expenditure of resources leading to the solution. Consequently, a rapid and well thought out response to calls for assistance begins to build the foundation for a positive relationship.

Before discussing the issue of maintenance of mechanical systems and it's effects on indoor air quality, it is important that the reader understand the variety and range of factors which can affect indoor air quality. This chapter will discuss the types of sources which can create a concern for indoor air quality and how pollutants are distributed and ultimately can be controlled. In addition we will discuss the impact that building occupants can have on indoor air quality. In addition, information regarding current legislation and proposed guidelines, as well as a table offering suggested preventive maintenance scheduling is included.

20.1.2   The Impact of Indoor Air Quality

Much information has been learned about the relationship of indoor air quality and the maintenance of mechanical systems in commercial buildings. Although it is now widely recognized by IAQ consultants, property owners and managers are only recently beginning to accept that:

- The expense and effort required to address indoor air quality in a pro-active manner is almost always less than the expense and effort required to address a problem after it has occurred.

- Many indoor air quality problems can be prevented by making sure that the facility's management staff and operation's personnel understand the many factors which can create these problems. In addition, they must be well trained in the operations of their equipment and must have budgets and support staff to allow them to remain diligent in preventive maintenance activities.

Indoor air quality problems which occur in high-rise buildings are very similar, in nature and cause, to those problems which occur in one-story strip centers and other similar buildings. Consequently, the design and use of the building does assume the same level of importance as the building's maintenance and operations activities.

20.1.3   Factors Affecting IAQ

The air quality found inside a building is the result of many factors which can affect it continually during a day's activities. There are factors that can change rather quickly such as occupant loading, a change in space conditioning, and even odor or contaminant releases which can occur in an uncontrolled manner. In addition, there are other factors which can develop into air quality problems over extended periods of time. Problems of this nature include the development of biological growths, the deterioration of building components, and even gradual modifications to the tenant makeup, affecting how and when the building and it's mechanical systems are being used.

Experience has shown however, that for an indoor air quality problem to occur there need to be four specific factors present. Those factors are:

- A contaminant source
- A pollutant pathway

- HVAC systems
- Building occupants

While any of these factors can play an important part in the overall indoor air quality complaint, all four must be present for a complaint to occur. Take for example a source such as an odor present in a building, away from any occupied spaces. The building's occupants, however, are located at the other end of the building, away from the source. In this instance, with the odor occurring and the building occupied, if there is no transport mechanism present (HVAC system) the chance of an indoor air quality complaint to occur is not likely. Given a second example in a building that clearly has a contaminant source and a fully functional HVAC system, but no building occupants. In this instance the chance for an indoor air quality complaint to occur is also limited. However, in any instance where an HVAC system, contaminant source and building occupants are present concurrently, the chance for an indoor air quality complaint to occur increases.

20.1.4   Contaminant Sources

The source is considered the area or product which is creating the indoor air quality contaminant. This can occur as an odor or fume brought into the building through the outside air supply. This can also occur within a building from deteriorating components, off-gassing of fixtures or furnishings, or can be generated by a building occupant activity (such as gluing, soldering, etc.).

As previously discussed, indoor air quality sources can occur *outside* of the building or can be generated *inside* of a building. Whenever these sources are uncontrolled, the potential for an indoor air quality problem to occur is increased dramatically. Recent indoor air quality training programs developed by the EPA have identified a variety of areas in which sources can occur.

20.1.4.1   Sources outside of the building

Sources found outside of the building can include outdoor air which has been contaminated by a variety of things both naturally occurring and man made. The naturally occurring ones are usually anticipated in the forms of pollen, dust, spores etc., which are generated by trees, grasses and other living things. Man made pollutants can include gasses, dusts and fumes from industrial activities.

In addition, outside air fans can bring in other outside sources, such as exhaust from vehicles parked at building loading docks, from idling traffic in parking lots and nearby congested freeways.

20.1.4.2   Mechanical equipment sources

Sources of indoor air quality contaminants related to the HVAC system can include dirt and dust which is found in duct work or on other components, but is most often seen as microbiological growth found in drip pans, humidifiers, ductwork coils and other moist portions of the system units. In addition, indoor air quality investigators commonly look for improper venting of combustion products such as carbon monoxide from water heaters, steam boilers and gas appliances.

20.1.4.3   Building occupant activities

Activities which can create contaminant sources in buildings come from a variety of areas. Building management firms must be apprised of the fact that housekeeping activities as well as maintenance activities can create potential indoor air quality sources. In addition, occupant activities have been shown as a source for a variety of indoor air quality contaminants. These contaminants include the obvious such as tobacco smoking, and may also include cooking, cosmetic odors, and even the use of certain office supplies which may affect the more allergic individual. Building housekeeping and maintenance crews must also be made aware of the fact that the materials used for cleaning, pest control activities, maintenance and lubricating supplies can also generate odors, vapors, and fumes which can be considered potential indoor air quality sources.

20.1.4.4   Building components and furnishings

Building components and furnishings can create indoor air quality problems. Such products can act as collectors for dust or fallen air-born fibers and other particles. These furnishings include actual hard-surface fixtures such as cabinets, desks, bookshelves, etc. In addition, carpeting has been found to be one of the largest collectors of dust and other indoor air contaminants. This problem increases as carpeting ages since the build-up of dirt and other contaminants can never fully be removed through cleaning processes. Other types of unsanitary conditions can be created through water damage which may occur in the occupied space. Water leaks or spills which come in contact with items such as cellulose ceiling tiles, plaster board walls and even the aforementioned carpeting, present a tremendous potential for the growth of microbio or biological concerns.

When products, fixtures, and furnishings are new, the potential for indoor air quality concerns also exists. Research has shown that the dyes and glues used in the manufacture and installation of items such as wallpaper, carpeting, and even drapery systems can off-gas harmful chemicals into a buildings environment. A method for "forcing" the premature off-gassing of these products has been developed, and is referred to as "bake-out". Although the bake-out processes has been shown to work in certain environments, building mechanical systems, certain wet components and other factors all play a part in determining the effectiveness of the bake-out process. (Bake-out should only be attempted under the proper guidance of experienced environmental consultants.)

20.1.4.5 Other unusual sources

There are many other sources of indoor air quality contaminants which are considered unusual, yet can be found in most any building. The most common of these is a small kitchenette area, found in many office suites. Odors are the primary cause of IAQ complaints where these are found. In addition, odors from a building or tenant cafeteria can create indoor air quality complaints.

Other areas which can generate unique though not unusual problems include: laboratory areas; print shops, gymnasium or health spa areas, beauty salons, smoking lounges, and even office areas with high volume copy machines.

20.1.5 Pollutant Pathways

The way in which air flows over, around and at times under a building, can create pollutant pathways. These pathways created by air pressure differences or air movement, can act as a carrier or transport mechanism for the pollutant, taking it from the source to previously unaffected areas of the building.

A fourth and sometimes critical factor affecting indoor air quality involves pollutant pathways and naturally occurring driving forces. Pollutant pathways can most commonly be referred to as pressure differentials which occur throughout the inside of a building.

These differentials can be caused from an oversupply of conditioned air being provided to one room while limited quantities of return air are being allowed. When this occurs, a positive pressure differential happens. If this room as described were a laboratory, kitchen, or other source of potential contaminants, those contaminants can then be delivered through the use of a pollutant pathway to other occupied areas in the building. Driving forces describes the activity which generally occurs through natural forces exerted on the outside of a building. These forces can include high winds on the northside creating a low-pressure area on the southside of the building or the phenomenon known as *stack effect* which can occur most dramatically during winter months. During stack effect, warm air rises inside of a building causing an accumulation of warm air into the upper floors while at the same time generating an obvious lack of warm air in the lower floors as heated air rises, it is replaced by cold air being "sucked" into the building at lower floors.

Even when a building is designed and maintained under a positive pressure, there is always an area or two which will occur to be periodically under a negative pressure. As this occurs so does the potential for a source to affect the low pressure areas. The interaction between pollutant pathways and intermittent or variable driving forces can lead to the problem where a single source is causing indoor air quality complaints in a variety and sometimes distant, areas within a building.

20.1.6   HVAC System

Although the HVAC system is not able to control air contaminants which are being generated at the source, it can be used to affect pressure relationships and even dilute (through the ingestion of outside air) indoor air quality contaminants. For the sake of our discussions here it's important to note that while the indoor air quality source can in fact be the HVAC system, it is most commonly found as the carrier of the source contaminants to the building occupants.

20.1.7   Building Occupants

Without a person or persons who are impacted by the presence of a contaminant source, there would not be an indoor air quality complaint.

20.1.8   Standards Governing Indoor Air Quality

Although at present there is no federal legislation which sets such standards for acceptable indoor air quality, there are a number of states which have enacted legislation requiring that. The scope and nature of indoor air quality regulation in those affected states varies on a state-by-state basis. However, a good rule of thumb is to include what legal experts refer to as "providing standard practice of care" when determining the level of indoor air quality.

However, during the design, construction, and even retrofit of commercial buildings there are recommended trade association standards which should be followed. The standards written by the American Society of Heating Refrigerating and Air Conditioning Engineers

(ASHRAE) were created to cover several different aspects of commercial building management which can affect indoor air quality.

20.1.8.1 ASHRAE standard 62-1989

The standard recommending guidelines for acceptable indoor air quality is referred to as standard 62-1989: "Ventilation for Acceptable Air Quality."

ASHRAE 62-1989 is intended to assist professionals in the proper design of ventilation systems for buildings. This standard presents two possible procedures for ventilation design: A "ventilation rate" procedure and an "air quality" procedure.

Whichever procedure is used during the design of mechanical systems, the standard states that the design criteria in assumption shall be documented and made available to those responsible for the operations and maintenance of the system. Some of the important features of ASHRAE 62-1989 include the following:

- A definition of acceptable air quality.

- A discussion of ventilation effectiveness.

- The recommendation of the use of source control of contaminants.

- Recommendations on the use of heat recovery ventilators.

- A guideline for acceptable carbon monoxide levels.

- Appendices listing suggested possible guidelines for common indoor air pollutants.

A proposed revision, ASHRAE 62-1989R, is currently being reviewed and considered. Readers are urged to obtain a copy from ASHRAE, 1791 Tullie Circle, N.E., Atlanta, GA 30329-2305.

20.1.8.2 ASHRAE standard 55-1992

The second of the two ASHRAE standards which can have an affect on indoor air quality is referred to as standard 55-1992: "Thermal Environmental Conditions for Human Occupancy."

ASHRAE 55-1992 covers several environmental perimeters including the temperature, radiation, humidity and air movement designed within a building's structure.

Since the perception of being too warm or cold, too humid or dry, can sometimes be misinterpreted as an indoor air quality issue, thermal comfort within the occupied space becomes important.

Some of the important features of ASHRAE 55-1992 include:

- A definition of acceptable thermal comfort.

- Recommendations for summer comfort zones and winter comfort zones clearly defined in graphs.

- Guidelines for making adjustments in air delivery depending on occupant activity levels.

20.1.9 Program Management

Once the construction of a building has been completed, the responsibility for preserving good IAQ rests with the building operations and maintenance functions. A comprehensive facilities program should exist within each occupying firm or building manager. This program will be used to manage and monitor acceptable indoor air quality including a scheduled testing and analysis phase, repairs, minor alterations to facilities, as well as to guide the reaction to unscheduled breakdown repairs and maintenance.

To implement an IAQ program, three elements must be provided:

- Management support and commitment from all levels of the organization.

- Staffing by trained and competent personnel;

- Budgetary allocations to provide the fiscal resources necessary to perform the comprehensive services required to maintain good indoor air quality.

The following pages cover areas that are shown to have an affect on the indoor air quality of a building.

20.1.9.1 Facility inventory

An audit of the condition of all buildings should be conducted as early as possible to establish a baseline for the physical plant. In addition to anticipating potential IAQ problems, an audit has value to all other aspects of operations and maintenance. Results of an effective (and periodic) audit can also be used as due diligence groundwork in instances of lawsuits related to building IAQ. Such audits can be performed by properly prepared and knowledgeable mechanical/environmental engineers.

The major building components that should be considered for evaluation include:

- Roof

- HVAC (including local exhaust systems in specialty areas)

- Plumbing

- Electrical

- Sub-floor/crawl space areas

- Other areas: Loading docks, food services, duplicating and copy rooms, laboratories, storage facilities, industrial and manufacturing areas, and so forth.

20.1.9.2 Systems description

Providing a description of each system is a particularly important, but often overlooked program element. It is important that both existing and new staff in the building understand the capacities and capabilities of the systems, to assure that the building is performing to it's optimum.

Review your existing equipment lists and mechanical plans. Compare the as-built drawings to the equipment installed. This helps to assure that components are receiving regular attention. Also, equipment that has been installed in inaccessible locations is often overlooked during routine maintenance. By verifying all equipment noted on the mechanical drawings, you are forced to confirm it's presence and condition.

20.1.9.3 Operations plan

An operations manual should describe how to run the building systems. The plan should provide a detailed step-by-step procedure covering operation from start up to shut-down. This document, an essential reference for day-to-day activities, will be a useful training document for new employees. A vital part of this plan should be a compilation of all available manufacturers literature and manuals.

20.1.9.4 Comprehensive preventive monitoring plan

Preventive monitoring is incorporated into all building IAQ Programs. The objective of this monitoring is to prevent small deficiencies from blossoming into major costly outages or repairs. For example, regularly oiling a bearing on a fan system will extend the useful lifetime of the unit, and prevent the potential loss of make-up or exhaust air that is necessary for proper IAQ. The table of Preventive Monitoring Tasks (Table 20.1.1) lists the tasks that are necessary in any program related to IAQ.

Scheduling periodic consultant audits can be an important determination of indoor air quality. Both frequency and timing are important. For example, biological growth of fungi and bacterial organisms

TABLE 20.1.1   Preventive Monitoring Tasks

Recommended Frequency

AIR HANDLING SYSTEM

HEAT/COOL/VENT

Clean or replace filters . . . . .Quarterly or as needed
Clean coils & spray with disinfectant . . . . .Annually
Clean & flush condensate pans with disinfectant as needed Quarterly or as needed

Check filters. Clean if needed . . . . .Monthly

Check condensate pan drains . . . . .Monthly

Check & clean air intake & intake louvers, etc . . . . .Monthly
Check air distribution boxes and overall duct system for
cleanliness, leaks, collapse . . . . .Quarterly
Clean heat & cool coils . . . . .Annually
Check & clean fan/blower blades of dirt & trash build-up . . .Quarterly or as needed

RETURN AIR FAN

Check & clean air intake screening & louvers, etc . . . . .Monthly
Check overall duct system for cleanliness, leaks, collapse . . .Monthly
Clean fan & blower blades . . . . .Annually

UNIT HEATERS

Check & clean coil & fan blades . . . . .Quarterly or as needed

FAN COOLING UNITS

Clean & flush condensate pans with disinfectant as needed Quarterly
Check & clean filters as needed . . . . .Quarterly
Check & clean coil with spray and disinfectant . . . . .Annually

CEILING FANS

Check & clean fan blades . . . . .Monthly

ROOF TOP UNITS

HEAT/COOL/VENT

Check operation of roll filter . . . . .Monthly
Clean or replace filters . . . . .Quarterly or as needed
Clean coils & spray with disinfectant . . . . .Quarterly or as needed
Clean & flush condensate pans with disinfectant as needed Quarterly or as needed
Clean all coils . . . . .Annually

DRAINAGE SYSTEMS

ROOF DRAINS

Check & clean roof or floor drains in area . . . . .Weekly

IAQ AUDIT

Independent testing & analysis of mechanical systems and
tenant areas for acceptable air quality levels . . . . .Quarterly

---

have often been found in condensation pans. In some environmental studies, these organisms have been shown to cause disease. Since HVAC systems run throughout the year, they will accumulate biological growth as the year progresses and may disseminate biologically active aerosols and odors, to the building's occupants all of the time. *If building management waits until a chosen annual date to clean and disinfect these units, they may pose a greater risk to building occupants than if they were inspected on a regular schedule and cleaned / disinfected as problems begin to develop.*

Scheduled maintenance and any preventive monitoring must also be tied to a building's utilization schedule. Painting, coil cleaning, pest control or other projects involving the use of volatile organic chemicals should be scheduled when there are few, if any, occupants in the immediate vicinity. The work area should be well ventilated, using fans or supplying large quantities of outside air wherever possible. In contrast, the monitoring phase must be scheduled when the building is occupied to allow for retrieval of "occupied" and/or potential exposure analysis figures.

Mechanical systems.   Pulleys, belts, bearings, dampers, heating and cooling coils, and other mechanical systems must be checked periodically. These are included in Table 20.1.1; consequently there is no reason that these should be the source of an indoor air quality problem if the staff is following the table schedules.

Pulleys and belts should be tightened as needed and changed prior to failure. Bearings should be lubricated or repacked to prevent major failure of vital system components. Air dampers and baffles should be cleaned and cleared of debris periodically. Failure to perform these activities will result in an increase in resistance, which causes a decrease in air supply.

The air conveyance system (ductwork) should have access ports available to allow for periodic sampling of IAQ at various points in the system. This data, when retrieved, can offer accurate information on the presence of mold, mildew, bacteria, dust, etc., allowing building management to address a problem before any concerns are raised. See Chapter 15.2: "Ductwork."

Building ventilation networks are systems that serve multiple locations. The common practice of arbitrarily adjusting the dampers or baffles to accommodate complaints from one area must be avoided. By changing the air flow in one area, the system balance is shifted and distribution throughout the entire network is affected. If there are distribution complaints, a test of the building's air handling system should be performed to confirm that the HVAC system and distribution network are in balance and are adequate.

Condensate pans.   Components that are exposed to water, such as condensation drain pans, require scrupulous maintenance to prevent biological growth and the entry of undesired biological contaminants into the indoor airstream. This is an item on Table 20.1.1. Therefore, there is no excuse if, or when a pan is found to be the source of an

IAQ problem. The need to use an algecide is an indication of inadequate maintenance. Refer to Chapter 22: "Condensate Control."

Filters.   Mechanical equipment for ventilation, heating and air conditioning contain filter media or screens that minimize particulate matter from collecting on the coils, fans and interior housings and ductwork. Originally, the primary consideration was to protect the system from contamination and loss of efficiency resulting in equipment shutdown and extensive cleaning. Recognition is now given to the role filters can play in improving indoor air quality.

Building systems are primarily concerned with mechanical filtration filters. Commonly used are replaceable fiberglass filter media, mechanical screen media that requires cleaning and recoating, and bag type filters. Filters are rated by efficiency. The method of particle entrapment is by impingement, which locks the particulates within the filter. Filters have a life expectancy rated in hours when used for a given air flow with expected contaminants. Severe dry spells that blow excessive dirt or, interior dust producing activities, or construction at a site nearby which sends dust towards the building's outside air intake, can increase contaminants reaching the filter.

Independent investigations have found that too often "permanent" metal filters are not being cleaned properly or often enough. Improper maintenance of these filters can have the following effects:

- Equipment efficiency is reduced and more energy is required to push the same quantity of air through the filter.

- Some contaminants pass through the overloaded filter, clog the coils, and enter the occupied spaces.

- Dirt collection on the coils diminishes the thermal transfer efficiency of the unit resulting in higher energy consumption.

- Contaminants on the coils become a breeding ground for bacterial and fungal growth.

Based on considerable experience, it is recommended that all "permanent" filters be replaced with the fibrous disposable type. The recommended period for checking and changing air filters is covered in the list of Preventive Monitoring Tasks, Table 20.1.1.

Vacuuming.   Normal industrial vacuums emit particles and fibers in their exhaust. An improvement in performance can be obtained if the vacuum can be fitted with a high efficiency particulate air filter (HEPA). HEPA vacuums should be used in areas that might have spores or microorganisms. HEPA vacuum filtration ensures that potentially toxic or harmful aerosols are not dispersed while responding to a problem. Dry sweeping in problem areas should be avoided.

Roofing. Poorly designed or improperly drained roofs may be a potential source of poor IAQ. Flat roofs will invariably collect water and cause pooling. Stagnant, standing water on roofs can support the growth of algae, bacteria, and possibly viruses that can be drawn into building air systems. Leaks in roofs result in water damage or accumulation in ceiling tile, carpeting or internal wall spaces. Fungi and bacteria that develop in this moisture have been found to be responsible for allergies and respiratory disease. Consequently, when roofs are sloped inadequately or roof repairs are postponed, IAQ can easily be compromised. Water damage materials must be removed and replaced in a timely fashion before they serve as a breeding ground for biological growth.

Pipe leaks. Pipe leaks can occur through corrosion, mechanical failure, or because of the age of the facility. In any case of leakage, repair or replacement of the damaged pipe section must be performed immediately and the contaminated water quickly removed and disposed of by pouring it into an appropriate drain. It is prudent to have wet vacuums, submersible pumps, squeeze brooms and mops available to handle water emergencies. Water damaged ceiling tiles, rugs, insulation or walls must also be moved and replaced in a timely fashion to prevent mold from growing. Following storms, it is good practice to inspect the building for discolored ceiling tiles or leaks as signs of water problems. Pipes that may be subjected to situations causing condensation, should be insulated to prevent the problem.

Drains. The antisiphon "P" traps in sinks must contain water to prevent noxious odors from the sanitary sewer lines from migrating back into the indoor air spaces. Sinks and drains that are used infrequently can dry out, allowing a path for sewer gases to enter. Cupsinks in laboratory fume hoods and on benches frequently dry out and have often been found to be the sources of odors. This problem can be resolved and prevented by periodically running water in these drains, plugging unused drains with a rubber stopper, or using a nontoxic liquid with a low vapor pressure to fill the P trap. Drains located in mechanical or air handling rooms are designed for condensate pan runoff. If no moisture is evident, these lines can dry out (this occurs most frequently in dry climates and during the winter months in humid climates). Dry P traps allow a path for noxious odors from sewer lines to enter. These odors, can then be unknowingly distributed throughout the building by the air distribution system.

Drains in laboratories must be kept clear and in working order. Sediment in drain traps can provide conditions that support the growth and accumulation of biological organisms.

20.1.9.5 Materials management program

It is important to understand that most chemicals used in housekeeping, maintenance, operations, pest control and cafeteria services can affect air quality. Their use in the building should be properly managed. Material Safety Data Sheets (MSDS) are obtained from the manufacturers or distributors and should be on file in the building where any potentially hazardous material is to be stored or used. MSDS's for all chemical products used should be kept in an easily accessible area for employee access. A master file should also be kept in the offices of the personnel responsible for the chemical's use. A building's population should not be subjected to unknown and potentially harmful effects of any material because of the absence of an MSDS.

The chemical composition of hazardous ingredients must be evaluated along with the hazardous reaction potential. Wherever possible, materials, chemicals and reagents that present the lowest toxic potential should be used. There are many authoritative sources available to help in this determination, including vendor salesperson, vendor chemists, and even the local or state health department.

Less toxic materials should be substituted for more toxic materials. In general, water soluble materials should be given preference to organic solvent systems. Materials that are higher in flash point (ignition) and/or have a lower vapor pressure are also preferred.

Building use changes. Special care must be exercised when building space utilization is changed. Renovation, redesign, or changes in building use can create situations that may lead to compromises in IAQ. For example, if a copy machine is installed in a small closet or other unventilated space, chemical emissions such as ozone or carbon black dust or ammonia may become problems. Further, the addition of blueprint processes, paper bursting activities, or even microwave cooking can add unexpected odors, dusts or gases to an unprepared or improperly engineered area. The effect on heat and noise levels must also be anticipated if new equipment is added to an area.

A common renovation problem arises when additional personnel need to be accommodated in a space. Office or instructional areas are often partitioned and additional furniture and equipment is installed. Anticipate the need to modify the air distribution in these situations. Conversely, when partitions are removed, creating new spaces, the ventilation distribution and balance must be revised. Care must be taken to ensure that, in the final layout, air supply grills are located far enough away from the returns so that complete air balance/mixing does occur. Otherwise, stagnant areas will develop.

Evaluation of building materials (prior to construction or renovation). New materials used during construction or renovation present the potential for occupant exposure to emissions (offgassing) from those materials. Many IAQ problems can be avoided by selecting building materials that are less likely to pollute. In cases where this is not possible, the responsibility to address the issue of offgassing products falls to building management. The most likely avenue to pursue is dilution of the concentrations caused by the source. Chemical content, chemical emission potential and the potential for toxicity and irritation, are all issues that should be considered in construction or renovation.

Most manufacturers are now supplying Material Safety Data Sheets (MSDS) with the delivery of their products. If the product is a carpet, wallboard, paneling or material, some vendors can now supply product "emission" statistics on these items. In cases where the product emission rates are not available, most vendors will employ staff or consultants who can relate information about the products' effects on IAQ.

The simplest approach to material evaluation is to identify materials that can raise IAQ concerns, and select the lowest-emitting and least toxic products.

Ventilation & isolation during construction or renovation. During renovation projects, special care must be taken to isolate the area being remodeled from all occupied areas. The contaminants can spread via the HVAC system, by airflow through corridors, up and down elevator shafts and through ceiling plenum areas. Contamination of occupied areas during remodeling can lead to tenant complaints ranging from dry eyes, itchy skin and burning throats to nausea and even vomiting. Individual effects almost always include lower productivity, which is used to support claims of tenant move-out.

There should be an effort to supply sufficient exhaust-only ventilation to areas that are involved in this construction or renovation process. Further, adjoining (occupied) areas should be kept under a slight "positive" pressure to avoid cross-contamination. This positive pressure in occupied areas, along with exhaust ventilation at the source of the contaminants will help to isolate and remove most contaminants.

After the construction or renovation project has been completed, other strategies fall into play that can have a major impact on the IAQ of the facility. Much has recently been written about "bake-out" or "flush-out" procedures.

Bake-out is a process of overheating a building or space to artificially age the materials that are sources of contaminants. Research on the effectiveness of this process is limited and any attempts at performing this process should be guided only by IAQ consultants.

Pesticides. Building management should be committed to providing the building with a pest managed environment through the implementation of preventive hygienic methods and chemical strategies when necessary. Integrated Pest Management (IPM) emphasizes the use of nonchemical techniques for the management of pests, relying on the use of pesticides only when nonchemical strategies are not effective.

Each building should receive a scheduled inspection for the purpose of identifying existing or potential problems that may contribute to harboring, feeding or population growth of pests. Inspectors should recommend nonchemical pest control measures whenever possible. Finally, the inspector will list options for chemical treatments should the nonchemical measures prove to be unsuccessful.

Chemical control treatment should be applied after building hours with the exception of emergency situations. All contract personnel will be required to possess a valid pest control applicators license, and such license must be on file in the building management office.

Building management has a responsibility to notify all tenants about the pending treatment for pests. Management should also assign personnel to accompany the application technician, and to monitor the use and location of all pesticides.

20.1.9.6   Record keeping

While there is a clear record keeping requirement for asbestos, other potential contaminants have no set procedure. Therefore, it is recommended that records on issues related to IAQ be kept on a scheduled, periodic basis. These records, should contain information related to the scheduled PM program, as well as engineering reports showing the IAQ health of the building. This IAQ survey will involve testing for specific contaminants related to Indoor Air Quality and should be performed every 3 months. If no IAQ problems are encountered during the first year, consideration can be given to extending the schedule to a 6-month basis.

The minimum testing required should include review of the operations of the HVAC system and testing and evaluations for the presence of CO (carbon monoxide), CO_2 (carbon dioxide), temperature and humidity (as a comfort indicator), pressure differential of various rooms and A/C zones, and in some instances samples for the presence of excessive levels of mold, mildew, fungi, and bacteria. Other tests which can be performed include sampling for the presence of: NH_3 (ammonia), NO_2 (nitrogen dioxide), HCHO (formaldehyde), O_3 (ozone), and RSPs (respirable suspended particles) such as glass fibers and inhalable dusts.

This record keeping should include records developed during an initial baseline audit, laboratory field tests (if any), reinspections (every 3 months), and other additional testing. Action plans and any maintenance or operation activities that have been used to mitigate problems, and results of work performed by any outside consultants or contractors that were used should be kept on file as well.

20.1.7 Preventive Monitoring Tasks Guidelines

The Preventive Monitoring Tasks guidelines (Table 20.1.1), are designed to offer guidance on daily, weekly or monthly procedures that can have a dramatic effect on indoor air quality. These items are offered for guidance only, and should not be treated as suggested requirements. Specific information for each building depends on a number of factors, including occupancy levels, location, number of heating days vs. cooling days, ambient air conditions, type of HVAC system, recommended manufacturer maintenance guidelines and other unique factors not addressed in this chapter. The purpose of these guidelines is to begin the thought process regarding items that can exhibit the potential for the development of an indoor air quality problem.

20.2

Coil Care*

Billy C. Langley, Ed. D., CM
Consulting Engineer, Azle, Texas

The two coils used in air conditioning and refrigeration systems are the evaporator and the condenser. In heat pump systems they are called indoor and outdoor coils, respectfully.

By knowing the finer points of their operation and maintenance, service technicians will help customers achieve improved operating efficiencies through their hvacr systems.

Evaporators. Evaporator coils are found inside the conditioned space. Their purpose is to absorb heat from the space and products stored inside, into the refrigerant as it flows through the system.

Several types of evaporator coils are in use. However, the most popular is the forced-air coil. They are generally more efficient than the other types.

Forced-air coils are further divided into blow-through and draw-through types. The draw-through type is the most efficient of the two. However, the blow-through is used extensively because of the extra space required by the draw-through type.

The draw-through type is more efficient because air flows through most of the coil surface in an equal amount. In the blow-through type, most of the air is forced through the coil directly in front of the blower, leaving the outside edges of the coil operating at less than the designed efficiency.

*Reprinted by permission from The Air Conditioning, Heating and Refrigeration News, Copyright 1996.

Other types include the flat-plate coil, and other types of fast-freeze evaporators used in food-processing plants.

The immersed type has the evaporator placed in a fluid bath of some type, with the fluid stored at a given temperature to be used later. This type is popular in some larger a/c applications that take advantage of off-peak electric rates.

Watch for low load

Efficient operation of any type of evaporator is due to the rate at which heat is absorbed into the refrigerant as it passes through the coil.

Probably the most common cause of reduced evaporator efficiency is low load. This may be caused by several factors.

1. Dirty evaporator surface—This dirt, which is carried with the air, contacts the wet coil surface and sticks to it. The dirt acts as an insulator, reducing the amount of heat absorbed. It also prevents the proper amount of airflow through the coil.

To bring the evaporator back to peak efficiency, it must be cleaned. Sometimes this can be done by brushing the fins in the direction of the fins with a wire brush. Sometimes it can be done by spraying the coil with a non-acid coil cleaner and then spraying it with a high-pressure water nozzle in the opposite direction of airflow.

In extreme cases, the coil will need to be steam cleaned to remove the oil and dirt clogging the fins. This cleaning sometimes requires that the coil be removed from the unit.

Be sure to use the proper procedures to remove the refrigerant. Make certain that the tubing is properly sealed to prevent moisture and dirt from entering the system. And protect the motors and electrical equipment from the water or steam to prevent damage.

2. When cleaning the coil, all bent fins should be straightened so that the air can flow freely through the coil. When the airflow is restricted by bent fins, the heat transfer rate is reduced because the air cannot contact the complete coil surface.

3. Frost on the evaporator can also cause the system to be underloaded. A thickness of ice or frost of $\frac{1}{8}$ in. (3 mm) can reduce the coil's efficiency approximately 25 percent.

Frost or ice on the coil is usually caused by:

a) Low load caused by a dirty coil, dirty filter, low refrigerant charge;

b) Low ambient outdoor temperatures;

c) Dirty blower; or

d) Blower that is running too slow.

A blower that is running too slow is usually indicated by a high temperature drop across the coil. Usually this temperature drop should be about 20°F (-6°C) when 400 cfm (11 m^3) of air per ton of capacity is moving through the coil.

Some higher-efficiency units have a temperature drop more than 20°F (-6°C), and an airflow greater than 400 cfm (11 m^3) per ton. Follow the equipment manufacturer's recommendations.

Sometimes, when proper piping practice has not been followed or when the coil was not installed level, oil will collect in some of the tubes and reduce the amount of coil surface used.

Sometimes when a compressor or condensing unit has been replaced, one that has too much capacity for the evaporator will be used. This causes the suction pressure to be too low, in turn causing frost or ice to build up on the evaporator.

To correct this problem, the proper size compressor or condensing unit must be installed.

Condenser coils

Condensers. The purpose of the condenser is to reject heat absorbed by the evaporator, plus the heat of compression created by compressing the refrigerant.

The condenser must be sized to fit the system for the circumstances under which it must operate.

Two types of condensers used are:

1. Air cooled—Both draw-through and blow-through types. Draw-through types are more efficient than blow-through types, because the air is pulled equally through the complete surface of the condenser coil. (See Fig. 20.2.1.)

Blow-through coils are less efficient because most of the air flows

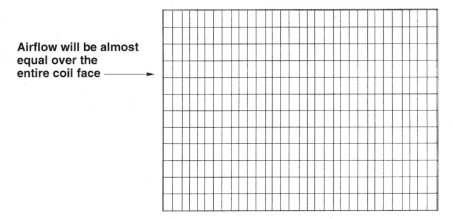

Airflow will be almost equal over the entire coil face ———▶

Figure 20.2.1 Draw-through coil.

through the condenser coil directly in front of the fan discharge. Thus, the outer edges of the condenser are not used to their full capacity. (See Fig. 20.2.2.)

In condensing units, the draw-through type is most popular because sufficient room is available for the extra space needed.

Air-cooled condensers are the most popular because they do not require special attention to prevent them from freezing during cold weather, especially when they are installed outdoors.

Air-cooled condensers must be installed away from shrubs, trees, flowers, and other plants whose debris can get into the condenser coil. Air-cooled condensers should not be installed near patios or under bedroom windows, where the noise could interfere with normal use of the adjacent area.

The prevailing wind should also blow in the direction of airflow through the condenser. This will aid in keeping the condenser coil cool, and not overload the condenser fan motor.

Usually the condenser requires about 1,000 cfm (28 m^3) of air per ton of capacity to remove the heat absorbed by the evaporator, plus the heat caused by compressing the refrigerant.

The temperature rise through the condenser should be from about 25° to 35°F ($-3°$ to 2°C). Some higher-efficiency units have different temperature rise requirements. Check the equipment manufacturer's recommendations.

The last two or three rows of condenser tubing are usually used for subcooling the refrigerant before it enters the liquid line. (See Fig. 20.2.3.) This is to help increase the efficiency of the unit by reducing flash gas at the flow-control device.

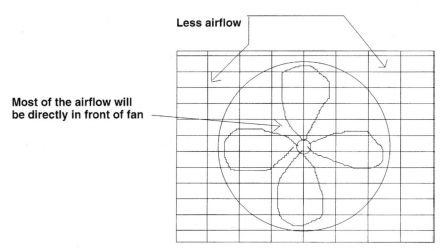

Figure 20.2.2 Blow-through coil.

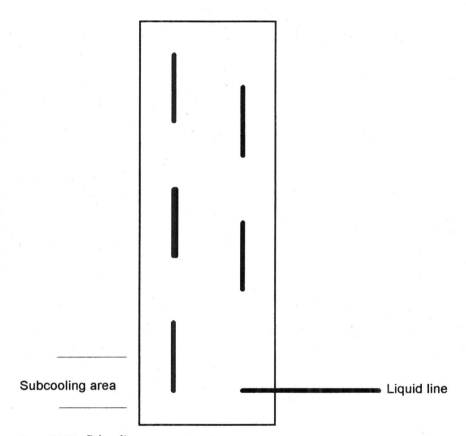

Figure 20.2.3 Subcooling area.

A subcooling of 10° to 15°F (−13° to −10°C) is generally used. Again, on some high-efficiency models the subcooling may be as high as 20°F (−6°C). Be sure to check the equipment manufacturer's recommendations.

Air-cooled condensers must be kept clean to prevent high head pressures and compressor overheating. And heat is the compressor's greatest enemy.

The coil may be cleaned with a high-pressure nozzle on a garden hose. Spray the water in the opposite direction of the normal airflow through the coil. Occasionally a non-acid cleaning agent may be needed to remove all the dirt and grease.

Unremoved, the dirt causes the heat transfer to fall below that required for the system. Also, it will prevent the proper amount of airflow through the coil for proper cooling and condensation of the refrigerant.

Always remove any debris left inside the condensing unit housing; it will only get back into the coil.

When flushing the condenser coil with water, protect the condenser fan motor and the electrical compartment to keep water out of them. They should be covered with a piece of plastic.

The fins on the coil must be straightened so that air can flow freely through the coil. When a large portion of the fins are bent over, the proper amount of air cannot flow through the condenser, causing high head pressure. (Fin combs are available at most supply houses that will help in straightening them.)

2. Water-cooled condensers—The type used usually depends upon the equipment manufacturer. However, they all operate on the same basic principle.

Water flows in the opposite direction of the refrigerant. Thus, the cooler entering water contacts the cooler leaving refrigerant first.

Water-cooled condensers require more maintenance than do air-cooled condensers. Maintenance generally depends upon the type of water used for cooling.

Water-cooled condensers will have an accumulation of scale on the tubes that slows the heat transfer from the water to the refrigerant. This scale must be removed so the condenser can operate as designed.

Care must be taken in removing the scale, because an acid-type chemical is used. Too much of the chemical also will remove some of the copper from the tubing inside the condenser shell.

When enough of the tubing has been removed, the tube will give way under the refrigerant pressure inside it. The refrigerant will leak out and the water will leak into the refrigerant system. This causes much trouble for the system—and for the service technician who finds it.

Water-cooled condensers are designed to operate properly with 3 gal (11 lit) of water/ton/minute flow through the condenser with a 10°F (5°C) rise. A temperature rise lower than this is usually an indication that the condenser is scaled and must be cleaned. A temperature drop higher than this indicates insufficient water flow through the condenser.

Insufficient water flow is usually caused by a plugged pump strainer or a pump that is not running at the proper speed. Also, the spray nozzles on the tower may be plugged.

Reduced water flow will generally cause higher-than-normal head pressure. A scaled condenser coil also causes higher-than-normal head pressure.

The amount of stoppage or scale determines how high the head pressure will rise. Normally, on water-cooled condensers using HCFD-22, the head pressure will be around 215 psig (1482 kPa) when everything is operating properly.

When a water-cooled condenser is subjected to freezing tempera-tures, it must be protected, usually by draining the water circuit, in-cluding the cooling tower, and blowing any water from the condenser tubes.

If the tubes should freeze and burst when the system is again placed in operation, water may be pumped into the refrigerant circuit. This condition requires much work and is very expensive to repair.

System Control Equipment

Terry Hoffmann
Johnson Controls, Inc.,
Milwaukee, Wisconsin

21.1 Introduction

In order to discuss the Maintenance and Operations of system control equipment, it is important to understand their purpose. Fundamentally, a control system does four things: (1) establishes a final condition, (2) provides safe operation of the equipment, (3) eliminates the need for ongoing human attention, and (4) assures economical operation.

Maintaining a set final condition is often considered the only purpose of the system because it has a direct impact on occupants of the facility. The comfort level of each individual is the end product of all control efforts, yet ASHRAE design standards are set to satisfy only 90 percent of all occupants. Well-designed systems that are operated correctly and maintained to their original delivered condition are capable of doing much better than this.

Safety pertains to both individuals and property. The most obvious safety-related HVAC controls are pressure and temperature cutout switches that shut off equipment to prevent damage to both the device and the operator. It is interesting to note that levels of safety consciousness vary throughout the world. While the general pattern would seem to indicate more concern for safety in economically developed countries, there are variances from country to country based on culture and operational practices. For example, it is quite normal for a motor control system installed in Germany to monitor each position of a hand-off-auto switch, as well as all electrical overload devices. As we develop into a completely global society based on a single set of standards, which are then tailored to the needs of a particular geographic area, it will be important to understand these differences and adapt our operational practices to meet them.

Elimination of the need for human attention is the cornerstone of the modern control system. In fact, we often call it an Automatic Temperature Control System or an Automated Building Control System. The impact here is that the whole concept of operation is less tangible because the device or system is designed to operate itself without the aid of a person. The first thermostat, dating back to the 1880s, was not designed for the *comfort* of occupants, but for their *convenience* so they did not have to notify the boiler room staff whenever they wanted more heat or less heat. A good automatic control system reduces human attention to making sure that commissioning has been correctly accomplished, and occasionally making setpoint adjustments to better meet the current situation if it is outside of normal design conditions. These variances might include extremes of heat and cold, as well as building upgrades and retrofits that affect control variables either temporarily or long term.

Assurance of economical operation has become another important factor due to two important global trends. First, the need to conserve our global natural resources places a responsibility on each individual and organization to utilize them in the most efficient manner. If these resources are depleted without the development of suitable alternatives, it would have a tremendous impact on the world economy. Second, the advent of global competition has made productivity a central piece of the strategic operation of both manufacturing and service industries. Automatic control systems provide the vehicle for this conservation and productivity. It is notable that many specifications refer to the controls as an Energy Management and Control System.

21.2 Fundamental Control Devices

By looking at a functional block diagram of the fundamental control loop, we can easily identify the individual pieces of equipment common to HVAC control systems (see Fig. 21.1):

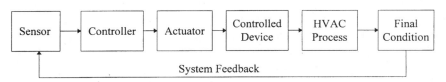

Figure 21.1 System feedback.

Sensors at the beginning of the diagram sense the controlled variable (i.e., the final condition we are attempting to impact). They send a signal to the controller's input. For HVAC controls, these devices usually sense temperature, humidity, flow (velocity), speed (percent or RPM), position or state (on/off, high/low).

The **controller** analyzes the sensor's input and makes a comparison against the desired final condition—the setpoint. The controller then outputs a signal to the controlled device assembly to maintain the setpoint value. In networked systems, it is also responsible for reporting information to a higher level controller on the network, which then aggregates the information and interfaces to a higher level information system. In turn, this information can be accessed on a personal computer interface or a dedicated display device.

The **actuator** is often part of a controlled device assembly, and is responsible for taking the output from the controller and transforming it—with the aid of some energy source—into a change of position in the controlled device. Actuators are usually powered electrically or pneumatically.

The **controlled device** regulates the flow of steam, water or air according to the commands of the controller as applied by the actuator. The regulation of these fluids helps to balance the load, gains or losses in the HVAC system. These devices include motors, valves and dampers.

The **HVAC Process** refers to the actual mechanical equipment, ducts, coils, fans, pumps, etc., that impact the *final condition* in accordance with mechanical design.

In summary, we are therefore concerned with the operation and maintenance of sensors, controllers, actuators, valves, dampers, motors, the infrastructure connecting them to the mechanical equipment, and finally, when installed, all related network components that feed into the human operator of a computerized management and control system.

21.3 Control System Types

HVAC control systems are usually categorized by energy source:

Self-contained control systems combine the sensor, controller and controlled device into one unit. In this system, a change in the controlled variable is used to actuate the controlled device. A self-contained pressure relief valve is an example of this type of control device. If the pressure in the line exceeds the maximum setpoint—which may be fixed or adjustable—it is vented to relieve pressure and protect the end devices, along with ensuring the safety of the operator and occupants.

Pneumatic control systems are compressed air to modulate the controlled device. In this system, the air is applied to the controller at a constant pressure, and the controller regulates the output pressure to the controlled device according to the desired rate of change. Typical air pressure for an HVAC control system is 20 PSI (136 kPa).

Electromechanical control systems use electricity to power a mechanical device. These systems may have two-position action (on/off) in which the controller switches an electric motor, solenoid coil or heating element. Or, the system may be *proportional* so that the controlled device is modulated by a motor.

Electronic control systems use solid-state components in electronic circuits to create control signals in response to sensor information. Greater reliability of electronic components and reduced power requirements have allowed electronic systems to supplant electric systems in all but the simplest applications — and where digital control is not required.

Digital control systems (direct digital control, or DDC) utilize computer technology to detect, amplify and evaluate sensor information. This evaluation can include sophisticated logical operations such as rule-based and fuzzy logic operations. The resulting digital output signal is then converted to an electrical or pneumatic signal, which is capable of operating the controlled device. Often this transducer is included in the actuator. Table 21.1 compares the different types of systems.

TABLE 21.1   Control system types.

Control System Types		
Classification	**Source of Power**	**Output Signal**
Self-Contained	Vapor Liquid Filled	Expansion as a result of pressure
Pneumatic	15 PSI (104 kPa) 20 PSI (136 kPa)	0 - 15 PSI (0 - 104 kPa) 0 - 20 PSI (0 - 136 kPa)
Electric	24 VAC 120 VAC 220 VAC	0 - 24 VAC 0 - 120 VAC 0 - 220 VAC
Electronic	5, 12, or 15 VDC 12 VAC 24 VAC	0 - 5 VDC (typical) 0 - 12 VAC 0 - 24 VAC
Digital	24 VAC 120 VAC 220 VAC	0 - 20 milliamps 0 - 10 VDC direct digital

21.4   Typical Applications

Since this book is about HVAC control, we are interested primarily in the following applications, or subsets, of HVAC control:

- Temperature Controls
- Flow Controls
- Pressure Controls

While the mechanical equipment may vary greatly in these applications, the control elements are largely the same. In the case of electric or electronic controls, the only difference between each application might be the sensing element. With regard to digital control systems, there is flexibility to handle any of these applications — and in many cases a single controller assumes responsibility for temperature, flow and pressure in order to control a single piece of mechanical equipment. For a detailed description of each application and system design fundamentals, refer to the *1995 ASHRAE Applications Handbook* or the *Handbook of HVAC Design*) (see Bibliography).

21.5   Impact of System Failure on System Performance

Failure of the control system has an obvious impact on the performance of the HVAC system. The most noticeable impact is on occupant comfort due to the loss of temperature, humidity or ventilation control. Older pneumatic and electronic systems are designed for a single "best alternative" failure mode. In the case of temperature control, failure usually defaults to a *Full Heat* setting to prevent freezeups in the northern climates, and to a *No Flow* setting where freezeups are not a problem.

Either alternative is concerned strictly with safety, not comfort. Modern digital control systems are designed for fail-safe or fail-soft operation to lessen the impact of such component failures. For example, if a sensor is determined to be out of range or non-operational, the digital controller can continue to operate using the last good reading that it had obtained prior to the failure — or it can be programmed to fail to a particular output. It might also be programmed to look at another input, such as outside air temperature, and make the best decision possible given that information. Any of the three alternatives are probably superior to the older "on/off" scenario.

Modern digital controllers, with their built-in failure modes, also have eliminated much of the domino effect that plagued less sophisticated systems. The chances of melting all of the chocolate bars in the cafeteria because temperature control failure defaulted to *Full Heat* over the weekend are greatly diminished. By the same token, it is less likely that any failure in a system will result in the failure or destruction of components in another, complementary system. That's because operators can preprogram and select a sophisticated array of checks and balances.

Another significant step toward the elimination of catastrophic failures is the growing trend of system manufacturers to include not only an adjustable alarm limit for critical values, but also a pre-alarm limit which notifies the operator of an impending failure. This form of predictive feedback fits well with the concept of both predictive maintenance and reliability-centered maintenance. For older systems with less built-in intelligence, it is the operator's responsibility to spot these anomalies.

21.6 Historical Experience Concerning Failure Rates

When comparing the Mean Time Between Failure (MTBF) of a digital HVAC control system to that of a pneumatic or electric system, it is appropriate to examine the system as having four levels of technology, each with its own characteristics (see Table 21.2).

Since the control and monitoring system requires all of the layers to be operating correctly, system reliability is equal to that of the least reliable component—which often is the **personal computer.** Therefore, it is important to make sure that a networked system is configured with multiple PCs having duplicate functionality. A recommended plan is to have access to a spare monitor, and to maintain a

TABLE 21.2 Mean time between failure (MTBF).

Layer	MTBF Characteristic
Sensors / Actuators	Very long MTBF for sensors (passive components)
	Medium MTBF for actuators (mechanical components)
Digital Controllers (Microprocessor based)	Long MTBF (failures usually relate to power supply)
Wiring Network	Very Long MTBF (passive components)
Personal Computers	Short MTBF (mechanical components in monitors and disk drives)

daily backup of the system on a second hard drive or a high-density removable backup device (tape or disk).

Actuators also present concerns because they are often the next least-reliable devices. Most buildings have older style actuators that are not very reliable, although many high-quality actuators being built today have one-year reliability rates of over 99 percent. It is reasonable to ask the system manufacturer for this information as part of the original purchase decision. The information also is valuable when determining which spares — and how many — to have on hand. There is a high probability that any manufacturer maintaining ISO 9000 certification of its design and manufacturing facilities will be able to provide this data.

The **wiring network** is not noted as having a high probability of failure. However, this is only true if the network does not rely on a PC-based server and if the network is not shared with other information technology organizations within the facility. Network maintenance is best handled by a computer network management specialist, so it lies beyond the scope of HVAC controls. The only way to make sure that network failures do not adversely impact system operations is to design and maintain a system with no single point of failure. In HVAC applications, the impact of a single failure can be minimized by using distributed control techniques and increasing the level of intelligence as far down the control ladder as possible. When adding devices to an existing system, always evaluate the impact of a controller failure on the operation of the system given its new tasks. In many cases, it is better to add a new controller than to overburden an existing one.

21.7   Operator Training Requirements

Depending on the size of the facility and the complexity of the HVAC control system, it is possible to have vast differences in the educational requirements for those who operate and maintain the equipment. Anyone with responsibility for the system should have a working knowledge of HVAC basics, as well as familiarity with the system itself and the technology behind it. In addition, the operator should be able to accomplish the following tasks:

- Program/change setpoints and schedules as required to meet the operational needs of the organization.
- Optimize energy usage and efficiency.
- Ensure occupant comfort.

- Understand the impact of system changes and maintenance on the core business functions of the organization.

- Assist with building systems safety, health and environmental standards, requirements and codes.

- Assist with planning for system upgrades and improvements designed to increase safety, improve reliability, and improve occupant comfort levels, thereby increasing productivity.

In order to accomplish these tasks, the operator will require training above and beyond that which is available on the job. The following recommendations are to be viewed as consecutive and incremental. Each level builds on the next, leading to the ultimate goal of an informed, productive control system manager.

HVAC basic training.   This course of study should include the following information:

- HVAC system types and piping systems
- Fan systems and fan characteristics
- Dampers and damper actuators
- Valves and valve actuators
- Boilers and related equipment
- Fundamentals of refrigeration
- Chiller types and applications
- Cooling towers
- Heat exchangers and miscellaneous equipment

Sources for this training include local technical colleges and universities, ASHRAE professional development seminars, mechanical systems and control systems manufacturers, and self/group teaching via books, videos and study guides available from a variety of sources.

Control systems fundamentals.   A typical course outline should include:

- Basic control theory
- Proportional control fundamentals
- Proportional plus integral control
- Proportional plus integral plus derivative control
- Economizer systems

- Mixed air systems control
- Variable air volume control systems control
- Building pressurization control strategies
- Fan capacity control methods
- Control fundamentals for water systems

Sources for this training are similar to those for HVAC systems, with a heavier emphasis on materials obtained from control systems manufacturers.

System-specific operation and maintenance (pneumatic). If pneumatic systems or devices are prevalent in the facility, it is important to understand:

- Pneumatic controls and air supply systems
- Air supply maintenance
- Thermostats: single and dual setpoint
- Controllers
- Valve and damper actuators
- Valve and damper actuator service
- Auxiliary pneumatic devices
- Temperature, pressure and humidity transmitters
- Pneumatic receiver/controllers
- Calibration and adjustment or receiver/controllers
- Safety devices and procedures

Hands-on courses are best suited for this training, and are available from local technical colleges and control system manufacturers.

System-specific operation and maintenance (digital). The differences between the pneumatic and digital courses are technology- and terminology-based:

- Control systems evolution
- Computers, programming, and direct digital control (DDC)
- Systems terminology
- Basic hardware, software and user input devices
- Analog and binary sensor types and applications

- Output types, actuators and transducers
- Software data types
- Control loop types
- Programming direct digital controllers
- Communication to automation systems

Training of this nature should be provided directly from the manufacturer of the digital controls or from an alternate source familiar with the equipment.

Energy management concepts. This course of study ensures the understanding of basic energy management concepts and features relating to heating, cooling, air distribution, electrical and lighting systems:

- Energy management fundamentals
- The building envelope
- Modeling the building environment
- Opportunities for conservation
  Cooling systems
  Heating systems
  Air distribution systems
  Water distribution systems
  Electrical systems
  Lighting systems
- Performing an energy audit

The Association of Energy Engineers (AEE) has courses available which lead to examination and granting of a Certified Energy Manager (CEM) certificate. This line of study is suitable for both operators and aspiring managers. ASHRAE also offers professional development seminars on this topic, as do local colleges and universities. Manufacturers of mechanical and control systems are likely to offer courses in applying their products to manage energy usage.

Building automation and control systems operation and maintenance. With a firm foundation of mechanical, systems, and energy knowledge in place, it is possible to comprehend and apply the features and functions of the building automation system to the needs of the enterprise:

- Basic system hardware configurations

- Communications network architecture
- Operator workstations and I/O devices
- System software
  On-line help, point types, menus
  Passwords, commands, summaries
  Reports and alarm management
  Scheduling points and summaries
  Point history, trending, operator transactions
  Energy management features
  Defining/creating systems and points
  Programming control loops and functions
- Hardware maintenance

It is strongly suggested that this course be taken on site using the installed equipment, or at a well-equipped lab operated by the control system manufacturer.

Given the breadth and depth of training required to properly maintain a complex HVAC control system, a decision must be made to provide training in-house or to augment the operation and maintenance team with an external control systems specialist. There are three ways to obtain this assistance:

- Include maintenance of the control system in a service contract with the mechanical contractor or controls company. This provides a limited amount of time each month to perform important tasks and observe system trends that may impact operation.

- Outsource a mechanical systems and controls specialist from an organization that has experienced staff people available. The advantage here is that the individual is full-time and blends into the organization, while not burdening the organization with the requirements for training and development.

- The operation and maintenance of the mechanical systems and controls can be handled by an outside supplier as a task under a total facility management contract. This allows management to concentrate on planning and developing strategies for improvements in operation and productivity, while ensuring competent operation and maintenance at a price that is constant for a fixed period of time.

21.8   Product Preventive/Predictive Maintenance Procedures

The following tasks are recommended for a comprehensive inspection and calibration of the HVAC control system. This may be accomplished on a rotating basis or several times throughout the year, depending on seasonal conditions, winter shutdown, spring changeover, etc. Many of the tasks are system-specific and may not be required in a particular facility.

It must also be noted that the type of maintenance program adopted in the facility may have a significant impact on the frequency of many tasks. For example, a program which includes predictive maintenance information may reduce the frequency of tasks requiring hardware checks for wear or accuracy. Likewise, a good Reliability Centered Maintenance (RCM) program (see Chapter 13) will reduce the frequency of most tasks while increasing the frequency of a lesser number of tasks performed on mission critical equipment.

Finally, the word *clean* is used sparingly in the following procedures to describe situations where dirt has a direct adverse effect on the operation of the device. It is good general policy to maintain all control equipment to a high standard of cleanliness. The interval between general cleaning can be evaluated on a yearly basis and extended if it is determined to be acceptable.

CONTROL AIR COMPRESSOR

- Drain the tank and check all traps
- Change compressor oil and check the oil pressure
- Check the belts, bearings and sheaves, visually or by vibration analysis
  Replace worn or damaged equipment
  Ensure proper sheaf alignment to avoid premature failures
- Change the suction filter on a regular basis
- Inspect the unloader and check valve
- Inspect the high-pressure safety valve
- Lubricate the motor(s) and analyze operating conditions
- Check the PE switch, starter and alternator as required
- Measure and record the compressor runtime under loaded conditions during building occupancy
- Where applicable, record the oil carry-over rate

REFRIGERATED AIR DRYER

- Check and record the refrigerant temperature
- Check and record the refrigerant pressure
- Clean all grills and coils as required
- Check and operate the drain trap and bypass valves

FILTER AND PRESSURE-REDUCING STATION

- Inspect the coalescent filter and change as required
- Inspect the charcoal filter and change as required
- Check and adjust the pressure-reducing valve, record all settings

BOILERS, CHILLERS, PUMPS AND ZONE CONTROL

- Check all controllers for correct calibration and adjust as necessary
- Calibrate all transmitters and set gauges as required
- Check all PE switches
- Check all control valves for operation and flow
- Check all pilot positions for accuracy
- Check auxiliary control devices

FAN SYSTEMS AND HVAC UNIT CONTROLS

- Review the sequence of operation for conformance to original specification and resolve differences
- Check operation and calibration of all dampers, lubricate as required
- Check all pilot positioners for accuracy
- Check all control valves and actuators
- Calibrate all controllers as required
- Calibrate all transmitters
- Set receiver gauges as required
- Check all solenoid valves, PE switches and air valves where installed and applicable
- Check all auxiliary controls

ROOM-TERMINAL UNIT CONTROLS

- Check all thermostats and calibrate as required
- Check all control valves for operation and leakage
- Check operation of unit coil steam traps
- Check operation of all dampers, lubricate
- Check all solenoid valves, PE switches and air valves where installed and applicable
- Check all auxiliary controls

TERMINAL UNITS

- Boxes — Mixing and Variable Air Volume
  Inspect connection to ductwork
  Lubricate and adjust dampers and linkage
  Verify operation of control

- Electric Duct Heaters
  Inspect coil for damage, clean if accessible
  Inspect isolators for damage or cracks
  Torque the heating terminals
  Verify operation and staging of control

- Induction Units
  Visually inspect coil and clean
  Check and clean drain pans
  Clean discharge grill
  Check and clean strainers
  Check steam traps and hand valves

- Reheat Coils
  Inspect coil for leaks/damage
  Dust coil, remove debris
  Check and clean strainers
  Verify operation of steam trap if applicable
  Verify operation of control

- Radiation
  Visually inspect fins/cast iron
  Check and clean strainers
  Check valve for leakage and operation
  Clean as required

NOTE: Always refer to the manufacturers' written documentation for calibration and maintenance procedures where available.

STANDALONE DIGITAL CONTROLLER, or APPLICATION SPECIFIC CONTROLLERS of a Building Automation System

On a Scheduled Basis

- Verify that device is being controlled at the appropriate setpoints and schedules

- Change one setpoint value; verify smooth transition and stable control at the new setpoint

- Return setpoint to original value

- Repeat for each additional control loop, if any

- Verify that controlled valves and dampers will stroke fully in both directions, sealing tightly where appropriate
- Verify the proper operation of critical control processes and points associated with this unit, and make any necessary adjustments

As Required

- Verify/calibrate other points associated with these units where the need for possible "Corrective Maintenance" is indicated

NETWORK LEVEL CONTROLLERS and INTERFACE DEVICES of a Building Automation System

On a Scheduled Basis

- Check meter and LED indications to verify proper AC and DC power levels, appropriate Transmit and Receive activity on the trunks, and to check for possible Error Code and diagnostic indications
- Inspect wiring for signs of corrosion, fraying or rapid discoloration
- Check voltage level of battery
- Cycle power to initiate Self-Test Diagnostic, if appropriate for device
- Remove excessive dust from heat sink surfaces
- Clean faceplate and input pad, if present
- Clean transparent window in door, if appropriate
- Clean enclosure exterior surfaces
- Verify the proper operation of critical control processes and points associated with this unit, and make any necessary adjustments

As Required

- Verify/calibrate other points and control processes, where the need for possible "Corrective Maintenance" is indicated

COMPUTERIZED OPERATOR WORKSTATIONS of a Building Automation System

On a Daily Basis

- Review workstation for CRITICAL, FOLLOW-UP and OFF-LINE status indications
- Review workstation for OVERRIDE, DISABLED and LOCKOUT status indications
- Review System Event Log
- Review Building Automation System operational concerns

- Perform on schedule "Corrective Maintenance" procedures as appropriate to resolve situations noted in the preceding Reviews

On a Scheduled Basis

- Check monitor for clarity, focus and color
- Clean read/write heads of removable disk drives(s)
- Cycle power, listen for unusual motor/bearing noise
- Verify proper system restart (check system date, time and hardware status)
- Clean exterior surfaces
- Save/copy workstation database, including graphics and resident controller databases
- Install appropriate workstation software refinement and problem correction revisions ("Minor Rev's") as they become available

NOTE. Major revisions to the Building Automation System workstation software which add new features and capabilities — or significantly enhance existing features — are not included in maintenance procedures outlined here.

NETWORK SERVICES for a Building Automation System

On a Scheduled Basis

- Reset the system diagnostic counters
- Allow data to tabulate in the diagnostic registers, if appropriate
- For each operator workstation and network connected unit:
  List all diagnostic statistics
  Analyze the error rate for each network node
  Analyze the transmission rate for each network node
  Determine the network performance ratio
- For each application specific device level trunk:
  List all diagnostic statistics
  Analyze the error rate for each network node
  Analyze the transmission rate for each network node
  Determine the network performance ratios
- Provide a report summarizing Network Analysis results

As Required

- Perform Network Analysis tasks as appropriate to verify or discount suspected communications or network throughput problems
- Perform Network Analysis tasks as appropriate to evaluate the impact on network performance of various configuration options — as part of a proposed system expansion or modification

21.9 Troubleshooting and Repair Tips

The following information is not designed to be a complete summary of troubleshooting techniques, but may be helpful in sorting out basic problems and identifying the cause of persistent failures.

The **control air compressor** is often the source of problems in pneumatic systems. Most compressors are designed for best performance when operating between 33 percent and 50 percent on-time. More than 50 percent on-time can result in objectionable levels of oil carry-over if proper filtration procedures are not followed. Oil in the control air lines can destroy certain pneumatic devices or make them inoperable. Excessive on-time also may result in a significant energy loss.

Common causes of excessive on-time include:

- Air leaks in the system

- Water in the compressor tank

- Extra control devices added to the system which require capacity above the original design specifications

- Inefficient compressor operation

Check the efficiency of the compressor is easy if you have the documentation that comes with the unit. Simply close off the compressor from the control lines, bleed off the air in the tank, and record the pump-up time to pressurize the tank to 90 PSI (620 kPa). Compare this to the manufacturer's specifications. If the number is too large, the compressor may require attention. If it is satisfactory, you must check the system for leaks or extra equipment.

Room controllers are basically maintenance-free devices as long as they receive accurate input signals, have clean uninterrupted power or supply air, and are not abused. Once installed, they usually require little calibration unless they have been tampered with or application requirements change. It is still a good idea to check the operation of a controller by reading the output pressure or voltage for a known condition in the space.

In actuality, the first evidence of a problem with an HVAC system is usually a complaint from an individual who is uncomfortable or bothered by drafts. Experience has shown that, in most cases, the problem lies in a misunderstanding of the capabilities and application of the control system as opposed to a malfunction of the equipment. There are several common causes of occupant discomfort that do not involve the system:

Sun load—Direct sunlight on a thermostat will cause overcooling or underheating in a zone. Also, direct sunlight on occupants will cause them to feel warmer than the actual space temperature.

Zone control—A person outside of a controlled zone may feel uncomfortable because the air distribution system is not delivering the full controlled medium to the space. Also, different zones may be controlling at different temperatures due to location or installed equipment requirements, so it may take a few minutes to adjust to the differences when moving from zone to zone. People working in an office atmosphere might not be comfortable in a high-tech manufacturing environment where constant air movement is required for cleanliness.

Covering of grills—Occupants frequently will cover part or all of a discharge diffuser, causing improper heating or cooling. The imbalance in the air distribution also can have an impact on other occupants of the same zone.

Occupant location—If occupants are located immediately adjacent to an outside wall or window, they may be subject to cold air leakage or to radiant cooling or heating from the walls.

People and equipment—Overheating may occur if more people or equipment occupy an area than it was originally designed for. This can occur when a meeting is held in an area not designed for the purpose, when an office area is converted to a makeshift laboratory, or when an area's computer usage increases significantly. In addition to the problems associated with overheating, it is also possible that the system is no longer delivering sufficient outside air to each person to ensure a healthy environment.

Drafts—To some people, even slight air movement is uncomfortable. Since most systems use air as a means of heating and cooling, there must be movement for the system to operate. This results in the need to relocate workstations or install new diffusers to solve minor problems. In extreme cases, the air delivery system may need to be rebalanced.

Even if none of these conditions exist, it is still possible for the controller to be operating properly. If the controller appears to be operating correctly, check this list before reprogramming or recalibrating the unit:

- Are the actuators operating correctly and coupled to the controlled device (damper or valve)?
- Is the control agent (water or air) sufficiently hot or cold enough to maintain the desired setpoint?

- Is the pressure sufficient to deliver enough of the control agent to maintain the desired setpoint?

- Are there unusual loads in the space which may overpower the efforts of the controller?

- Is there leakage in water lines or ductwork?

If the answer to all of the above is No, it is time to refer to the manufacturers instructions and recalibrate or reprogram the controller.

Valves are another occasional source of problems in HVAC control systems. The most common valve failures are sticking, leakage and failure to close completely. Leakage around the valve stem is easily identified and corrected by replacing the packing. If there is persistent leaking of multiple valves, it is possible that the water supply is dirty and causing premature failures.

Valves that are not controlling correctly may be sticking. To check this, stroke the valve from one extreme position to the other, and check for travel that is smooth and even. If the stem sticks, replace the packing. Failure of a valve to shut off tightly also can cause comfort problems. To check for leakage through the valve, stroke it fully closed and check the temperature of the outlet piping. If the outlet remains hot on a heating valve or cold on a cooling valve, there is some leakage through the valve. Check with the control valve manufacturer to see if a repair kit is available for the disk, or replace the entire valve.

Networked systems present an entirely different set of challenges to the operator. Far and away the most likely scenario for a network failure which can be treated by a layman is a faulty connection. Check all connections by carefully disconnecting, inspecting and reconnecting the network to the device that is not responding, and then to the next device in line. Pay special attention to shields and grounds, as a faulty shield may cause a segment of wiring to act like an antenna. Once the connections have been checked and reconnected, power down and re-power the device that is not responding — and check again for a response.

Beyond these simple troubleshooting measures, it is time to bring in a network specialist with the correct hardware and software tools to thoroughly test the network. *Always perform a complete backup of all information on computer hard drives before attempting to fix any software problems on the system — or allowing anyone else to touch the hardware.* Rebuilding a network from scratch can take weeks — and all because someone neglected to back up the system or erased the existing backup media to use it for another purpose. It is not too severe to recommend keeping the backup under lock and key.

Additional suggestions for troubleshooting and repair of HVAC control systems are available directly from manufacturers and are specific to their products. Ask for a copy of the operating manual for the type of system or equipment you are operating, as well as any publications dealing with maintenance and troubleshooting.

21.10   Criteria for Repair vs. Replacement

Replacement parts for most HVAC controls are available directly from the manufacturers or their authorized distributors. The decision regarding repair vs. replacement of the device is easier for controls than it is for mechanical equipment in an HVAC system because controls usually are much less expensive—and the troubleshooting process often recommends replacement. In general, there are three scenarios for inoperable or damaged equipment:

- If the manufacturer, or third party, offers a repair kit, it is usually safe to assume that repair is cost-effective and easy to accomplish.

- If the manufacturer offers a repair and return service, or exchange credit, there is benefit to taking advantage of the reduced pricing. This is usually limited to large controllers or complex electronic devices where a repair station has been established to handle the exchange process.

- If a repair parts kit is not available and there is no repair and exchange policy, the part must be replaced. Controllers and auxiliary equipment should not be repaired locally, as testing and adjustment of most items cannot be adequately accomplished without sophisticated laboratory testing.

If the troubleshooting and replacement process is unsuccessful and there is no service contract in place with the controls vendor, then a suitable third party must be called in to determine the next step in the repair process.

21.11   Summary

The importance of well-maintained and operated HVAC system control equipment to the delivery of comfort for building occupants cannot be overstated. Insufficient attention to controls can undermine the efforts—and the cost advantages—of keeping the building's mechanical equipment in peak condition. Control equipment failure can cause extensive damage to HVAC systems, which may in turn create unsafe conditions.

Effective maintenance and operations of HVAC system control equipment depends on a number of factors, including the type of equipment installed in the facility, the maintenance system that is being employed, and — most importantly — well-trained people who understand the complexities of building controls.

21.12   Bibliography

1. Alliance Planned Service, Johnson Controls, Inc. Milwaukee, Wisconsin, 1992.
2. ANSI/ASHRAE 55-1992 ASHRAE Standard, American Society of Heating, Refrigerating and Air-Conditioning Engineers, Inc., Atlanta, Georgia, 1992.
3. *1995 ASHRAE Applications Handbook,* Chapters 8 and 42, American Society of Heating, Refrigerating and Air-Conditioning Engineers, Inc., Atlanta, Georgia, 1995.
4. *Building Environments — HVAC Systems,* Chapters 16 and 17, Global Learning Services, Johnson Controls, Inc., Milwaukee, Wisconsin, 1997.
5. Grimm, Nils R., and Rosaler, Robert C. *Handbook of HVAC Design,* McGraw-Hill Publ. Co., New York, New York, 1990.
6. Installation and Service Manual for PureFlow™ Air Compressors, First Edition, Johnson Controls, Inc., Milwaukee, Wisconsin, 1993.
7. Johnson Controls Institute Training Programs, Johnson Controls, Inc., Milwaukee, Wisconsin, 1996.

Condensate Control

Warren C. Trent, M.S., P.E., CEO
C. Curtis Trent, Ph.D., President
Trent Technologies, Inc.
Tyler, Texas

22.1 Introduction

Suitable condensate control is achieved and an acceptable HVAC maintenance program can be implemented only when condensate is confined to (a) surfaces of the cooling coil, (b) a small and properly sloped condensate drain pan, and (c) a well drained system through which condensate flows freely and never stands nor stagnates.

Confining condensate to these three areas allows the system to operate virtually free of excessive maintenance, property damage and health-threatening biological growth. When condensate is confined in this manner, the required system maintenance consists of rather simple periodic scheduled procedures: inspecting, cleaning, and flushing the drain system (pan, seals and lines).

Unfortunately, far too many systems now in operation are not designed to restrict the spread of condensate, and are not amenable to reasonable maintenance.

A successful maintenance program for these systems must begin with an assessment of the capacity of each to *confine condensate to the cooling coil, drain pan and drain system.* Anytime condensate spreads beyond these areas, system modification is necessary to ensure that condensate is properly confined, under all operating conditions.

System deficiencies that allow the spread of condensate beyond the cooling coil and the drain system include the following:

- Condensate carryover from cooling coils
- Condensate drips onto internal HVAC system components
- Unsuitable drain pan designs
- Very low supply air temperatures
- Improper fan position inside the air handlers
- Ineffective seals on condensate drain lines
- Unsuitable installation of condensate drain lines

When these deficiencies are present, no amount of system maintenance can prevent equipment damage, surrounding property damage and health threatening biological growth. *Scheduled maintenance has only limited value.* It occurs after property damage has been done and biological growth has had its effect. Moreover, the damaging effects begin all over as soon as system operation is resumed. The design considerations necessary to avoid these conditions in future system are provided in the *McGraw-Hill HVAC Systems & Components Handbook* (Ref. 1).

Whenever the above deficiencies appear in any system, they must be remedied before a meaningful maintenance schedule can be defined and implemented. The following paragraphs suggest suitable remedies and define a drain system that can be maintained with reasonable routine and preventive maintenance programs.

22.2   System Deficiencies and Remedies

Specific system deficiencies that preclude the implementation of a feasible maintenance program, along with remedies to these deficiencies are reviewed below.

22.2.1   Condensate carryover from cooling coil

Condensate carryover in any observable quantity is incompatible with a practical and acceptable system maintenance program. Damage and contamination begin when carryover occurs, neither will wait for the next scheduled maintenance action.

Condensate carryover occurs when the velocity of the air passing through the cooling coil is sufficient to entrain condensate and blow it off the coil. Any time the system components or other surfaces downstream of the cooling coil become wet, condensate carryover is a possible cause that must be assessed. The presence of carryover can best be established by visual observation (portholes, fiber optics, etc.)

downstream of the coil, during the cooling operation when the latent heat load (water removal) is high. The entire surface of the coil must be viewed in order to determine the cause and location of the deficiency. Uniform carryover indicates one deficiency, carryover in local areas indicates other deficiencies. The three most common causes of condensate carryover are (a) unsatisfactory coil design, (b) dirty cooling coils and (c) distorted air velocity profile entering the coil.

22.2.1.1 Unsatisfactory coil design. Coil design is unsatisfactory when condensate carryover appears somewhat uniformly over the entire face of a *clean* cooling coil. The following design parameters determine the cooling coil condensate carryover characteristics: airflow of the air handler; height and width of the cooling coil; size and spacing of the coil tubes; and thickness and spacing of fins on the tubes. Fig. 22.1 illustrates, for a typical coil design, the relationship among these parameters. The air velocity at the coil face, shown in this figure, is determined by dividing the air handler airflow by the face area of the coil.

Problem definition. When condensate carryover, due to coil design, occurs in a particular system, it can be remedied only by changing one or more of the parameters included in Fig. 22.1. Generally, in existing

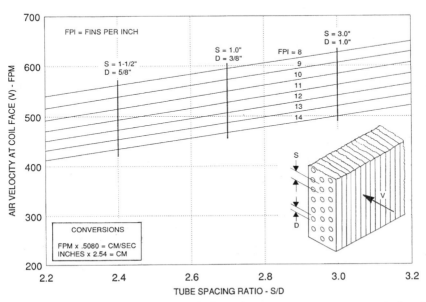

Figure 22.1 Coil face velocity above which condensate carryover occurs, typical cooling coil.

systems, it is not practical to make significant changes in coil geometry. Thus, the most practical way to eliminate condensate carryover is to reduce air velocity at the coil face; that is, reduce airflow.

Remedies. Reduced airflow can be achieved most effectively by changing fan speed, which involves either a change in size of pulleys or a change in the motor speed, if the motor speed is variable.

It is often possible to reduce airflow without causing system problems. Many times HVAC systems are oversized. In such cases, reduced fan speed introduces no penalty in cooling performance. Moreover, total cooling capacity is relatively insensitive to airflow. For example, a reduction in airflow of 20 percent typically reduces total cooling capacity by about 5 percent at the rated point. Sensible cooling capacity is reduced more, but latent cooling capacity is increased. Even in those instances where the total cooling capacity is compromised by reduced airflow, the best choice may be to accept this compromise and eliminate the serious property damage and health problems associated with condensate carryover.

In installations where the coil design is similar to that shown in Fig. 22.1, the reduction in air velocity needed to eliminate carryover can be approximated as follows: Compute air velocity at the coil face in feet per minute by dividing the airflow (CFM) by the face area (square feet — height times width) of the cooling coil. With the coil face velocity, enter Fig. 22.1 at the spacing ratio and fin spacing determined from coil measurements. The difference between this point and the point where carryover is indicated by Fig. 22.1, indicates the reduction in velocity and, therefore, the airflow reduction required to eliminate carryover. At best, however, this process provides only a starting point. The proper airflow reduction is that which eliminates condensate carryover, a condition that must be determined by visual observation or other suitable means.

The installation of moisture "eliminators" downstream of the coil is not a viable method for preventing carryover. Eliminators add significantly to the pressure loss in the system and, therefore, reduce airflow. In addition, they introduce another potential growth place for biological contaminates.

22.2.1.2 Dirty cooling coils. Condensate carryover may sometimes be observed in systems where the cooling coil design is entirely satisfactory. The cause may be dirty coil surfaces. Because, dirt does not always collect uniformly on the cooling coil, carryover may occur in limited local areas of the coil.

Problem definition. Dirt and other foreign material deposited on the surfaces of cooling coils can reduce the area for airflow and increase the air velocity sufficiently to effect condensate carryover. Unrelated to condensate carryover, dirty coils have other adverse consequences. They reduce heat transfer and decrease system efficiency.

Remedies. Within the industry, the most widely endorsed solution to dirty coil conditions is a maintenance program that involves periodic coil cleaning. When properly defined and performed regularly, coil cleaning can be adequate to avoid carryover resulting from dirty coils. But, coil cleaning is a costly and time consuming process. Furthermore, in many existing systems, the cooling coils are so inaccessible and cleaning is so difficult that it is often deferred until a major problem arises.

Probably the most dependable and cost effective way to maintain coils in an acceptably clean condition is to utilize filters with adequate capacity to remove particles that accumulate on the coil; thereby, avoiding the need to perform frequent cleaning. Available data (Ref. 2) indicate that filters with dust, spot efficiency ratings (as defined in Ref. 3,) of 25 percent or higher can virtually eliminate dirty coils and the airflow problems they cause. Of course, to be effective a filter must be placed in an essentially air tight holder, otherwise by-passed particles can reach the coil and be deposited on the surfaces.

22.2.1.3 Distorted air velocity profile entering the cooling coil. Air entering the cooling coil with a nonuniform velocity profile can cause carryover even when the average face velocity is below where carryover would occur, as indicated in Fig. 22.1. Carryover caused by this condition can be defined by visual observation. And, the location and source of the distorted airflow can be identified and corrected.

Problem definition. In draw-through systems, distorted airflow can be generated; for example, when air enters a short-coupled return plenum at right angles to the coil. In blow-through systems the coil may be subjected to a very adverse velocity profile created at the fan discharge. Carryover of this type is evidenced by concentration in local areas of the coil, where airflow distortion occurs.

Remedies. Carryover caused by distorted airflow may be remedied by installing longer plenums and/or turning vanes. Properly installed turning vanes and longer more efficient diffusers not only improve the velocity profile they can significantly reduce pressure losses. Turning vanes and longer plenums require more space, additional hardware, and added costs. Nevertheless, in certain cases, major changes may be necessary to eliminate the detrimental spread of condensate if an effective maintenance program is to be achieved.

22.2.2   Condensate drips onto internal HVAC system components

Any HVAC system that allows condensate to drip onto internal surfaces and components is subjected to internal damage and the growth of contaminating organisms. Sloped cooling coils and noninsulated coolant lines (refrigerants or water) are often the source of condensate drips. Systems that exhibit these qualities must be modified, because routine scheduled maintenance does not protect against these conditions, nor does it remedy the causes.

22.2.1.4   Drips from sloped cooling coils.
Sloped cooling coils included in some HVAC systems are prone to drip condensate onto surfaces outside the condensate drain pan.

Problem definition.   Condensate that drips from a slanted coil onto surfaces outside the drain pan creates destructive and contaminating conditions. At small slope angles, surface tension may be adequate to retain the condensate and allow it to drain into the condensate pan. However, foreign deposits on the coil can easily destroy the effects of surface tension and cause dripping to occur.

Air velocity entering the coil tends to reduce dripping from a coil that is sloped rearwardly (from the top), when the fan is operating. But it is not a reliable force for preventing condensate dripping.

Remedies.   Extending the condensate drain pan to catch condensate drips is not a viable solution. It introduces another equally serious condition, large drain pans, discussed in Section 22.2.3. ("Unsuitable drain pan designs"). In order to ensure adequate condensate control and a dry system that is free of harmful contaminates, sloped coils must be kept clean at all times and the fan should not be turned off during the cooling period. The air filter must never be placed below the cooling coil, where dripping condensate creates conditions conducive to the growth of health threatening organisms. Filters mounted below the coil must be moved upstream to eliminate a major source of system contamination. Because slanted coils increase the susceptibility of systems to internal wetness and contamination; as a minimum, a more stringent and high frequency maintenance program is necessary, for these systems.

22.2.1.5   Drips from noninsulated coolant lines.
All too frequently, existing HVAC units have noninsulated coolant lines that pass through the cooling airflow passage, causing condensate drips along their paths.

Problem definition.   Condensate that drips on surfaces outside the drain pan cause damage to the HVAC system and create conditions conducive to health threatening biological growth.

Remedies. This condition, much too common in current systems, is simple to remedy. It can and must be eliminated by applying suitable insulation to all bare coolant lines.

22.2.3 Unsuitable drain pan designs

Condensate drain pans in many HVAC systems now in the field are so configured that the necessary maintenance effort varies between very difficult to impractical to perform. Among the most troublesome features are (a) large drain pans, (b) primary drain port location, and (c) internal baffling. Systems that exhibit these characteristics must be modified before a reasonable maintenance program can be implemented.

22.2.3.1 Large drain pans. Systems that make use of large condensate drain pans (often extended downstream of the cooling coil to protect against condensate carryover)—cannot confine the spread of condensate within the boundaries necessary to permit successful maintenance.

Eliminating condensate carryover, as discussed in paragraph 22.2.1, will not keep large condensate pans dry and free of damage and biological growth.

Problem definition. As condensate forms and drains into the pan, it will stand there at some finite depth. If the drain pan is level—as is common for systems now in the field—condensate will cover the entire pan. The precise depth at which condensate stands depends upon the pan geometry, and the rate at which condensate is drained from the pan. Typically, the depth varies between about 1/8-in (3 mm) and 1/2-in (12 mm) (or greater). See Fig. 22.2.

During system operation, condensate in the drain pan will flow from the area below the cooling coil to the drain port, leaving the remainder—in fact most—of the condensate in a stagnant state. There, it becomes a growth haven for contaminating organisms, as illustrated in Fig. 22.3. A photograph of one such result is shown in Fig. 22.4.

Remedies. For HVAC units now in the field, sloping the drain pan by tilting the HVAC unit is one way to reduce the pan area covered with stagnant condensate. Fig. 22.5 illustrates the effect of sloping a drain pan in one direction. Sloping the pan in both directions, of course, further reduces the area of stagnant condensate. Hence, with sufficient slope in two directions, it is possible to virtually eliminate stagnate condensate in the pan.

Sloping the pan, however, by no means makes a large pan an acceptable remedy for condensate carryover. Carryover droplets depos-

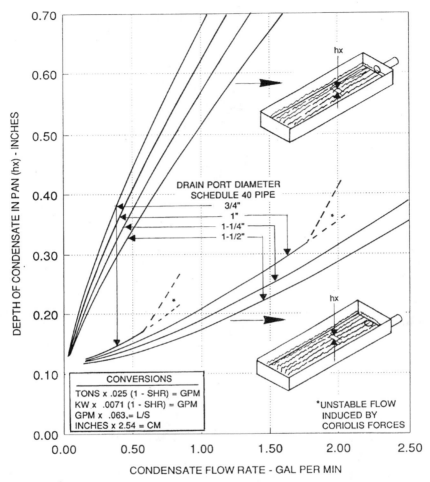

Figure 22.2 The effect of drain port size and position on condensate flow rate under the force of gravity.

ited on the pan will not drain readily to the drain port. Instead, they will be held in place by surface tension providing another potential source of biological contamination. In reality, the large drain pan serves no useful purpose. It is not a solution to condensate carryover, a condition that must be prevented, as discussed in par. 22.2.

The equipment damage and contamination problems caused by large drains pans now in the field, can be remedied by simply reducing the pan size to that required to catch the condensate and accommodate the flow in the drain pan. The length of the pan must be sufficient to cover the base of the cooling coil. The pan area is then fixed by pan

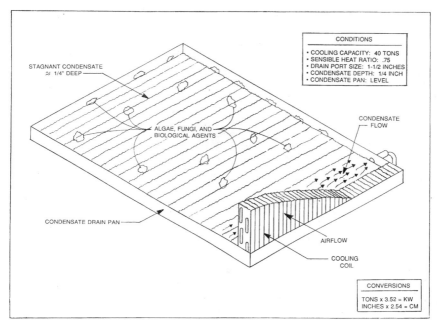

Figure 22.3 Contamination problems created by wide drain pans.

width—the distance the pan extends away from the cooling coil. The width of the drain pan must be sufficient to accommodate the maximum condensate flow rate, yet not so wide as to allow condensate to stagnate. Pan widths considered acceptable for systems with various cooling capacities and condensate drain sizes are shown in Fig. 22.6.

The most effective way to reduce large pans to a suitable size depends upon the specific system involved. Where possible, the most desirable pan is one constructed of durable nonmetallic material or stainless steel. Often, however, the most practical way to effect pan size reduction is to install a wall inside the current drain pan, as illustrated in Fig. 22.7. In some cases, it may be necessary to relocate the drain port and place it at the end of the pan as indicated.

If the pan and attachments already in place are constructed of ferric metals, they must be replaced or treated with durable (long-life) protective coatings. This is because under some conditions the presence of iron accelerates the growth of certain harmful bacteria. (See Ref. 4.)

Condensate drainage can be enhanced and protection against the formation of water puddles can be realized by tilting the HVAC unit toward the drain port, as discussed above. A slope of 1/4-in./ft

Figure 22.4 Photo showing contamination in wide drain pan.

(2 cm/m) is usually adequate. However, the unit should be tilted only with the approval of the equipment manufacturer.

Large pans modified as defined above will permit condensate to be confined within boundaries that allow practical maintenance and ensure a minimum of property damage and biological contamination.

22.2.3.2 Primary drain port location. Proper drain port location is essential to adequate drainage and condensate control.

Problem definition. Drain ports located above the bottom of the drain pan inherently prevent complete drainage of condensate. Water standing in the pan below the drain port level collects and retains debris, supports the growth of biological agents, and contaminates the system. Frequent cleaning and maintenance are required to prevent serious property damage and human health problems. Moreover, location of the drain port above pan floor level makes cleaning and scrubbing unduly difficult and time consuming. Maintaining such a pan free of contaminates is not realistically feasibility.

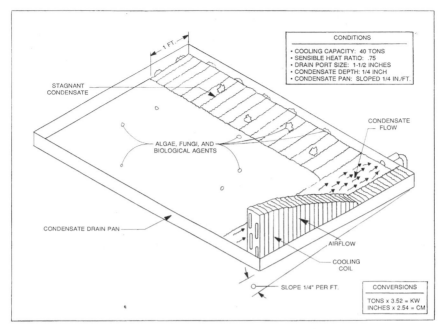

Figure 22.5 Effects of pan slope on condensate flow and contamination.

Remedies. An acceptable maintenance program dictates that drain ports be flush with the bottom of the drain pan. Whenever feasible, the drain port should be placed in the bottom floor of the drain pan, to further improve drainage. See Fig. 22.2.

22.2.3.3 Baffles in condensate drain pans. The drain pans in some draw-through HVAC packaged units now in the field, are equipped with internal baffles. Evidently, these are intended to prevent condensate from blowing into the system where it can cause damage to internal components and promote health threatening biological growth. In most draw-through systems—in the field today—condensate blowing is a serious problem. The problem arises when no seal has been installed or when the seal depends upon a trap that is dry; for example, during initial system start-up, or start-up for summer cooling. (Traps become dry in winter due to evaporation and/or freeze-plug expulsion).

Problem definition. During the above operating conditions, baffles can reduce, although they rarely eliminate, condensate blowing when no drain seal is present. But they present a significant maintenance problem. More surface area is exposed to condensate and the potential for

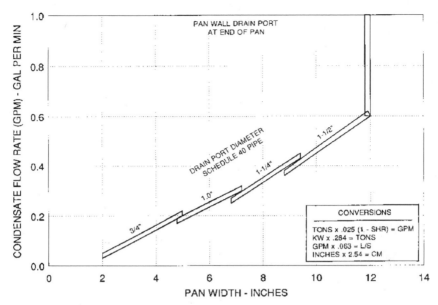

Figure 22.6 Pan widths suitable for units with various flow rates and for commonly used port sizes.

system contamination is magnified. And because of the small air and water passages in baffles, condensate pans seldom drain well and condensate flow is often blocked by debris. Frequent cleaning of baffled systems is therefore imperative. Yet, the interior of baffle arrangements are so inaccessible that reasonable maintenance procedures are usually not feasible.

Remedies. The implementation of a practical maintenance program requires the removal of troublesome baffles and the use of a reliable and effective drain seal that eliminates the condensate blowing, for which the baffles were initially installed. See par. 22.2.6 "Ineffective seals on condensate drain lines" in the following pages.

22.2.4 Very low supply air temperatures

Low supply air temperatures present a special problem in condensate control, which occurs too frequently to be ignored. The problem appears in the form of damage to and biological contamination of air supply grilles.

22.2.4.1 Air supply grilles In buildings where the latent heat load is high, the dry bulb temperature of the air leaving the cooling coil is usually below the dew point temperature of air in the conditioned room.

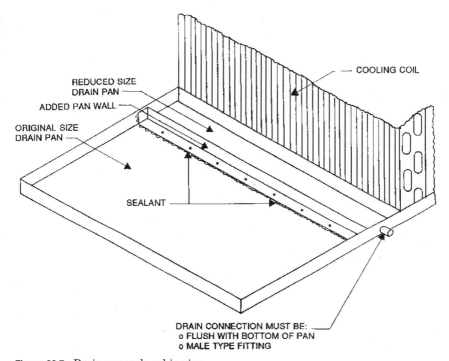

Figure 22.7 Drain pan reduced in size.

Problem definition. Whenever supply air reduces the supply grille temperature to the dew point temperature inside the room, condensate can form on the surfaces of the grille. Deposited there, it causes the formation of molds and other fungi that discolor, damage and contaminate surfaces of the grille. Systems that allow formation of condensate on supply grille surfaces require constant attention and must be modified to be amenable to a realistic maintenance program.

Remedies. There are three possible solutions to this problem: (a) change in grille design, (b) increase in cooling airflow rate, and (c) addition of cooling air reheat.

The internal and external geometry as well as the material of construction determine how resistant a supply grille is to condensate formation. Some grille manufactures are aware of the problem and can be helpful in selecting designs that are the least susceptible to condensate formation.

An increase in airflow rate will increase the supply air temperature and reduce moisture formation on supply grilles. Many times it may be possible to increase airflow sufficiently to avoid this problem and still provide adequate system performance. This possibility should always be considered, anytime condensate is found on air supply grilles. (The effect of increasing airflow on condensate carryover from the cooling coil must be evaluated when increased airflow is considered.)

Providing reheat to cooling air is an effective way to avoid condensate formation on supply air grilles, but it is usually an inefficient and a costly process. There are, however, other reasons for using reheat. If needed for other purposes, adding reheat may offer an acceptable method of eliminating condensate on supply grilles.

More details on what can be done to avoid condensate formation on supply grilles are provided in Ref. 1.

22.2.5 Improper fan position inside air handlers

Fans improperly positioned within an air handler often cause condensate problems that cannot be remedied by routine maintenance. Poorly located fans can (a) cause condensate to be entrained directly from surfaces of the cooling coil or (b) generate vortices that ingest condensate from either the drain pan or the cooling coil.

22.2.5.1 Condensate entrainment. Air entering the fan inlet is accelerated to a velocity much higher than that in the coil plenum. This condition generates a high velocity core that extends well beyond the fan inlet.

Problem definition. If fans are placed too close to cooling coils the high velocity core will extend into the coil, entrain condensate and blow it onto downstream components.

Remedies. In instances where the fan draws condensate directly from the coil, the separation distance must be increased in order to alleviate the problem. Ref. 1, provides information on the separation distance necessary to avoid condensate entrainment.

22.2.5.2 Vortex ingestion. Airflow distortion developed upstream of fan inlets can interact with fan blades and generate troublesome vortices. The presence of vortices, is most readily confirmed by visual observations.

Problem definition. Vortices touching the coil surfaces or condensate in the drain pan can carry streams of condensate into the fan. From there, it is spread onto downstream components causing damage to equipment and promoting the growth of biological contaminates.

Remedies. Vortices can usually be eliminated by installing properly designed turning vanes. Reference 1, provides additional information on the use of turning vanes.

22.2.6 Ineffective seals on condensate drain lines

Ineffective seals on condensate drain lines of draw-through systems is a major cause of equipment damage, surrounding property damage, and system contamination. In fact, dysfunctional drain seals appear to be the primary cause of biological growth in HVAC systems.

An ineffective drain seal can be identified in a number of ways. (Table 22.1)

The consequences of operating a draw-through HVAC system with an ineffective drain seal are summarized in Fig. 22.8. These consequences are clearly unacceptable and dictate that every draw-through HVAC unit be equipped with an effective and reliable condensate drain seal.

Within the industry, three different types of devices are being used to form condensate drain seals for draw-through HVAC systems: (a) condensate (water) traps, (b) condensate pumps, and (c) a fluidic flow control device — the latter a recent technological development.

TABLE 22.1 Determining Seal Effectiveness

Symptom	To Determine Seal Effectiveness
Outside air is drawn into HVAC unit during normal operation.	Place soap bubbles or a small flame at the drain exit, or at any open port (downstream of the drain seal). If either the bubbles or flame is drawn into the system, the seal is ineffective and unsatisfactory.
Outside air is drawn into HVAC unit when the fan is operating with the drain pan, trap,* and drain system dry. * (unless the seal is equipped with a water replenishing system)	Place soap bubbles or a small flame at the drain exit, or at any open port (downstream of the drain deal). If either the bubbles or flame is drawn into the system, the seal is ineffective and unsatisfactory.
Condensate stands in the drain pan during the cooling operation.	Determine the depth of condensate in the drain pan. If it exceeds the level shown in Fig. 22.2 the seal is ineffective and unsatisfactory.
Condensate is blown from the drain pan during cooling system start-up, or normal operation, and condensate is spread onto internal surfaces.	View conditions of the drain pan and cooling coil through port holes or other means.

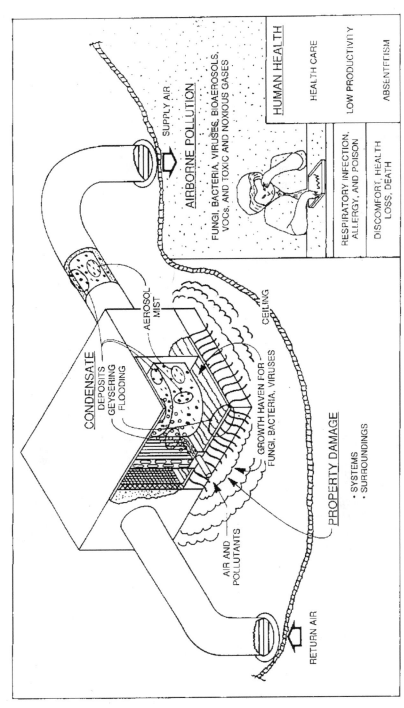

Figure 22.8 Consequences of no seal or a dysfunctional seal on draw-through HVAC units.

Each of these devices exhibits unique physical and operating characteristics and provides a different level of effectiveness and reliability.

22.2.6.1 Condensate traps. The condensate trap is widely used as a seal in condensate drain lines. It is usually mounted outside the HVAC unit, as indicated in Fig. 22.9. The seal is formed by gravitatoinal forces acting on trapped water and a water column.

Problem definition. Properly configured, as described in Ref. 1, the conventional condensate trap can provide a seal under *ideal* circumstances. Even when configured properly, however, the trap exhibits numerous failure modes and is so unreliable that it is ineffective and is unsuitable for use on a draw-through HVAC system.

Some failure modes are inherent with the condensate trap: (a) flow blockage—pan overflows; (b) trap freeze damage (in outside locations)—seal destroyed; (c) evaporation of condensate—seal destroyed. The causes of these failures and the destructive consequences are summarized in Fig. 22.10.

In addition to inherent deficiencies, traps are susceptible to design deficiencies and unwise field practices, which add failure modes and further decrease trap reliability and effectiveness.

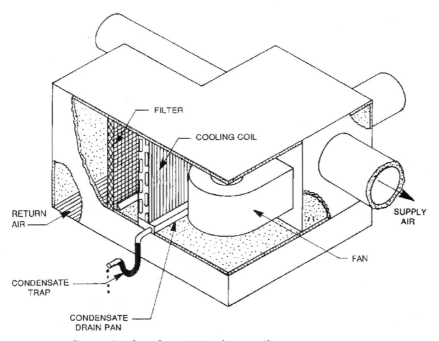

Figure 22.9 Conventional condensate trap in operation.

<table>
<tr>
<th>BLOCKED TRAP</th>
<th>FREEZE DAMAGED TRAP</th>
<th>EMPTY TRAP
(Winter operation)</th>
<th>EMPTY TRAP
(Cooling start-up)</th>
</tr>
<tr>
<td>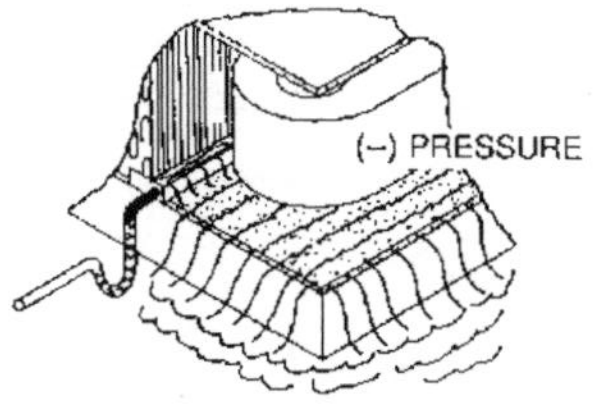
</td>
<td>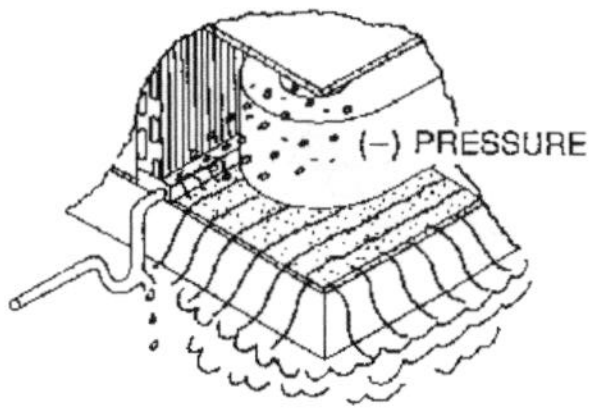
</td>
<td>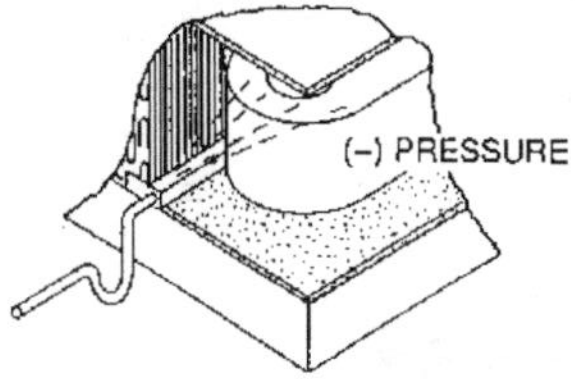
</td>
<td>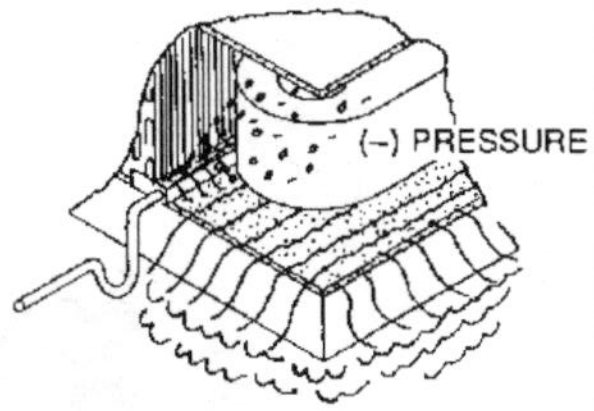
</td>
</tr>
<tr>
<td>Cause: Algae growth and trapped debris</td>
<td>Cause: Installed in low temperature environment</td>
<td>Cause: Evaporation during non-cooling periods - no seal present</td>
<td>Cause: Evaporation during non-cooling periods - no seal present</td>
</tr>
<tr>
<td>Likely Consequences:
• Overflow of condensate
• Damage to unit and surroundings
• Pollution of system and conditioned air
• User Costs: Property danage, health care, excessive maintenance and litigation</td>
<td>Likely Consequences:
• Seal destroyed
• Ingestion of outside air which may be polluted
• Overflow of condensate
• Blowing of condensate into unit and ductwork
• Formation of aerosol mist
• User Costs: Property damage health care, excessive maintenance and litigation</td>
<td>Likely Consequences:
• Ingestion of outside air which may be polluted
• User Costs: Health care and litigation</td>
<td>Likely Consequences:
• Ingestion of outside air which may be polluted
• Overflow of condensate
• Blowing of condensate into unit and ductwork
• Formation of aerosol mist
• User Costs: Property damage, health care, excessive maintenance and litigation</td>
</tr>
</table>

Figure 22.10 Some trap failures, causes, and consequences.

Reference 5 presents a comprehensive summary of failure modes that plague the condensate trap, and torment building owners/users.

Remedies. The methods available for improving trap effectiveness and reliability are limited.

Trap flow blockage is virtually impossible to avoid. Because traps trap water, they likewise trap debris, which eventually blocks flow. In addition, the cool water in the trap promotes the growth of algae, a condition that almost ensures periodic flow blockage. Annual cleaning of traps is one possible way to overcome flow blockage. However, because of prolific growth of algae and the drying-out and hardening that takes place during winter months, trap cleaning is not always effective. The best approach is to replace traps annually.

Trap freeze-up and seal destruction, in outside locations, can be avoided by applying heat to the condensate during freezing periods. Although such a system requires a heating element and a sensor device, it can be effective and reasonably reliable, but not cost free. The use of freeze plugs, often employed to prevent trap damage, is an unsuitable choice because when expelled they destroy the seal and render traps ineffective until the plugs are replaced. Further, in many applications, it is not feasible for maintenance personnel to replace freeze plugs after each freeze-up. Thus, after thaw-out the seal is lost until plugs are replaced and traps are filled, often months later. Evaporation of condensate from traps, which begins as soon as the cooling operation and condensate-flow ceases, can be overcome by using a water replenishing system. The control of water flow may be effected by a continual drip arrangement, an on/off timer, or level sensor. In any case, a water replenishing system adds cost and decreases reliability. For successful operation in outside locations (where freezing occurs), a water replenishing system must be equipped with methods for keeping the water above freezing temperature.

Condensate traps should be used as drain seals only as a last resort. Even when water replenishing and condensate heating systems are employed, systems must be monitored carefully—a very expensive procedure.

22.2.6.2 Condensate pumps. Condensate pumps are usually installed in the condensate drain pan, inside a suitable water sump. They are used primarily where there is insufficient depth for the installation of a gravity dependent drain system or where condensate must be discharged to a level above the drain pan. Pumps provide a positive and effective seal during all phases of operation. External power (usually electrical) is required. To accommodate variations in condensate flow, an on-off switch, which is usually operated by a float, must be provided.

Problem definition. Although they are well developed and have been used successfully in special applications, the long term reliability of condensate pumps must be questioned. They depend upon moving parts that operate in a cool and humid environment. Pumps must handle condensate that often carries significant quantities of debris, which can interrupt pumping action and cause flooding. The on-off switch used to control pump operation is exposed to a somewhat hostile environment, and is subject to frequent failures.

Condensate pumps may be damaged when exposed to freezing conditions. Freezing, generally, does not destroy the seal, but it can result in pump failure and condensate overflow whenever cooling operation begins. Because they usually require considerable space inside the drain pan, pumps are not often used as replacements for condensate traps.

Remedies. Acceptable pump reliability can be achieved by implementing an aggressive maintenance and monitoring program. It must include periodic cleaning of debris from the sump, and replacement of aging and deteriorating electrical and mechanical components. In outside locations, freeze damage can be avoided by providing a heating system to maintain water in the sump at a temperature above freezing.

22.2.6.3 Fluidic flow control device. The fluidic flow control device is a recent advancement in drain seal design. It was developed specifically for use on the drain lines of draw-through HVAC units, to be free of the failure modes common to traps and pumps. The device is connected to the condensate drain port much like a conventional condensate trap. The desired seal is formed by a unique combination of hydraulic and pneumatic forces, available in the HVAC unit. It has no moving parts. A key feature of the device is that it uses air as a seal instead of water. Thus, it negates the problems associated with a water seal. The operating principles of the fluidic flow control device are illustrated in Fig. 22.11. During both heating and cooling operations, the drain seal is formed as follows:

Fresh air from the fan discharge is supplied to point (a) at a pressure slightly above atmospheric. Some of the air flows away from the HVAC unit; thus, preventing ingestion of outside air. A portion of the fresh air returns to the HVAC unit, passing through points (b) and (c). The quantity of air returning to the unit is minimized by the high pressure loss in the mitered elbows. This pressure loss plus the air flowing through the bypass connected at point (c), assures that the air entering the condensate drip pan will not produce blowing and geysering and an aerosol mist.

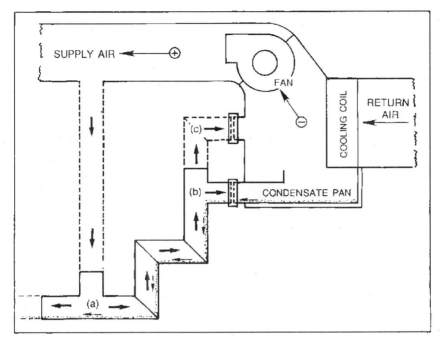

Figure 22.11 Operating principles of the fluidic flow control device.

Condensate flows through the device without being trapped. At the same time, the counterflow of condensate and air creates a pulsing action that ensures free passage of debris. Hence, the potential for freeze-up and flow blockage—common problems with traps—are nil.

A typical field installation of the fluidic flow control device is shown in Fig. 22.12.

The device is available in sizes suitable for HVAC units up to 100 tons of cooling capacity and −5.0 in. (12° cm) water column at the condensate drain outlet. Fig. 22.13 shows how the physical dimensions of the device vary with system cooling capacity and drain pan pressure. Reference 6 provides more technical information about the device and explains how it works. (The product is manufactured and marketed by Trent Technologies, Inc., of Tyler, Texas.)

Problem definition. Installation procedures for the device are new to HVAC service personnel, and the effort required is slightly more than that required to install a conventional condensate trap (although, less than that required for installing a trap with a condensate heater and a water replenishing system). Not all existing systems are adaptable to installation of the new device. For example, units that are too close

Figure 22.12 Typical field installation of the fluidic flow control device.

to the floor to allow installation of a trap, cannot accommodate the fluidic control device either, without major changes.

Remedies. Detailed step-by-step installation procedures, including a pictorial guide, are available to assist service personnel with their first installation. Subsequent installations become routine. Once installed, the device operates virtually void of maintenance effort and free of the many problems caused by the condensate trap, in that it:

1. Allows condensate to flow freely and unimpeded from the HVAC unit;

2. Prevents air (which may be contaminated) from being drawn into the system through the condensate drain pipe during heating operations, cooling operations, and cooling system start-up operations (when p-traps are usually empty);

3. Prevents condensate in the drain pan from being blown into the air conditioning unit and the duct work (during both normal and start-up operations);

4. Removes the condensate drain system as a source of an aerosol mist;

5. Eliminates condensate overflow caused by trap blockage and negative pressure inside the system;

6. Is not affected by algae growth;

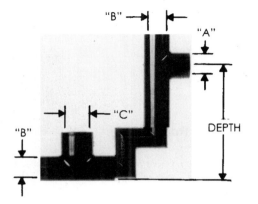

NOMINAL RATED TONS	-- PORT SIZE -- NOMINAL DIAMETER		
	"A"	"B"	"C"
1 to 20	3/4" or 1"	3/4"	1"
1 to 100	3/4" to 1-1/2"	1"	1-1/4"

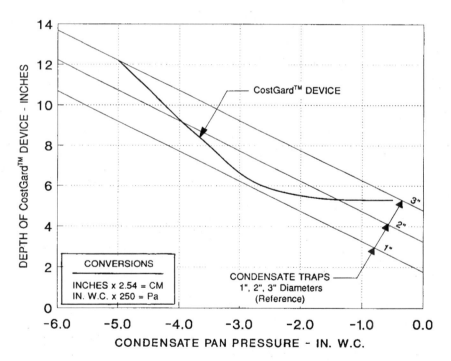

Figure 22.13   Physical dimensions of the CostGard® condensate control device.

7. Is not affected by condensate evaporation (as are traps);

8. Precludes damage from freezing temperatures;

9. Includes no moving parts; and

10. Is self-cleaning and self-regulating.

22.2.7 Unsuitable installation of condensate drain lines

Condensate drain lines, which extend from the drain seal to the condensate disposal place, can be — and frequently are — the source of serious condensate problems. Drain lines receive less design, installation, and maintenance attention than any component of the HVAC system. As a result, line failures are significant contributors to property damage, equipment damage, and system contamination.

An acceptable drain line is simply one that has adequate flow capacity and offers minimum potential for flow blockage. The critical design factors are basic geometry and the support system for the drain line.

22.2.7.1 Drain line geometry. Unsatisfactory drain line geometry is a common cause of drain line flow blockage that results in condensate overflow problems.

Problem definition. Small diameter, long and meandering, non-sloped, and deflected drain lines are major causes of drain line blockage. Lines that are too small are easily blocked by debris and algae growth. Long lines are more susceptible to flow blockage, because there are simply more places for blockage to occur. Meandering lines with elbows are highly prone to flow blockage. Non-sloped drain lines retain condensate along with debris. Retained or standing condensate supports the growth of algae, which restricts and blocks flow. The effect of drain line blockage, which occurs all too often, is flooding and the overflow of the condensate pan. The consequences are damage to equipment and surrounding property plus contamination of the HVAC unit and the surroundings.

Remedies. The diameter of the drain line must be equal to or greater than the exit diameter of the drain seal device. When more than one HVAC unit drains to a common line, the area of the common line must be increased proportionally, at the downstream point where the additional HVAC drain line enters. The line length should be the minimum possible, following the shortest path to the condensate disposal place (the shorter, the better). And it should include the least possible number of elbows. The line must be sloped away from the drain seal at a rate of *no less* than 1/8-in per ft (1 cm/m).

22.2.7.2 Drain line support

A firm fixed drain line support system is essential to ensure satisfactory condensate drainage.

Problem definition. Condensate drain lines are often located in areas of high maintenance activity; for example, building roof tops. In this environment, if lines are not securely fixed they can be broken or damaged, by careless personnel, and permit condensate to be drained to unwanted places. Damage to drain lines frequently destroys the condensate drain seal (a very common occurrence). A destroyed condensate drain seal results in all the property damage and contamination problems discussed in paragraph 22.2.6: "Ineffective seals on condensate drain lines".

A fixed support system is also required to ensure that the drain line maintains a satisfactory slope and prevents line defections sufficient to create harmful secondary traps. Any amount of line defection allows condensate to collect at the low point, where it promotes the growth of algae and increases the potential for flow blockage. Deflections greater than one diameter of the drain line creates a much more detrimental situation. In systems with a conventional condensate trap, a trap formed by a dip in the drain lines forms an airlock, which will block condensate flow and cause the pan to flood and overflow.

Remedies. To avoid excessive deflection of the drain line, prevent shifting of position, and ensure a satisfactory slope, fixed supports should be provided at distances no less than the following: PVC (Schedule 40 pipe), 2 to 3 ft (0.6 to 0.9 m); copper, 4 to 6 ft (1.2 to 1.8 m), and steel 8 to 10 ft (2.4 to 3 m) intervals.

22.3 Condensate Control and Maintenance

HVAC systems with the above deficiencies cannot avoid internal wetness caused by uncontrolled spreading of condensate. Under these circumstances, no level of maintenance effort can prevent damage to and contamination of system interiors.

However, once these HVAC system deficiencies are remedied as describe above, the necessary condensate control can be realized. And, only the condensate drain system—consisting of the drain pan, drain seal and drain line—requires any maintenance effort.

The drain system features and characteristics required for acceptable condensate control and maintenance are summarized below for each drain system component.

Condensate drain pan.   The condensate drain pan must be:

1. Wide enough (extended downstream of the cooling coil) to accommodate the maximum condensate flow rate, but not so wide as to allow condensate to stagnate

2. Sloped toward the drain outlet port

3. Equipped with a male drain connection that is flush with the bottom of the drain pan

4. Constructed of non corrosive material — such as plastic or stainless steel and

5. Accessible for easy cleaning.

Condensate drain seal.   The drain seal must:

1. Allow condensate to flow freely and unimpeded from the HVAC unit;

2. Prevent air (which may be contaminated) from being drawn into the system through the condensate drain pipe during winter heating operations, summer cooling operations, and cooling system start-up operations (when p-traps are usually empty);

3. Prevent condensate in the drain pan from being blown into the HVAC unit and the duct work (during both normal and start-up operations);

4. Remove the condensate drain system as a source of an aerosol mist;

5. Eliminate condensate overflow caused by trap blockage and negative pressure inside the system;

6. Not be affected by algae growth;

7. Not be affected by condensate evaporation (as are traps);

8. Preclude damage from freezing temperatures;

9. Include no moving parts; and

10. Be self-cleaning and self-regulating.

Condensate drain line.   The drain line must be:

1. Equal to or greater in diameter than the outlet of the drain seal;

2. Sloped toward the condensate disposal place at least 1/8-in./per ft (1 cm per m);

3. Installed with fixed supports at proper intervals to prevent dips that retain condensate, promote algae growth, and form line traps;

4. The shortest length possible, with a minimum of bends, turns, and elbows; and

5. Insulated or equipped with heating provisions, when required to operate in freezing temperatures.

22.4   Routine and Preventive Maintenance Programs

Properly configured drain systems, with the features and characteristics defined above, are amenable to effective and economical routine and preventive maintenance programs. Program details, of course, differ according to system usage; i.e., where and how heating and cooling systems are used. Accordingly, specific programs depend upon whether the system is required to provide:

1. Summer cooling and winter heating, (where outfdoor temperature is below freezing)

2. Summer cooling and winter cooling, (where outdoor temperature is below freezing);

3. Summer cooling and winter cooling (when outdoor temperature is above freezing)

Routine and preventive maintenance programs under these various conditions are outlined in Table 22.2, for the effective and economical condensate drain system defined above.

In these programs, the projected maintenance effort is small. The only effort for which a finite time is stated is that for scheduled inspections. Experience shows that the time required for servicing these systems is minimal. The estimated time for maintenance service varies from zero to a few minutes. Accordingly, in any particular installation, maintenance experience may indicate that the inspection and service frequencies stated in the table can be decreased significantly, thus reducing the effort and cost.

The key to achieving this highly maintainable drain system is the use of an effective drain seal (in this case the fluidic flow control device). The conventional condensate trap simply cannot be adequately maintained and is, therefore, unsuitable as drain seal.

TABLE 22.2 Routine and Preventive Maintenance Program for Properly Configured Condensate Drain System

Drain pan

1. Frequency and Time of Inspection and Service:

   (a) For Systems That Provide Summer Cooling and Winter Heating
   - Annually—during cooling operation, when condensate is flowing

   (b) For Systems That Provide Summer Cooling and Winter Cooling
   - Semi—annually—during summer and winter operations, when condensate is flowing

2. Maintenance Effort Required:

   (a) If biological growth and/or debris are present in the drain pan:
   - Physically remove all contaminates
   - Flush with water and
   - Treat with EPA approved biocide

   (b) Otherwise, no effort is required.

3. Equipment and Material Needed:

   (a) Surface scraper
   (b) Scrub brush
   (c) Water hose
   (d) Biocide

4. Estimated Time Required:

   (a) Less than 5 min per inspection + (25 min travel time to and from maintenance shop and system site)
   (b) 0 to 60 min per time serviced + (25 min travel time to and from maintenance shop and system site)

Drain seal

1. Frequency and Time of Inspection and Service:

   (a) For systems that provide summer cooling, winter heating and cooling
   - Annually—during cooling operation, when condensate is flowing

2. Maintenance Effort Required:

   (a) If condensate is not flowing freely during cooling operation and/or condensate is standing in the pan more than the operating level indicated in Fig. 22.2, at drain outlet:
   - Check for debris inside the device. If present, physically remove and flush inside with water
   - Check operating pressures per manufacturer's instructions

   (b) Otherwise, no effort is required.

3. Equipment and Material Needed:

   (a) Water hose
   (b) Pressure gauge

4. Estimated Time Required:

   (a) Less than 5 min per inspection + (25 min travel time to and from maintenance shop and system site)
   (b) 0 to 30 min per time serviced + (25 min travel time to and from maintenance shop and system site)

TABLE 22.2   Continued

Drain line

1. Frequency and Time of Inspection and Service:
   - (a) For Systems That Provide Summer Cooling, Winter Heating and Cooling
     - Annually—during cooling operation, when condensate is flowing
   - (b) For Systems That Provide Summer Cooling and Winter Cooling
     - Semiannually—during summer and winter operations, when condensate is flowing
2. Maintenance Effort Required:
   - (a) If condensate is not flowing freely through the drain line during cooling operation and/or lines are deflected more than one diameter between supports—
     - Physically remove flow blocking algae and/or: debris
     - Flush with water
     - Treat with EPA approved biocide
     - Service drain line supports and remove deflectons and
     - Service drain line insulation and/or drain line heating system when so equipped
   - (b) Otherwise, no effort is required
3. Equipment and Material Needed:
   - (a) Water hose
   - (b) Biocide
4. Estimate Time Required:
   - (a) Less than 10 min per inspection + (25 min travel time to and from maintenance shop and system site)
   - (b) 0 to 60 min per times serviced + (25 min travel time to and from maintenance shop and system site)

22.5   Maintenance Limitations of the Condensate Trap

Table 22.3, prepared in the form of a routine and preventive maintenance program, shows why it is not possible for the conventional condensate trap to provide a satisfactory drain seal. In summary, this is because the trap:

1. Requires frequent periodic cleaning to remove algae and debris, in order to prevent condensate flow blockage and overflow, resulting in system contamination and property damage
2. Must be filled with water frequently,
   - (a) During non-cooling periods—e.g. winter—to prevent the ingestion of potentially toxic and noxious gases, and
   - (b) Prior to each cooling system start-up, to prevent drenching and/or flooding that contaminate and damage the system interior and surroundings

TABLE 22.3   Routine and Preventive Maintenance Program for Conventional Condensate Trap

Trap located indoors or outdoors, with outdoor temperatures above freezing

1. Frequency and Time of Inspection and Service:

   (a) For Systems That Provide Summer Cooling and Winter Heating
   During Cooling Operation:
   - Annually—at initial syhstem start-up for cooling
   - Semiannually—at initial system start-up and at second system start-up if facility is shut down annually for a week or more, e.g. schools
   During Heating Operation:
   - Biweekly, between cooling system shut-down and the beginning of winter heating

   (b) For Systems That Provide Summer Cooling and Winter Cooling
   - Semiannually—at 6-mo: intervals (one inspection must be made at system start-up, following an annual shut-down of facility for a week or more, e.g., schools

2. Maintenance Effort Required:

   (a) At each annual inspection (and semi-annually if need is indicated)
   - Physically remove flow-blocking algae and/or debris, or replace trap
   - Flush with water
   - Treat with EPA approved biocide and
   - Fill trap with water and add biocide tablets

   (b) At each biweekly inspection
   - Fill with water and add biocide tablets if need is indicated.

3. Equipment and Material Needed:

   (a) Internal pipe scraper
   (b) New trap
   (c) Water hose
   (d) Biocide

4. Estimated Time Required:

   (a) Annually and Semiannually:
   - 5 min per inspection + (25 min travel time to and from maintenance shop and system site)
   - 0 to 60 min per time serviced + (25 min travel time to and from maintenance shop and system site)

   (b) Biweekly:
   - 5 min, per time serviced + (25 min travel time to and from maintenance shop and system site)

Trap located outdoors, with outdoor temperatures below freezing

1. Frequency and Time of Inspection and Service:

   (a) For Systems That Provide Summer and Winter Cooling and Winter Heating
   During Cooling Operation:
   - Not possible to maintain drain seal with a trap during winter cooling under these conditions:—flowing condensate will freeze in trap, block flow, and damage trap
   During Heating Opertion:
   - Not possible to maintain drain seal with a trp during winter heating under these conditions:
     —Unless the trap is filled with water, it will not hold a seal and when filled, water will freeze and block condensate flow

3. Is not suitable for use during winter cooling in outdoor locations where the temperature is below freezing, because of flow blockage and trap destruction.

Although it is possible by routine and preventive maintenance to minimize the damage caused by trap blockage, the procedure is costly; because, traps must be cleaned or replaced frequently, requiring a significant maintenance effort.

Overcoming evaporation which destroys the drain seal, during non-cooling periods and at system start-up time for cooling, places an enormous burden on maintenance personnel. Filling each condensate trap with water, frequently, to avoid gas ingestion, drench problems, and condensate flooding places impractical demands on maintenance organizations, even in indoor locations.

For example, using information in Table 22.3, it has been estimated that the maintainance effort for a typical system located indoors — used for both heating and cooling — is about 6 hr per yr per unit.

In outdoor locations, where outdoor temperatures are below freezing, it is virtually impossible to implement an effective trap filling program. Freezing condensate will destroy the trap, rendering it ineffective until it is replaced and refilled with water. Freeze plugs are of little value, even if they protect the trap, they destroy the seal.

The maintainability of the condensate trap can be made feasible by utilizing a water replenishing system and a condensate heating system, to ensure a water-filled trap and prevent freezing in outside locations. This does not, however, eliminate all the maintenance problems of the conventional trap and it introduces others. The system is subject to the same trap blockage from algae growth and debris, as is the conventional trap. In addition, both the water replenishing and the condensate heating systems require some type of control and usually utilize moving parts. Such systems require appreciable maintenance, which is often much too great for a practical drain seal.

22.6   References

1. N. R. Grimm, and Rosaler, R. C. *Handbook of HVAC Systems and Components*, McGraw-Hill Publishing Company, New York, New York, 1998.
2. T. C. Ottney, "Particle management of HVAC systems," *ASHRAE Journal, 35*, American Society of Heating, Refrigerating, and Air Conditioning Engineers, Inc., Atlanta, Georgia, July 1993, pp. 26–34.
3. *1996 ASHRAE Handbook: HVAC Systems and Equipment*, American Society of Heating, Refrigerating, and Air Conditioning Engineers, Inc., Atlanta, Georgia, 1996, p. 24.3
4. G. W. Brundrett, *Legionella and Building Services*, Butterworth-Heinemann Ltd., London, 1992, p. 4

5. W. C. Trent, "The Condensate Trap: A Costly Failure," *Air Conditioning, Heating and Refrigeration News*, February 21, 1994, p. 3. Business News Publishing Co., Troy, Michigan.

6. W. C. Trent, and Trent, C. C. "Considerations in Designing Drier, Cleaner HVAC Systems," *Engineered Systems*, August 1995, p. 38. Business News Publishing Co., Troy, Michigan.

Metric Conversion Tables

NOTE: Most of the text in this handbook provides SI metric conversions. However, editorial factors necessitated omissions in some tables and graphs. Listed below are conversions most likely encountered in HVAC.

In the table below the first two digits of each numerical entry represents a power of 10. For example, the entry "-22.54" expresses 2.54×10^{-2}. Standard abbreviations are used as appropriate.

To convert from	To	Multiply by
	AREA	
acre	m^2	$+03\ 4.046$
circular mil	m^2	$-10\ 5.067$
in^2	m^2	$-04\ 6.4512$
ft^2	m^2	$-02\ 9.290$
mi^2	m^2	$+06\ 2.589$
$yard^2$	m^2	$-01\ 8.361$
	DENSITY	
lb/in^3	kg/m^2	$+04\ 2.768$
lb/ft^3	kg/m^2	$+01\ 1.602$
	ENERGY	
Btu (mean)	joule	$+03\ 1.056$
Btu/lb	J/kg	$+03\ 2.324$
ft · lb	joule	$+00\ 1.356$
	ENERGY / AREA TIME	
$Btu/(ft^2 \cdot s)$	W/m^2	$+04\ 1.135$
	FORCE	
lb	N	$+00\ 4.448$
oz	N	$-01\ 2.780$
	LENGTH	
ft	m	$-01\ 3.048$
in	mm	$+01\ 2.54$

To convert from	To	Multiply by
LENGTH (cont.)		
mil	m	−05 2.540
mi (U.S. statute)	m	+03 1.609
yd	m	−01 9.144
MASS		
oz	kg	−02 2.835
lb	kg	−01 4.536
ton (short)	kg	+02 9.072
POWER		
Btu/h	W	−01 2.931
(ft · lb)/hr	W	−04 3.766
hp [550 (ft · lb)/s]	W	+02 7.457
tons (cooling capacity)	kW	00 3.52
PRESSURE		
bar	Pa	+05 1.00
ft H_2O	Pa	+03 2.989
inHg	Pa	+03 3.386
lb/in^2 (psi)	Pa	+03 6.895
SPEED		
ft/s	m/s	−01 3.048
mi/h	m/s	−01 4.470
TEMPERATURE		
°F	°C	(°F −32)/1.8
VISCOSITY		
cSt	m^2/s	−06 1.00
cP	(Pa · s)	−03 1.00
VOLUME		
barrel (42 gal)	m^3	−01 1.590
fluid oz (U.S.)	m^3	−05 2.597
ft^3	m^3	−02 2.832
gal (U.S. liquid)	m^3	−03 3.785
gal (U.S. liquid)	L	00 3.785
in^3	m^3	−05 1.639
ppm	mg/L (of H_2O)	00 1.000

Index

ABOUT THE EDITOR

Robert C. Rosaler is a consulting engineer with over 30 years of experience in HVAC and plant engineering operations. He is editor in chief of McGraw-Hill's *Handbook of HVAC Design, HVAC Systems and Components Handbook*, and *Standard Handbook of Plant Engineering*. He is a member of the American Society of Mechanical Engineers and the Association for Facilities Engineering.